AF400756

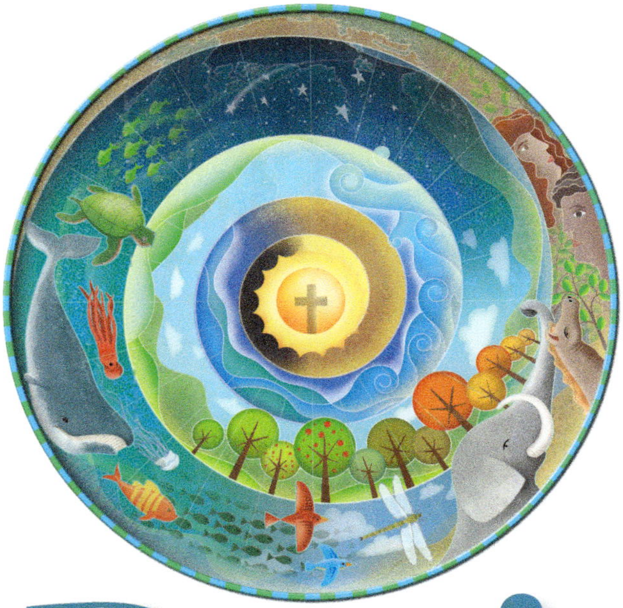

By Design

A Journey to Excellence through Science

Kendall Hunt Religious Publishing

A Division of Kendall Hunt Publishing

Developed in Collaboration with the
Seventh-day Adventist NAD Office of Education and
Kendall Hunt Religious Publishing
A Division of Kendall Hunt Publishing Company

Acknowledgments

SDA/NAD Office of Education

Larry Blackmer, Executive Editor/Vice-President of Education
Carol Campbell, Director of Elementary Education
Debra Fryson, Director of Education, Southern Union

Program Development and Publisher

Kendall Hunt Religious Publishing, a Division of Kendall Hunt Publishing

Program Consultants Lead Team I

Dan Wyrick, Director of Nature by Design
Lee Davidson, Associate Professor of Teacher Education/Dept. Chair, Andrews University
Jerrell Gilkeson, Associate Director of Education, Atlantic Union
Ed Zinke, Biblical Researcher
Tim Standish, Geoscience Research Institute

Associates Representing Lead Team II

Betty Bayer, Associate Director of Education, SDA Church in Canada
Diane Ruff, Associate Director of Education, Southern Union
Ileana Espinosa, Associate Director of Education, Columbia Union
James Martz, Associate Director of Education, Lake Union
Keith Waters, Associate Director of Education, North Pacific Union
LouAnn Howard, Associate Director of Education, Mid-America Union
Martha Havens, Associate Director of Education, Pacific Union
Mike Furr, Associate Director of Education, Southwestern Union
Patti Revolinski, Associate Director of Education, North Pacific Union

Contributing Writers and Reviewers

Carol Raney, Freelance Writer
Katia Reinert, NAD Director of Health Ministries
Alayne Thorpe, President of Griggs University
Rick Norskov, MD, Southern Adventist University
Lucinda Hill, MD, Southern Adventist University
Raul Esperante, PhD, Geoscience Research Institute
Mitch Menzmer, PhD, Southern Adventist University

Contributing Multi-Grade Classroom Writers and Reviewers

Karen Reinke, Freelance Writer
Mark Mirek
Heidi Jorgenson
LaVona Gillham
Cindy French Puterbaugh

www.kendallhunt.com
Send all inquiries to:
4050 Westmark Drive
Dubuque, IA 52004-1840
1-800-542-6657

ISBN 978-1-4652-0112-6

1 2 3 4 5 6 7 8 9 10 19 18 17 16 15 14

Science Classroom and Lab Safety

It is important to be safe when you perform hands-on investigations. Always look for the safety symbol in your textbook. It tells you how to be safe as you perform your investigations. Below are some important safety rules to follow for most investigations. Remember to be alert and listen to your teacher for any additional safety rules.

Prepare for laboratory work

- Review classroom laboratory procedures with your teacher.
- Never perform unauthorized experiments.
- Keep your lab work area organized and free of clutter.
- Know the location of the safety fire blanket and first aid kit.

Dress for laboratory work

- Tie back long hair.
- Do not wear loose sleeves.
- Wear shoes that completely cover your feet.
- Wear lab apron, goggles, and gloves as required.

Avoid contact with chemicals

- Never taste or sniff chemicals.
- Never draw materials in a pipette with your mouth.
- When heating substances in a test tube, point the mouth away from people.
- Never carry dangerous chemicals or hot equipment near other people.

Avoid hazards

- Keep combustibles away from open flames.
- Use caution when handling hot glassware.
- When diluting acid, always add acid slowly to water. Never add water to acid.
- Turn off burners when not in use.
- Keep lids on reagent containers. Never switch lids.
- Use sharp equipment and objects as they were designed to be used and in a safe manner. For example, always cut away from yourself rather than toward yourself, and be careful to study which edges and points of an object are the sharp ones.

Clean Up

- Consult the teacher for proper disposal of materials.
- Wash hands thoroughly following experiments.
- Leave laboratory work area clean and neat.

In case of accident

- Report all accidents and spills immediately.
- Place broken glass in designated containers.
- Wash all acids and bases from your skin immediately with plenty of running water.
- If chemicals get in your eyes, wash them for at least 15 minutes with an eyewash or clean water.

Scavenger Hunt

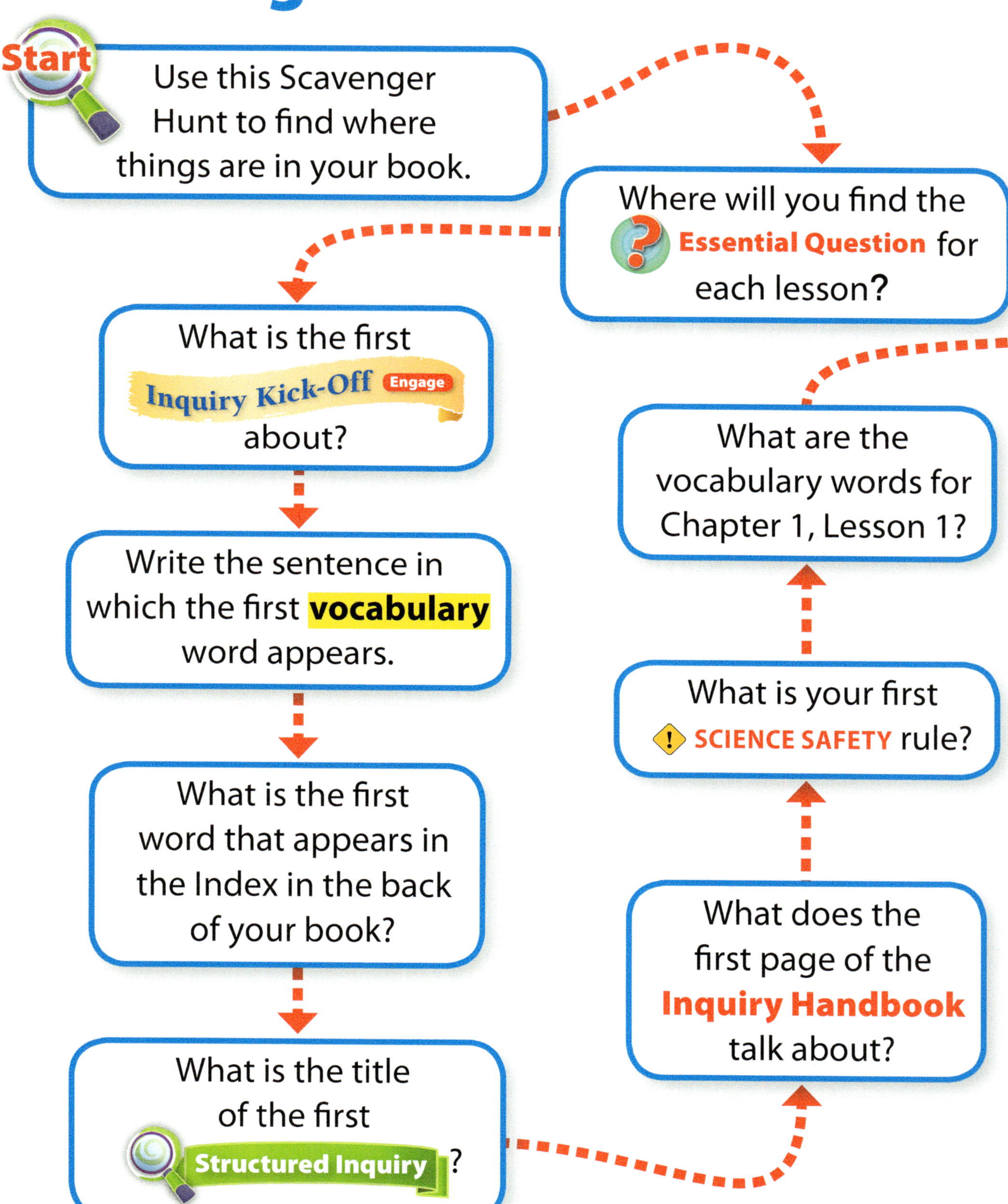

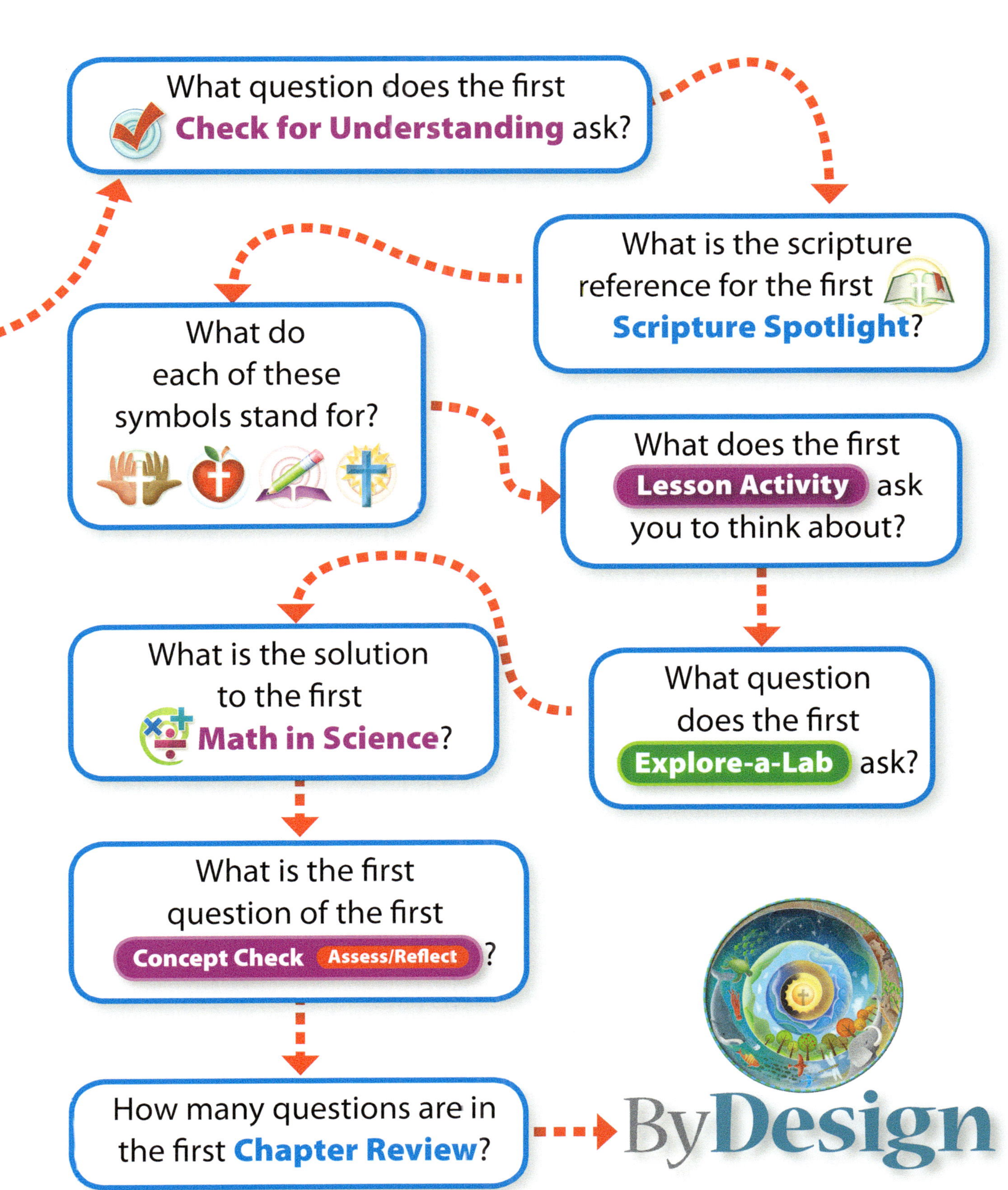

What question does the first Check for Understanding ask?
What is the scripture reference for the first Scripture Spotlight?
What do each of these symbols stand for?
What does the first Lesson Activity ask you to think about?
What question does the first Explore-a-Lab ask?
What is the solution to the first Math in Science?
What is the first question of the first Concept Check Assess/Reflect?
How many questions are in the first Chapter Review?
ByDesign

Table of Contents

Science

How would you define science? Is it all the knowledge accumulated from disciplines such as biology, chemistry, and physics? What other disciplines might it include? Does it include the latest technology? Science includes all of these things, but we usually focus most on science as a process of discovery—a systematic search for understanding of the natural world.

In order to systematically search for the truth, a scientist would have to be true about the natural world. What are some truths about the natural world that a scientist would consider? Can you think of any other assumptions that would need to be true for scientists to proceed with this process?

Scientists make several assumptions that provide the foundation for science:

- Living things and the physical world can be studied and understood.
- The processes of nature follow predictable laws.
- By observation and experimentation, we can learn what these laws are.

Where do you think scientists got the idea that nature follows predictable laws?

Sometimes we distinguish between empirical science and historical science.

Empirical Science

You are probably most familiar with empirical science, which is the basic scientific process that you have been learning about for several years.

Process of Empirical Science	
Observe	Use your senses to make observations.
Question	Ask a question that you would like to have answered.
Hypothesize	Provide an explanation for an observation or problem that can be tested by experimentation. Hypotheses are also known as educated guesses.
Experiment	Design a test that will confirm or falsify your hypothesis.
Record data	Carefully gather and record data. (Data might include pH, time, temperature, distance, growth, or other measurements.)
Analyze data and draw conclusions	Determine whether the data confirms or falsifies your hypothesis. This may lead you to ask more questions and start the process all over again.

Scientists do not always follow these processes in the same order. New questions may be raised at every step in the process, additional observations along the way may lead to changes in the experimental procedure, and multiple experiments may need to be done before a hypothesis can be confirmed or rejected.

Testable

It is important to emphasize that in empirical science, a hypothesis must be testable. This means that a scientist must be able to design an experiment that can test whether a hypothesis is true.

Repeatable

It is also important for an experiment to be repeatable. If one scientist performs an experiment and leaves a detailed description of the exact procedure followed and the data gathered, another scientist should be able to repeat the experiment and verify the results.

Historical Science

Unlike empirical science, where observations can be made and experiments can be performed in the present, historical science is more like trying to solve a mystery about something that happened in the past. How many of these examples do you think would be considered historical science?

- A detective solving a crime
- An archaeologist piecing together what happened at the ruins of an ancient city
- A paleontologist studying fossils buried in a certain order in the geologic column

Did any of these scientists witness the event in person? Can they design an experiment that could repeat the event? How might the scientist go about solving the mystery? After gathering as many clues as possible, each scientist can reason based on what he or she knows about cause and effect and can suggest a hypothesis to explain the data. Scientists often compare multiple hypotheses to see which one can explain the most data. Although reasonable suggestions can be made, it is impossible to directly verify a hypothesis related to historical science. How is this different from the process of empirical science?

Activity

Form a group of two or four students. On a large sheet of paper, create a two-column chart. Label the columns *Evidence* and *Hypothesis*. Your teacher will present evidence, and your group will form a hypothesis based on this evidence. Record the evidence and the corresponding hypothesis on your chart. After you have recorded the hypothesis for the first evidence, write *Evidence 2* on your chart. Your teacher will present additional evidence. Record the new evidence and revise your group's hypothesis based on this. Follow the same procedure until all the evidence has been presented and work together to create a final hypothesis. Present your final hypothesis to your class and explain how your group came to this conclusion. Were some of your group's ideas different from those of other groups? If so, why do you think they were different?

When presented with the same evidence, how different are our hypotheses?

When scientists search for explanations for events from history, their search often leads them to ask whether the event is best explained as the result of natural processes like the ones we see occurring around us today or whether the event is better explained as the result of a catastrophe. How might they be able to figure that out? Geologists used to think that the Channeled Scablands in the state of Washington formed as the result of slow geological processes. A scientist named J. Harlen Bretz suggested that a catastrophic flood from glacial Lake Missoula could better explain the evidence. After much data was gathered over many years, other scientists finally agreed that the data was, in fact, best explained by catastrophic processes.

In addition to regular natural processes and catastrophes, another kind of event—a nonrepeatable, supernatural event—would be considered part of historical science.

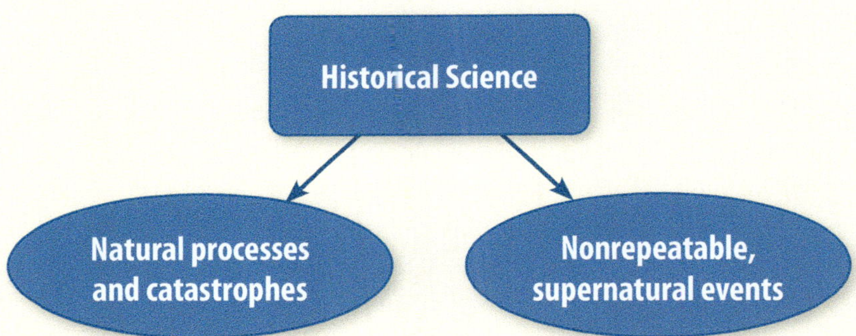

The creation of the Universe and the creation of life on Earth would be examples of nonrepeatable, supernatural events that would need to be approached as historical rather than empirical science. Why would this be?

What about the global Flood described in **Genesis**? Would that be considered historical science? Would it be a natural catastrophe or a nonrepeatable, supernatural event?

To understand historical science, it is necessary to examine more assumptions that scientists make about science. Before you read the chart on the following page, predict how you think each worldview will react to the idea of intervention by God in history.

Assumptions		
Biblical Worldview Only	**Both Worldviews**	**Naturalistic Worldview Only**
God does exist.		God may or may not exist. If He exists, He did not create and does not interfere with nature.
Divine revelation may reveal information about nature that human beings might never discover by themselves.	Living things and the physical world can be studied and understood.	Human beings are capable of understanding nature without divine revelation.
God created the Universe by a process we don't understand and sustains nature through the laws He has established.	The processes of nature follow predictable laws. By observation and experimentation, we can learn what these laws are.	Nature operates on its own according to these laws.
	Change has occurred in organisms and in the physical Universe.	
Scientists are willing to consider the possibility of intervention by God at various times in history.	Scientists cannot do experiments to test the supernatural. Scientific hypotheses must be testable using only criteria accessible to the five senses.	Scientists are only willing to consider theories that do not imply or require any supernatural activity at any time in history.

Both creationist and naturalistic scientists would agree that supernatural events themselves (like Creation or the Flood) are not testable directly using the scientific method. Why would this be? Scientists with different worldviews differ dramatically in their willingness to consider those theories that imply or require any supernatural activity. Naturalistic scientists believe that since Creation and the Flood are not directly testable, they are outside of the realm of science. Scientists with a biblical worldview, however, believe that even though the supernatural events themselves are not directly testable, they can inspire genuine scientific research.

Informed interventionism is a view of historical science that recognizes the intervention of God at various times in history as well as God's divine revelation to us through the Scriptures. It is broader than creationism, because, in addition to recognizing God's intervention at Creation, it also recognizes His intervention in geological history at the time of the Flood. Based on their belief in events described in Scripture, scientists with this view of historical science design experiments that can be done using the scientific process. Which worldview—naturalism or informed interventionism—considers more hypotheses? Which do you think would allow the most data to be gathered?

God and Science

Have you ever thought about the fact that science is a creation of God? You have learned that science includes both a product (knowledge and technology) and a process. As you know, the things we study in science—like the Universe, Earth, the oceans, the laws of nature, plants, animals, and people—were created by God. When God created us in His image, He gave us curiosity and the desire to explore His Creation and learn new things. He also gave us the ability to think and be creative. Nature, the process we use to accumulate knowledge about nature, as well as the ability to apply that knowledge and invent new things all come from God.

With these gifts from God come the responsibility to use them in ways that glorify Him.

Gifts from God	To Think About
Nature	How should we care for nature?
Scientific process	How can we participate in the scientific process in a way that glorifies God?
Ability to apply knowledge/ technology	How might responsibility to God affect how we apply knowledge and use technology?

Did you know that God was the first science teacher? In the book *Patriarchs and Prophets,* Ellen White describes how Adam and Eve received instruction directly from the Creator, Himself. Imagine having the laws of nature explained to you by the One who created them! They enjoyed studying the mysteries of the heavens, the orbiting planets, day and night, light and sound. They were familiar with the habits of the animals, from the massive leviathan to the tiny insects. They saw the handiwork of God.

Scripture Spotlight

What do **1 Corinthians 10:31**, **Ecclesiastes 9:10**, and **Ephesians 6:7** have in common? What do they have to do with the scientific process?

Sharing, Critiquing, and Changing

Whether a scientist is using observation and experimentation in the present or seeking explanations for past events, a very important part of the scientific process is sharing ideas with other scientists, verifying experimental results, and critiquing one another's conclusions. In this way, scientists help one another evaluate ideas, find and correct mistakes, and work together to increase the body of accurate knowledge about the natural world. Can you think of a time when working with other people helped you to learn more or be more accurate in your work? What kind of communication skills do you think would be necessary for scientists during this process?

As data is gathered, a hypothesis may be verified or it may be falsified. If scientists accumulate evidence to verify a hypothesis or a group of hypotheses, a scientific theory may be formulated. Theories are only formulated when a significant amount of data supports a particular idea. Although science is a search for knowledge about the natural world, scientists are hesitant to claim that any theory is absolute truth. Instead, the idea that any theory may be discarded as new data is gathered is a valued and foundational characteristic of science.

Hypothesis (plural: hypotheses)	A hypothesis is a reasonable guess based on something you know or can observe. The goal of the scientific process is to verify or falsify hypotheses. It is easier to disprove a hypothesis than to prove that it is true because the possibility always remains that additional data may disprove it later.
Theory	A theory is a hypothesis that is generally accepted as true because it is supported by a significant amount of data. A theory can change over time if new evidence suggests that it was incomplete. A theory can also be disproved and discarded if evidence is discovered that disputes the theory.
Law	A law describes something that always happens, with no exceptions, under specific circumstances (such as the law of gravity). A law does not explain how or why it happens.

In many ancient civilizations, people believed that the Sun, the Moon, and the planets circled Earth. This was based on their observations that Earth appeared to be unmoving while the other celestial bodies appeared to revolve around Earth each day. Over time, other scientists gathered data that conflicted with and eventually disproved this model. In time, it was replaced by a heliocentric model that recognizes that Earth and the planets of our Solar System revolve around the Sun instead.

In the same way, the accumulation of scientific evidence is challenging another idea that has been around for a while in scientific circles—that Earth is an average planet spinning around an average star in an average galaxy. According to this idea, there is nothing exceptional about Earth, and there is likely to be life on any number of other similar planets throughout the Universe. Scientists have gathered and analyzed dozens of pieces of evidence that point to the conclusion that Earth is anything but ordinary, that it has finely tuned conditions that allow intelligent life to exist here, and that those very conditions also make it perfectly suited for scientific discovery.

The Privileged Planet—Designed for Science

Two scientists named Guillermo Gonzalez and Jay Wesley Richards wrote a book called *The Privileged Planet*. In it they describe how Earth's location in the Universe, in our galaxy, and in our Solar System provides the resources necessary for life to exist on Earth while protecting us from many of the dangers that exist elsewhere in the Universe. These conditions, along with things such as the size and rotation of Earth, the mass of the Moon and the Sun, and a host of other factors, conspire together to make Earth a habitable planet.

Interestingly, they suggest that the same conditions that make life possible here also make Earth strangely well suited for viewing and analyzing the Universe. One example of this is that solar eclipses can be viewed better from Earth than from any other planet in our Solar System. A total eclipse is possible because the Sun is four hundred times larger than the Moon, but it is also four hundred times farther away. The fact that eclipses can be viewed best from the only place where there are observers to see them seems like more than just a coincidence. Being able to see these eclipses has resulted in important scientific discoveries that would have been difficult, if not impossible, to make from places where eclipses don't happen.

In addition, Earth's location provides us with the best platform for observing both nearby and distant stars. We are also in an excellent position to detect cosmic background radiation, which helped scientists confirm that the Universe had a beginning. Gonzalez and Richards do not believe that there is any reason to assume that the very same properties that allow for life on Earth would also by chance or natural law provide the best setting in which to make scientific discoveries about the world and the Universe. Instead they suggest that the evidence points to Design—not only that God designed our Earth to be inhabited, but that He also designed it specifically so that people could engage in scientific discovery. What reasons can you think of to explain why God might have designed the Universe for discovery?

That Earth was created to be inhabited is consistent with what the Bible teaches. *"For thus says the Lord, who created the heavens (He is the God who formed the Earth and made it, He established it and did not create it a waste place, but formed it to be inhabited.)."* **Isaiah 45:18**

That God designed the Universe for discovery also seems reasonable. Humans have the capacity to be curious, to think, and to explore. Several verses in the Bible mention trying to figure out the mysteries of God. If God did design the Universe for discovery, what do you think He most wants us to discover? Paul describes in **Romans 1:20** that we can understand God's invisible attributes, His eternal power, and His divine nature from the things He has made. In **Psalm 19:1**, David says that the heavens can teach us about Gods glory.

If God specifically positioned our Earth in such a way that we could engage in scientific discovery, what implications might that have for how we do science?

Christian Roots of Science

Do you know of any scientists in the Bible? You probably know that God gave Solomon ***"wisdom and very great discernment and breadth of mind, like the sand that is on the seashore"*** **and that his wisdom** ***"surpassed the wisdom of all the sons of the east and all the wisdom of Egypt."*** You can read about it in **1 Kings 4:29–34**. But did you know that Solomon was a scientist? In addition to his 3000 proverbs and his 1005 songs, the Bible says that he could speak about ***"trees, from the cedar that is in Lebanon even to the hyssop that grows on the wall; he spoke also of animals and birds and creeping things and fish."*** People came from all over to listen to his wisdom, which included what he knew about science.

Leonardo da Vinci was an artist and scientist, best known for his painting of the Last Supper, and for his inventions.

The wisest man who ever lived studied the natural world from the perspective of one who believed in God as Creator and recognized the authority of scripture. Since that time, thousands of scientists have followed in Solomon's footsteps. Many of the greatest scientists of all time were Bible-believing Christians. In fact, Christianity itself actually provided the environment in which modern science developed. Look at the chart on the next page to learn more about some Christian scientists.

Biblical Christianity believes that nature was created and is sustained by a law-giving God. By discovering predictable patterns in nature, it is possible to figure out the natural laws established by God. Because the laws of nature do not change, it is possible to make predictions about what will happen in certain circumstances. This belief that natural laws existed and could be studied provided the rational foundation for modern science.

Scientist	Contribution to Science	Christian Beliefs
Leonardo da Vinci	Founder of modern science	Believer in Christ and the Scriptures; painted *The Last Supper*
Johann Kepler	Founder of physical astronomy; discovered the laws of planetary motion	He stated that in his research, he was merely "thinking God's thoughts after Him." He also said, "Since we astronomers are priests of the highest God in regard to the book of nature, it befits us to be thoughtful, not of the glory of our minds, but rather, above all else, of the glory of God."
Francis Bacon	Formulated the scientific method	Devout believer in the Bible. Said: "There are two books laid before us to study, to prevent our falling into error; first, the volume of the scriptures, which reveal the will of God; then the volume of the Creatures, which expresses His power."
Robert Boyle	Father of modern chemistry	Diligent student of the Bible; profoundly interested in missions; funded Bible translation
Isaac Newton	Discovered the law of gravity	Believed in Jesus as Savior and in the Bible as God's word; defended Creation and the worldwide biblical Flood; said, "We account the scriptures of God to be the most sublime philosophy. I find more sure marks of authenticity in the Bible than in any profane history whatsoever."
Michael Faraday	One of the greatest physicists ever, who developed the sciences of electricity and magnetism	Humble and sincere Christian who had an abiding faith in the Bible and in prayer
Louis Pasteur	Established the germ theory of disease; disproved spontaneous generation; developed vaccines for rabies, diphtheria, anthrax, and others; developed pasteurization and sterilization	Strongly religious man who opposed Darwinism. When asked about his faith, he said, "The more I know, the more does my faith approach that of the Breton peasant. Could I but know all, I would have the faith of a Breton peasant woman."

In addition to these and other scientists from history who believed in God as Creator and in the authority of Scripture, many scientists today still hold these beliefs and study science to learn more about God as He is revealed in nature.

Unit 1 Life Science

Unit Overview

What living organisms have you encountered today? What do all of these organisms have in common? Living organisms are all around you. They vary in size and shape, but all living things are made of the same small units—cells. In this unit, you will learn more about cells and the cell parts that work together inside the cell. You'll also find out how cells differ in structure and function.

Many living things are made of just one cell. Larger organisms are made of trillions of cells! In this unit, you will take an in-depth look at the variety of organisms God designed during Creation. Why do you think God wanted so much variety? You'll find out about the features of many groups of living things, as well as how the organisms survive in their environments. You will learn how organisms are classified, or grouped, according to their features.

For an organism to survive, it must reproduce. When organisms reproduce, they pass on their traits to their offspring. In this unit, you will learn why organisms often resemble their parents, and you'll find out about the mechanism that allows parents to pass on their traits.

A young elephant, which is called a calf, can weigh more than 120 kg. For the first two years of its life, a calf drinks milk from its mother as it learns to eat grasses and leaves.

ADVENTISTS InStep for Life

Your teacher may assign an Open Inquiry lab and a Lifestyle Challenge activity. Use your ***Science Journal*** to record your work.

Science Journal

Bacteria, Protists, and Fungi

Scripture Spotlight

When you think about the things God created, do you think about the things you need a microscope to see? Most bacteria and protists and some fungi are microscopic. By studying these organisms you can better understand God and the world He created for us. You will read the following passages in this chapter.

Isaiah 43:1 (p. 19)
Leviticus 14:33–45 (p. 40)
Numbers 19:15 (p. 27)
Exodus 7:14–25 (p. 37)

Listeria, shown growing on an agar plate, is a bacterium found in soil, water, and some animals, such as poultry and cattle. It can be present in raw milk and foods made from raw milk. *Listeria* is one of the harmful bacteria that can cause food poisoning.

The Big Idea

All living things are classified into groups according to the characteristics they share. Microscopic organisms such as bacteria, protists, and fungi are alike because most cannot be seen without magnification. Each can be either beneficial or harmful.

How are the bacteria in this picture like you? How are they different from you?

Inquiry Kick-Off Engage

The live bacteria in yogurt provide many health benefits, such as preventing infections. What do you think bacteria look like? In your ***Science Journal,*** you will investigate what one type of bacteria found in yogurt looks like.

Essential Question

How Are Living Things Classified?

Objectives

- Describe how scientists name and classify living things.
- Explain the structure of scientific names.
- Describe the classification system used by scientists.

Vocabulary

common name
scientific name
species
binomial nomenclature
taxonomy
kingdom
domain

Look out a school window. Make a list of things you see. Which things are living, and which are nonliving? How are the living things—called organisms—alike? How are they different? What characteristics would you use to arrange these organisms into groups? To efficiently study the amazing variety of organisms on the planet, scientists have developed different ways to sort them into similar groups. How would this approach help scientists study and classify organisms? Using the characteristics that you just considered for the organisms outside, how would a lizard fit into the grouping? How would a cactus fit into the grouping? What about a blue whale or a paramecium? In this chapter, you will learn about the different considerations scientists use to group organisms and how science benefits from performing these organizational tasks on the diverse organisms that inhabit the planet.

Studying the Living World Explain

Where in a grocery store would you look for canned corn? Where would you look for fresh carrots? Items in grocery stores are sorted and grouped to make them easy to find. What difficulties would you have shopping in a grocery store where the items weren't sorted and grouped in a way that you understand?

Just as a store arranges similar items together, a system of classification places organisms into groups. These groups are based on similar characteristics. To form these groups, scientists study how organisms are structured and how they function. Scientists look at the following characteristics.

Classifying Organisms

Characteristic	Defining Question
Body type/Cell type	Is the organism single celled or multicellular?
Cell structure	Are the cells of the organism prokaryotic or eukaryotic?
Method of getting energy	Is the organism an autotroph or a heterotroph?

Naming Organisms

The Bible tells us in **Genesis 2:20** that Adam named all the animals at Creation. When speaking, we usually use a **common name** when describing living things. You may use common names to refer to a robin, an oak tree, or a deer. However, the same common name may mean different things to different people. For example, the bird that you call a robin is very different from a robin in Great Britain. In Australia, yet another kind of bird is called a robin. How can three different birds all be called a robin?

To avoid confusion, scientists give each kind of organism a unique, two-word Latin name. For example, the species known as the polar bear is named *Ursus maritimus*. The two-word name assigned to each kind of organism is referred to as its **scientific name**. The first word of a scientific name is the genus name. A *genus* (plural, *genera*) is a group of closely related species. The second word in the scientific name identifies the species. A **species** is a group of organisms that can mate to produce fertile offspring. The genus name of the polar bear is *Ursus*, and the species name is *maritimus*. A genus name is always capitalized, while the species name is always lowercase. Notice that a scientific name is always in italics. Scientific names are underlined if being handwritten. This two-part Latin naming system, called **binomial nomenclature**, is used to name all living things. Do you know the scientific name for humans?

A species is the narrowest category. A genus may contain several species. For example, the genus *Ursus* contains several species of bears. But only polar bears are given the scientific name *Ursus maritimus*.

Swedish naturalist Carolus Linnaeus developed this two-part Latin naming system in the 1750s. Scientists use Latin because it is standardized, consistent, and no longer changing. Why do you think Latin is no longer changing? What issues might occur by using a language that is still changing? People speaking any language can understand Latin scientific names. For example, a Spanish scientist, a French scientist, and a Russian scientist will each use the same Latin name for the same kind of organism. A Latin genus or species name often tells you something about the organism. For example, consider the species name for the polar bear, *maritimus*. What does this name suggest to you? Why do you think *maritimus* is a good species name for the polar bear? The binomial nomenclature for humans is *Homo sapien*. What is the meaning of the word *sapien*?

Each of these birds is called a robin. Yet they are different species.

How are these robins different from the common American robin? What could be some reasons these birds have the same common name?

Record your work for this inquiry. Your teacher may also assign the related Guided Inquiry.

Classifying Seeds

How do characteristics help you to classify seeds?

Materials

- 10 different kinds of seeds
- hand lens
- metric ruler

Procedure

1. **Observe** your seed collection closely, and use a hand lens if necessary. Divide your collection of seeds into two groups based on an observable characteristic. Use the characteristic to identify each group of seeds in the chart in your ***Science Journal***. Label one of the groups *A* and the other group *B*. Draw and color the seeds in each of the two groups.
2. Take the seeds in group *A* and divide this group into two more groups based on a new observable characteristic. Use this characteristic to identify each group of seeds in the chart in your ***Science Journal***. Draw and color the seeds in each of the two groups.
3. Take these two groups of seeds and divide them one more time into two groups based on a new observable characteristic. Use this characteristic to identify each group of seeds in the chart in your ***Science Journal***. Draw and color the seeds in each of the two groups.
4. Repeat Steps 2 and 3 with the group *B* seeds from Step 1.

Analyze Results

Compare and discuss your systems of classification with those made by your classmates. Then use your classification system to develop a dichotomous key to explain your classification process in the activity. A dichotomous key is a series of questions, each with only two answers. The answer to each pair of questions will either identify the unknown organism or lead you to another set of questions that lead to an identification.

Create Explanations

1. How do characteristics help you to classify seeds?
2. What characteristics are useful in distinguishing between seeds?
3. How might a scientist use the seed of a plant to help classify the plant?

Many species's names describe the region where the species lives, its color, size, or shape. The sugar maple, *Acer saccharum*, is the tree from which we get sweet-tasting maple syrup. *Saccharum* is the Latin word for sugar. Sometimes the species name includes the name of the scientist who first described it. In 2012, a new species of hermit crab, *Aeropaguristes tudgei*, was named after Dr. Christopher Tudge, the scientist who discovered it on the barrier reef off the coast of Belize. What does the name of the black bear, *Ursus americanus*, tell you about the species? What are the similarities between a black bear and a polar bear that group them into the same genus?

Scripture Spotlight

Names are important. The Bible says in **Isaiah 43:1** that God has redeemed us and calls us by name.

Classifying Organisms Explain

The science of classifying living things is called **taxonomy**. A *taxonomist* classifies living things by placing organisms into increasingly larger groups based on shared features. The smallest, or narrowest, classification category is a species. Similar species are grouped to form a genus. Similar genera form a family, while similar families make up an order. Taxonomists group similar orders to form a class and similar classes to form a phylum (plural, phyla). Similar phyla form a **kingdom**. Hamsters and other animals form the Animal kingdom. Plants make up the Plant kingdom. Fungi and organisms such as molds and mushrooms are placed in the Fungi kingdom. All living things are classified in the same way. Latin names are used throughout taxonomy. Why do you think taxonomists need to use the same type of classification? What might happen if taxonomists in the United States and Canada classified the groups differently?

Classification of a Dwarf Hamster

Taxonomic Group	Group Name	Group Members
Domain (largest group)	Eukarya	Eukaryotic organisms
Kingdom	Animalia	Animals
Phylum	Chordata	Animals with a nerve cord; many also have a backbone
Class	Mammalia	Animals with fur or hair; also produce milk to feed their young
Order	Rodentia	Mammals that are characterized by a single pair of continuously growing incisors in the upper and lower jaws that must be kept short by gnawing
Family	Cricetidae	Hamsters, including voles, lemmings, and New World rats and mice
Genus	*Phodopus*	A lineage of small hamsters native to central Asia that displays unusual adaptations to extreme temperatures
Species (smallest group)	*P. campbelli*	Hamster named for W.C. Campbell, who collected the first specimen

Domains

A discovery in the 1970s made scientists rethink the use of kingdom as the largest taxonomic category. Research revealed new differences in cell structure and DNA among all of the kingdoms. Scientists found a unique group of organisms called *archaea*. Like bacteria, they are single celled and lack nuclei. However, their cell walls are not made from the same substance as those of bacteria. Also, their DNA structure is similar to that of eukaryotes, not bacteria.

Scientists decided to place the archaea into a new category. They developed a new taxonomic category called a domain. A **domain** is a taxonomic category above the kingdom level. There are three domains: Archaea, Bacteria, and Eukarya.

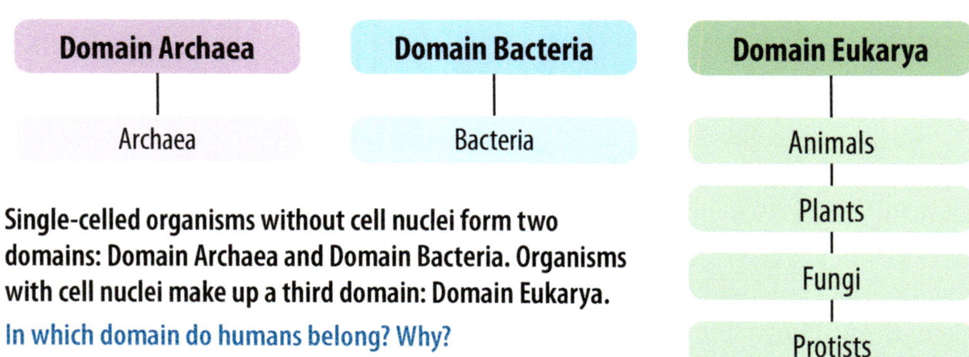

Single-celled organisms without cell nuclei form two domains: Domain Archaea and Domain Bacteria. Organisms with cell nuclei make up a third domain: Domain Eukarya.

In which domain do humans belong? Why?

Classifications May Change **Explain**

Compare pictures of a shark and a dolphin. How are they alike? How are they different? How do scientists classify the two animals? Sometimes organisms that look similar are not as alike as they may seem. Can you think of other animals or plants that look similar but aren't classified in the same group?

Red panda

Animals we know as bears are species in the bear family (family Ursidae). The bear family includes the black bear, brown bear, and polar bear. How about pandas? Unlike other bears, the red panda and the giant panda feed primarily on bamboo, but they occasionally eat other plants and even meat. Both species have a small, bony projection on their wrists that helps them grip bamboo stalks. Try to grasp something without using your thumb. Then pick up the same object using your thumb. How does having an opposable thumb help you pick up things?

The giant panda looks more like a bear than the red panda. The red panda shares some characteristics of raccoons. As a result, scientists have debated their classification. After studying the DNA of pandas, scientists learned new information. They discovered that the giant panda is more similar to a bear than to the red panda. Today the giant panda is usually classified as a member of the bear family. What characteristics does the giant panda share with other bears? Many taxonomists classify the red panda, however, in its own family.

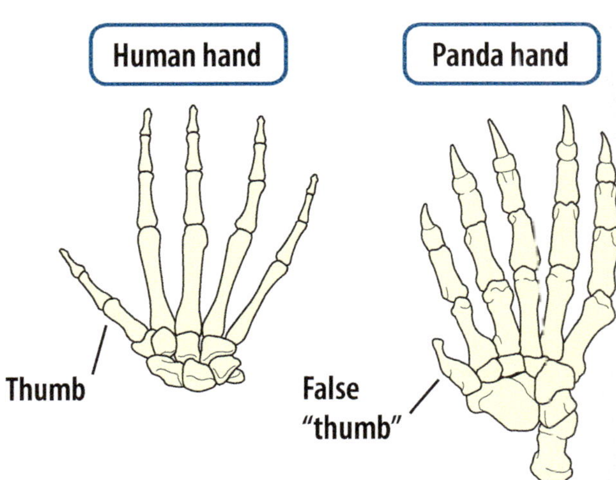

The false thumb allows pandas to grip bamboo stalks.

How are the panda's unique wrist bones similar to a human's thumb?

Giant panda

Finding New Species Explain

Faith Connection

Scientists continue to find new species of animals in modern times. What does this tell you about God's Creation?

Scientists continue to find new species. Many new species are found in remote areas that have not been previously explored. Off the coast of Indonesia in 2013, scientists researching the coral reef discovered a new type of shark. This unique species of epaulette shark, now named *Hemiscyllium halmahera*, frequents the shallow waters near large coral reefs and tide pools. Like many epaulette sharks, this shark has small fins on the bottom of its body that it uses to wriggle from side to side to move along the ocean floor. For this reason, it is often called a "walking shark." This new species of shark can grow to be 70 cm long and is covered in large, dark spots or wide, dark stripes. How do you think such a large animal remained hidden for so long?

In 2010, researchers working in Madagascar identified a new species, Durrell's vontsira (*Salanoia durreli*). Scientists think this small, ferret-like animal is the first new meat-eating species seen in more than 25 years. It was first spotted swimming in Madagascar's Lac Alaotra in 2004. Since 2004, scientists have spotted just four Durrell's vontsiras. Two were captured briefly, while another two have only been photographed. Why do you think the two were captured and then released? How do you think new, rare species should be handled? Should they be captured, or should they live in the wild? Why do you think this?

Newly discovered species are often few in number. Many are at risk of extinction because the areas where they live are being threatened by development. Biologists explore these areas, hoping to identify new species before their habitats are destroyed so that they can protect the animal and its habitat.

Epaulette sharks feed on small fish and crustaceans that live on coral reefs.

What are some examples of human activities that could threaten this newly found species?

Math in Science

There are more than 1.5 million identified species of animals. More than 1.3 million of these are known species of invertebrates. What percentage of animals are invertebrates? Why are scientists unable to say definitively how many species of animals there are?

Lesson Activity

Use the Internet or other resources to learn about one of the organisms listed below or another organism discovered since 2000. Create a display that includes a picture of the organism; its scientific name; when, where, and how it was discovered; and who discovered it. Present your poster to the class.

- Cypriot mouse
- Walter's duiker
- walking shark
- Darwin's bark spider
- Olinguito
- red-bearded titi
- Pinocchio frog
- Cambodian tailorbird
- Hero shrew
- legless lizard
- epaulette shark

 **What organisms have been discovered recently?**

Concept Check Assess/Reflect

Summary: How are living things classified? Scientists classify living things by shared characteristics. They name living things using a two-part system developed by Carolus Linnaeus that uses Latin. The first part of an organism's scientific name is the genus to which the organism belongs. The first letter of the genus name is capitalized. The second part of the scientific name identifies the species and is lowercased. An organism's scientific name is italicized in type and underlined when handwritten. The classification system used by taxonomists becomes more precise from the largest to the smallest group. The system progresses from domain to kingdom, to phylum, to class, to order, to family, to genus, and to species.

1. Why is it necessary for scientists to use scientific names rather than common names when describing an organism?
2. Classify the following animals into four groups based on their similarities: gorilla, butterfly, chimpanzee, lobster, mosquito, mouse, wasp, crayfish, lobster, chipmunk, and crab. Describe the characteristics that you used to group the organisms.
3. Life scientists estimate that millions of living species are undiscovered. What might be some reasons so many species are undiscovered?
4. How can new information about DNA help taxonomists?

Lesson 2

Essential Question

What Are Bacteria?

Objectives

- Identify characteristics and structures of bacteria.
- Describe the growth of bacteria.
- Describe the types of bacteria.
- Describe diseases caused by bacteria.
- Summarize the roles of bacteria.

Vocabulary

microorganism
autotroph
heterotroph
binary fission
aerobic bacteria
anaerobic bacteria
antibiotic

Did you know that bacteria (singular bacterium) are the most abundant organisms on Earth? Bacteria are microscopic organisms found in almost every environment on Earth, no matter how extreme. Bacteria have been found on the bottom of the ocean floor, in volcanoes, at the poles, and even in your intestinal tract. Did you know that there are 10 times more bacteria cells living in your body than there are human cells? What other places do you think you could find bacteria if you were looking through a microscope? How can bacteria survive in such diverse environments?

Characteristics of Bacteria Explain

Have you ever gone to the doctor and been told that you have a bacterial infection? Bacteria are **microorganisms**, organisms too small to see without magnification. Bacteria can thrive almost anywhere. What precautions can you take to prevent infections? In what areas of your home are these precautions most important to take? Most of us think of bacteria as tiny, single-celled organisms that cause disease. Because bacteria are small, it is tempting to think of them as simple. Bacteria are more complex and varied than you might think, and not all bacteria are bad for humans. Can you think of any useful bacteria? What are some useful bacteria you have heard of?

What could happen if bacteria got into your food? How can you prevent that from happening?

Bacteria are prokaryotes. Recall that a *prokaryote* is a single-celled microorganism that does not have a nucleus or membrane-bound organelles. As you know, *eukaryotes* have cells that have nuclei surrounded by a nuclear membrane. Bacteria are smaller than many eukaryotic cells. Like plants and fungi, bacteria have cell walls. But how are bacterial cell walls different? They contain *peptidoglycan,* a tough compound made of sugars and protein woven together, which is evidence of its complexity and God's Design. Why do you think a bacteria cell needs a tough wall?

Escherichia coli

The bacterium *Escherichia coli* (better known as *E. coli*) normally lives in the lower intestines of people and other warm-blooded animals.

Why would the presence of *E. coli* in food be a concern?

Food Sources

Recall that organisms that can make their own food are called **autotrophs**, or *producers*. These organisms make their own food using energy from sunlight or from inorganic compounds. Photosynthesis is the process by which autotrophic organisms harvest sunlight to produce energy. Some bacteria, called *cyanobacteria*, are photosynthetic. These bacteria have substances called *pigments* that capture sunlight. Chemosynthesis is the process by which autotrophic organisms harvest energy from chemicals in the environment. The bacteria in the soil that convert nitrogen compounds into a form that plants can use are chemosynthetic bacteria.

Organisms that cannot make their own food must obtain energy from other sources. They are called **heterotrophs**, or *consumers*. Most bacteria are heterotrophs. Heterotrophic bacteria obtain energy in many different ways. Some obtain nourishment from dead organisms. Other bacteria get nutrients from another living organism, called a *host*. These bacteria, called *parasites*, can cause disease as they feed on their hosts. What are examples of parasites? Most strains of *E. coli* bacteria are harmless heterotrophs. However, some strains can cause serious food poisoning in humans.

Reproduction

Bacteria usually reproduce asexually by **binary fission**, a form of reproduction common in single-celled organisms. During binary fission, a single cell first grows in size. Then it divides to form two cells of the same size. Binary fission is considered asexual reproduction because a single parent cell produces offspring that are identical to the parent cell. This type of reproduction allows bacteria to grow quickly in number. Under ideal conditions, a single bacterium can divide roughly every 20 minutes. How long do you think it would take one bacterium to grow to one million? What factors might affect how quickly a bacterium cell grows?

Record your work for this inquiry. Your teacher may also assign the related Guided Inquiry.

Bacterial Growth Rates

Where do bacteria grow at school?

Procedure

Materials
- safety goggles
- lab apron
- disposable gloves
- 4 Petri dishes with nutrient agar
- wax pencil or permanent marker
- water, sterile distilled
- sterile cotton swabs
- tape

1. Working with a partner, obtain four Petri dishes from your teacher. Do not open your dishes until you are told to do so. Use a wax pencil or permanent marker to write your team's name on the top of the dishes.
2. Select three school locations to test for the presence of bacteria. Write the name of each location on one of your Petri dishes. **Record** the locations in your ***Science Journal***. Write *Control* on the remaining Petri dish.
3. Moisten a sterile cotton swab with distilled water. Wipe the swab against the surface of the first location that you are testing. Open the dish and gently rub the swab across the agar on the appropriate side of the dish. Place a piece of tape on each side of the Petri dish to attach the top of the dish to the bottom.
4. Repeat Step 3 using a new cotton swab at a second and third location. Be sure to use a clean cotton swab for each sampling.
5. Store your dishes as directed by your teacher. **Observe** your dishes and those of other groups after 24 hours. **Compare** growth rates for different areas. **Record** what you see in your ***Science Journal***.
6. Return the Petri dish to your teacher. After another 24 hours, observe how the bacterial colonies changed. Record what you see in your ***Science Journal***.

Analyze Results

Bacterial growth will appear as shiny, round dots on the agar. **Display your data** for each location in a table summarizing your findings.

Create Explanations

1. Where do bacteria grow at school?
2. Which of the locations produced the greatest amount of bacteria?
3. How did the growth of bacteria change over time? What factors affected the rate of growth of your samples?

Kinds of Bacteria Explain

Why are bacteria so widespread? Bacteria are adapted to live in many different conditions. **Aerobic bacteria** are bacteria that require oxygen for growth. **Anaerobic bacteria** grow only in the absence of oxygen. Some bacteria live at temperatures just above freezing, while others do well at temperatures near boiling. Other bacteria have adapted for living in the bodies of animals, including humans. These bacteria can thrive only at specific body temperatures.

Most bacteria are one of three shapes: spherical, rod, or spiral. Scientists use special stains to make bacteria visible. When a scientist studies an unknown bacterium under the microscope, the shape is the first clue to its identity.

The bacterial cell wall offers a second clue. Scientists classify most bacteria into two groups based on composition of the cell wall. The researchers use a technique called Gram staining, which reveals differences in cell wall structure. How does Gram staining work? The Gram staining procedure distinguishes between Gram-positive and Gram-negative groups by coloring the cells red or violet. Gram-positive bacteria stain violet due to the presence of a thick layer of peptidoglycan in the cell walls. This molecule retains the crystal violet that stains these cells. However, Gram-negative bacteria have a thinner peptidoglycan wall. The thinner wall does not retain the crystal violet stain, producing a red color. What challenges would scientists have if they didn't use Gram staining? How do you think scientists first discovered the unique chemical nature of bacterial cell walls?

Scripture Spotlight

Why do you think that, in **Numbers 19:15**, God said that any open vessel would be unclean?

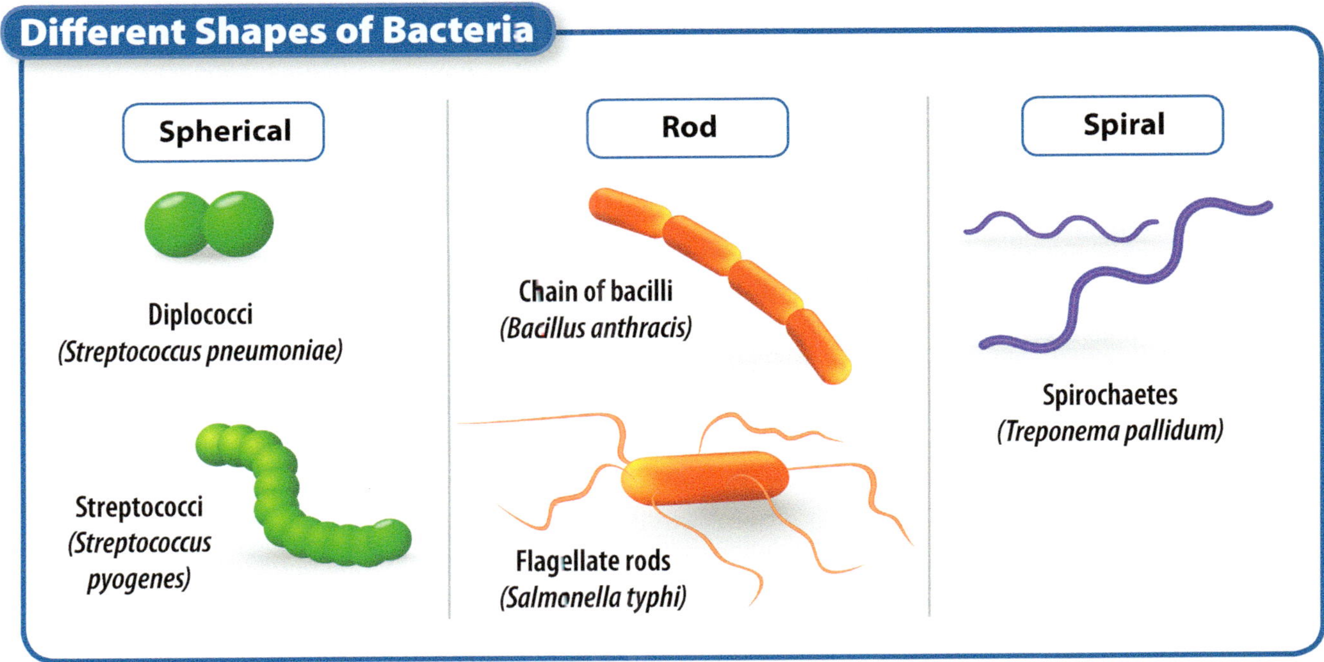

Because of their outer membrane, gram-negative bacteria are often more resistant to host defenses. They are also harder to kill with antibiotics. An **antibiotic** is a drug that can kill bacteria in the body. Different antibiotics kill bacteria in different ways. Many antibiotics, such as penicillin, stop bacteria from dividing. They do this by interfering with the peptidoglycan in the cell wall. Human cells do not contain peptidoglycan, so antibiotics do not stop human cells from dividing. Why is it important for a doctor to know the species of bacteria causing an illness before prescribing an antibiotic? Why might a doctor want to confirm that bacteria cause an illness before prescribing an antibiotic? Antibiotics are not effective on viruses. Why do you think this is so?

Roles of Bacteria Explain

Bacteria are vital organisms in all ecosystems. Did you know that some bacteria in the human body keep our immune systems healthy? Other bacteria produce chemicals that help harness energy in our food. What other vital functions do bacteria perform?

Bacteria as Nutrient Recyclers

Perhaps the most important function of bacteria is to serve as Earth's natural recyclers. Many kinds of bacteria decompose, or break down, dead organisms and waste products. This decomposition releases carbon, nitrogen, and other elements back into the environment. Other living things will use these elements again. Without the actions of decomposers, these elements would remain locked up and unavailable. What other problems would result if there were no decomposers?

Photosynthetic bacteria use light energy to make sugars, releasing oxygen needed by other living things. Some kinds of bacteria capture, or fix, atmospheric nitrogen. These nitrogen-fixing bacteria change nitrogen into forms that plants can use. Bacteria are very important partners in the ecology of all of the places where they live.

Bacteria have been found deep in the ocean. Here, they harness compounds such as hydrogen sulfide to generate energy. This energy supports all of the other organisms in the area. Chemosynthetic bacteria living inside tubeworms get energy from chemicals released in the hot water of hydrothermal vents. The tubeworm protects the bacteria, and the bacteria provide a food source for the tubeworms. Some of these bacteria are decomposers, which provide the same function as they do on land. What would happen if the source of hydrogen sulfide disappeared from a tubeworm community?

Lesson Activity

Observe the labels of several yogurt brands. Look for the words *live and active cultures*. These words refer to the living bacteria, *Lactobacillus bulgaricus* and *Streptococcus thermophilus,* that form yogurt. These bacteria convert pasteurized milk to yogurt during fermentation. Compare three or more brands of yogurt. Read the yogurt labels. Gather and record data that identifies the names of bacteria.

Which brands of yogurt list the scientific names of bacteria? What species of bacteria are listed?

Commercial Uses of Bacteria

Did you eat yogurt, cheese, pickles, buttermilk, or sour cream today? Bacteria contribute to the production of these and many other foods. Dairy foods such as yogurt and many cheeses are made when bacteria break down the lactose in milk, producing lactic acid. Other kinds of bacteria are used in industry to make substances such as acetone or vinegar. Some bacteria can even digest the petroleum released during oil spills.

Check out your ***Science Journal*** for a Guided Inquiry that explores how to test homemade yogurt for live bacteria.

Extend

Tubeworms, the red and white structures in the picture, live near hydrothermal vents deep in the ocean. They contain chemosynthetic bacteria that support these unique ecosystems.

Harmful Bacteria

Bacteria can sometimes cause disease in humans and other organisms. Recall that a disease-causing organism is called a pathogen. Bacterial diseases can be transmitted in the air, in water, or in food. Each kind of pathogenic bacteria is transmitted in a characteristic way. Some pathogenic bacteria invade food or water supplies. These bacteria can grow in food and cause food poisoning if the contaminated food is eaten. What are examples of illness or disease caused by bacteria?

Pathogenic bacteria harm a host because they damage the host's cells. Some pathogenic bacteria destroy cells directly. Others release harmful substances called toxins, which damage the cells. Each species of disease-causing bacterium affects a specific area of the body. Tooth decay, for example, is caused by a bacterium that lives on teeth. Another species of bacteria lives in the stomach and can cause stomach ulcers. What other harmful bacteria have you heard of?

Refer to the table on bacterial diseases. Botulism is a dangerous foodborne illness that can occur when food is not canned properly. Botulism is the illness that occurs when a person is exposed to the toxins produced by the bacterium *Clostridium botulinum*. How do you think companies prevent botulism from affecting products bought by consumers? What are the symptoms of botulism poisoning? What can you do if you believe you have been exposed to botulism?

Focus on Health

Leprosy is a bacterial disease that was common in Bible times. Why is leprosy not a huge concern in most parts of the world today? What other bacterial infections has modern medicine all but eliminated?

Bacterial Diseases

Disease	Bacterium Shape	Area Affected	How Transmitted
Botulism	Rod	Nervous system	Improperly canned foods
Cholera	Rod	Intestines	Contaminated water
E. coli 0157:H7	Rod	Intestines	Contaminated vegetables or raw or undercooked beef
Lyme disease	Spiral	Joints, heart, nervous system	Infected tick
Salmonella food poisoning	Rod	Intestines	Contaminated food or water
Methicillin-resistant *Staphylococcus aureus* (MRSA) infection	Spherical	Skin, lungs, blood	Touching contaminated object or the skin of infected person
Tuberculosis	Rod	Lungs	Breathing in air droplets from an infected person's cough or sneeze

The deer tick, which carries the bacterium that causes Lyme disease, is very small and can be easily overlooked.

Lesson Activity

Separate into groups. Each group should select one of the diseases (except botulism) from the Bacterial Diseases table. Using the Internet, each group should answer at least the following questions: What causes the disease? What are the symptoms of the disease? How is the disease spread? What are the precautionary measures to prevent people from becoming sick? Prepare a three-minute presentation with accompanying poster-board illustrations to present to the class. After all of the presentations, regroup as a class and discuss whether more can be done by society to prevent these illnesses.

How can you help prevent illnesses?

Concept Check Assess/Reflect

Summary: What are bacteria? Bacteria are prokaryotic microorganisms. Some bacteria are heterotrophs, while others are autotrophs. Bacteria usually reproduce asexually by binary fission, often very quickly. Aerobic bacteria require oxygen, while anaerobic bacteria grow only in the absence of oxygen. Bacteria can be round, rod, or spiral shaped. Some bacteria recycle nutrients in ecosystems. Other bacteria are used to produce foods, make chemicals, or clean up pollution. Some bacteria cause diseases. Antibiotics are used to treat bacterial diseases.

1. Explain the difference between a heterotroph and an autotroph. Describe one type of beneficial autotrophic bacteria and one type of beneficial heterotrophic bacteria.
2. How do bacteria differ from plants and animals?
3. Explain the connection between bacterial growth rates and the ability of bacteria to cause disease.
4. Predict some of the negative consequences if all bacteria disappeared.

Lesson 3

Essential Question

What Are Protists?

Objectives

- Identify the characteristics and structures of protists.
- Describe the three main groups of protists.
- Explain the environmental importance of protists.
- Describe helpful and harmful protists.

Vocabulary

protist
protozoan
phytoplankton
pseudopodia
flagella
cilia

Imagine an amoeba, a paramecium, and an alga on the table in front of you. How would you classify these organisms? They have different life cycles, trophic levels, forms of locomotion, and cell structure. Which characteristic would you use to classify these organisms? Are they like plants? Are they like animals? What if other classmates disagreed with your classification? How would you classify the organisms then? Would it help to know that these organisms do not fit into any of the other kingdoms? Scientists are often faced with this same dilemma. They resolved this conflict by creating a brand new kingdom. Meet the protists!

Characteristics of Protists Explain

A **protist** is a eukaryotic organism that cannot be classified as a fungus, a plant, or an animal. If you think this is a broad definition, you are correct. Protists include an astonishingly wide variety of organisms. Some protists resemble plants or fungi. Others look and act more like animals. These single-celled eukaryotes that move and eat food were once called **protozoans** and classified in the Animal kingdom.

Protists have different ways of obtaining nutrition. Some protists are autotrophs. Autotrophic protists include algae, seaweeds, and diatoms, for example. What characteristic do these organisms share? How do these protists get the energy they need? Other protists are heterotrophs. Heterotrophic protists include amoebas, paramecia, and the organisms that cause the disease malaria. Some heterotrophic protists are parasites, while others are decomposers and live on decaying organisms.

Marine diatoms, as viewed with a microscope, float in the surface water of marine environments.

How do you think these diatoms obtain energy?

Protists include organisms of many different sizes. Many are single celled. But some, such as brown seaweed, are more than 60 m in length. Many protists can move, sometimes extremely fast. Other kinds of protists stay in one place.

As scientists learn more about the structure and function of protists, they continue to find new ways to classify them. Many of these classifications are based on scientists' understandings of their similarities and differences. Scientists are also still discovering new species of protists. Why do you think the scientific classification of protists is frequently updated? Do you think protists will ever have a clear and concise classification? Explain.

Faith Connection

As scientists study and discover more data, they may change their thinking. How might your life change the more you study God's Word?

Protist Diversity Explain

The easiest way to learn about protists is to study them in three general groups that are based on the shared characteristics of the members of each group. The three groups are plant-like protists, animal-like protists, and fungi-like protists.

Plant-Like Protists

These protists are autotrophs that need sunlight energy to perform photosynthesis. What other living thing uses sunlight for photosynthesis? Algae are primarily aquatic protists. Some algae resemble plants. But, unlike plants, algae lack roots, stems, and leaves. Red algae and brown algae live in the oceans. Brown algae are commonly called *seaweeds* and can be many meters in length. Green algae include both single-celled and multicellular varieties. Green algae are found in marine water, in freshwater, and living on land.

Phytoplankton are small photosynthetic organisms that float and drift near the ocean's surface. These organisms, which include photosynthetic bacteria as well as algae and other protists, produce oxygen through photosynthesis. They also provide food for other organisms, such as heterotrophic protists, animals such as fish, and ultimately consumers who eat the fish. Why do you think phytoplankton aren't classified as plants? How much oxygen do these tiny organisms contribute to the planet?

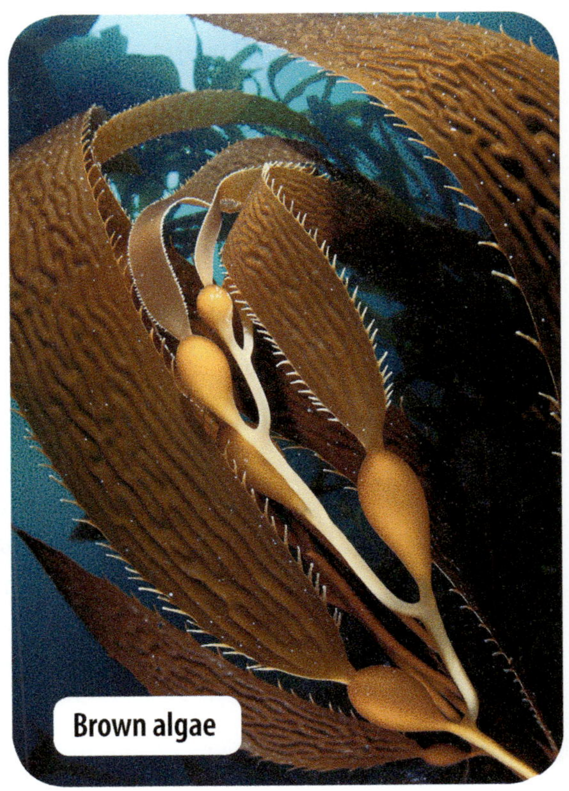

Brown algae

Giant kelp is a brown alga that grows abundantly off the Pacific coast of North America. It can be as much as 100 m long.

How do other organisms and humans use giant kelp?

How is this amoeba plant-like? How is it animal-like?

Animal-Like Protists

Early scientists used the simple microscope to study water droplets. They identified some protists as animals. Why do you think they thought these protists were animals? The cells of most protists have adaptations enabling them to move through water. For example, an amoeba, a single-celled protist, moves by using extensions of its cell called **pseudopodia**. It also uses its pseudopodia to capture and engulf food. How do you think an amoeba determines the location of food?

Some animal-like protists, such as the euglena, have whiplike structures called **flagella**. Euglena are single-celled photosynthetic protists. They swim, using a flagellum to move toward light. Flagella rapidly move back and forth, causing the cell to move forward. Other protists, such as paramecium, have cells covered with short, hairlike structures called **cilia**. These protists move and search for food by beating their cilia. Can organisms control the speed and direction of their movement with cilia or flagella? The discovery of Euglena and similar organisms in the mid-1800s created the need for the addition of the third kingdom Protista to the existing two kingdoms, Animalia and Plantae.

Fungi-Like Protists

Fungi-like protists absorb food from their environments and release spores to reproduce, much like fungi do (which will be studied in the next lesson). Organisms called slime molds are not molds at all, but protists. Slime molds can form large, colorful masses. They often live on the damp forest floor, where they absorb nutrients from decaying vegetation. Did you know that a slime mold could negotiate a maze to find food? What do you think happens to a slime mold when conditions become unfavorable? When food is scarce, slime molds produce spores that can travel through the air.

Explore-a-Lab

Structured Inquiry

How do the protists in a drop of water move about?

Observe, under a microscope, a drop of pond water or a culture your teacher provides. Look for the three different methods protists use to move—cilia, flagella, or pseudopods. Can you find amoebas, euglenas, and paramecia? Did you find any that look like they are colonies made from many cells? Do any appear to have green coloring matter within their cells as plants do? Make detailed, labeled diagrams of each organism you see at low power and at medium or high power. Identify each as an autotroph or a heterotroph.

Record your work for this inquiry. Your teacher may also assign the related Guided Inquiry.

Observing Slime Molds

How does a slime mold change as it grows?

Materials

- plastic container with lid
- paper towel
- wax pencil or permanent marker
- water
- piece of slime mold
- oatmeal flakes (not instant)
- medicine dropper

Procedure

1. Line a plastic container with a moist (not wet) paper towel. Label the container with your name.

2. Your teacher will give you a piece of the slime mold *Physarum polycephalum*. Place the slime mold on the paper towel. Drop in a few flakes of oatmeal that have been moistened with a drop or two of water. Observe the slime mold for 5 minutes.

3. Put the lid on the container and place it in a cool, dark place.

4. **Observe** the slime mold each day for three days. **Record** your observations and make drawings of the slime mold in your ***Science Journal***. Add a few more oatmeal flakes to the container each time you make observations. Return the container to a cool, dark place after each observation.

5. On days 4–7, **observe** but do not feed the slime mold. Make sure the paper towel is kept moist. Use the medicine dropper to add water, if needed. **Record** your observations and make drawings of what you see. Return the container to a cool, dark place after each observation.

Analyze Results

Compare your drawings and observations with those of other class members.

Create Explanations

1. How does a slime mold change as it grows?

2. What conditions are necessary for the slime mold to grow?

3. Where do you think you would find slime molds in the natural environment?

Importance of Protists Explain

You may not know it, but protists can be found in many of the products people use every day. Some of these products are agar, medicines, soaps, and food products. Protists affect people and other living organisms in another way—they can cause diseases.

Environmental Roles

Protists play a vital role in global photosynthesis. Scientists believe that phytoplankton contribute between 50 to 85 percent of the oxygen in Earth's atmosphere. They aren't sure because it's a tough thing to calculate. In the lab, scientists can determine how much oxygen is produced by a single phytoplankton cell. The hard part is figuring out the total number of these microscopic organisms throughout Earth's oceans. Phytoplankton populations increase and decrease with the seasons. Phytoplankton bloom in spring when there is more available light and nutrients. The sugars and other carbon compounds made by the phytoplankton provide food for many organisms. Phytoplankton form the base of the food chains in the ocean. What do you think would happen to ocean ecosystems if phytoplankton didn't exist?

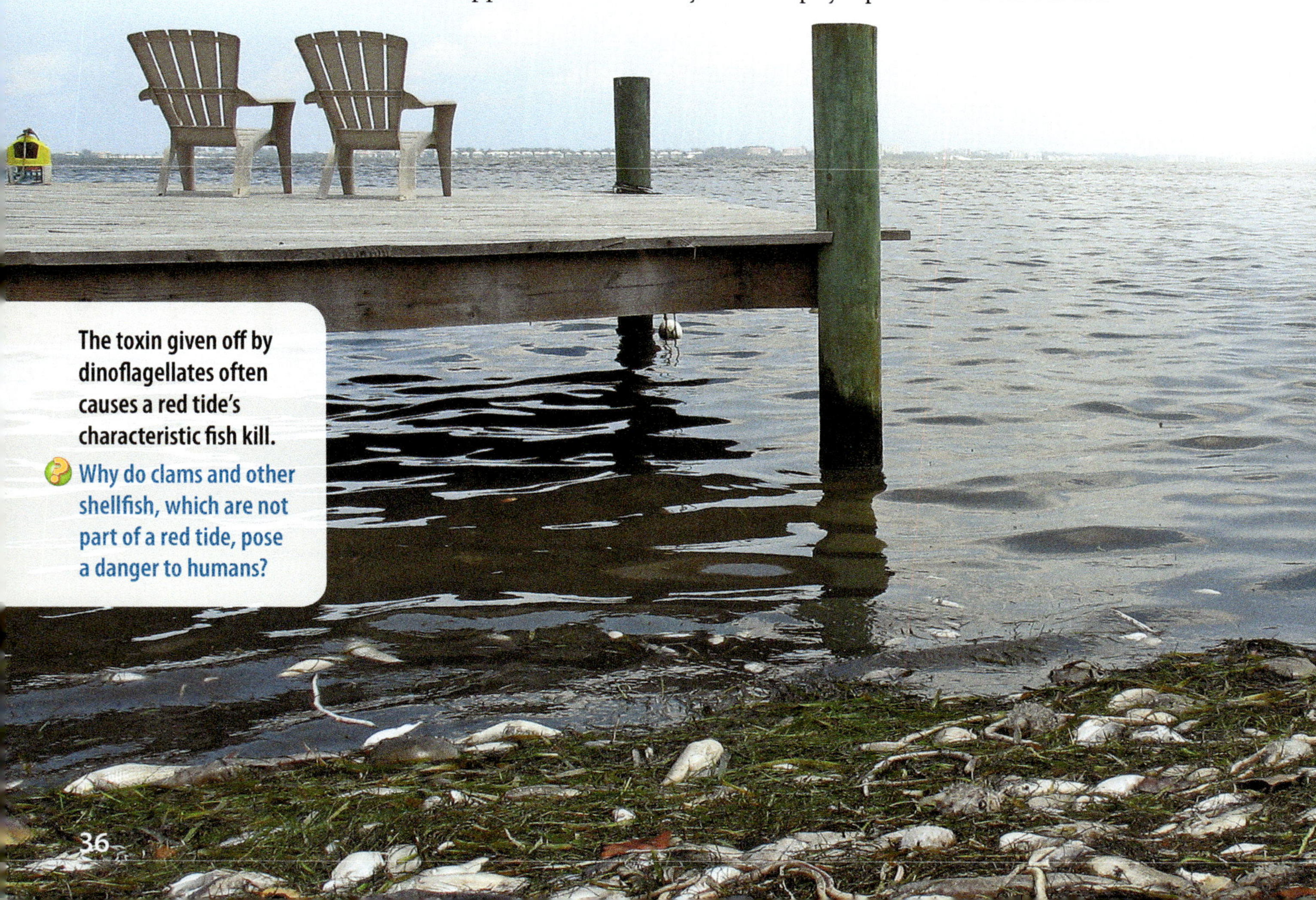

The toxin given off by dinoflagellates often causes a red tide's characteristic fish kill.

Why do clams and other shellfish, which are not part of a red tide, pose a danger to humans?

Some kinds of algae have a major impact when they suddenly increase in number. A large, rapid growth in the population of algae is called an *algal bloom*. When the algae die, they decompose. In freshwater lakes and ponds and even in the ocean, the decomposition of large algal blooms reduces the oxygen level in the water. How might this drop in oxygen levels change the relationships of other living things in the ecosystem?

In coastal areas, large populations of photosynthetic protists called *dinoflagellates* can cause harmful algal blooms known as *red tides*. Some species of dinoflagellates produce a toxin that kills fish. The toxin affects the fish's nervous system so that it is paralyzed and cannot breathe. When humans eat seafood affected by the organisms living in these waters, they can become very sick. The toxin can damage the human nervous system, causing illness and even death. Red tides can reduce fishing in affected areas and harm tourism.

Scripture Spotlight

Read **Exodus 7:14–25**. Some people think the Nile turning to blood was God using a phenomenon similar to red tides. What protist creates conditions such as those described in this text?

Diseases Caused by Protists

Protists that are parasites can sometimes cause diseases. For example, *Phytophthora infestans* causes potato blight, a disease that has had great economic and historic impact. Soon after *P. infestans* infects potato plants, the stalks and the stems become a slimy, black mass. Entire potato crops are ruined. When potato blight swept through Ireland in the late 1840s, it created widespread famine. During the great potato famine, about one million people died of famine and related diseases. As a result, many thousands of Irish citizens immigrated to the United States.

Lesson Activity

Use the Internet or other resources to learn about two additional diseases caused by protists—African sleeping sickness and giardiasis. Make a poster that details how the diseases are spread, symptoms of the diseases, treatments, and methods of prevention. Present your poster to the class.

How do protists cause disease?

Cycle of Malaria

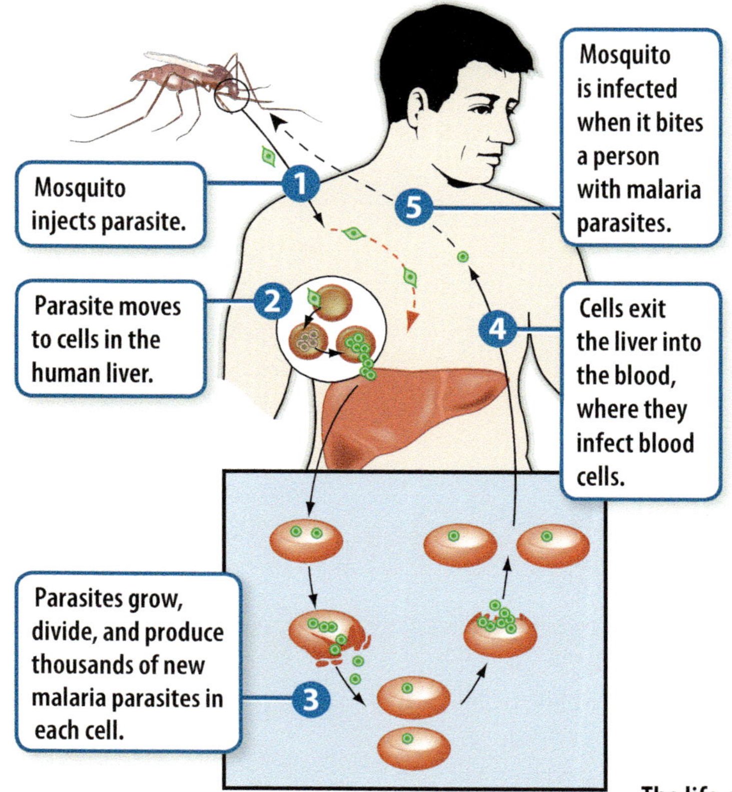

Malaria is a life-threatening human disease caused by parasitic protists that are transmitted through the bites of infected *Anopheles* mosquitoes. Each year, malaria causes nearly one million deaths, mostly among African children. Malaria causes fever, headache, chills, and vomiting. These symptoms can become severe and may result in death. The disease can usually be treated and often cured, but no vaccine yet exists. Vaccines are in the testing stage. By the time you read this, there may be a vaccine that is effective against malaria.

The most effective way to prevent people from getting malaria is protection against mosquito bites, which usually occur at night when the *Anopheles* mosquito is active. Insecticide-treated mosquito nets protect people from mosquito bites while asleep, reducing the chance of developing malaria. What other ways can humans protect themselves against mosquito bites?

The life cycle of the parasitic protist that causes malaria involves the host *Anopheles* mosquito before it infects a human or animal host.

 What are some reasons why malaria is so difficult to eradicate?

 Concept Check Assess/Reflect

Summary: What are protists? Protists are eukaryotic organisms that are not classified as fungi, plants, or animals. Many protists are single celled, while others are multicellular. The three main groups of protists are plant-like protists, animal-like protists, and fungi-like protists. Scientists continue to debate the classification of protists based on new research. Autotrophic protists form the basis of many food chains. Algal blooms reduce oxygen in water, causing other organisms to die. Protists also cause diseases, such as potato blight and malaria.

1. Why are protists considered both helpful and harmful organisms?
2. How are photosynthetic protists and heterotrophic protists alike? How are they different?
3. What would happen if most of the world's ocean phytoplankton died?
4. What three steps would you recommend to reduce the cycle of malaria? Why might communities be slow to adopt these changes?

Lesson 4

Essential Question

What Are Fungi?

Did you know that one gram of forest soil can contain more than one million microscopic fungi? In forest soils, other than plant roots, fungi are the dominant life forms. How can fungi be so prevalent but seemingly invisible to us?

In Oregon's Blue Mountains, a gigantic fungus occupies about 965 hectares of soil. This fungus would cover 1665 football fields, or nearly 10 square kilometers. But you cannot really see this fungus because, except for the mushrooms it forms on the surface, it is entirely underground! It is one of the world's largest organisms. What is the purpose of mushrooms in the life cycle of a fungus?

Objectives

- Identify the characteristics and structures of fungi.
- Explain the role of fungi in decomposition.
- Describe the different kinds of fungi.
- Describe how fungi are harmful and useful.

Vocabulary

hyphae
mycelium
saprophyte
mycology
parasite

Omphalotus olearius **is more commonly called the jack-o'-lantern mushroom. It is orange in color and glows at night due to bioluminescence.**

Why shouldn't people eat the wild mushrooms they find in the forest?

Scripture Spotlight

God gave the Israelites very specific instructions for dealing with mold in a building in **Leviticus 14:33–45**. What were these instructions? What health problems may result from inhabiting a building infested with mold?

Characteristics of Fungi Explain

The organisms known as fungi (singular, fungus) are so unique that they are placed in their own kingdom. The Kingdom Fungi includes mushrooms, molds, mildews, and yeasts. A mold is a fungus that appears as a fuzzy film growing on surfaces and foods, such as fruits and bread. Where are some places you have seen a fungus growing? Do fungi always look alike?

Many strange and beautiful life forms populate the Kingdom Fungi. Like protists, plants, and animals, fungi are eukaryotes and have cells with a nucleus surrounded by a membrane.

Have you ever opened a bag of bread and found green, white, or black mold growing on the bread? How did it get on the bread? What did the mold look like? Fungi may be single celled or multicellular. Multicellular fungi, such as mushrooms and molds, have threadlike bodies made of thin filaments. These filaments are called **hyphae** (singular, hypha). The hyphae may be tightly packed, as in the familiar mushroom. In other species, the hyphae are single strands, like the moldy fungal threads that reach deep into bread or overripe fruit. The hyphae together form a tangled mass called a **mycelium**. However, not all fungi produce mycelium. Yeasts are single-celled sac fungi. Unlike other fungi, yeasts do not have a mycelium.

Because mushrooms do not move, they once were thought to be plants. However, unlike plants, they are not able to photosynthesize and are not autotrophs. Fungi must get their food from other sources and are considered to be heterotrophs, like animals. Like animals, fungi require oxygen and release carbon dioxide. But unlike animals, mushrooms and other fungi do not ingest, or eat, their food. Instead, they release digestive enzymes outside their bodies. These enzymes break down the food into smaller molecules, which are then absorbed by the fungus. The fungal mycelium gives a fungus a very large surface area with which to absorb nutrients. However, fungi dry out easily because of this large surface area. What happens to a fungus when it dries out?

The thick, white structures are the spore-producing parts of this mold.

How are the spores dispersed?

Record your work for this inquiry. Your teacher may also assign the related Guided Inquiry.

Activity of Yeast

How does sugar affect the growth of yeast?

Procedure

1. Use a wax pencil or permanent marker to write your name on each bag. Label the first bag *A*. Label the second bag *B*.
2. Use a thermometer to measure the water temperature. **Record** the temperature in your ***Science Journal***.
3. **Measure** and pour 100 mL of warm water into the plastic cup. Add one packet of yeast to the cup and stir gently until the yeast is mixed thoroughly in the water. Add the water and yeast mixture to bag *A*. Squeeze most of the air out of the bag and seal it. Place the bag flat on top of three layers of paper towels. Do not move or touch the bag for 30 minutes.
4. Rinse out the plastic cup and dry it.
5. Repeat Step 3 with bag *B* but add one spoonful of sugar to the water and yeast mixture. Note: Make sure that all of the sugar has dissolved in the cup before adding the mixture to the bag.
6. **Observe** the bags at the end of 30 minutes. **Record** your observations.

Materials

- safety goggles
- lab apron
- 2 large, zip-top bags
- wax pencil or permanent marker
- thermometer
- warm water (40°C to 45°C)
- 100 mL graduated cylinder
- 2 packets active dry yeast
- plastic spoon
- granulated sugar
- 3 paper towels
- metric ruler

Analyze Results

Measure the height that each bag has risen above the tabletop.

Create Explanations

1. How does sugar affect the growth of yeast?
2. How can you explain the difference in the appearance of the bags?
3. Why is yeast useful in baking?
4. What cell process is going on inside the yeast cells that produces the gas that filled the bag?

Fungi have characteristics similar to both plants and animals. Recall that a plant cell is surrounded by a cell wall made of cellulose, a tough, rigid material made of carbohydrates. What is the function of the cell wall in plants? Like plants, fungi have cell walls. The cell walls of fungi are made of chitin, a tough carbohydrate similar to cellulose. Chitin is the same substance that forms the hard, outer coating of insects. Why might fungi require a tough exterior? Do fungi cell walls serve the same function as plant cell walls? Fungal proteins are also more like the proteins found in animals than those found in plants. Both animal and fungal cells contain chemicals called *sterols*. These compounds play an important role in biological messenger systems. Human bodies contain the sterol cholesterol, while fungi contain the sterol ergosterol. However, both animals and fungi contain lanosterol. None of these chemicals are found in plants. How might fungi use ergosterol?

Along with bacteria, fungi are important decomposers, breaking down the bodies of dead organisms. Such decomposers are called **saprophytes**, organisms that absorb carbon and other nutrients from dead organisms. Saprophytes recycle nutrients that would otherwise remain unavailable to the living world. How would the forest floor be different without saprophytes, such as fungi?

Studying Fungi

The study of fungi is called **mycology**. Scientists who study fungi examine fungal cell structure and reproduction. They use these characteristics to classify the Kingdom Fungi into several distinct groups. What characteristics might scientists use to group fungi?

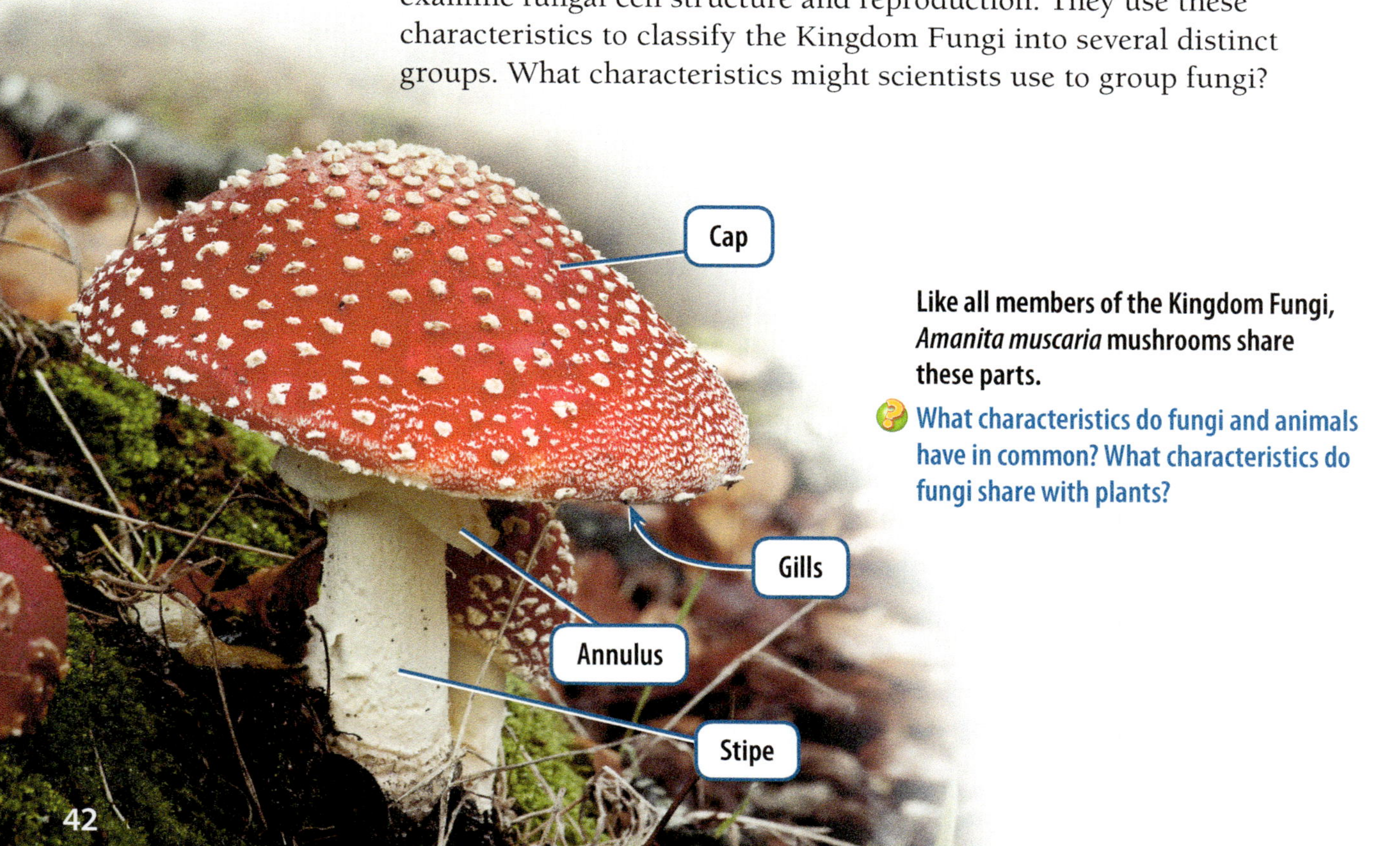

Like all members of the Kingdom Fungi, *Amanita muscaria* mushrooms share these parts.

What characteristics do fungi and animals have in common? What characteristics do fungi share with plants?

Kinds of Fungi Explain

Many people think of fungi as nothing more than molds and mushrooms. But fungi are far more diverse. The major groups that make up the Kingdom Fungi include an astonishing variety of organisms.

Fungi reproduce both asexually and sexually. All fungi produce spores. A *spore* is a thick-walled, asexual reproductive cell that is very effective for reproduction. Spores are lightweight so that they can be carried great distances by the wind, and they are resistant to drying. Why do you think resistance to drying is an advantage to the spore?

The sexual stage of fungi reproduction generally occurs underground. Fungi are placed into taxonomic groups according to the type of spore and spore-bearing structures they have. Study the table to learn about different fungi and their characteristics.

Black bread mold *(Rhizopus stolonifer)*

How should you handle foods with molds on them?

Fungi

Type	Characteristics	Examples
Club fungi	• Organisms are primarily an underground network of hyphae. • Mushrooms are reproductive organs. • Mushroom has umbrella-like cap lined with gills that release asexual spores.	• Mushroom
Sac fungi	• Asexual reproduction is primarily through budding. • Spores are located in microscopic, saclike structures. • Some sac fungi are parasites, organisms that feed on other organisms.	• Yeast • Chestnut blight • Dutch elm disease
Zygote fungi	• Asexual reproduction is by producing spores in spore cases atop stalks. • Spore cases give molds their fuzzy, dusty appearance. • Sexual reproduction is by producing reproductive structures that form zygotes (fertilized eggs) inside a tough capsule.	• Black bread mold
Aquatic fungi	• Organisms live in water. • They produce spores that swim during sexual reproduction. • They can be parasitic. • They can cause disease in fish.	• Frog fungus

Faith Connection

Yeast often symbolizes evil or corruption in the Bible. When the Jews celebrated Passover, why did they prepare unleavened bread?

Explore-a-Lab

Structured Inquiry

What can you learn from a spore print?

Carefully remove the stems from two mushrooms and lay the mushroom caps with the gills down on an unlined sheet of white paper. Cover the mushrooms and paper with a container to keep them moist. Let the mushrooms sit undisturbed overnight and remove them from the paper the next day. Observe the paper with a hand lens. Draw and label a sketch of what you see on the white paper. Describe the marks on the paper and what made them. With classmates, discuss ways you could estimate the number of new mushrooms that could be produced from one mushroom cap.

Lichens come in a multitude of colors: green, gray, black, white, yellow, orange, and even red.

Why are lichens not classified as plants or fungi?

Fungal Partnerships

Remarkably, some fungi form lifelong symbiotic partnerships with other organisms. Lichen is an association between a fungus and a partner, such as a photosynthetic cyanobacterium or a green alga. The photosynthetic partner produces sugars, which are shared with the fungus. The fungus, on the other hand, protects the photosynthetic partner and keeps it from drying out. Both organisms benefit from the partnership. Other fungi, called mychorrihizae, form an association with plant roots. These fungi help the plant roots absorb minerals and aid in the uptake of water. In return, the fungus receives nutrients from the plant. Why do you think some plants have a symbiotic relationship with mycorrhizae while others do not? What kinds of plants do you think are most dependent on a relationship with these fungi?

Economic Impact of Fungi Explain

Fungi are versatile and seem to grow everywhere. What benefits come from being able to grow almost anywhere? What disadvantages come from this capability? Fungi are important in the building of soil. Would you be surprised to learn that fungi have been used for centuries to produce dye, especially brown, yellow, and gray colors for fabrics? Fungi have even been used to enhance the color of food!

Plant Diseases

Why don't people like fungi? Many fungi produce plant diseases that cause millions of dollars of crop damage every year. Besides this being a problem for farmers, what other problems come with a loss of crops? How do agricultural scientists try to combat plant pathogens that can destroy crops? Most economically important plants have one or more fungal parasites. A **parasite** is an organism that feeds on another organism. Fungal hyphae burrow into plant tissues, damaging leaves, stems, and roots, resulting in plant death and reduced harvests. Rusts, so named because they cause powdery, orange spots on infected leaves, destroy wheat and many other grasses. How do you think these fungi spread from plant to plant? Powdery mildews infect the leaves of grasses, vegetables, trees, and shrubs. Chestnut blight and Dutch elm disease have destroyed these once-common North American trees. Why would a cool, wet summer be of concern to farmers? Are plant diseases only a concern when plants are alive, or can they also affect plants, such as grains, after harvest? If so, how?

Fungal diseases can affect crops around the world, especially in regions where dews are frequent and temperatures are warm, between 18°C and 30°C.

Explore-a-Lab

Structured Inquiry

 **Which type of bread molds fastest?**

Obtain two zip-top bags from your teacher. Into one bag, place a slice of white bread from a commercial grocery store. Into the other bag, place a slice of bakery white bread or homemade bread of equal size. Label each bag with the type of bread and your name. Moisten the bread in each bag with five drops of water. Seal the bags and place them in a warm, dark place. Do NOT open the bags! After several days, observe the bread through the bags using a hand lens. Compare the two slices of bread.

Human Diseases

In humans, some fungi are parasites that infect body tissues. Athlete's foot and ringworm are common skin diseases caused by fungi. Although the name *ringworm* suggests that a worm is the cause, it is actually a fungus that leaves round, itchy, red, raised, scaly patches on the skin. If ringworm occurs on the scalp, round bald patches may form. Athlete's foot is a fungal infection that develops in the moist areas between the toes. What are the symptoms of athlete's foot? Similar fungal infections can affect other regions of the body. Why do you think both of these fungi grow best in warm, moist environments?

The yeast-like fungus *Candida* causes the infection candidiasis, known as thrush when it affects the mouth. This fungus is a normal internal resident of the body, but when the body is weakened, *Candida* can grow in large numbers, causing an infectious disease. Patients with undeveloped or weakened immune systems, such as babies, people with AIDS, or those undergoing cancer treatment, are at higher risk of such infections.

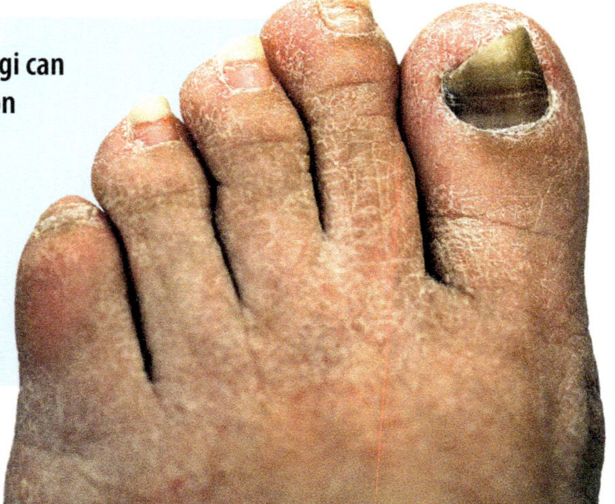

Athlete's foot and other fungi can damage the skin and nails on feet. Most athlete's foot infections can be treated with over-the-counter creams.

 **How could someone avoid the fungus athlete's foot?**

Commercial Uses of Fungi

Fungi have many commercial uses. Some sac fungi can actually feed on paint and jet fuel and are used to clean up contaminated waste sites. Another species of sac fungus produces citric acid that is used to add flavor to candy and soft drinks. Mushrooms are grown as a source of food. The bluish-green mold *Penicillium* is the original source of the antibiotic penicillin. Yeasts are vital for baking. What other commercial uses of fungi can you think of?

These molds belong to the genus *Penicillium*. They are used to make the antibiotic penicillin.

Concept Check Assess/Reflect

Summary: What are fungi? Fungi are eukaryotic organisms that can be single celled or multicellular. Yeasts are single celled, and mushrooms are multicellular. Fungi are heterotrophic and have cell walls made of chitin. Multicellular fungi have bodies made of hyphae, which together form a mass called a mycelium. Fungi can be both helpful and harmful. Types of fungi include club, sac, zygote, and aquatic. Fungi can reproduce by asexual and sexual reproduction.

1. Why have scientists placed fungi in their own kingdom?
2. How are fungi important to the environment?
3. How are fungi both helpful and harmful to other living things?
4. List some animal-like and plant-like characteristics of fungi. Do fungi appear to be more like an animal or more like a plant? Explain.

People in Science

Extend

Get to Know Carolus Linnaeus

Carolus Linnaeus was born Carl von Linné in 1707 in Sweden, but he is more often known by the Latinized version of his name. He is recognized as the father of modern taxonomy. Linnaeus's system for classification, naming, and ranking of organisms is still widely used today.

Although Linnaeus entered the University of Lund to study medicine in 1727, he fell in love with botany instead. He collected and studied plants as part of his medical training. Doctors at that time needed to learn to prepare and prescribe medicinal plants.

Linnaeus is particularly known for his work *Systema Naturae* ("The System of Nature") which was originally only pages long and described a taxonomy of three kingdoms of nature: stones, plants, and animals. Each kingdom was then subdivided into classes, orders, genera, species, and varieties. Linnaeus continued to revise his *Systema Naturae* throughout his life from his 11-page pamphlet to a multivolume system. Although today's biologists have added additional rankings, such as family, this system is largely unchanged.

Called to Serve

Carolus Linnaeus was filled with wonder over nature's beauty and intricacies. He spent his life devoted to learning God's order of living things.

Linnaeus was a deeply devout man, and this was reflected in his love of living things. Linnaeus felt that only by studying and classifying nature could humanity understand God's Divine order of Creation. It was humanity's task, and perhaps duty, to reveal this natural order.

Concept Check

1. In what ways does Linnaeus's *Systema Naturae* resemble today's classification of organisms?
2. How did Linnaeus express his belief in God and love of nature?

Careers in Science

Extend

Mycologist

Just as botanists study plants, mycologists study the mysteries of fungi. Mycologists begin their education by getting a four-year college degree in certain areas of biology, such as microbiology. Then most continue their college education for two or more years. There are more mycology-related jobs than there are qualified mycologists to fill them. Careers in mycology can range from applied agriculture to academic research and laboratory work.

Fungi have many uses in numerous industries. The most obvious positive application is as a food source. Most of us are familiar with mushrooms, such as the button, portobello, and morel varieties in our local markets. Identifying new species that are safe to eat is an important field. David Arora is an American mycologist. He has authored several books on amateur mushroom hunting and conservation to arm mushroom hunters with information on which species are safe to eat. Identifying poisonous mushrooms and treatments for eating toxic mushrooms are two of the tasks a mycologist might undertake. Fungi are also a constant concern in agriculture. Fungal infections can destroy crop plants, and plants tainted with even a small amount of certain fungi can poison people who unwittingly consume them. We know that 25 percent of the human population is sensitive to fungal molds, so this is an important area of research.

How fungi relate to human health is the main focus of mycology. Fungal molds in particular can affect many aspects of human health, whether they are ingested in tainted food or inhaled. Mold in homes is well known for creating respiratory illnesses for the people who live there. However, some molds heal. The fungal mold *Penicillium* has saved millions of lives. This common bread mold is now an antibiotic. Certain strains are also used to produce cheese, including blue cheeses, Gorgonzola, Brie, and Camembert.

Concept Check

1. In what ways may mycologists contribute to human health?
2. Fungi can be harmful or beneficial. Create a two-column chart to compare ways fungi can have a positive or negative effect on humans. Then identify how mycologists use their knowledge of fungi to help people.

Chapter 1

Review

Assess/Reflect

Study Guide

Lesson 1

1. Scientists classify living organisms based on their characteristics. The organisms are placed into increasingly smaller and smaller groups until the organism is given its unique scientific name.
2. Scientists use a naming format called *binomial nomenclature* to give each organism a scientific name. Scientific names consist of two Latin names—genus and species.
3. Organisms are organized into eight levels of classification. The order of the levels used in classification is domain, kingdom, phylum, class, order, family, genus, and species.

Lesson 2

1. Bacteria are microscopic, single-celled, prokaryotic organisms. Bacteria belong to the Domain Bacteria and can be autotrophic or heterotrophic.
2. Bacteria can grow in aerobic and anaerobic environments. Most bacteria reproduce.
3. Bacteria are described by shape, spherical, rod, and spiral, and Gram-staining, negative or positive.
4. Bacteria cause many diseases, including botulism, cholera, *E. coli*, lyme disease, salmonella, and tuberculosis.
5. Bacteria decompose material at Earth's surface and are used by industry.

Lesson 3

1. Protists are eukaryotes that cannot be classified as fungi, plants, or animals. They can be autotrophic or heterotrophic. They can be very small or very large.
2. There are three main types of protists. Plant-like protists are autotrophs that photosynthesize. Animal-like protists move using cilia, flagella, or pseudopods. Fungus-like protists absorb food from their environment.
3. Photosynthetic protists form the base of freshwater and marine food chains and produce much of Earth's oxygen.
4. Helpful protists are important in making products such as agar, soaps, and certain medicines. Harmful protists cause many diseases, including malaria, African sleeping sickness, and giardiasis.

Lesson 4

1. Fungi are eukaryotic heterotrophs with cell walls composed of chitin. They are made up of thread-like structures called hyphae. The hyphae form a mass of tangled threads called a mycelium.
2. Fungi are decomposers. They release digestive enzymes outside their bodies that digest organic material that is then absorbed by the fungi.
3. There are four types of fungi: club fungi, which include mushrooms that reproduce by releasing spores; sac fungi, including yeast and morels, which reproduce by budding and by releasing spores; zygote fungi, such as bread mold, which reproduce sexually by producing zygotes and asexually by releasing spores; and aquatic fungi, such as frog fungus, which reproduce sexually using spores.
4. Fungi are both helpful and harmful. Helpful fungi break down paints and other contaminants and are also used in food and medicine production. Harmful fungi cause spoilage of food, contamination in homes and businesses, and many diseases, including chestnut blight and athlete's foot.

Vocabulary Check

Explain how each pair of items is related.

1. scientific name—species
2. domain—kingdom
3. autotroph—heterotroph
4. hyphae—mycelium

Multiple Choice

Choose the best answer.

5. Which is the largest category of classification?
   A. domain
   B. family
   C. kingdom
   D. phylum
6. Which is the correct way to write the scientific name for a dwarf hamster?
   A. *phodopus campbelli*
   B. *Phodopus Campbelli*
   C. *Phodopus campbelli*
   D. *phodopus Campbelli*
7. How are bacteria and fungi alike?
   A. Both have cell walls.
   B. Both are autotrophic.
   C. Both are multicellular.
   D. Both are eukaryotic.
8. Which disease is caused by a protist?
   A. botulism
   B. malaria
   C. Dutch elm disease
   D. ringworm
9. What two characteristics do all fungi share?
   A. pigments that capture sunlight, flagella
   B. cells without nuclei, spores
   C. cell walls of chitin, absorb food
   D. binary fission, cilia

Check Point

Answer the following questions.

10. What characteristics do scientists use when classifying organisms?
11. Why do scientists rely on scientific names rather than common names to classify organisms?
12. Antibiotics kill certain bacteria but do not harm human cells. Explain.
13. A student discovers that fungi usually grow only in damp places. Why might fungi thrive in moist environments? Explain.
14. **Develop a hypothesis** about the effect of temperature on the growth of yeast, and **design an experiment** to test your hypothesis.

Chapter 2: Characteristics of Invertebrates

Scripture Spotlight

As you learn about invertebrates, look for evidence of God's Creativity and Design. You will read the following passages in this chapter.

John 7:24 (p. 58)	Proverbs 14:27 (p. 85)
Proverbs 6:6–9 (p. 74)	Deuteronomy 7:17–21 (p. 87)
Colossians 3:8–14 (p. 81)	Joshua 24:12 (p. 87)
Proverbs 13:14 (p. 85)	1 Corinthians 2:9 (p. 89)

Sea jellies are invertebrates. They have no bones in their bodies.

The Big Idea

God made the majority of animal species without backbones. These invertebrates are found in diverse environments and exhibit fascinating variety.

How could not having bones in its body benefit a sea jelly?

Inquiry Kick-Off Engage

Why do invertebrates move differently than vertebrates? What structures do invertebrates have that enable them to move? Use your ***Science Journal*** to compare the movements of vertebrates with those of invertebrates.

Lesson 1

Essential Question

How Are Invertebrates Classified?

Objectives

- Describe the difference between vertebrates and invertebrates.
- Describe the types of skeletal structures present in invertebrates.
- Explain how animals are classified by body symmetry.
- Describe common invertebrate phyla.

Vocabulary

symmetry
bilateral symmetry
radial symmetry
asymmetry

Look at the earthworm in the picture. How would you describe it? How does it move? How do you think it breathes? Earthworms inhabit every continent except for Antarctica. The world's largest earthworm, at nearly two meters long, is the giant Gippsland earthworm of Australia. Earthworms are soil-eating machines. They are nature's cultivators and recyclers. As these worms move through the soil, they mix and aerate it. Their burrows allow water to reach deep underground, and the digested soil that they produce provides important nutrients for plants. How is the earthworm similar to you? How is it similar to a clam or a mosquito?

Earthworms are familiar invertebrates.

Which body parts can you identify on this earthworm?

Classifying Animals Explain

God created thousands of kinds of animals, giving us a great deal of diversity to enjoy. Each one fills a special niche in nature. The fact that there are so many animals can make them difficult to study. Recall that scientists devised a universal method of classifying animals to make studying them easier. Scientists agree that all animals

- are eukaryotes (cells have a nucleus and organelles);
- are multicellular, and most have bodies with differentiated tissues;
- have cells that lack rigid cell walls;
- respond to stimuli; and
- are heterotrophs.

In addition, most animals are capable of movement and undergo sexual reproduction (some, such as sponges, are capable of asexual reproduction). Hydra and some sea anemones reproduce by asexual budding. You may recall that scientists sort animals into groups, with each grouping composed of similar animals. Animals belong to the Eukarya domain. The domain is further organized into kingdoms, phyla (singular, phylum), classes, orders, families, genera (singular, genus), and species. Think of several animals you are familiar with. Do you know the phylum each belongs to? How about the class, order, or family of the animal?

When classifying animals, one of the characteristics that scientists use is the presence or absence of a backbone made of vertebrae. Vertebrates are those animals that have vertebrae; invertebrates are those animals that lack vertebrae. What physical similarities would you expect to see between these two groups of animals? What differences would you expect to see between the groups?

Animal Classification	
Domain	• Highest level of classification • Three domains: Archaea, Bacteria, Eukarya
Kingdom	• A group that makes up a domain • Six kingdoms: Archaebacteria, Eubacteria, Protists, Fungi, Plants, and Animals
Phylum	• A group that makes up a kingdom • Plural form of *phylum* is *phyla.* • This level is called *division* in Plants and Fungi.
Class	• A group that makes up a phylum
Order	• A group that makes up a class
Family	• A group that makes up an order
Genus	• A group that makes up a family • Plural form of *genus* is *genera.*
Species	• A group that makes up a genus • Species are able to reproduce and have offspring that can reproduce.

Record your work for this inquiry. Your teacher may also assign the related Guided Inquiry.

Dissect an Earthworm

What are the structures of an earthworm?

Materials

- 5 paper towels
- Petri dish
- water
- balance
- dissecting kit and pan
- living earthworm
- preserved earthworm
- magnifying lens
- metric ruler

Procedure

1. Tear a paper towel in half and wet one half thoroughly. Fold the wet towel so that it will fit in the bottom half of the Petri dish. Place the living worm on the wet towel and place the cover on the Petri dish.
2. Place the preserved worm on a paper towel and cover it with another paper towel to keep it from drying.
3. **Observe** the living worm inside the Petri dish. **Identify** the external structures identified in the earthworm dissection diagram in your ***Science Journal***.
4. Take the worm out of the Petri dish. **Measure** and **record** its mass and length. Record the number of segments in your earthworm. Then make a labeled drawing of your living earthworm.
5. Repeat Step 4 with your preserved earthworm.
6. Carefully place the preserved worm in the dissecting pan dorsal side up. Place a pin in the worm at the anterior and posterior ends. Locate the clitellum and insert the tip of a scissors about 3 cm posterior. Carefully cut all the way up to the top of the head. Be sure to cut only through the skin.
7. Gently spread apart the skin. Use pins to hold the skin to the pan. Identify all structures and make a labeled drawing of the earthworm.
8. Locate the digestive system, beginning with the mouth and including the pharynx, esophagus, crop, gizzard, intestines, and anus.
9. Look for the worm's tiny, whitish brain near its far anterior end. Remove the intestines and find the ventral nerve cord.
10. Examine the other structures you see within the worm. Determine what other parts you can identify. Discuss the functions of each body part.

Analyze Results

Compare the data you found for the living and preserved earthworms.

Create Explanations

1. What are the structures of an earthworm?
2. Based on your observations, explain how an earthworm moves.
3. Determine the phylum of the earthworm.

Vertebrates Versus Invertebrates

Many of the larger animals you are familiar with are vertebrates. The backbone inside an animal's body protects the spinal cord and supports the body. However, most animals are invertebrates. Some invertebrates lack any kind of hard skeleton and are quite soft, or jellylike. Squid have a flexible structure that supports their body, while earthworms and sea jellies use body fluids under pressure for support. Other invertebrates, such as sea stars and insects, have a rigid skeleton that gives them shape, protects them, and allows them to move.

Most invertebrates have one of four main types of these skeletal structures:

- Skeleton made of calcium, silica, and organic fibers (for example, many sponges)
- Skeleton of interlocked calcium-carbonate plates and spines (for example, sea stars and sea urchins)
- Skeleton made of chitin (for example, arthropods)
- Skeleton made of a calcium-carbonate shell (for example, many mollusks)

There are many more types of invertebrate animals than there are vertebrate animals. In fact, invertebrates make up about 97 percent of all animal species. Scientists group invertebrates into eight major phyla, including sponges, cnidarians, worms, mollusks, echinoderms, and arthropods. Many invertebrate phyla may be divided into more than one class. You will learn more about these animals later in this lesson and in Lesson 2. What evidence of God's Design can you see in the body forms of animals?

Red rose sea anemone

Why might someone incorrectly think this anemone is a plant?

Classifying Animals by Symmetry Explain

In addition to classifying animals by whether they have a backbone, scientists classify animals by studying symmetry. **Symmetry** describes the balanced distribution of an animal's parts around an axis. Animals can have bilateral symmetry, radial symmetry, or no symmetry, known as asymmetry.

Bilateral Symmetry

Most animals have **bilateral symmetry**, which means their body parts are arranged in the same way on their left and right sides. You have bilateral symmetry, with one arm and one leg on each side of your body. Bilateral animals have a front, a back, and two sides. Many animals with bilateral symmetry have a head with centralized nerve tissue, or a brain. Vertebrates have bilateral symmetry. The backbone typically divides the vertebrate's body into two halves that are mirror images of each other.

Imagine a pet dog. Where could you draw a line on the dog to get two equal halves? How does bilateral symmetry help this dog survive in its environment? To what other animals did God give this design?

Scripture Spotlight

What adjectives might you use to describe different invertebrates? While some may look graceful and beautiful, others may appear plain, slimy, prickly, or even menacing. What does **John 7:24** say about appearances, and how might that relate to the people around you?

This peacock butterfly has a hard, outer shell covering its body and is an example of an invertebrate with bilateral symmetry.

Where is the butterfly's line of symmetry?

Radial Symmetry

The body parts of some animals are arranged in a circle around a center point, like a bicycle wheel with spokes. This arrangement is called **radial symmetry**. For example, a sea jelly has a circular shape. It is the same all the way around, with no front or back or sides. The radial design God gave these animals allows them to reach out in all directions to obtain food. They can also easily sense danger in all directions. Can you think of other creatures that have radial symmetry?

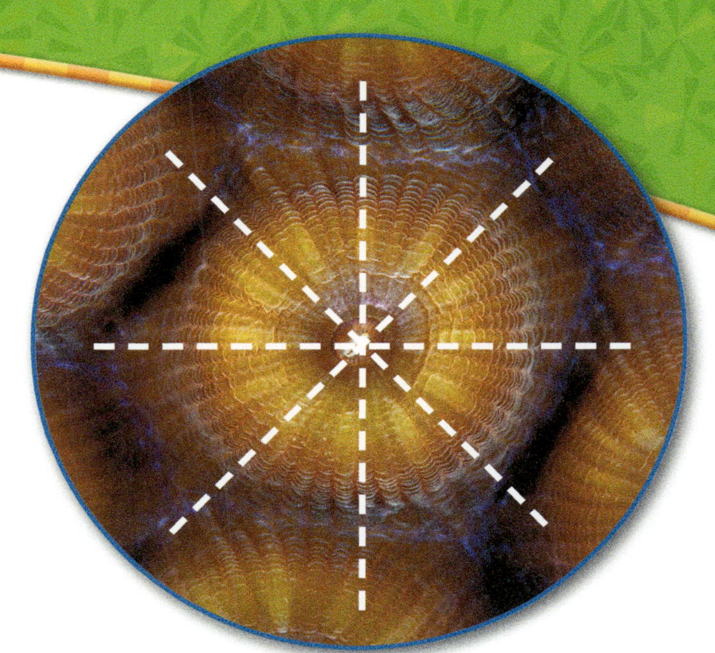

These coral polyps have radial symmetry.

If you had radial symmetry, how would life in your classroom be different?

Asymmetry

Some animals, including most sponges, lack any symmetry and have **asymmetry**. Their bodies are shaped irregularly, and they cannot be split into equal parts. How many organisms can you think of that have asymmetry?

Asymmetry is a sign that the animal's body parts aren't very specialized. The more specialized the body parts of an organism are, the more symmetrical it generally is. Sponges, for example, do not have specialized tasks for various parts of the body. Each part must absorb food on its own. Sea stars with their radial symmetry have some specialized parts to absorb nutrients and excrete waste within each segment. How many specialized parts does your bilaterally symmetrical body have? Would your body function as well if it had radial symmetry? Explain.

Faith Connection

Notice the order and beauty of bilateral and radial symmetry and the simplicity of asymmetry. Can you think of any spiritual applications of symmetry to a Christian's personality, character, or relationship with Jesus?

Explore-a-Lab

Structured Inquiry

How can you identify the lines of symmetry in invertebrates?

Working with a partner, take turns searching the Internet or other sources for 10 images of invertebrates. Print the images. Swap your set of images with your partner's. Use a ruler to define the lines of symmetry in each image. Label each image with the type of symmetry you observed. Then compile the images and review your results. Do you concur with each other's classification?

Invertebrate Phyla Explain

Invertebrates	
Phylum Description	**Examples**
Porifera (Sponges) Sponges have a central cavity and live underwater. They may seem like plants, because they are attached to the sea bottom. Sponges have no true body cavity and no specialized organs. Sponges are asymmetrical. Animals that you are more familiar with, such as insects and birds, have some type of body cavity in which we find organs for specialized body functions.	Sea sponges are often harvested for people to use for bathing. Freshwater sponges live in lakes and slow-moving water.
Cnidaria (Cnidarians) Cnidarians are made of a jellylike substance with connecting tissues but have no specialized organs. Tentacles surround the opening to the central cavity. Cnidarians have radial symmetry.	Examples of cnidarians include sea jellies, often mistakenly called jellyfish. Corals and sea anemones are also cnidarians.
Platyhelminthes (Flatworms) Platyhelminthes have flattened bodies with no body cavity, lack body segments, have distinct heads and tails, and may be free living or live as parasites. Platyhelminthes have bilateral symmetry.	The tapeworm is a common flatworm that is a parasite in many animals, including humans. Planaria and trematodes are other platyhelminthes. One type of trematode causes schistosomiasis, the second-most devastating parasite caused disease affecting humans.
Nematoda (Roundworms) Nematodes have cylindrical bodies, lack body segments, are often microscopic, and may be free living or live as parasites. Some parasitic nematodes cause disease in humans. Roundworms have bilateral symmetry.	One parasitic nematode is the trichina worm, which enters the human body through undercooked meat (usually pork). It causes the disease trichinosis. Hookworm is another parasitic nematode.
Annelida (Segmented Worms) Annelids have cylindrical bodies divided into many segments, well-developed organ systems, and may be free living or live as parasites. The body segments of annelids contain specialized organs for body functions, such as blood circulation, blood purification, and breathing. Segmented worms have bilateral symmetry.	The common earthworm is an annelid as are leeches.

Invertebrates (cont.)	
Phylum Description	**Examples**
Mollusca (Mollusks) Mollusks have a soft body covered in a cloak-like mantle. They often use shells to protect themselves. They have three main body parts: the brain, the visceral mass containing the internal organs, and the foot. Mollusks have bilateral symmetry.	Oysters, snails, and squid are all members of the Mollusca phylum.
Echinodermata (Echinoderms) Adult echinoderms have rays or arms, often in multiples of five. The large majority of echinoderms live in marine environments. They use tube feet for movement and food collection. Adult echinoderms have radial symmetry.	Sea urchins and sea stars, sometimes erroneously called starfish, are echinoderms. Sea cucumbers are also echinoderms.
Arthropoda (Arthropods) Members of this category have exoskeletons and may have segmented bodies and multiple jointed legs. As arthropods grow, they must shed, or molt, the exoskeleton to produce a larger one. Arthropods have bilateral symmetry.	Arthropods include insects, spiders, centipedes, and crustaceans.

Concept Check Assess/Reflect

Summary: How are invertebrates classified? An invertebrate is an animal withouwt a backbone. A vertebrate is an animal with a backbone. Most vertebrates have bilateral symmetry. Invertebrates can have bilateral or radial symmetry, or they can be asymmetrical. Bilateral symmetry means that the animal is arranged in the same way on both sides. Radial symmetry means the body parts are arranged in a circle around a central point. Asymmetry means that the body shape is irregular and cannot be divided equally. Invertebrates are classified into several phyla, including Porifera, Cnidaria, Platyhelminthes, Nematoda, Annelida, Mollusca, Echinodermata, and Arthropoda.

1. What are two characteristics that animals classified as vertebrates might share? What are two characteristics that invertebrates might share?
2. Would an animal with radial symmetry be able to move better on land or in water? Explain your answer.
3. If an animal is asymmetrical, what can you tell about its body parts?
4. How do bilateral symmetry, radial symmetry, and asymmetry affect the feeding habits of an animal?

Lesson 2

Essential Question

What Are Sponges, Cnidarians, and Worms?

Objectives

- Describe the structure and characteristics of sponges, cnidarians, flatworms, roundworms, and segmented worms.
- Explain each group's role in its ecosystem.

Vocabulary

sponge
spongin
regeneration
cnidarian
annelid

When God created Earth's creatures, He designed each with what it needed to do the job for which it was designed. Can you describe some ways in which animals carry out the design for which God created them?

For some types of invertebrates, that means a very simple body plan. For other types of animals, the body plan is more complex. Regardless of the body plan, all creatures have the features they need to survive in their natural environment. As you learn about invertebrates in this lesson, ask yourself the question, "How does this animal show God's Design?"

Phylum Porifera (Sponges) Explain

A **sponge** is an invertebrate with a central cavity, many small pores, and a large opening at the top. All sponges live underwater, attached to a solid surface. Sponges range in size from one or two centimeters wide to a width of one to two meters. The largest known sponge reaches a width of nearly three meters. Most sponges live in saltwater and can be brightly colored—red, orange, yellow, or blue. Freshwater sponges are often dull brown or green. Why do you think freshwater sponges have dull colors and marine sponges have bright colors?

The body of a sponge is asymmetrical and is made of just a few types of cells. Cells of the inside layer use flagella, or whiplike tails, to move water through the sponge. As the water passes through the sponge, other cells filter out microscopic bits of food. Sponges are such efficient filters that some are able to filter 20,000 times their volume in a day. Through the act of eating, sponges are able to benefit their ecosystems. The sponges eat bits and pieces of decomposing plants and animals. Sponges also collect bacteria from the water as it filters through, creating and releasing nitrogen. Furthermore, sponges play an important role in nitrogen cycles within coral reefs. What would happen to Earth's oceans if sponges stopped doing their jobs?

The flexible, fibrous substance that makes up the body of a sponge is called **spongin**. Some sponges have sharp, glass-like structures in their body for protection. Sponges are made up of specialized cells that filter water, form the external skin of the sponge, engulf and digest food particles, transport nutrients, and form the sponge's skeletal fibers. All of the sponge's cells are able to change function as necessary. Some animals have a symbiotic relationship with sponges, using the sponges as their home and as protection from predators. Why would a sponge make a good hiding place for other animals? What other benefits do the sponges provide to their ecosystems?

Sponges can reproduce in two ways. They can reproduce sexually by producing egg cells and sperm cells. They also reproduce asexually by budding and regeneration. During budding, a bud forms on the parent sponge, drops off, and grows into a new sponge. Pieces of sponges that are broken up or torn apart by natural forces or predators also grow into new sponges. This process of growing new body parts is called **regeneration**. Why do you think God created sponges with the ability to reproduce in two distinct ways? Can you think of parts of the human body that are able to regenerate when damaged?

For centuries, people have harvested and used sponges from the sea. The natural sponges you can buy at the store were once living animals. Like many other organisms that humans use, harvesting too many sponges could cause them to become extinct. What can be done to protect sponges?

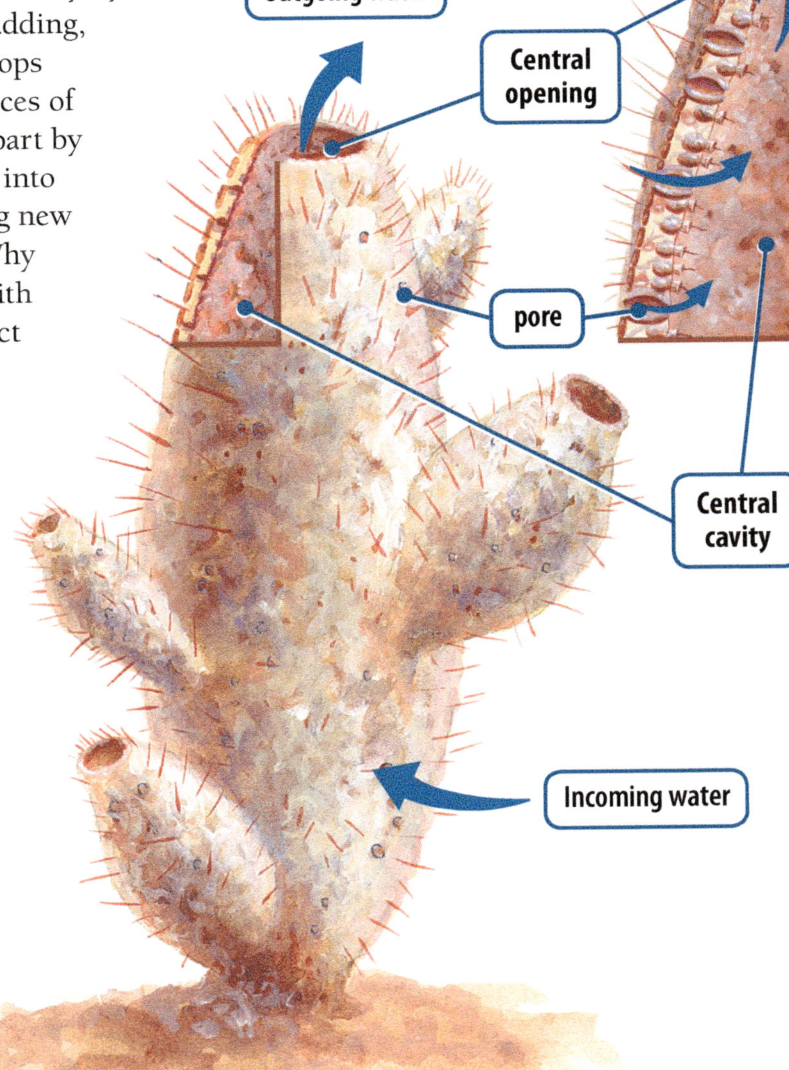

People have harvested and used sea sponges for thousands of years.

In what ways do people use natural sea sponges?

Record your work for this inquiry. Your teacher may also assign the related Guided Inquiry.

Comparing Sponges

Which is more absorbent—a natural sponge or a synthetic sponge?

Materials

- hand lens or dissecting microscope
- natural and synthetic sponges cut into several 3 cm cubes (3 pieces of each type for each group)
- balance
- 250 mL beaker
- water

Procedure

1. Use the hand lens or dissecting microscope to **observe** the structure of each kind of sponge. Draw what you see.
2. Use the balance to **measure** and **record** the mass of one of the natural sponge samples to the nearest 0.1 gram.
3. Fill the beaker with water and place the sponge sample in it.
4. After the sponge is completely saturated, take it out and let the excess water drip off. **Record** the mass of the saturated sponge.
5. Repeat Steps 2–4 with the other two natural sponge samples.
6. Repeat Steps 2–4 with the three synthetic sponge samples.
7. Graph your data in your ***Science Journal***.

Analyze Results

Calculate the percentage of absorbency for each sample you tested. To do this, subtract the dry mass of the sponge from the wet mass of the sponge to find the mass of the water absorbed. Divide the mass of the water absorbed by the mass of the dry sponge. To get the percentage of absorbency, multiply the result of the division by 100. Repeat for each sample of the sponges and then calculate an average for each type of sponge.

Create Explanations

1. Which is more absorbent—a natural sponge or a synthetic sponge?
2. Do you believe there is a connection between the size of a sponge's holes and its ability to hold water? Use evidence from your observations to support your answer.
3. Which type of sponge do you think is typically found in most homes for cleaning? Why do you think more people use this particular type of sponge?

Phylum Cnidaria (Cnidarians) Explain

Cnidaria is the group of invertebrates that includes corals, sea anemones, sea jellies, hydras, and the Portuguese man-of-war. Most cnidarians live in saltwater. Hydras, however, are tiny freshwater organisms. More than 9000 species of cnidarians live on Earth today.

Faith Connection

The Portuguese man-of-war is an excellent example of God's plan for cooperation among animals. It is made of four individuals that live together in a colony. Each individual has a different form and serves a highly specialized function. The four individuals depend on each other for survival.

Characteristics of Cnidarians

Most of the cnidarian body is a jellylike substance. Tissue-forming cells surround this substance. What are some benefits to having a jellylike body? Cnidarians have a simple digestive cavity and nerves. Signals from these nerves allow cnidarians to respond to their environment.

Cnidarians have radial symmetry. They have one body opening, or mouth, that is surrounded by tentacles. Cnidarians use their stinging tentacles to stun and capture prey. Touch triggers their specialized stinging cells, called cnidocytes, which inject venom into their prey.

Most cnidarians have a two-stage life cycle: the polyp and the medusa. The polyp stage is vase shaped. Its tentacles reach upward and wait for food to float by. Sea anemones, corals, and hydras are examples of polyps. The medusa stage is free swimming and is shaped like a bell. Its tentacles dangle downward and wait for prey to swim by. Sea jellies and the Portuguese man-of-war are examples of the medusa stage. What do you think is the advantage to cnidarians of having a two-stage life cycle?

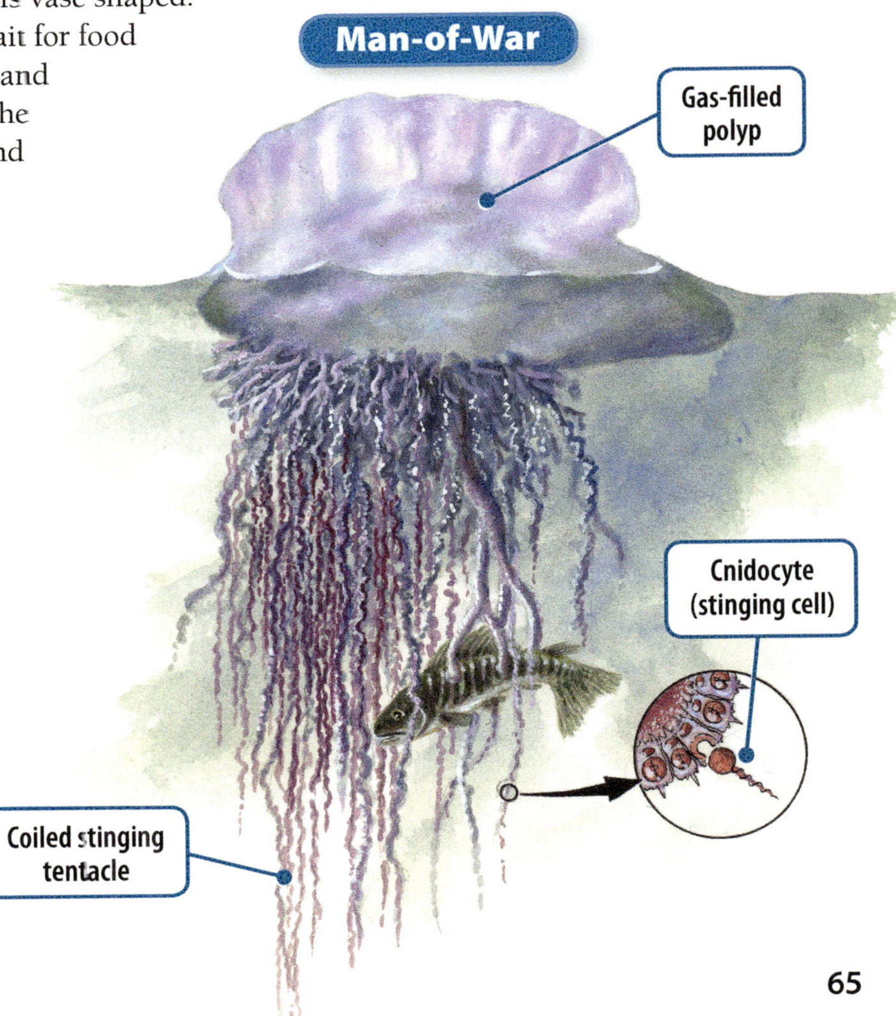

The Portuguese man-of-war floats on the surface of warm, tropical waters, trailing its stinging tentacles behind it. This example shows the medusa stage.

How might this floating behavior cause the Portuguese man-of-war to be dangerous to humans?

Explore-a-Lab

Structured Inquiry

What are the characteristics of a hydra?

Obtain a living hydra from the teacher and place it with a drop of water on a depression slide. Using a hand lens, observe and then record the characteristics of the hydra. Use a microscope on low power to observe the hydra in more detail. Leave the hydra undisturbed for several minutes and record its behavior again. Use the microscope on medium power to observe the hydra and then draw a sketch of it, including specific parts, such as the tentacles. Gently use a toothpick to touch the hydra and observe its behavior after it has been touched. Return the hydra to the teacher and obtain a slide with a budding hydra. Observe and record your observations.

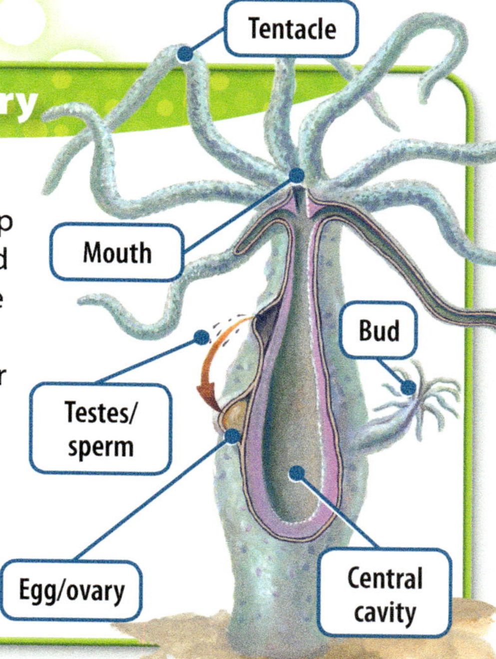

Importance of Cnidarians

Cnidarians are an important part of the ecosystems they are in. The many species of sea anemones, sea jellies, and corals add to the diversity of the ocean environment and help maintain its health. Cnidarians are an important part of the ocean food chain, acting as predators as well as prey. Hydra are just as important to freshwater ecosystems. What animals do you think eat the cnidarians? What do the cnidarians eat?

The large reef structures created by the corals provide an important habitat for thousands of species of marine life. Cnidarians, especially sea anemones, are potential sources for new medicines and anticancer compounds. Researchers have found that the venom within the sea anemone's tentacles can be used as treatments for various medical conditions and to battle obesity. Coral reefs are important to the economy as they attract and increase tourism. These reefs reduce the energy of waves, thus reducing beach erosion and helping to preserve this environment. And in some areas, coral is mixed with concrete to produce strong building materials. How would destruction of coral reefs affect cnidarians and thus the environment?

Faith Connection

How is the cnidarian like things you have experienced in your life?

A piece of coral is made of hundreds of identical individuals called polyps. Each polyp makes a cuplike "shell" outside its body, and its tentacles extend upward into the water.

What is the function of the "shell" of a coral polyp?

Worms Explain

A worm is an invertebrate with a long, soft body and no limbs. Many worms are parasites, while others improve the soil or have medical uses. All worms have bilateral symmetry. Worms make up three phyla of invertebrates: Platyhelminthes, Annelida, and Nematoda.

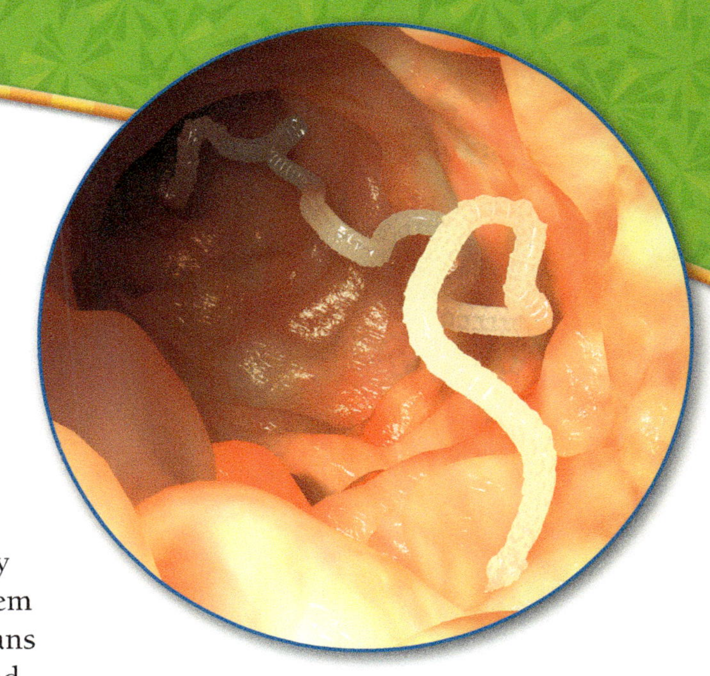

This tapeworm has attached itself to the inside of a human's intestines. Tapeworms can grow to be one to two meters long or even longer if they are not treated.

What negative health effects would you expect to see in humans infected by tapeworms?

Phylum Platyhelminthes (Flatworms)

As the name suggests, flatworms have flattened bodies. Their bodies are soft, and they have eyespots on top of their head allowing them to sense light. Tapeworms, flukes, and planarians are examples of flatworms. They are often found living in shallow water, but some parasitic flatworms live inside a host, infecting the blood, liver, lungs, eyes, or other organs.

Tapeworms are flatworms that attach themselves to the intestines of their host and absorb the host's digested food. Tapeworms reproduce by growing new body segments filled with as many as 100,000 eggs. When a segment passes out of the host's body, it can contaminate food or water. Another animal can become a host if it eats undercooked meat or drinks water containing tapeworm eggs or larva. What precautions can be taken against these flatworms?

Planarians are flatworms, but they are not parasites. These free-living worms thrive under rocks and in freshwater. They eat small organisms and the remains of dead organisms.

Explore-a-Lab

Structured Inquiry

How do planaria regenerate?

Obtain three planaria in a Petri dish with some water from your teacher. Observe the worms with a hand lens and draw your observations. Use a plastic slip cover to cut each of the worms in half. Try to cut one just behind the head, another in the middle, and the third closer to the tail. These worms move fast, but adding ice to the water can slow them down. Removing some of the water from the Petri dish may also help, but the worms must have some water to live. You may want to look through the hand lens when doing your cutting. Use a medicine dropper to transfer the tails to a second Petri dish filled halfway with water that has not been chlorinated. Place a lid on each Petri dish and label heads and tails. Keep the worms in a warm, dark place. Observe the worms daily and record how their appearance changes. What parts grew the fastest? Did the location of the cut affect how quickly parts regenerated?

Phylum Annelida (Segmented Worms)

A common segmented worm, or **annelid**, is the earthworm. The word *annelid* means "little rings," which describes the tube-shaped bodies of these worms. Their bodies, with as many as 100 segments, are not as smooth and slippery as they look. Instead, four pairs of bristlelike structures on each segment help the worm grip the soil as it moves along. In what ways do earthworms benefit gardeners and farmers? In what ways are they harming some forests?

Segmented worms are different from other kinds of worms. They have two body openings, a small brain, nerves, an intestine, blood vessels, and a circulatory system. These worms also have two types of muscle tissue. One type runs the length of their bodies, and the other muscles circle the segments. The bristles and these two sets of muscle tissue allow the earthworm to move forward. Have you ever lain flat on the floor and tried to move without using your arms or legs? What happened?

Focus on Health

One reason to avoid eating pork is that undercooked pork can transmit trichinosis, a disease caused by a roundworm. There are many references in the Bible where God says not to eat pork. Why do you think He does not want people to eat pork?

The earthworm's brain and nerves enable it to respond to light, temperature, and moisture. In this way, it can seek the dark, warm, moist environment that will help it survive. Why do you think you find so many earthworms on sidewalks after an overnight rain?

Some segmented worms reproduce asexually by budding, while other species reproduce sexually. These segmented worms are hermaphroditic, meaning they have both male and female parts. Following fertilization, a cocoon is formed around the eggs until they hatch.

Leeches are also segmented worms, but they are flatter and shorter than earthworms and do not have bristles. Instead of feeding on plant material, leeches use a sucker at the front end of their bodies to suck blood from their hosts. The rear sucker is used for leverage and movement. One type of parasitic leech produces a chemical that prevents blood from clotting and widens blood vessels, improving the blood flow. How can leeches be used for medical purposes?

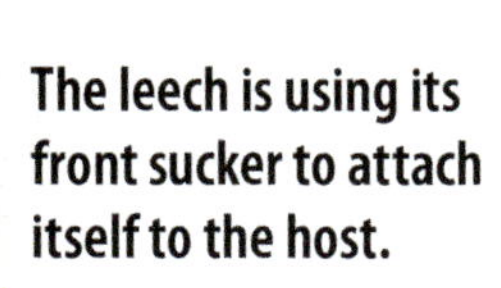

The leech is using its front sucker to attach itself to the host.

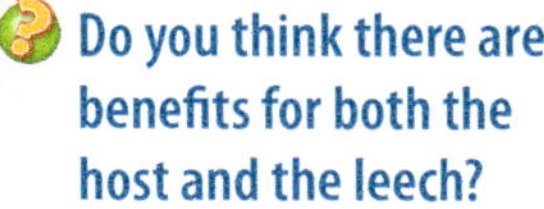

Do you think there are benefits for both the host and the leech?

Phylum Nematoda (Roundworms)

There are more than 500,000 species of roundworms. Two common roundworms are parasitic heartworms and nematodes. Heartworms can infect dogs and can cause painful knots in body tissues. How can heartworm infections be prevented? Nematodes live in the soil and can cause millions of dollars in damage to crops.

Trichina are parasitic worms that enter the body when undercooked meat (usually pork) is eaten. An infection of trichina worms causes the disease trichinosis, a potentially fatal disease. In what ways can trichinosis be prevented?

Some roundworm species can be beneficial. They feed on insects that damage crops. Roundworms also help with the nitrogen cycle that adds nutrients to the soil by breaking down dead plants and animals.

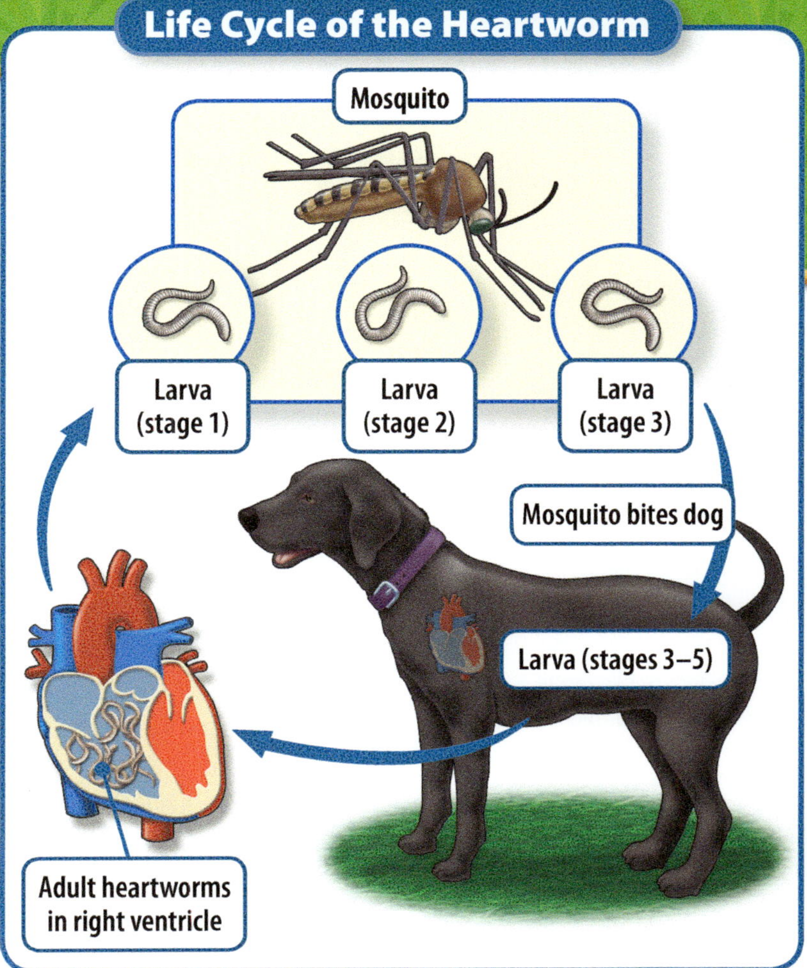

Heartworms are transmitted by mosquitoes to dogs. The larval worms travel to the dog's heart and continue growing.

What important steps can be taken to prevent the contraction of heartworms in a dog?

Concept Check Assess/Reflect

Summary: What are sponges, cnidarians, and worms? Sponges, cnidarians, and worms are all invertebrates with characteristic features. Sponges are asymmetrical, and the majority of them live in saltwater. Sponges have cells with flagella that move water through them, helping to filter the water. Cnidarians have radial symmetry and can be found mainly in saltwater. They use their tentacles to stun and capture prey. Worms have bilateral symmetry, and many of them live on land. All of these invertebrates use their unique body structures to help them move and function in their ecosystems.

1. How do sponges reproduce?
2. Why are cnidocytes important to a sea jelly?
3. Compare and contrast the methods used by tapeworms and leeches to obtain food.
4. Why do you think worms are relatively rare in the fossil record?

Lesson 3

Essential Question

What Are Mollusks and Echinoderms?

Objectives

- Describe the characteristics that help identify mollusks and echinoderms.
- Explain the specialized structures that equip mollusks to thrive in their environment.
- Explain the specialized structures that equip echinoderms to thrive in their environment.

Vocabulary

mollusk
mantle
radula
bivalve
echinoderm
endoskeleton
water-vascular system
tube foot

Mollusks and echinoderms are two different phyla categories of invertebrates. The body structures of each animal have specific functions that help the animal survive. The wonderpus octopus (pictured below) has several characteristics that make it unique. This octopus emerges at dawn and dusk to feed on small crustaceans. Why do you think the wonderpus hunts at these times? The wonderpus also has unique markings that allow researchers to distinguish individuals. How can this feature help scientists study this newly discovered species? Why do you think it took scientists so long to discover the wonderpus?

Phylum Mollusca (Mollusks) Explain

Think about the last time you saw a garden snail or slug. What do you remember about how it looked? How did it move? Snails and slugs are two common examples of animals that are classified into the phylum Mollusca. Squid, octopuses, clams, and scallops are also part of this phylum. Other members of this phylum include the eight-plated chitons and the beautiful chambered nautilus. Scientists have named and described nearly 100,000 species of mollusks worldwide.

The wonderpus octopus, discovered in the 1980s, lives in shallowwaters around the islands of Southeast Asia. The pattern of spots and stripes on each wonderpus is unique, so scientists and photographers can use these patterns to identify individuals.

How does an octopus differ from a squid?

A mollusk is an animal that has a soft, unsegmented body. In addition to the soft body, mollusks share other important characteristics described in the table below.

Phylum Mollusca	
Type of symmetry	Bilateral
Body parts	Head—the part that includes the mouth and sense organs Foot—a muscular structure that extends from the body and allows the animal to move, attach to a surface, or to capture food **Mantle**—fleshy tissue that covers and protects the mollusk's internal body parts; secretes the shell in those mollusks that have shells
Type of covering	One or more shells, or no shell
Examples	Snail, slug, squid, clam, scallop

Mollusks have a variety of adaptations that enable them to find and eat food. Many mollusks, such as clams, mussels, and scallops, simply filter plankton out of the water. Other mollusks, such as snails and slugs, are equipped with a toothed, or file-like, structure called a **radula** that is used in feeding. The octopus, squid, and chambered nautilus have specialized legs that capture food. Do any of these feeding habits surprise you? Why or why not?

All mollusks have these parts, but the parts have different shapes in different types of mollusks.

What is the function of the mantle?

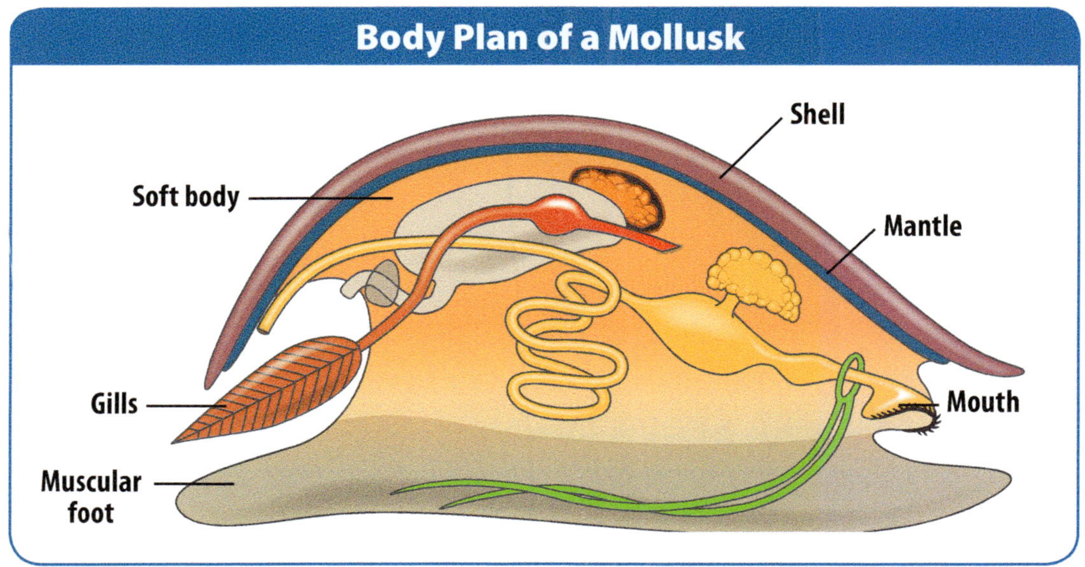

The mantle is one of the important body parts of the mollusk; its role is to protect the internal parts. In water-dwelling mollusks, the mantle covers the animals' gills. As water moves across the gills, oxygen diffuses into the mollusk and carbon dioxide diffuses out. Terrestrial mollusks, snails and slugs, have a breathing pore located beneath the mantle.

Among the seven classes of mollusks, the most common are Class Polyplacophora (chitons), Class Gastropoda (gastropods), Class Bivalvia (bivalves), and Class Cephalopoda (cephalopods).

Class Polyplacophora (Chitons)

Chitons and similar animals belong to the class Polyplacophora. Chitons range in size from a few millimeters long to the giant gumboot chiton of the Pacific coast that can grow to more than 30 centimeters. Chitons are boat-shaped animals that have shells made up of eight overlapping plates. The plates are held together by tough, leathery flesh called a *girdle*. Most members of this class feed on the algae and plankton that cover intertidal rocks. However, some, such as the lip chiton, are predators that chase down potential prey. The chiton's large muscular foot is perfectly adapted to holding tight to rocks. This protects the chiton from being pried off by predators and from being washed about when large waves crash on top of it. If a chiton does come off the rock, it rolls itself into a tight ball for defense, much like an armadillo does. Are there other characteristics of the chiton that resemble an armadillo?

The chiton is holding on tightly to the rock with its large foot.

What other creatures does the chiton remind you of and why?

Record your work for this inquiry. Your teacher may also assign the related Guided Inquiry.

Take-A-Part Clam

What are the external and internal features of a clam?

Allow your teacher to open the clam for you. Wash your hands and materials thoroughly after handling the clam.

Materials

- balance
- clam
- clam diagram
- dissecting kit
- dissecting pan and pad
- disposable gloves
- hand lens
- metric ruler
- string

Procedure

1. Obtain a clam.
2. **Observe** the clam carefully. In the data table, **record** the clam's color and shape. Make a sketch of the external view of the clam in the box provided. Label the following parts, using the diagram in your ***Science Journal*** for reference:
   - Anterior end
   - Dorsal side
   - Hinge
   - Posterior end
   - Umbo
   - Valve
   - Ventral side
3. Use the hand lens to look closely at the umbo. Find the tiny shell that represents the size of this clam when it started out as an adult clam.
4. Measure the length, width, girth (distance around), and mass of the clam. Use the string to help measure the girth. **Record** these measurements.
5. Determine the approximate age of your clam by counting its growth lines. Use the hand lens if needed.
6. Follow Steps 7–9 in your ***Science Journal*** to study the internal features of the clam.

Analyze Results

Compare the clam with the drawing of the clam. Identify any parts that are in the clam but not shown in the drawing. Determine what these parts might be and the purposes they serve.

Create Explanations

1. What are the external and internal features of a clam?
2. How does a clam eat food, and what body part is used for collecting food?
3. Explain how a clam breathes and which body parts are used for breathing.

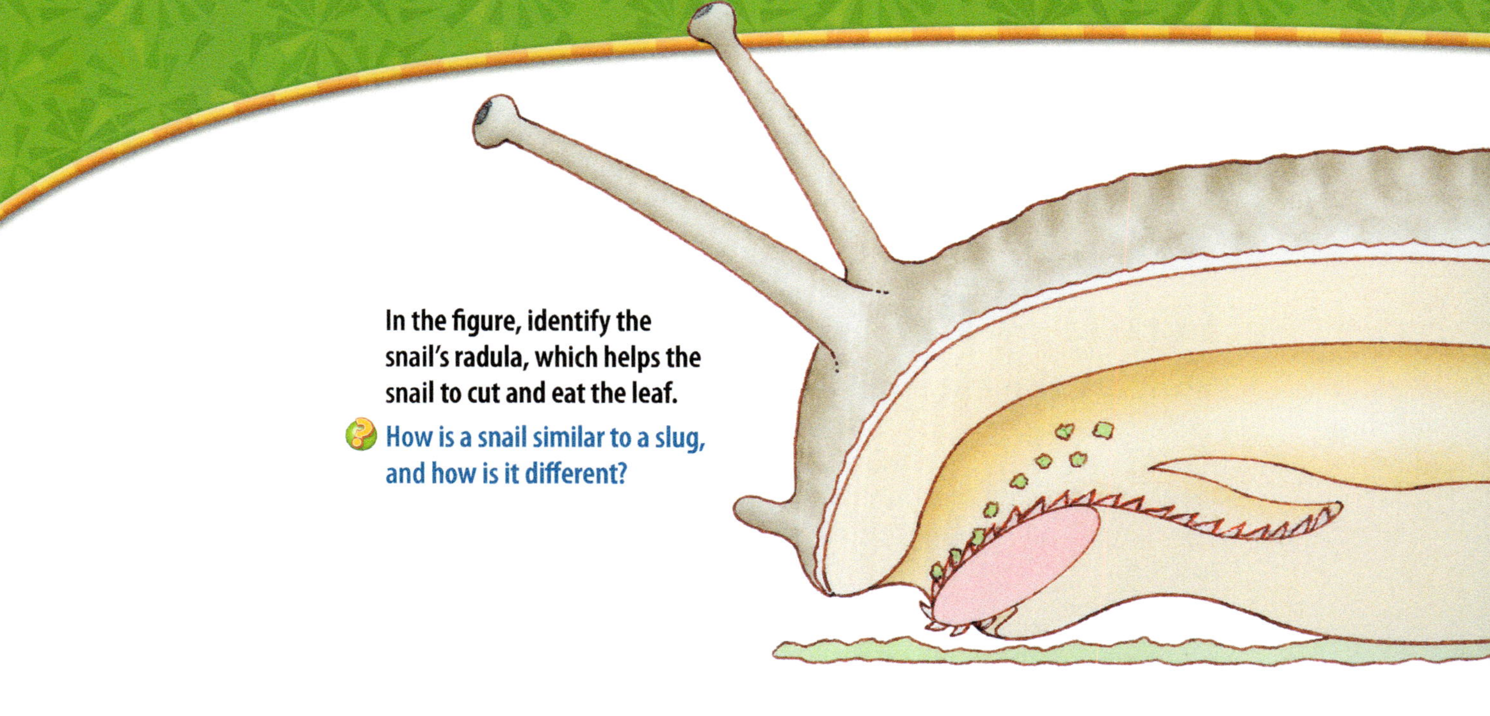

In the figure, identify the snail's radula, which helps the snail to cut and eat the leaf.

How is a snail similar to a slug, and how is it different?

Class Gastropoda (Gastropods)

Snails and slugs are gastropods. The word *gastropod* means "stomach-footed." The muscular foot of these mollusks runs along the underside, or belly, of the animal. Gastropods have a well-developed head, often equipped with sensory tentacles used to locate food and alert the animal to danger. At the rear of the foot, many snails have a lid-like structure called an *operculum* that closes when the snail retreats into its shell. How do you think this structure benefits the snail?

Gastropods have a hard and rigid *radula* that works like a saw or file to cut or scrape food from the rocks or to munch away at seaweed. Some snails have a radula that acts as a harpoon to capture prey. In other predatory snails, the radula is modified to bore through the shell of other mollusks. Once the snail bores the hole through the prey's shell, how do you think the snail eats its prey?

While many marine and aquatic snails and slugs breathe with gills, some land-dwelling gastropods have a lung-like structure that allows oxygen to diffuse across its thin walls into the bloodstream, while at the same time allowing carbon dioxide to pass from the bloodstream through the thin walls and out of the snail.

Scripture Spotlight

Read **Proverbs 6:6–9**. The term *sluggard* comes from *slugs*, which move like snails. What is the message of this passage?

Explore-a-Lab

Guided Inquiry

What are some characteristics of a garden snail?

Closely observe a garden snail. What external features do you see? Make a drawing and label as many parts as you can. Next, place some small pieces of vegetable leaves, such as lettuce or cabbage, near the snail. Observe the snail. Does it move toward the food? How do you think the snail senses the food? How does the snail move?

Class Bivalvia (Bivalves)

Bivalves are mollusks that have a shell of two halves; each half is called a *valve*. (The term *bivalve* literally means "two shells.") The valves are mirror images of each other. A flexible band of tissue joins the two halves and functions as a hinge. Most bivalves have toothlike structures at the hinge of each shell. A muscle that is attached to the shell's inner surface allows the bivalve to open and close its shells. Were you able to observe these features when you dissected the clam in the Structured Inquiry earlier in this lesson? How does this muscle help the bivalve protect itself from predators?

The mantle lines the valves. This structure secretes calcium carbonate that builds the valves from the inside out. In many mollusks, the mantle also secretes a smooth, shiny layer called nacre, or mother-of-pearl, that is designed to protect the mantle should small sand grains or other debris get into the shell and create an irritation. It is this material that forms the pearls sometimes found in certain bivalves.

The majority of bivalves live in the ocean, although a few varieties of clams and mussels live in freshwater. Bivalves have a thick foot that they use to dig into the sand on the ocean floor and bury themselves. The bivalve's foot can also be used to secure it to a rock. This is accomplished through a bundle of fibers attached to its foot that pass through the gap between the two shells and anchor the animal. However, bivalves are not secured to one location for their entire lives. Some bivalves, such as the scallop, swim. How do you think a scallop can swim without fins or a tail?

Unlike other mollusks, bivalves do not have a head or a radula. They feed by filtering food particles from the water. To do this, special tubes pump water through the bivalve's digestive system. One tube, the incurrent siphon, brings water into the clam. As the water passes through the mantle cavity oxygen is absorbed by the gills. Cilia on the gills capture the particles of food drifting in the water and bring the food to the clam's mouth. The water then flows out of the clam through the other siphon.

Giant clams (*Tridacna gigas*) live in shallow (up to 20 meters) areas of the Indian and Pacific Oceans. They are the largest of all the mollusks and can live for 100 years or more.

Algae live in the mantle of the giant clam. How do you think the algae benefit? How do you think the mollusk benefits?

Faith Connection

Even though squid are invertebrates, their eyes have many similarities to human eyes. Their complex eyes are evidence of God's Design.

Class Cephalopoda (Cephalopods)

Squid, octopuses, cuttlefish, and nautiluses are cephalopods. The term *cephalopod* means "head-footed." What most of us call the arms, or tentacles, of an octopus or squid are actually feet. They are attached to the part of the body that contains the eyes (the "head").

Except for the nautiluses that live in the South Pacific and Indian Oceans, cephalopods do not have external shells. They breathe with gills. They have well-developed eyes and excellent eyesight. The giant squid has the largest eyes of any animal on Earth. Why do you think it needs such large eyes? The squid has a pen, which is a slightly flexible, internal shell that supports the mantle and that has muscles attached to it.

When the octopus hunts for food, it uses both its excellent eyesight and its sensitive tentacles to find its prey. Once it finds a crab, a fish, or other suitable prey, the octopus quickly uses its tentacles, equipped with powerful suction cups, to capture the prey and bring it up to its mouth where its beak can tear the food into small pieces for swallowing.

Cephalopods can also move by jet propulsion. They push water from the mantle cavity through a siphon. The sudden squirt shoots them forward. Squid and octopuses also shoot out a dark liquid, or ink. The ink confuses predators and gives the squid or octopus time to move away from danger.

An octopus is able to camouflage the coloring and texture of its outer mantle.

What advantage do you think the coloring of the octopus provides?

Phylum Echinodermata (Echinoderms) Explain

Sea stars, sand dollars, brittle stars, sea cucumbers, and sea urchins are **echinoderms**, which means "spiny skin." Why is this a good name for these animals? How are these animals similar to mollusks? How are they different? Scientists have identified about 6000 types of echinoderms. All known echinoderms live in the world's oceans.

Phylum Echinodermata	
Type of symmetry	Bilateral symmetry (larva) Radial symmetry (adult)
Body parts	Endoskeleton—internal skeleton made of hard plates that may protrude through the skin Water-vascular system—system of water-filled canals that extend through the body and attach to tube feet Tube feet—allow the animal to move, attach to a surface, or capture food
Type of covering	Spiny skin
Examples	Sea star, sand dollar, sea urchin

Adult echinoderms have radial symmetry. Many have rays, or arms, in fives or multiples of five that radiate outward from the center of their body. In their immature (larval) forms, some echinoderms have bilateral symmetry. Scientists don't know exactly why this is. Why do you think echinoderms have different forms of symmetry as larval and adult forms? Echinoderms have an **endoskeleton** made of branching crystals of calcium carbonate ($CaCO_3$) and that often includes spines. How else do animals use the mineral calcium carbonate?

Echinoderms are also characterized by a system of water-filled canals that form a **water-vascular system**. These canals radiate out from a central ring canal that encircles the gut of an echinoderm. The canals extend down the body of the echinoderm and attach to **tube feet**, which allow the echinoderm to move.

Faith Connection

An echinoderm uses its tube feet and water-vascular system to grasp and hold its food. How is this like your relationship with God?

Starfish

This diagram shows the water-vascular system of a sea star.

Sea stars are sometimes called starfish. Explain why that name is misleading.

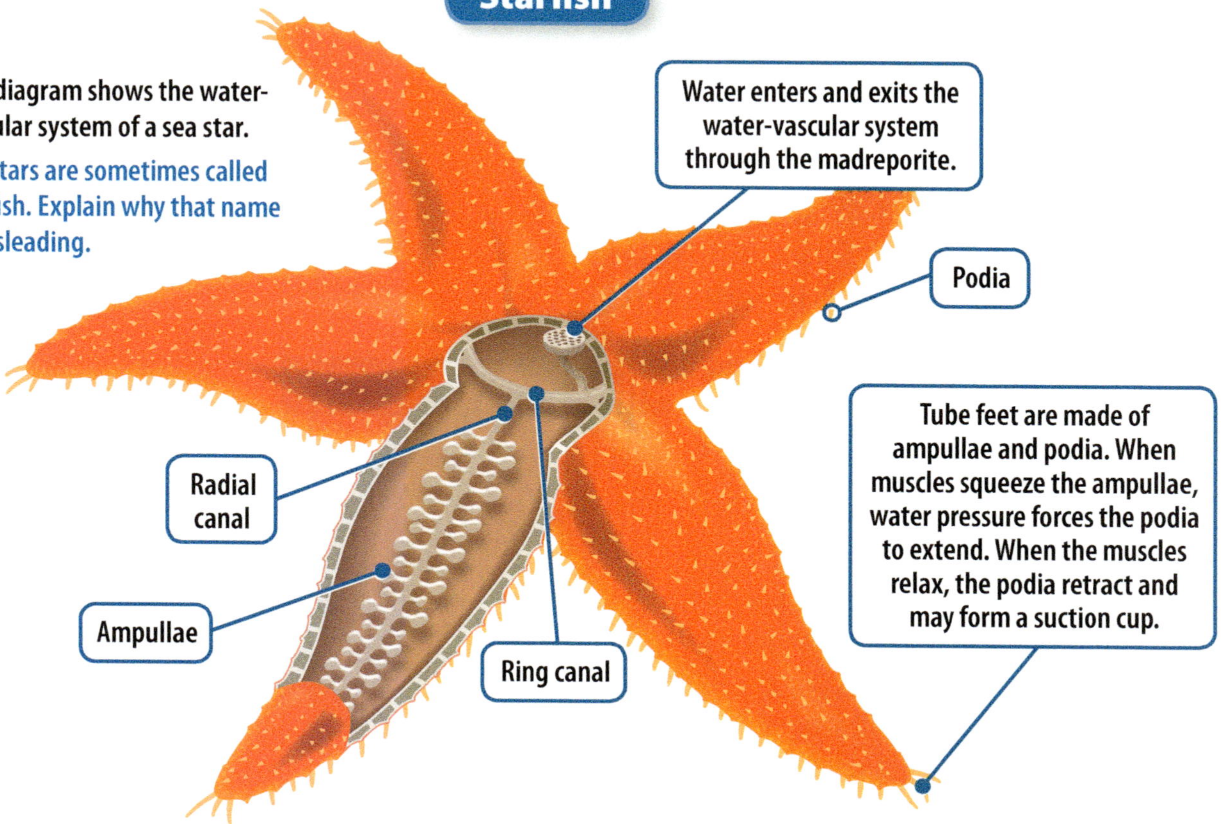

Water enters and exits the water-vascular system through the madreporite.

Podia

Tube feet are made of ampullae and podia. When muscles squeeze the ampullae, water pressure forces the podia to extend. When the muscles relax, the podia retract and may form a suction cup.

Radial canal

Ampullae

Ring canal

The water-vascular system gives echinoderms an ability to move in a unique way. Both the canal system and the tube feet are filled with seawater. By expanding and contracting chambers within the water-vascular system, the echinoderm can force water into certain tube feet to extend them. The animal has muscles in the tube feet, which are used to retract them. By expanding and contracting the right tube feet in the proper order, echinoderms can walk along the sea floor, hold on tightly to rocks, and capture their food. Like many marine invertebrates, echinoderms breathe by using gills. Many echinoderms live in the ocean tidal zone. What do you think might happen if an echinoderm is out of water for too long?

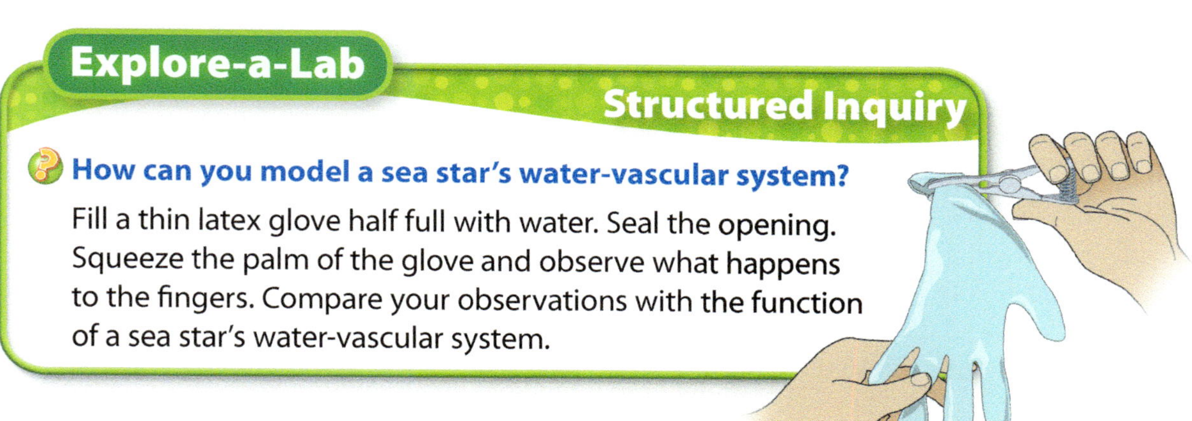

Explore-a-Lab

Structured Inquiry

How can you model a sea star's water-vascular system?

Fill a thin latex glove half full with water. Seal the opening. Squeeze the palm of the glove and observe what happens to the fingers. Compare your observations with the function of a sea star's water-vascular system.

Scientists classify echinoderms into four main classes: sea lilies (crinoids), sea urchins and sand dollars, sea stars, and sea cucumbers. Each class has its own unique characteristics.

Echinoderms			
Class	**Spine**	**Body Type**	**Tube Feet**
Sea lilies	Absent	A stem with long arms stretching upward	Long, tentacle-like tube feet located on arms and used for capturing food
Sea urchins and sand dollars	Present	Globe- or disk-shaped body	Used for movement
Sea stars and brittle stars	Present	A central disk with arms radiating outward	Used for movement and capturing prey
Sea cucumbers	Absent	Elongated, soft body; no legs	Used for movement

Concept Check Assess/Reflect

Summary: What are mollusks and echinoderms? Mollusks and echinoderms are two phyla of invertebrates. Mollusks have soft, unsegmented bodies and possess bilateral symmetry. Each mollusk has three body parts. Some mollusks have a shell that protects their body. Examples of mollusks include slugs, octopuses, and clams. Echinoderms have an endoskeleton and possess radial symmetry. All echinoderms live in the world's oceans. Examples of echinoderms include sea stars, sea cucumbers, and sea urchins. Tube feet help echinoderms move, hold on to rocks, and capture food. Muscles and a water-vascular system work together to extend and contract the tube feet.

1. What body parts do all mollusks have?
2. How do mollusks and echinoderms breathe?
3. How do bivalves eat without a head?
4. Explain how echinoderms use the water-vascular system.
5. You have just learned about the structure of mollusks and echinoderms. What is one of these structures that you think provides evidence of God's Design? Explain why.

Lesson 4

Essential Question

What Are Arthropods?

Objectives

- Describe the characteristics and structures of arthropods.
- Identify the characteristics of the major classes of arthropods.
- Describe the importance of arthropods to people.
- Describe the specialized structures of arthropods.
- Explain the difference between complete and incomplete metamorphosis.

Vocabulary

arthropod
exoskeleton
antenna
molt
complete metamorphosis
incomplete metamorphosis

Look at the crab below. It is a member of the same phylum as spiders and insects. Why do you think scientists would place a unique creature such as the crab in the same phylum? What do crabs have in common with spiders? There are more than 6000 species of crabs known in the world. How do you think the crab helps sustain our Earth? What role do the spiders and insects play? As you read, think about how these animals help maintain homeostasis on Earth.

Characteristics of Arthropods Explain

Spiders, centipedes, millipedes, mites, ticks, lobsters, crabs, shrimp, crayfish, barnacles, and insects are all members of the same phylum, Arthropoda. They may look very different from one another, but there are similar features shared among these creatures. **Arthropods** are invertebrates in the phylum Arthropoda. God designed the body structures of each animal to have specific functions that help the animal survive. Arthropods are found all over the world, and you most likely see at least one on a daily basis.

Of all the animal phyla, Arthropoda is the largest. More than three-fourths of all known organisms are arthropods. Why do you think there are so many arthropods? Scientists have identified more than one million species, but the true number of arthropod species is still unknown. Why do you think scientists don't know exactly how many arthropods there are? Arthropods live on land, in the water, and in the air. They range from the giant king crab that can have a leg span of 3.6 meters to the microscopic, aquatic copepod that is an important part of ocean food webs.

Crabs are classified as arthropods and are an example of Crustacea.

What characteristics of arthropods can you identify on the crab's body?

Three Major Characteristics

Arthropods share the following three major characteristics:

- Segmented bodies that form units, such as a head, thorax, and abdomen
- An external skeleton, called an **exoskeleton**, made of chitin; What do you think are the advantages and disadvantages of such a skeleton?
- Jointed appendages, which include legs, mouthparts, and antennae

Antennae are jointed, movable, sensory structures on the head of arthropods. Some arthropods have a single pair of antennae, while others have two pairs of antennae. Arthropods also have a number of different types of mouthparts, each suited to a specific style of feeding. Arthropod mouthparts are designed for cutting and chewing, piercing and sucking, siphoning, and filtering. Name an arthropod that might have each kind of mouthpart.

Find the appendages, segments, body units, and exoskeleton of this arthropod.

 Why are antennae considered appendages?

In addition to these characteristics, arthropods share other features.

Arthropods				
Type of Symmetry	**Body Parts**	**Type of Covering**	**Circulation**	**Examples**
Bilateral symmetry	Head, thorax, abdomen	Exoskeleton made of chitin	Open circulatory system	Spider, lobster, flea, centipede, insect

Arthropods have an open circulatory system. The blood of an animal with an open circulatory system is not enclosed in a network of blood vessels. How do you think bilateral symmetry and an open circulatory system help arthropods survive?

Shedding the Exoskeleton

The exoskeleton of an arthropod is rigid and inflexible, and it does not grow as the animal grows. Eventually the exoskeleton gets too tight, making it necessary for arthropods to **molt**, or shed their exoskeleton and replace it with a new one. When this process occurs, the animal wiggles out of the old exoskeleton and emerges with a new, soft exoskeleton that has been growing underneath. The animal expands quickly before the new exoskeleton hardens. What other animals replace their body coverings by molting?

Scripture Spotlight

Read **Colossians 3:8–14.** How is the molting of an exoskeleton like conversion? How is it different?

Structured Inquiry

Discover

Record your work for this inquiry. Your teacher may also assign the related Guided Inquiry.

Behavior of Arthropods

How do mealworms respond to different stimuli?

Procedure

1. Cut a circle out of paper towel. Draw a line down the middle of the paper-towel circle. Put it inside the Petri dish.
2. Drop a few drops of water onto one half of the circle. Make it damp but not soaking wet. Leave the other half dry.
3. Place 5 mealworms in the dish. Watch for 5 minutes or more, and **record** your **observations** every 60 seconds. Draw a diagram showing the position of each worm for each minute.
4. Remove the worms and dry the Petri dish. Cut a half circle out of the black paper. Tape it over the top of one half of the Petri dish so that half of the dish is dark.
5. Place 5 mealworms in the dish. Watch for 5 minutes, and **record** your **observations** every 60 seconds. Draw a diagram showing the position of each worm for each minute.

Materials
- paper towel
- scissors
- Petri dish
- dropper
- water
- 5 live mealworms
- black construction paper
- stopwatch or clock that indicates seconds

Analyze Results

What did the mealworms do in wet and dry environments? What did they do in light and dark environments? Explain how you know.

Create Explanations

1. How do mealworms respond to different stimuli?
2. Why was it important to record your observations every 60 seconds?
3. How might the behaviors you observed help these arthropods survive in their environment?

Metamorphosis

Recall that insects go through a transformation during their life cycle. Scientists have found that some arthropods experience a **complete metamorphosis**, and some experience an **incomplete metamorphosis**. The majority of insects experience a complete metamorphosis. This involves four stages: egg, larva, pupa, and adult. The transformation from a caterpillar to a butterfly is a common example of a complete metamorphosis. When the larva emerges from the egg, it is often a wormlike creature that will grow and molt many times. The larva looks very different from an adult of this species. The larva will then become inactive as a pupa and eventually emerge from its cocoon as an adult fully transformed. What changes are taking place while the organism is in the pupa stage? Bees and flies are two types of insects that experience a complete metamorphosis.

In an incomplete metamorphosis, there are only three stages: egg, nymph, and adult. In this process, the nymph, which resembles a small adult of its species, emerges from the egg. The nymph will grow and molt many times until it is an adult. Grasshoppers and dragonflies are two types of insects that experience an incomplete metamorphosis. Which type of metamorphosis is more similar to changes that mammals undergo?

Check out your ***Science Journal*** for a Guided Inquiry to observe how mealworms change over time.

Extend

When an insect hatches, how can you tell what type of metamorphosis it will experience?

Incomplete Metamorphosis

Adult

Nymph

Egg

Classes of Arthropods Explain

Scientists have identified a large number of arthropod classes. The table below shows five of the major arthropod classes.

Arthropods		
Class	**Major Characteristics**	**Examples**
Crustacea	• Fused head-thorax (cephalothorax) and abdomen • Five pairs of appendages on the head, including two pairs of antennae • Five pairs of legs	Crayfish, crab, shrimp, lobster
Arachnida	• Two body regions (cephalothorax and abdomen) • Four pairs of legs	Spider, tick, mite, scorpion, harvestman (daddy longlegs)
Chilopoda	• One pair of legs per body segment	Centipede
Diplopoda	• Two pairs of legs per body segment	Millipede
Insecta	• Body with three regions (head, thorax, abdomen) • Five pairs of appendages on head, including one pair of antennae • Three pairs of legs	Silverfish, fly (many types), lice, grasshopper, cockroach, termite, beetle, flea, bee, moth, butterfly

Millipede

Centipede

Both of these animals are arthropods, but they are different classes. Centipedes belong to the class Chilopoda, while millipedes belong to the class Diplopoda.

What feature can you **see** here that distinguishes one class from another?

Lesson Activity

Arthropods have an exoskeleton that is replaced as the creature grows. Choose an arthropod and make a flipbook showing how the animal molts. On the back of the book, write a description of each stage of the process.

How does the arthropod get a new shell?

Structure and Function in Arthropods Explain

Arthropods have many structures that allow them to live, grow, and reproduce in different environments. For example, arthropods must take in oxygen and get rid of carbon dioxide, but they do not have lungs. Instead, they exchange gases with the environment in other ways.

- Crabs breathe with gills, using a mechanism similar to that of octopuses.
- Some arachnids have "book lungs," which are not lungs at all. They are a series of thin-walled, air-filled plates surrounded by blood. The plates are stacked like the pages of a book. Diffusion occurs across the plates, moving oxygen into the blood and carbon dioxide out.

Scripture Spotlight

Spiders have fangs that inject venom into their prey. Some spiders actively hunt their prey, while others spin webs to ensnare unsuspecting creatures. What things around us—obvious or subtle—may endanger us spiritually? What do **Proverbs 13:14** and **14:27** say will protect us from Satan's snares?

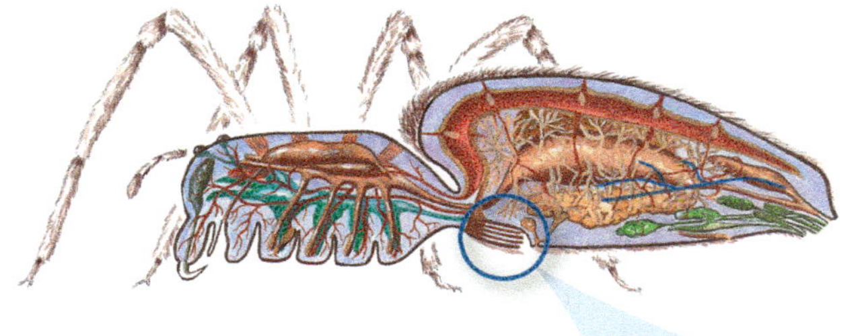

The book lung of a spider

How is this structure suited for the exchange of gases between the spider and the environment?

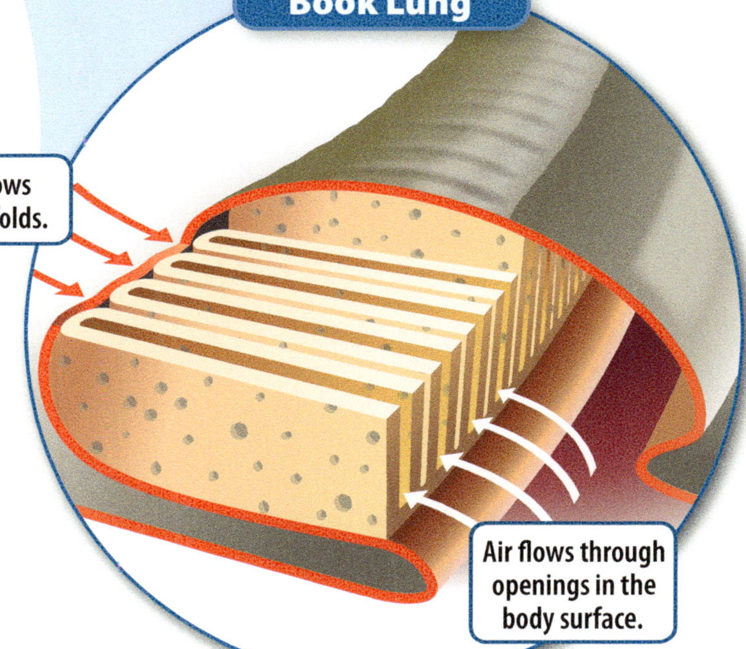

- Insects have *spiracles* on their abdomen. The spiracles are pores that open into tubes, called *tracheae*. Air enters the tracheae through the spiracles. Gases diffuse into and out of the blood through the tracheae. Many aquatic insects also have gills.

Why do you think different arthropods exchange gases in different ways?

Pictured are the spiracles and tracheae of a grasshopper.

Through which body segment does the air enter the insect?

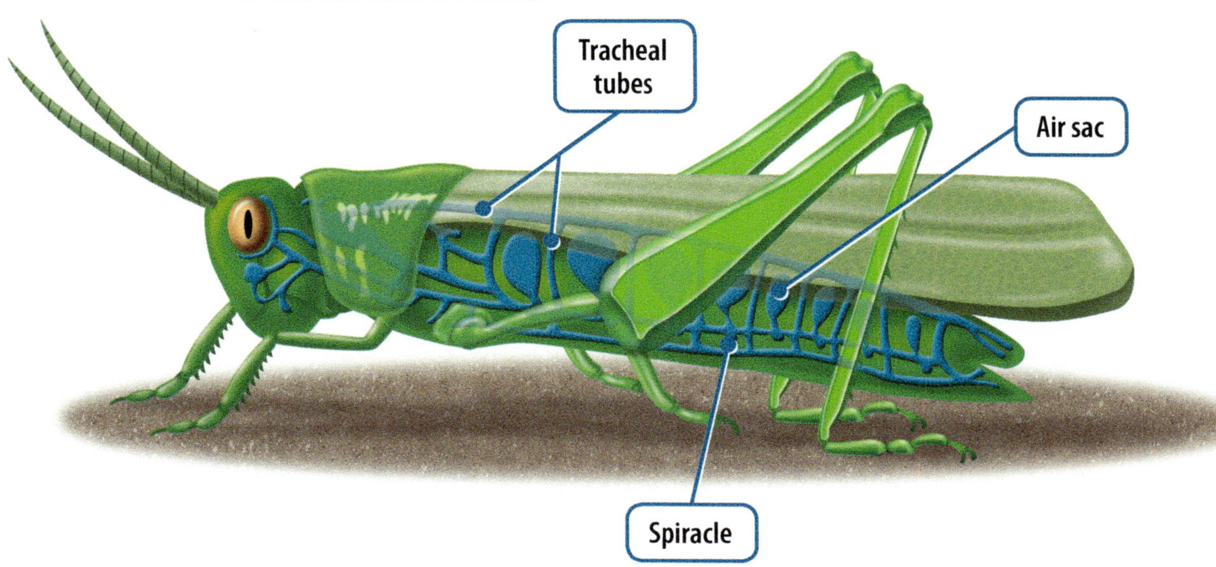

Importance of Arthropods

Many people view insects and spiders as pests because some bite or sting, carry disease, and destroy food crops. However, insects are vital to food chains in every ecosystem on Earth, and their activities help maintain balance in ecosystems. What are some ways that arthropods can be useful to humans?

- Bees, ants, beetles, and butterflies pollinate flowering and fruit-bearing plants. The fresh fruit section in grocery stores would be empty if it were not for insect pollinators.
- Spiders eat mosquitoes, flies, and many other harmful insects.
- Burrowing insects create channels in soil that allow water to reach plant roots.
- The praying mantis eats caterpillars and aphids, two insects that can do significant damage to plants.
- Insect droppings also help fertilize soil.

Some arthropods produce products that humans use. Bees produce honey that we enjoy as food. They also make the beeswax that we use in candles, furniture polish, and medicinal ointments. The caterpillar of the silk moth produces a strong fiber. Humans use silk fiber for making yarns and fabrics. What products do you use that might have come from insects?

Many cultures around the world use arthropods as a source of food because they are rich in protein. In Africa, grasshoppers and the Goliath beetle are commonly eaten.

In Africa and Australia, termites are valued as food because they have higher amounts of protein and fat than other insects. The termites may be fried and eaten or made into oil used for cooking. Locusts are eaten in South Korea, and scorpions are eaten in China. Caterpillars, grubs, and other insect larvae have been eaten all over the world. Campers, hikers, and military personnel are also taught to eat insects to survive if they are stranded in the wilderness. Some insects are even dipped in chocolate for a sweet treat! Would you eat one?

Scripture Spotlight

Compare **Deuteronomy 7:17–21** with **Joshua 24:12.** What arthropod did God use to defeat Israel's enemies?

Concept Check Assess/Reflect

Summary: What are arthropods? Arthropods are invertebrates that make up the majority of all animal species on Earth. There are several classes of arthropods, including the Crustacea, Arachnida, Chilopoda, Diplopoda, and Insecta. Arthropods share several characteristics and undergo either a complete or incomplete metamorphosis once during their life cycle. They have an exoskeleton that provides them with support and protection. As arthropods grow, they molt the exoskeleton, and a larger one replaces it. They exchange gases with the environment in different ways, including the use of gills, book lungs, and spiracles and tracheae. Arthropods are a vital and beneficial part of an ecosystem.

1. What three major characteristics do all arthropods have?
2. How do the numbers of legs differ among the classes of arthropods?
3. Explain how the structure of a book lung allows it to perform its function.
4. How can an exoskeleton be compared with a suit of armor?

Essential Question

How Do Invertebrates Use Their Senses?

Objectives

- Describe ways in which invertebrates can sense their environments.
- Identify ways in which invertebrates communicate.

Vocabulary

ocelli

compound eye

olfaction

pheromone

chromatophore

Have you ever seen a movie that featured talking animals? That's usually a good indicator that the movie is fictional; animals do not talk. Animals, both vertebrates and invertebrates, do communicate with one another; it just looks and sounds differently than a conversation between humans. Invertebrates communicate with others of their species and sense their surroundings. They communicate with other species as well. They see, hear, taste, touch, and smell, but not always in the same ways that you do. Look at the butterfly in the picture. What body parts do you think the butterfly might use to help it communicate? As you read, think about how you sometimes communicate nonverbally with others.

These monarch butterflies use their antennae to help them fly in the right direction when they migrate.

How would the ability of butterflies to sense direction help them during migration?

Sensing the Environment Explain

A sponge has no nervous system, but it contracts its body when touched. Earthworms have receptor cells along their body surface that let them respond to light, touch, vibration, and chemicals. Sea jellies have a series of interconnected nerve cells, a nerve net. They also have sense organs called *rhopalia*. The rhopalia contain special cells that let the animal detect chemicals, pressure, gravity, and light. The antennae of arthropods do many jobs, too. They sense touch, motion, temperature, sounds, and chemicals (odors and tastes). Mollusks use their sense of smell and chemical signals to detect food and predators at long distances. Echinoderms have a nerve that runs the length of each arm to the tube feet, which are sensitive to touch. Other cells scattered throughout the arms are sensitive to touch and to other stimuli in the water the echinoderms live in. This allows the animals to gather information about their environment, such as the temperature or amount of light present.

Some sea stars have specialized cells in their tube feet that detect chemical scents left behind by potential prey; they are able to follow the scent to hunt their prey. This same sense ability is used to make them aware of potential predators so that they can escape or prepare to defend themselves. All of these different creatures are able to sense their environment using different body parts.

How Invertebrates See

Invertebrates have their own ways of seeing. Many invertebrates have simple, light-sensitive cells called **ocelli**. Flatworms, sea jellies, and sea stars have ocelli called pigment-spot ocelli, which have pigment distributed randomly and have no additional structures such as a cornea or lens. These eyes can detect only light and shadow. The color of these eyes is either red or black. Why do you think these eyes are one of those colors?

Many snails and slugs also have ocelli located either at the tips or at the bases of the tentacles. Other gastropods, such as the conches, have much more sophisticated eyes. Giant clams have ocelli that allow light to penetrate their mantles, and scallops may have as many as 100 blue eyes that ring the edge of the mantle. Why do you think these invertebrates don't need more complex eyes?

Ocelli are also part of the vision system of arthropods. In addition to ocelli, however, arthropods such as crabs, shrimp, spiders, and insects have many individual, specialized receptors fused together to form a **compound eye**. A single compound eye can contain thousands of these receptors. Each receptor works like a lens; it responds to light from only a small part of the visual field. The insect's brain puts the parts together to form an image. Some insects can detect wavelengths of light that people can't detect—the short waves of ultraviolet light and the long, infrared wavelengths. What advantage does this give invertebrates?

Scripture Spotlight

Read **1 Corinthians 2:9.** Have you ever wondered whether our sight will be any different in heaven? Will we be able to detect wavelengths that we can't see now? Will we see colors we've never seen before?

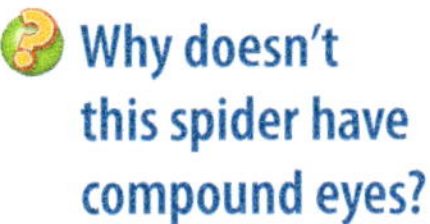

This spider has an ocellar eye.

Why doesn't this spider have compound eyes?

Explore-a-Lab

Structured Inquiry

How would you describe a compound eye?

Use a hand lens or dissecting microscope to examine the compound eye of a housefly or other insect. Make drawings to show the structure of a compound eye. Record your work for this inquiry.

Record your work for this inquiry. Your teacher may also assign the related Guided Inquiry.

Ant Foods

What foods do ants communicate about?

SAFETY: Do not touch the ants. Some ants can sting or bite.

Materials
- 3 flat lids
- brown sugar
- cheese
- cracker
- hand lens

Procedure

1. Label each lid with a letter: *A*, *B*, or *C*.
2. Put brown sugar in lid *A*, cheese in lid *B*, and a cracker in lid *C*.
3. Place the lids close together outside in an area accessible to ants.
4. **Observe** the lids when you first place them and then again every 5 minutes for 30 minutes. Use the hand lens to observe any ants that are attracted to the lids and any types of communication between them.
5. **Record** your observations in the chart in your ***Science Journal***, including the number of ants on each lid at each interval.
6. Repeat Steps 2–4 every day for one week in the same area, recording the results.

Analyze Results

Compare your observations about the three lids during the week. Infer the meaning of the differences in the ants' communication based on your observations. Analyze how the ants are communicating with one another. Use your observations and recorded notes to help you determine a conclusion.

Create Explanations

1. What foods do ants communicate about?
2. How do you know the ants communicated with other ants?
3. How does communicating benefit ants?
4. What can we learn about communication from the ants? How could this help people?

Although the compound eye is an amazing structure, the eyes of octopuses and squid are the most complex eyes of all the invertebrates. The eyes of these animals are very similar to the eyes of vertebrates, such as your own eyes. Like the vertebrate eye, the eye of a cephalopod is filled with a clear fluid and includes a cornea, a lens, an iris, a pupil, and a retina. Because of this structure, the eyes of cephalopods are a wonder to scientists. Why do you think scientists look at these eyes as a wonder? How do creationists and evolutions each interpret the eyes of octopuses?

Communicating with Chemicals Explain

The chemical senses of invertebrates are taste and smell, which we collectively call **olfaction**. Olfaction helps invertebrates find food, choose a mate, and escape predators.

Insects use their antennae and some other body parts for olfaction. An insect may have as many as 100,000 hairs on each antenna. Each hair is an olfactory organ. When stimulated by a chemical, the hair generates a nerve impulse that travels to the insect's brain. The hairs can send signals in many different patterns, each of which is unique. It allows the insect to distinguish thousands of chemicals.

Many insects use olfaction to communicate. They do this with pheromones. A **pheromone** is a chemical released by one individual that affects the behavior of another individual of the same species. Pheromones communicate many different messages. Some signal danger. Aphids fly away from a predator when other aphids release a warning pheromone. Ants wage war in response to a pheromone that is released when a predator attacks. Do you think that insects have to learn what the pheromones mean, or are they born with this knowledge?

Faith Connection

When God designed animals, He gave them all the body parts they need to survive. Some parts may seem odd to us. Some butterflies, for example, taste with their feet! What other examples of God's Design can you see in this lesson?

Some ants use pheromones to help the rest of their colony find food.

How does using pheromones help ants exploit food sources?

Some pheromones help invertebrates find mates. A female silkworm moth, for example, releases a pheromone that male moths can detect up to 11 kilometers away. Can you think of something that people use that acts similar to a pheromone?

Farmers and foresters sometimes use pheromones to control pests that damage forests or destroy food crops. Traps baited with a female pheromone attract males, but the males find no mate. Unable to reproduce, the pest population declines in number. Pheromone control has been used successfully in many places and on many pests worldwide. Have you ever seen any examples of pheromones used by insects?

Communicating with Behavior Explain

Invertebrates also communicate with behavior. Some invertebrates can communicate with sound, such as crickets. Male fruit flies vibrate their wings to make sounds. One kind of sound vibration attracts females. Another warns male competitors away.

Many cephalopods can change color to communicate. In their skin, they have specialized cells called chromatophores. Each **chromatophore** is a tiny bag of pigment encircled by muscles. When the muscles contract, the bag opens, revealing the color inside. Relaxing the muscles closes the bag and hides the color. What do you think the different colors might mean?

This cuttlefish, a cephalopod mollusk, opens and closes chromatophores to create many different colors on its body.

What message might this cuttlefish be sending?

Squid and octopuses can open and close chromatophores on different parts of their body at the same time. In this way, they can produce a variety of colors and patterns. What benefit does this have? The colors can also send messages to others of the same species, especially during courtship. Males change their colors to warn away other males or attract a female. The female's colors can signal males that she is ready to mate. Do humans do this? Does this have implications for how we dress?

Movements can also send messages. The Austrian zoologist Karl von Frisch (1886–1982) won a Nobel Prize in 1973 for his studies of communication in insects. He found that scout bees "dance" in the hive to tell other bees where food can be found. The circular and "waggling" motions of the bees communicate the direction, distance, quality, and quantity of the food. In what ways do you use body language to communicate?

Faith Connection

The dance of the bees, as well as the complex designs and behaviors of other invertebrates, is evidence of God's Design.

This honey bee has found a source of nectar.

 How is bee communication like a grocery store sales ad?

Concept Check **Assess/Reflect**

Summary: How do invertebrates use their senses? Invertebrates communicate in a number of nonverbal ways. Invertebrates communicate with chemicals, using pheromones to signal danger or find a mate. Some invertebrates have ocelli that are light sensitive and are used to help them see, while others have compound eyes to detect wavelengths of light. Some invertebrates use sound, such as vibrations of wings or legs, as a method of communication. They also communicate with behavior, such as octopuses changing colors during courtship and bees dancing to share information about food. Some invertebrates have chromatophores, which open and close based on muscle contraction to reveal or hide different colors, sending messages and warnings to others. Through sound, chemicals, and movement, invertebrates can be observed "conversing" with each other in many ways.

1. How do arthropods use their antennae to sense their environment?
2. How are ocelli like eyes?
3. What are pheromones, and how do they work?
4. What are three ways invertebrates communicate with behavior?

People in Science

Extend

Get to Know
Jean-Henri Casimir Fabre

Born in France in 1823, Jean-Henri Casimir Fabre is best remembered for his work in the field of entomology. Fabre's interest in insects began as a child. He was fascinated with the butterflies and dragonflies he observed around his home in Le Malaval, a village in France.

As a young adult, Fabre began teaching while pursuing studies in physics, chemistry, botany, and entomology. Although he was well known for his excellent teaching abilities and passion for science, he is best remembered for making entomology, the science of insects, popular.

A keen sense of observation allowed Fabre to observe, study, and experience live insects and their behaviors in their natural habitats. Studying the life and habitats of insects became his greatest passion. During experiments, Fabre's thorough observations and meticulous note-taking became the basis for countless articles, books, and letters dedicated to entomology. His most well-known writings, the 10 volumes of *Souvenirs Entomologiques,* took Fabre almost 30 years to complete.

When writing *Souvenirs Entomologiques,* Fabre did not follow the clinical scientific writing approach used by many of his counterparts. Instead he chose to write in an informal style so that people without scientific backgrounds would find the writings easy and enjoyable to read. Fabre's writings about his experiments and observations were so interesting that entomology became popular among scientists and nature enthusiasts.

Jean-Henri Fabre's lifelong love of science, his acute observation skills, and his ability to write for all people enabled him to popularize entomology. To learn more about entomology while peering into the amazing life of various insects and to learn about their unique habits and their instincts, read some of Fabre's most popular experiments and observations, such as *Bramble-Bees and Others, The Mason-Bees, The Life of the Caterpillar, The Life of the Fly,* and *The Wonders of Instinct: Chapters in the Psychology of Insects.*

Concept Check

1. What skills did Jean-Henri Fabre use when collecting information about insects?
2. Observe an insect for at least 15 minutes and write a brief essay about the insect.

Cuttlefish— Chameleons of the Sea

Found mainly in tropical, temperate, shallow ocean waters, cuttlefish are amazingly complex creatures. Because they have soft bodies, they are placed in the phylum Mollusca along with octopuses, squid, snails, and clams. The cuttlefish shows an impressive collection of adaptations that enable it to survive in its environment.

Cuttlefish have well-developed heads, beak-like jaws, and large eyes that are similar to the human eye. Their bodies are flat with an oval section and a fin that runs around the body from the back of the head.

Like octopuses, cuttlefish have eight arms, each with many suckers. Unlike octopuses, cuttlefish have two long tentacles they use to capture prey. The internal shell of the cuttlefish is called a cuttlebone. The cuttlebone has several chambers; some are filled with gas, and some are filled with water. The cuttlebone allows this cephalopod to change its density so it can float at different levels in the ocean.

Cuttlefish have tremendous camouflage ability. They are able to change their color to match their surroundings, hiding in plain sight. Scientists believe that cuttlefish have tiny light sensors throughout their skin that aid the cuttlefish in changing color. Some scientists are researching this ability to create a material that can camouflage clothing, buildings, and vehicles.

There are many applications for new technology like this. This new material could make tanks and submarines seem to disappear. It can also act as a three-dimensional camera, changing in real time. One day you may be able to walk right into a movie with images on every wall. Such technology could also make significant advances in medical imaging.

Concept Check

1. How does the structure of the cuttlebone aid in its function?
2. How do you think observing nature can help scientists improve technology?

Chapter 2 Review Assess/Reflect

Study Guide

Lesson 1

1. Animals can be divided into two groups—vertebrates and invertebrates. Vertebrates have a backbone inside their bodies that protects the spinal cord. Invertebrates do not have a backbone.
2. Many invertebrates have skeletal structures, such as plates, spines, and shells. These structures may be made of calcium, silica, organic fibers, calcium carbonate, or chitin.
3. Body symmetry is the balanced arrangement of an animal's body parts along an axis. Animals can have bilateral symmetry or radial symmetry, or be asymmetric.
4. Invertebrate phyla include Porifera (sponges), Cnidaria, Platyhelminthes (flatworms), Nematoda (roundworms), Annelida (segmented worms), Mollusca, Echinodermata, and Arthropoda. Sponges and cnidarians have no specialized organs. Annelids have organs for blood circulation, blood purification, and breathing. Arthropods have segmented bodies, jointed legs, and exoskeletons.

Lesson 2

1. Sponges have a hollow body with many small pores and a hollow opening at the top. Cnidarians are mostly made of a jellylike substance, with simple digestive cavity, nerves, and tentacles. Flatworms have long, soft, flat bodies and eyespots that detect light. Segmented worms may have as many as 100 segments, each with small bristlelike structures that help the worm move. Roundworms have cylindrical bodies.
2. Sponges help to purify water in their environment and play a role in reducing nitrogen levels. Cnidarians act as both predators and prey. Some flatworms are parasites, while others eat dead organisms. Many segmented worms eat plant material, but others are parasites. Although many roundworms are parasites, they play an important role in the nitrogen cycle.

Lesson 3

1. Mollusks have a muscular foot and a soft, unsegmented body surrounded by a mantle. Echinoderms have spiny skin and an endoskeleton.
2. Many mollusks, including snails and clams, secrete a shell for protection. Some mollusks have tentacles that help them find, capture, and eat prey.
3. Echinoderms have a water-vascular system that attaches to their tube feet and allows them to move.

Lesson 4

1. An arthropod has a segmented body that may include a head, thorax, and abdomen. The exoskeleton of an arthropod is made of chitin. The legs, mouth, and antennae are jointed.
2. Arthropods include crustaceans, arachnids, chilopods, diplopods, and insects. Crustaceans have a fused head and thorax called a cephalothorax, as well as five pairs of legs. Arachnids have a cephalothorax, an abdomen, and four pairs of legs. Each of the many body segments of chilopods has one pair of legs. Diplopods have two sets of legs per body segment. All insects have three body parts, a head, thorax, and abdomen. Insects have three pairs of legs.
3. Arthropods pollinate plants, eat harmful insects, fertilize soil, and provide food. Some arthropods make useful materials, such as beeswax and silk.

4. Crabs have gills to help them breathe underwater. Insects have spiracles on their abdomens that allow gases to enter and leave their bodies.
5. Many insects go through complete metamorphosis during their life cycle, changing their appearance dramatically. Complete metamorphosis has four stages: egg, larva, pupa, and adult. Incomplete metamorphosis is less dramatic. It has three stages: egg, nymph, and adult.

Lesson 5

1. Many invertebrates have antennae, which can detect touch, motion, sound, and chemicals. Ocelli help some invertebrates detect light and shadow. Some invertebrates have compound eyes, while others have eyes similar to human eyes.
2. Some invertebrates communicate using chemicals called pheromones. Different pheromones communicate different messages. Some invertebrates communicate with sounds, colors, or movements.

Vocabulary Check

Explain how each pair of terms is related.

1. Nematoda—invertebrate
2. exoskeleton—molt
3. arthropod—appendage
4. water-vascular system—tube feet

Multiple Choice

Choose the best answer.

5. What function do pheromones serve?
   A. communication
   B. locomotion
   C. digestion
   D. molting

6. How are roundworms and annelids different?
   A. Roundworms lack antennae.
   B. Roundworms lack segments.
   C. Roundworms live only in water.
   D. Roundworms have bilateral symmetry.

7. What are two characteristics of echinoderms?
   A. endoskeleton, two shells
   B. radula, mantle
   C. radial symmetry, pores
   D. tube feet, spiny skin

8. Which is an example of a bivalve?
   A. snail
   B. clam
   C. octopus
   D. sea urchin

9. What is the major difference between millipedes and centipedes?
   A. type of symmetry
   B. presence of paired antennae
   C. fusion of head and thorax
   D. number of legs per body segment

Check Point

Answer the following questions.

10. Crabs breathe with gills both in and out of the water. How might they achieve that?
11. What is the easiest way to tell the difference between an insect and a crustacean?
12. What features do adult sea anemones and sea jellies share? What features are different?
13. How would you design an **experiment** to test whether a species of nematode affects crops?
14. How do spiders **compare** with insects?
15. What features did God give sea jellies that help them survive?

Chapter 3 Characteristics of Vertebrates

All vertebrates have backbones. The backbone of this snake is very long, with many ribs attached.

Scripture Spotlight

God's Design is evident in the many amazing vertebrates He created. There are many Bible stories that involve vertebrates. You will read the following passages in this chapter.

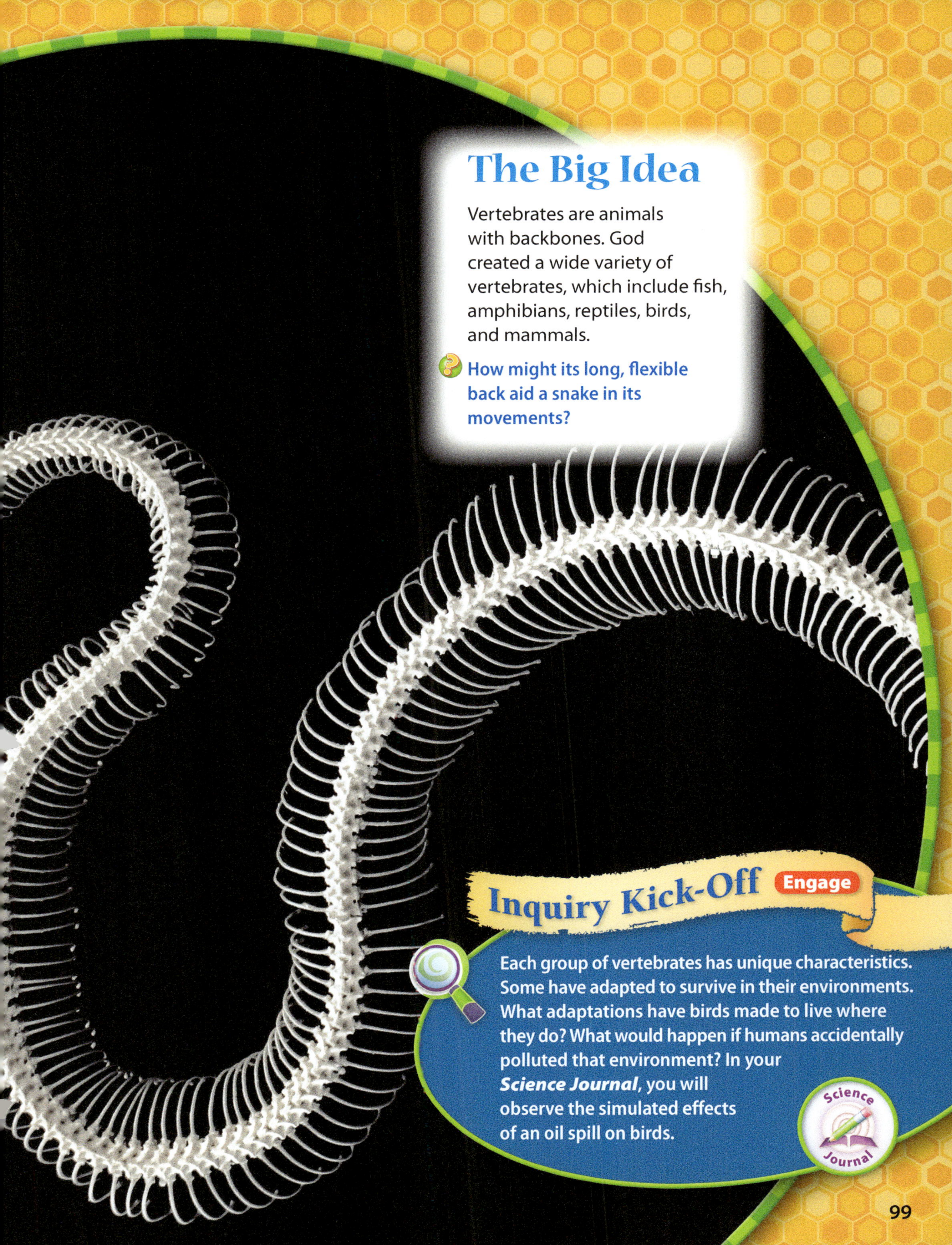

The Big Idea

Vertebrates are animals with backbones. God created a wide variety of vertebrates, which include fish, amphibians, reptiles, birds, and mammals.

How might its long, flexible back aid a snake in its movements?

Inquiry Kick-Off

Engage

Each group of vertebrates has unique characteristics. Some have adapted to survive in their environments. What adaptations have birds made to live where they do? What would happen if humans accidentally polluted that environment? In your ***Science Journal***, you will observe the simulated effects of an oil spill on birds.

Lesson 1

Essential Question

What Are Fish?

Objectives

- Describe the major characteristics of fish.
- Relate a fish's external and internal structures to their functions.
- Describe the classification of fish.
- Compare and contrast bony fish and cartilaginous fish.

Vocabulary

scales

spawning

operculum

How are you similar to a fish? How are you different? Like you, fish are vertebrate animals, but almost all types of fish spend their entire lives in the water. You breathe with lungs, but a fish breathes with gills. You walk on your legs, while fish swim by means of fins. Scientists have found an abundance of interesting facts about fish. For example catfish have more than 27,000 taste buds. Lungfish can breathe by means of not only gills but also lungs. A lungfish can live out of water for several years by burrowing under the earth and secreting a mucus cocoon. Sharks are the only type of fish that have eyelids, but they do not blink. Why do you think catfish have so many taste buds? Why do lung-fish have lungs while other fish don't? Why do you think sharks have eyelids and other fish don't?

Exterior Parts of a Fish

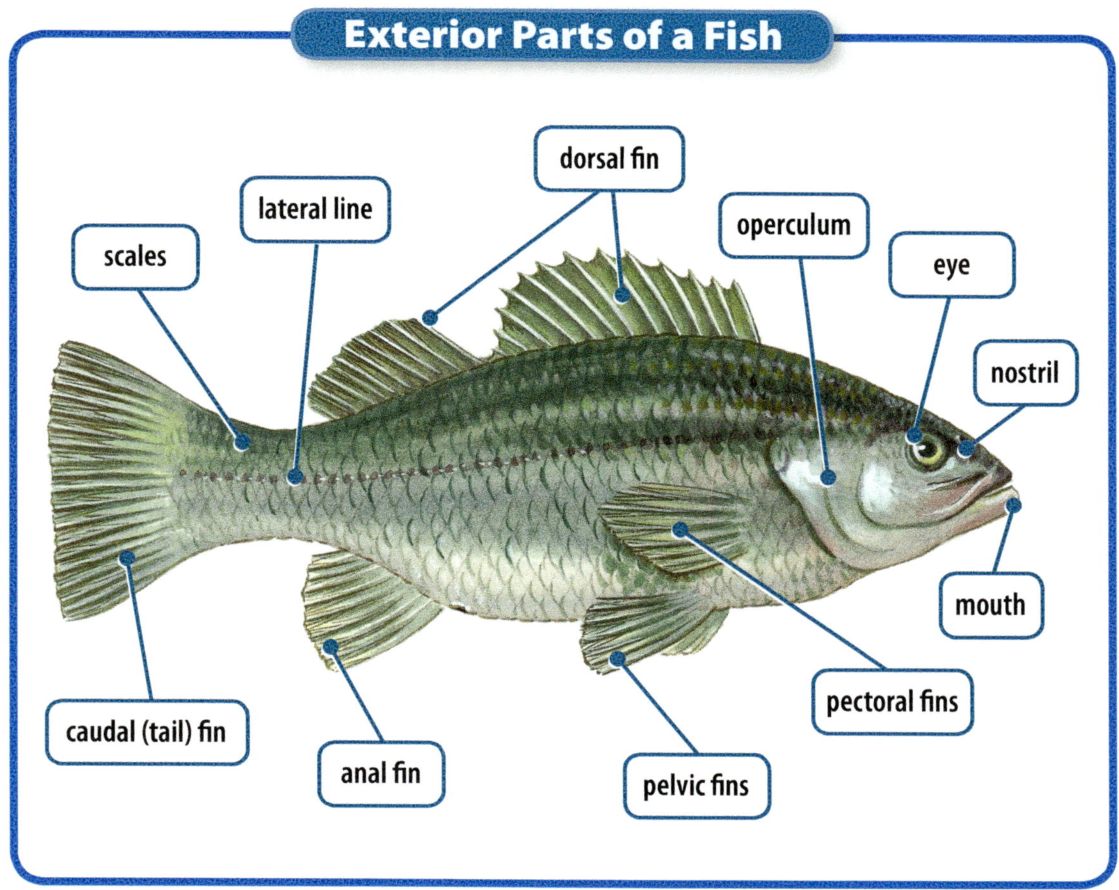

Record your work for this inquiry. Your teacher may also assign the related Guided Inquiry.

Fishy Behavior

What types of behavior do fish exhibit?

Procedure

1. Place the goldfish in a jar of cool, spring water that is well oxygenated. **Observe** the fish for 2–3 minutes. **Record** your observations, including how it moves in your ***Science Journal***.
2. Identify the different external features of the fish. In your journal, draw a detailed diagram of the goldfish and label these features.
3. Let the fish rest undisturbed for 2–3 min. Then gently tap the side of the jar with a pencil. Observe the fish's response. Record your observations.
4. Place a sheet of paper against the side of the jar. Use tape to hold the paper in place. Allow the fish to rest undisturbed for 2–3 min. Gently tap the paper on the side of the jar. Observe the fish's response. Record your observations. Remove the paper from the side of the jar.
5. Let the fish rest undisturbed for 2–3 min. Sprinkle a small amount of fish food into the water. Observe the fish's response. Record your observations.

Materials

- 1 L jar filled with spring water
- goldfish
- sheet of paper
- tape

Analyze Results

Compare and contrast the behaviors of the fish observed in this inquiry.

Create Explanations

1. What types of behavior do fish exhibit?
2. How did the fish's response differ in the three parts of the experiment?
3. Which sense—sight, motion, or hearing—does the goldfish rely on to detect danger? Explain.

Characteristics of Fish Explain

The internal and external structures of fish equip them to survive and thrive. What structures do you see in the picture below and on page 100 that provide the best evidence of the Creator's Design? Other characteristics of fish include:

- Fish absorb dissolved oxygen from the water through their gills.
- Most species of fish protect their bodies with hard plates called **scales**. The edges of scales can be jagged or smooth. How do scales benefit the fish?
- Mucus covers their body to help protect against infections. What is another way that mucus benefits fish?
- Fish move by means of fins. What function might a single fin have compared with paired fins?
- Fish are *ectothermic*, which means they regulate their body temperature by means of external sources. A few fish, such as the tuna, are able to keep their body temperature above the temperature of the water. How might fish stay warmer than the surrounding water? How do you think this is an advantage to tuna?
- Fish reproduce by **spawning**, a process by which male and female fish release sex cells near each other in the water. Is this method of reproduction efficient for animals that live in water? Why?

A fish's swim bladder lies well protected beneath its bony spine.

Would you expect a fish's swim bladder to change in size? Why or why not?

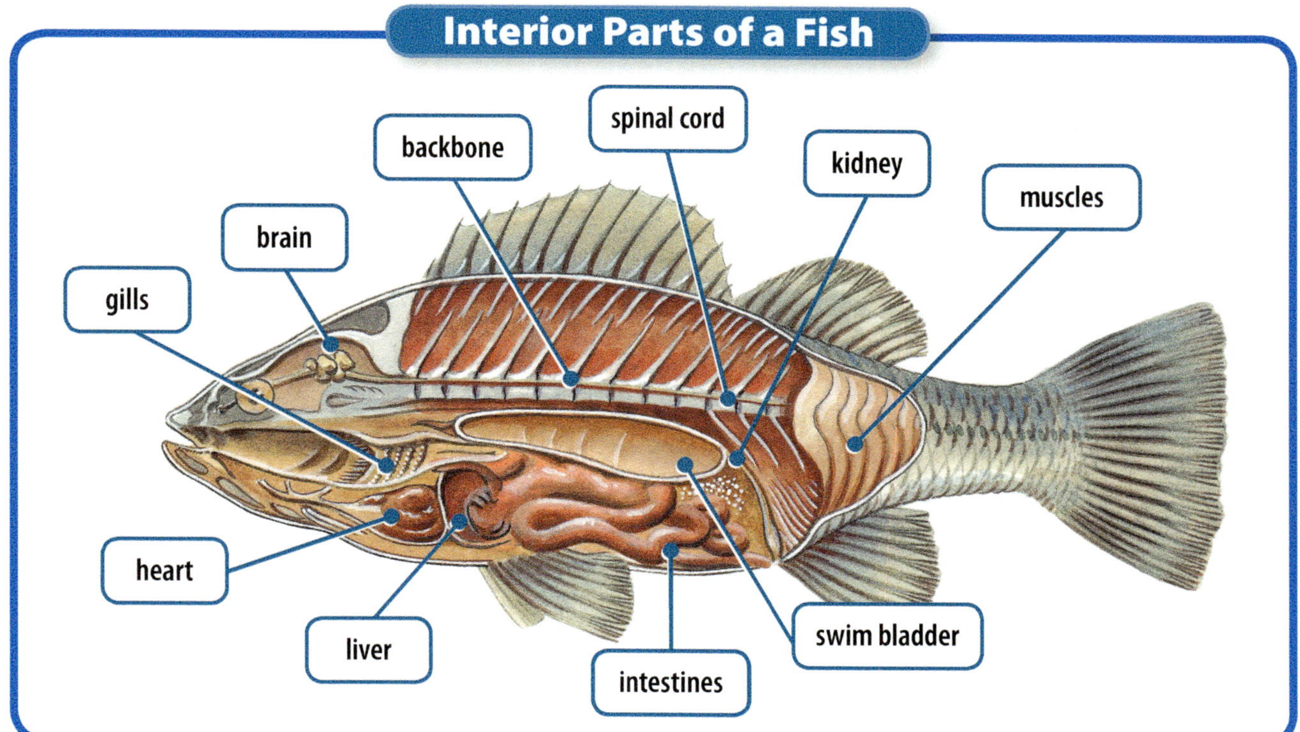

Reproduction

During spawning, some fish fertilize their eggs externally. The eggs then develop and hatch without any help from either parent. This is called *oviparous* development. In this method, the female lays eggs in a nest called a *redd*, and the male then fertilizes them. With *broadcast spawning*, the female releases her eggs and the male releases his sperm simultaneously into the water. Some fish also reproduce by means of internal fertilization, which is called *ovoviviparity*. In this case, the female retains the fertilized eggs within her body until they hatch internally, and the new fish are born live. What is the advantage of broadcast spawning? What are the disadvantages? What is the value of live birth to a fish species?

Scripture Spotlight

Matthew 17:24–27 tells the story of Peter finding money in a fish. Given what you have learned about fish so far, what is special about Jesus's miracle in this story?

Fins

As you saw in the diagram on page 100, fish have a wide array of fins that they use to move, swim, and steer. The caudal fin pushes the fish forward. The fish steers, hovers, and stops using the pectoral and pelvic fins along its sides. The single dorsal and anal fins keep the fish upright in the water. In some fish, sharp spines support the fins. How might these spines serve as a defense against predators?

Sense Organs

How do you think the senses of fish compare to those of other aquatic or marine animals? Like you, fish have several sense organs that provide information about their environment. Bony fish have eyes that have lenses that move in and out to focus and that can see color. Fish use their sensitive nostrils, or *nares*, to detect odors in water. Do you think fish use their nose to help with breathing? Why or why not?

As you know, fish also have a *lateral line*, a series of fluid-filled sacs from which tiny hairs protrude, producing the line that can be seen along the side of the fish. The lateral line senses movement, pressure, currents, and sound vibrations in the water. Why do fish need to respond to differences in pressure and currents? Fish do not have external ears, but they have *otoliths*, internal structures that are sensitive to sound and vibration. What might these structures look like?

Fish have a sense of taste, and many will spit out nonfood items before swallowing them. What kinds of taste do you think fish prefer—sweet, salty, sour, or something else?

There are about 54 species of sea horses. They range in size from 1.5 to 35.5 cm.

Swim Bladder

How do fish stay at a certain depth in the water? Recall that bony fish have a *swim bladder*, a hollow, gas-filled organ inside their bodies. It lies just below the backbone and above the other internal organs. It helps the fish keep its balance. It also saves energy. The fish can remain motionless, suspended in the water, due to the buoyancy of its swim bladder. What do you think might happen to a fish with a damaged swim bladder? Some species of fish, including the jawless fish and the cartilaginous fish, do not have swim bladders; they sink when they stop swimming.

Scripture Spotlight

Read the most famous Bible fish story in **Jonah 1–2**.

Gas Exchange in Fish

When you inhale, your lungs expand as they fill with air. When you exhale, your lungs contract as the air flows out of them. How are lungs like gills? How do gills obtain oxygen? Fish use their gills to get dissolved oxygen from water and excrete carbon dioxide as waste. A protective bony cover called an **operculum** lies over the soft, feathery surfaces of the gills. Why is the operculum important? The bright red color of the gills is due to the many blood vessels that lie close to the surface of the gill. Why is important for the blood vessels to lie close to the surface of gills?

When a fish opens its mouth and closes the operculum, water moves into the mouth. As the fish closes its mouth and opens the operculum, water is forced over the gill surfaces.

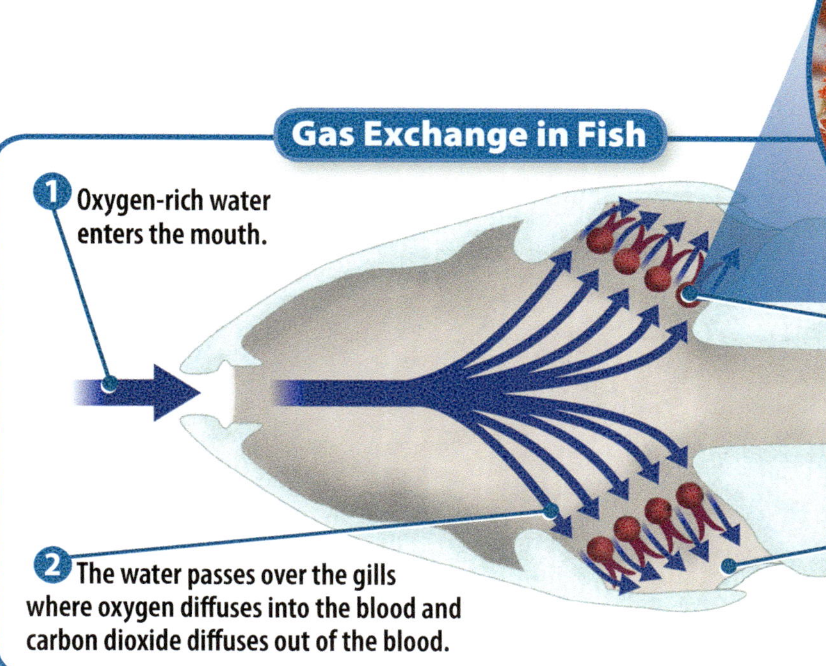

The exchange of gases—oxygen and carbon dioxide—occurs in the gills. How is this process similar to the gas exchange that takes place in your lungs?

Gills are not the only means of gas exchange in fish. Some eels can absorb oxygen through their skin, and some fish can breathe air either through their mouths or their stomachs. Lungfish gulp fresh air through the mouth and pass it out over the gills, but they also have lungs for gas exchange. They can live out of water for months by burrowing underground and living in a mucous cocoon when a river or lake dries up. How does this provide evidence of God's Design?

Explore-a-Lab **Guided Inquiry**

 **How are gills well-suited to their function?**

Using a whole fish or a fish head obtained from a market, examine the gill structure of a fish. Find the operculum, gills, and gill rakers. (Rakers are stiff projections along the inner edges of the gills that keep food out of the gills.) **Observe** a gill filament with a hand lens or dissecting microscope. Make drawings of your observations.

Diversity Among Fish **Explain**

With more than 27,000 species of fish known, fish are the most diverse and the largest group of all the vertebrates. Scientists classify fish into two main superclasses:

- Superclass Gnathostomata (jawed fish)
  - Class Chrondrichthyes (cartilaginous fish)—sharks and rays
  - Class Osteichthyes (bony fish)—trout, bass, tuna, and other familiar fish
- Superclass Agnatha (jawless fish)—hagfish and lamprey

Check for Understanding

How are fish's bodies and sense organs especially suited for their life in the water?

Jawed Fish

Most fish have jaws. These jawed fish can be found in two classes—bony fish and cartilaginous fish. Bony fish account for 96 percent of all known fish species and are incredibly diverse. They range in size from 12 mm to 4.5 m or more in length. Examples of bony fish include bass, trout, catfish, sturgeon, eel, carp, and salmon. What are some other species of bony fish? How do you think their bony skeletons help these fish move? How do their bones affect their speed, strength, and agility? They have a swim bladder, which controls their buoyancy at various depths and helps them stay in place when necessary. Because they can live at different depths and in different environments, they have a wide range of adaptations. What might be some special adaptations in bony fish?

There are far fewer cartilaginous fish than bony fish. The cartilaginous fish include sharks, rays, and skates. These fish do not have true bones. Instead, their skeletons are made of cartilage. Feel the tops of your ears and the tip of your nose. Are these places on your body hard and bony? Are they soft? Are they somewhere in between? These areas are formed of cartilage. How do they compare with bony areas, such as the femur or rib cage? How would your body be different if all your bones were made of cartilage, like a shark?

Explore-a-Lab

Structured Inquiry

How sensitive is a shark's sense of smell?

Pour 500 mL of clean water into each of 6 large beakers. Label the beakers *A1, A2, B1, B2, C1,* and *C2.* In beaker A1, add 10 drops of red food coloring. Stir gently. In beaker B1, place 5 drops of red food coloring. Stir. In beaker C1, place 1 drop of red food coloring. Stir. Place beakers A2, B2, and C2 in different parts of the classroom. Place A1 beaker beside the A2 beaker. Repeat with the B and C beakers. In beaker A2, place 10 drops of cologne. Stir. In beaker B2, place 5 drops of cologne. Mix. In beaker C2 place 1 drop cologne. Mix. Sniff each beaker and note any differences. Also observe the beakers with red food coloring. Record your observations. A shark's minimum sensitivity to smell blood is about 1 part per million. Beaker A1 represents 1000 parts per million. If the red food coloring represents blood, in which beakers could a shark smell the blood? How does this compare to your ability to smell the cologne?

The whale shark is the largest fish in the world, yet it feeds on microscopic plankton.

Cartilaginous fish, as the name suggests, have cartilage instead of bone. Unlike bony fish, the cartilaginous fish lack the swim bladder. As a result cartilaginous fish must keep swimming to maintain buoyancy in the water, so they cannot stop and stay in one place in the water. Their scales are also different from those of bony fish. Bony fish scales are called cycloid scales. Cartilaginous fish scales are called placoid scales. Placoid scales have a rough, sandpaper-like texture. The members of this class lack the diversity of the bony fish due to the limitations of their physical structures.

Jawless Fish

The jawless fish, as their name suggests, lack any true jaws. Instead, these fish are equipped with a mouth designed for sucking. The jawless fish are a small superclass that includes several subgroups found only in the fossil record. The two living subgroups of this group are the hagfish and lamprey. Hagfish live in deep water and feed primarily on dead and decaying organisms. A few lamprey species are parasitic and feed on fish and marine mammals. These species spend their adult lives in the ocean but move to freshwater streams and rivers to spawn.

Math in Science

You have learned that there are at least 27,000 known species of fish and that 96 percent of them are bony fish. Of the 27,000 known species, how many of the fish are bony? Why do you think there are so many more bony fish? How many of the fish species yet to be discovered do you think could be bony fish? Justify your answer.

Concept Check Assess/Reflect

Summary: What are fish? In general, fish are ectothermic vertebrates that live in water and use gills to obtain the oxygen that is dissolved in water. Unique features of fish include the operculum, lateral line, swim bladder, and fins. Fish are classified by the presence or absence of a jaw. Most fish have a jaw, and scientists classify these fish in two groups: bony fish and cartilaginous fish. There are only two subgroups of jawless fish alive today, although several examples exist in the fossil record.

1. What are the three major characteristics of fish.
2. How do the two main groups of jawed fish differ in their internal structure?
3. What senses do fish possess, and how do they use them to survive in their watery environment?
4. Compare and contrast your lungs with a fish's gills.

Lesson 2

Essential Question

What Are Amphibians and Reptiles?

Objectives

- Describe the major characteristics of amphibians.
- Explain the process of metamorphosis.
- Describe the major characteristics of reptiles.
- Identify three ways in which snakes capture and consume their prey.

Vocabulary

herpetology

scute

constriction

venom

How far can you jump? One meter? Two meters? More? Some species of tree frogs can jump nearly 50 times their body length. If you had this ability, how far could you jump? Flying frogs have huge webbed feet that act like parachutes and enable the frogs to glide more than 14 meters between treetops.

In the last lesson, you learned the characteristics of fish. Which of those characteristics do amphibians and reptiles also share? How are amphibians and reptiles different from fish? As you study amphibians and reptiles in this lesson, think about how they compare and contrast with fish and with each other.

Characteristics of Class Amphibia (Amphibians) Explain

How do you change as you get older? How would you feel if your legs disappeared or you grew a tail? Amphibians are smooth-skinned vertebrates and are classified as salamanders, frogs, toads, and caecilians. In most species, the juvenile form differs significantly from the adult form. These juvenile forms are suited for life in the water, and the adult forms are better suited for life on land. Although adults live on land, they must reproduce in or near freshwater.

Why do you think no amphibians live in the ocean? Most amphibians live in warm, humid places close to lakes, streams, and ponds. A few have adaptations that allow them to live in drier climates. For example, the desert spadefoot toad can tolerate long periods of drought by burrowing into the sand using its large front feet. The spadefoot toad returns to the surface at night or after heavy rains when the air is moist to gather food. Why would the spadefoot toad return to the surface only when the air is moist?

The feet of this flying frog are webbed. The webs catch air as the frog falls, which enables the frog to glide long distances.

The axolotl is an amphibian native to Mexico that does not develop adult amphibian characteristics but retains its gills, fins, and a tail throughout its life.

How do these characteristics help the axolotl live in water?

Except for caecilians, which are a limbless group that resembles worms or snakes, adult amphibians have four limbs that they use for moving around. The amphibian's skin is coated with mucus. What do you think the function of the mucus is? Amphibians are ectothermic and cannot regulate their body temperature internally. What other ways do amphibians regulate their temperature?

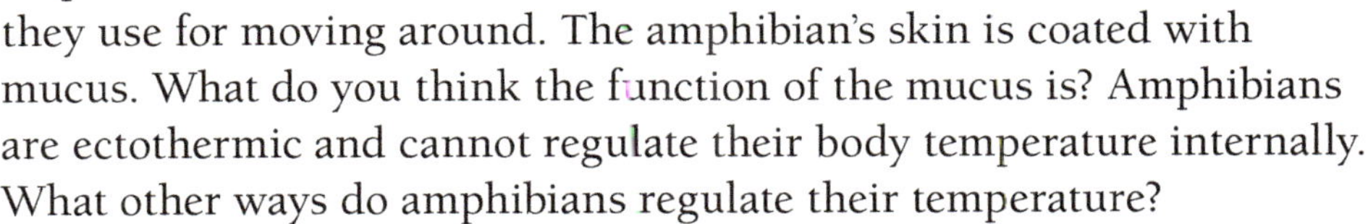

As discussed earlier, amphibians reproduce in freshwater. In some amphibians, fertilization occurs inside the female body. In others, it is external. For external fertilization, the eggs are laid in freshwater, and a male of the species releases sperm over them. After fertilization, the eggs develop into embryos. The embryos hatch into larval, or juvenile, forms. What is the larval stage of a frog called? As the juveniles mature into adults, their bodies undergo metamorphosis. What other types of animals experience metamorphosis? What dangers would the larva face as they matured?

Scripture Spotlight

What kind of amphibian is mentioned in **Psalm 78:45** and **105:30**? To what story do these verses refer?

An important part of amphibian metamorphosis is a change in respiratory structures. As juveniles, amphibians live in water and breathe with gills. The gills allow the juvenile amphibians to absorb needed oxygen from the water, as fish do. As adults, however, they can develop a variety of structures for gas exchange. Many breathe with lungs. Some amphibians exchange gases with the environment through their skin, and some use the lining of the mouth. Many amphibians use a combination of these methods. How do their methods of gas exchange explain why amphibians are particularly sensitive to environmental pollutants?

Math in Science

The gray tree frog lays its eggs in ponds. A single female can lay as many as 2000 eggs in one year. If all of the eggs of a single female survived, and half of them were females that also produced 2000 eggs, how many frogs would there bein three years? Suggest a reason why this does not happen.

Record your work for this inquiry. Your teacher may also assign the related Guided Inquiry.

Temperature Regulation in Ectotherms

How do ectotherms regulate their body temperature?

SAFETY: The light bulb may become dangerously hot. Do not touch it while it is lit.

Materials

- thermometer
- plastic container
- lamp with bulb that gives off heat
- masking tape
- clock or stopwatch
- ruler
- water
- sand

Procedure

1. **Record** the temperature of the room. Then put the thermometer in a plastic container.
2. Place the container near the lamp, mark the spot where you placed the container with a piece of masking tape, and turn on the lamp. Wait until the container reaches 30°C. Record how long it took to reach this temperature.
3. Maintain a temperature between 30°C and 35°C for 14 min. Record the temperature every 2 min. Record what you did to maintain the correct temperature range.
4. After 14 min, turn off the lamp and remove the thermometer from the container. Wait for the thermometer to return to room temperature.
5. Place the thermometer back in the container. Pour water into the container to a height of 8 cm. Repeat Steps 2 through 4. Place the container at the same marked spot.
6. Dispose of the water. Place the thermometer back in the container. Pour sand into the container to a height of 8 cm. Repeat Steps 2–4. Place the container at the same marked spot.

Analyze Results

How successful were you in maintaining the correct temperature?

Create Explanations

1. How do ectotherms regulate their body temperature?
2. For which container was it easiest to maintain the correct temperature?
3. How do the conditions in the lab mimic the conditions that amphibians face in the environment?
4. How do amphibians control their body temperature in nature?

Orders of Class Amphibia Explain

Scientists who study amphibians and reptiles study the field of **herpetology**. Herpetologists have described thousands of different amphibian and reptile species. If you were a herpetologist, what characteristics would you study? The three orders of amphibians are the Caudata, Anura, and Apoda. The table below shows the common names, characteristics, and habitats of each order.

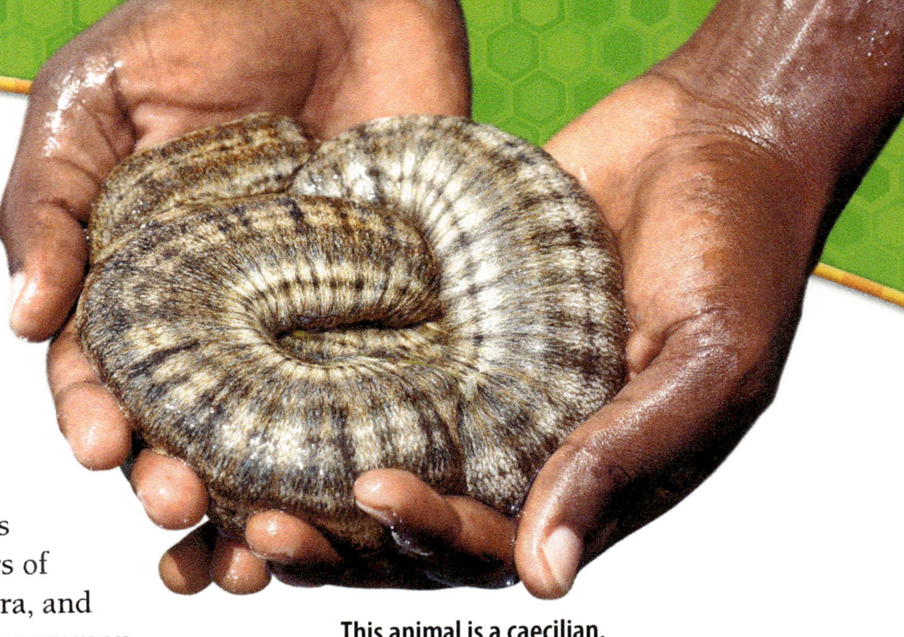

This animal is a caecilian.

How is it similar to and different from a snake?

Amphibia			
Order Name	**Characteristics**	**Habitat**	**Examples**
Caudata	• Adults have tail • No scales • Elongated body • Four limbs, usually of equal size • Smooth, glandular skin (may have poison glands) • Carnivores and predators—none eat plants	Live in soil or water; reproduce in ponds.	Salamander, Congo eel (not a true eel), mudpuppy, newt, siren, axolotl
Anura	• Most adults lack tail • No scales • Adults have four limbs, usually of unequal size; hind limbs modified for jumping • Webbed feet (in some)	May live in trees, on the ground, in burrows, or in water	Tailed frog, spadefoot toad, tree frog, narrow-mouthed toad
Apoda	• Scales (some species) • Wormlike body shape • Pair of tentacles on head (believed to be olfactory organs)	Live underground	Caecilian

Lesson Activity

Investigate the life cycle of an amphibian other than a frog. Make a flow chart or some other visual device to show what happens in each stage of its life cycle. Share what you learn with your class.

What is metamorphosis like in an amphibian?

Characteristics of Class Reptilia (Reptiles) Explain

Scripture Spotlight

Read **Genesis 3:1; 14–15**. Think about the story of the fall of man. Why does the snake not have any limbs like other reptiles?

Crocodiles, turtles, tortoises, snakes, and lizards are reptiles. Because they do not need a watery environment to reproduce, they can live in drier places than amphibians. What are some examples of environments where reptiles live? How are their bodies adapted to live in each environment? Reptiles have a scaly, tough outer skin. It slows the loss of water from their bodies. Many reptiles have four limbs with claws, but some, such as snakes, have no limbs.

Would it be easier to capture a reptile in the morning or the afternoon? Why? Like amphibians, reptiles are ectotherms, so their body temperature is about the same as the environment. In some ways, this is an advantage. Because ectotherms do not need to use as much energy to maintain their body temperature, they require less food to survive. This does not mean, however, that they do not have ways to control their internal temperature.

Reptiles do not perspire. How do they stay cool? At the hottest times of the day, reptiles retreat to the shade, under a rock, into a shallow burrow, or into shallow water. How might shallow water help the reptile cool down? In the mornings when the air is cool, reptiles will often sit on a rock to absorb the heat from the Sun. They also expand their rib cages while sunning. How would increasing the size of the rib cage affect body temperature?

The size of a reptile is related to how well it stays warm. Larger reptiles can absorb more energy from the environment and hold heat longer, although they warm more slowly than smaller reptiles. Using this comparison, which reptile would warm up faster in the morning, an alligator or a salamander?

Fertilization in reptiles occurs inside the female's body, similar to the process in ovoviviporous fish. In some reptiles, such as garter snakes, the eggs develop in the mother, and the young hatch alive. How might live birth benefit the offspring? Some reptiles lay fertilized eggs that are protected by a hard or leathery shell. Why are reptile eggshells thicker than bird eggshells?

The python belongs to the order Squamata. The U.S. Department of Agriculture considers the python to be an invasive species.

Why is the python considered an invasive species, while other snake species are not?

Comparing and Contrasting Orders of Reptiles

The table below lists the three orders of reptiles. In what environments would you anticipate finding a rattlesnake? Why? What characteristics of a sea turtle would make it difficult for it to live in the environment inhabited by a Komodo dragon? What characteristics do the box turtle, skink, and crocodile share? Explain why scientists classify all three as reptiles.

Reptilia			
Order Name	**Characteristics**	**Habitat**	**Examples**
Chelonia	• Shell made up of plate-covered bone (can be either hard or soft) • Shell in two parts: carapace (top part) and plastron (on the lower, belly side)	Live in terrestrial and aquatic environments, coastal areas, and the open ocean	Box turtle, snapping turtle, mud turtle, tortoise, terrapin, sea turtles
Squamata	• Skin with scales • Legs (most lizards) • No legs (snakes)	Live in a variety of environments from the desert to warm, wet rainforests	Glass lizard, whiptail lizard, skink, boa, python, cobra, Komodo dragon, chameleon, iguana, king snake, rattlesnake
Crocodilia	• Distinct head, trunk, and long tail • Four toes on front feet, five toes on back feet	Live in tropical lowlands in the brackish water of estuaries, mangrove swamps, and salty lakes	American alligator, crocodile, speckled caiman

Explore-a-Lab

Guided Inquiry

What are some characteristics of lizards?

With your class, go to the playground or local park to hunt for a lizard. Use a net to capture the lizard and place it in a box to observe. What color is its skin? What is the texture of its skin like? How long is the tail? How does the lizard respond to a loud noise? How does the lizard respond when you place leaves or a rock in the box? Record answers to these observations and other details you notice about the lizard. When you are finished, release the lizard into its natural habitat in a safe location.

This desert tortoise is native to the Mojave and Sonoran Deserts of the southwestern United States.

Why might this tortoise thrive in the desert, while other members of the Chelonia order live in moister environments?

Chelonia

How do scientists distinguish turtles from tortoises or from terrapins? The answer is by where they live. Turtles spend most of their lives in water. Many have webbed feet for swimming. Sea turtles have flippers and streamlined bodies that let them speed through ocean water. Tortoises are land dwellers. They walk on round, stout legs. Terrapins live near freshwater sources. They spend some time on land and some time in the water. How are their legs different from turtles and tortoises?

All three reptiles have protective shells. The shells are made of bones covered by plates called **scutes**. The shell is part of the animal's skeleton, attached to its spine and ribs. A turtle cannot crawl out of its shell. Some tortoises can pull their heads, legs, and feet inside their shells and can close their shells completely when danger threatens. Others cannot retreat in this way, but they can tuck their head sideways against the shell. The horned turtles cannot retract their heads at all. They have only bony knobs and horns to protect them.

Squamata

Although snakes and lizards belong to the same group of reptiles, they differ in some important ways. One way is obvious. Lizards have limbs, and snakes do not. If you look closely at lizards and snakes, you will see some other differences. Lizards have movable eyelids. They close the eyelids to protect and lubricate their eyes. Snakes have no movable eyelids. A snake's eyes are always open. Transparent scales protect their eyes. Snakes have a thin, forked tongue, while most lizards do not. What is the purpose of a forked tongue? Snakes have a double-hinged jaw. What is the benefit of this feature? Lizard jaws are fixed, which limits the size of the food they can swallow. While lizards have ears on either side of their head, snakes lack external ears.

Check for Understanding

How are the life cycles of amphibians and reptiles similar? How are they different?

Crocodilia

As for alligators and crocodiles, how do you tell the difference? What characteristics do you think scientists use to tell the difference between the two? Need a hint? Look at the snout. In general, the alligator's snout is wide and rounded. The crocodile's snout is narrow and pointed. When an alligator closes its mouth, its upper jaw covers its lower teeth. In crocodiles, both upper and lower teeth remain visible. There's a difference in the skin, too. The scales of a crocodile have a small spot or dimple close to the edge. The scales of an alligator do not.

Which of these animals is a crocodile? How do you know?

Feeding in Snakes **Explain**

How does a snake capture and eat food without the use of limbs? Snakes are predators. They eat mice, rats, gophers, and bird eggs. Large snakes eat even larger prey, such as small deer. Farmers like having certain species of snakes around because they eat the pests that destroy valuable crops. Other snakes can be dangerous to humans because of the venom they inject if they bite. What are the benefits of snakes to the ecosystem? Snakes feed in three ways:

- Some species take the simple approach: they seize their prey with their mouth and swallow it whole.
- Others kill their prey before they eat it. The gopher snake, for example, is a constrictor. **Constriction** means wrapping the snake's body around an animal of prey and squeezing until the prey suffocates.
- Other snakes, such as rattlesnakes, produce **venom**, a poison, to immobilize or kill their prey.

Explore-a-Lab

Guided Inquiry

Are all reptiles alike?

Divide into groups and use the Internet to research the length of the following reptiles: blue gecko, panther chameleon, green basilisk, paradise flying snake, water monitor, eastern diamondback rattlesnake, Komodo dragon, and green anaconda. Share your information with the class.

Divide into two groups. One group should gather a tape measure, a ball of yarn, and a scissors. Measure a length of yarn to represent each reptile. The second group should gather labels and pens and decide which string represents which animal. As a class, discuss the diversity in reptiles.

Venomous snakes produce different types of venom. Some snakes produce venom that contains a neurotoxin that attacks the nervous system of the prey. How do you think this action affects the respiratory function of the snake's victims? Other snakes produce a hemotoxin that breaks down proteins in the tissues of the blood and other tissues. How do scientists use these venoms in medicine?

Sinaloan milk snakes *(Lampropeltis triangulum sinaloae)* are not venomous, so some people keep them as pets.

Concept Check

Assess/Reflect

Summary: What are amphibians and reptiles? Amphibia and Reptilia are two classes of vertebrates. Like fish, amphibians and reptiles are ectotherms. Three orders of Amphibians are Caudata, Anura, and Apoda. Amphibians hatch from eggs laid and fertilized in water, live in the water as juveniles, and live on land as adults. They must be in or near freshwater to reproduce and most undergo metamorphosis. Reptiles are organized into three orders: Chelonia, Squamata, and Crocodilia. Reptiles have scaly, tough outer skin. Many have four limbs with claws, but some, such as snakes, do not have limbs. Snakes, a common type of reptile, are efficient predators, seizing their prey and swallowing it whole, suffocating prey by constriction, or using venom to subdue their prey.

1. What are the main characteristics of amphibians?
2. What happens when amphibian metamorphosis occurs?
3. In what ways are amphibians and reptiles alike and different?
4. In your opinion, which snake is better adapted for survival—one that seizes its prey live and swallows it whole, or one that paralyzes its prey with venom before consuming it? Explain your reasoning.

Lesson 3

Essential Question

What Are Birds and Mammals?

What do you do to regulate your body temperature when the temperature of your environment changes? Animals live in almost every environment on the planet, and, like people, they must maintain a certain temperature for their body processes to take place. How do animals regulate their body temperature in these diverse environments? How do they regulate body temperature as their environment becomes warmer or cooler?

Ectotherms and Endotherms Explain

Birds and mammals are endotherms. *Endotherms* are animals that are capable of internal generation of heat. These animals have large numbers of mitochondria in their cells. Recall that cells contain organelles called mitochondria, which power the cell. In the mitochondria, food molecules are broken down, and the energy in those foods is released. Some of that energy is released in the form of heat. Because endoderms have an increased number of mitochondria in their cells, they generate more heat than the bodies of ectotherms.

Compared with ectotherms, endotherms need larger amounts of fuel to generate heat to maintain their body temperature. A *homeotherm* is an endotherm that can maintain a stable body temperature by metabolic activity. Most birds and mammals are homeotherms. Why do you think God designed these animals with the ability to maintain their body temperature?

Not all endothermic animals maintain a stable temperature. Many endotherms alter their temperature based on environment and season. A **poikilotherm** is an endothermic animal whose internal temperature varies considerably. For example, hummingbirds drop their temperature at night to save energy. What other animals are poikilotherm? Do you think this adaptation helps them survive in their environments? Why?

Objectives

- Explain the difference between ectothermic and endothermic animals.
- Describe the types of feathers and their functions.
- Explain how the respiratory system of birds is designed for flight.
- Identify the distinguishing characteristics of common orders of birds.
- Identify the distinguishing characteristics of common mammal orders.

Vocabulary

poikilotherm
semiplume
filoplume
barbule
air sac

Characteristics of Class Aves (Birds) Explain

How many species of birds have you seen? How many different colors do these birds have? What is the largest bird? What is the smallest bird? Despite these vast differences, all birds have characteristics in common that make them birds. What are some of the common features of all birds?

All birds are equipped with a beak or bill to help them obtain food. In what other ways to birds use their beaks?

Also as part of the digestive system, birds have a large muscular stomach and a crop that grinds food. What body part do you have that does a job similar to that of the crop?

Another characteristic common to all birds is the hard-shelled eggs that they lay. Their eggs have a large yolk surrounded by albumin. The yolk and albumin nourish the growing embryo. How do birds protect their eggs? Do they all use the same strategies?

Most birds, but not all, have strong, lightweight skeletons made of hollow bones. This lightweight structure allows birds to take flight. Many bones, including the collarbone, which is sometimes called a wishbone, are fused to make the skeleton rigid to brace the bird's body during flight. The sternum (breastbone) of birds is large to provide increased surface area for sturdy attachment points for the large and powerful wing muscles. Some birds lack this lightweight skeleton. Why might this be an advantage to these birds? How might the lack of this skeleton affect how the bird lives?

In addition to the lightweight skeleton, birds also have a specialized respiratory system designed to absorb extraordinary amounts of oxygen. Why do you think birds need this extra oxygen?

Bird Skeleton

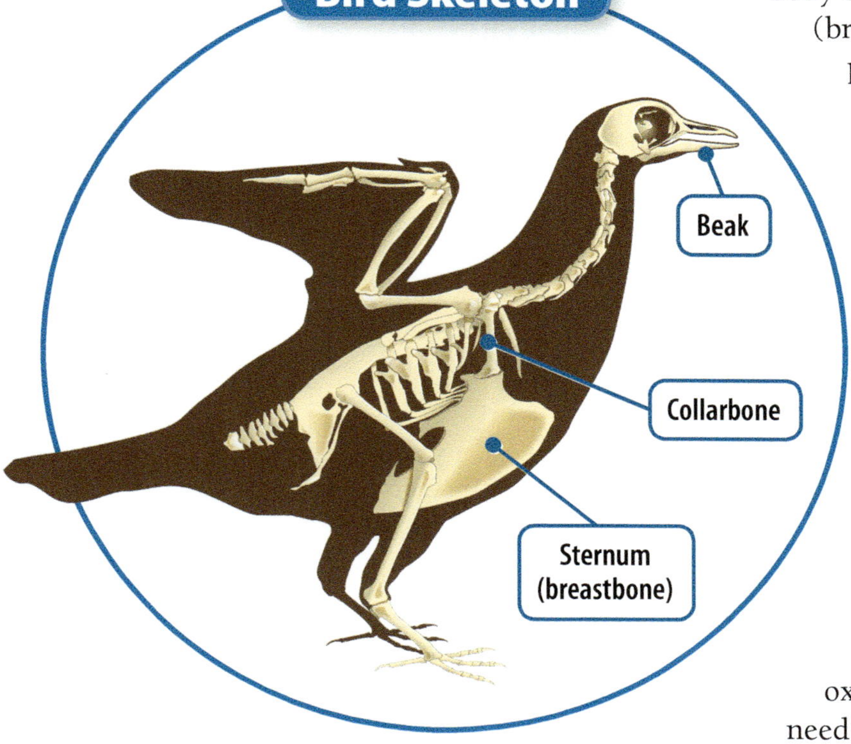

Record your work for this inquiry. Your teacher may also assign the related Guided Inquiry.

How Birds Use Feathers

How does a feather's structure relate to its function?

Materials

- feathers (A–F)
- hand lens
- tweezers
- medicine dropper
- water
- metric ruler
- stopwatch

Procedure

1. Get one of the sample feathers provided.
2. Using a hand lens, look closely at the construction of a feather. Pull gently to separate the barbs, and then smooth them back together with your fingers, or tweezers (if necessary). In your ***Science Journal***, draw the feather and any features that you **observe**. Label the parts of the feather.
3. Use the medicine dropper to place three drops of water on different areas of the feather. Observe whether the feather repels the water. **Record** your observations in the table provided in your journal.
4. Drop the feather from a height of 1 m and time how long it takes the feather to fall to the ground. Record the time in your journal. Repeat this process two more times.
5. Examine the feather again and decide what the function(s) of the feather is. Record the function(s). See the list below for possible functions. Record the function or functions of the feathers.

camouflage	visual signals
flight	breeding
floating/swimming	warmth
nesting material	weather proofing

6. Repeat Steps 2 and 3 for the remaining feathers.

Analyze Results

Compare the features of the feathers and describe their similarities and differences.

Create Explanations

1. How does a feather's structure relate to its function?
2. What is the structure of each feather? Why do feathers have different structures?
3. Use the Internet to research the functions of different types of feathers. Were your predictions of those functions correct? What are the advantages of each type of feather structure?

Feathers

Have you ever looked carefully at a bird feather? In what ways is a feather unique? What characteristics of feathers make them suitable for flight? Feathers are outgrowths from the skin of birds and have a variety of forms. What do you know about different types of feathers?

Type of Feather	Definition	Question
Contour	Cover the bird's outer body	How do birds care for their feathers? Why do birds sometimes shed their feathers?
Down	Soft, fluffy feathers under the outer feathers that help to maintain the bird's body temperature	How do people use down feathers?
Semiplume	Provide an insulating effect and provide shape for the bird	Where on a bird would you expect to find this type of feather?
Filoplume	Small, hairlike feathers with few, if any, barbs that have sensory receptors next to their bases; grow along down feathers	How do the filoplume feathers provide the bird information?

The contour feathers of the wing and tail are most important for flight. They stiffen when they push down against air, providing lift. These feathers also help the bird steer. Study the diagram as you read about the structures of a contour feather. The feather consists of these structures:

- A main shaft called a *rachis* that attaches by a *quill* to the bird's skin
- *Barbs* that are branches that make up the soft *vane*
- **Barbules** that have tiny hooks that hold the barbs together

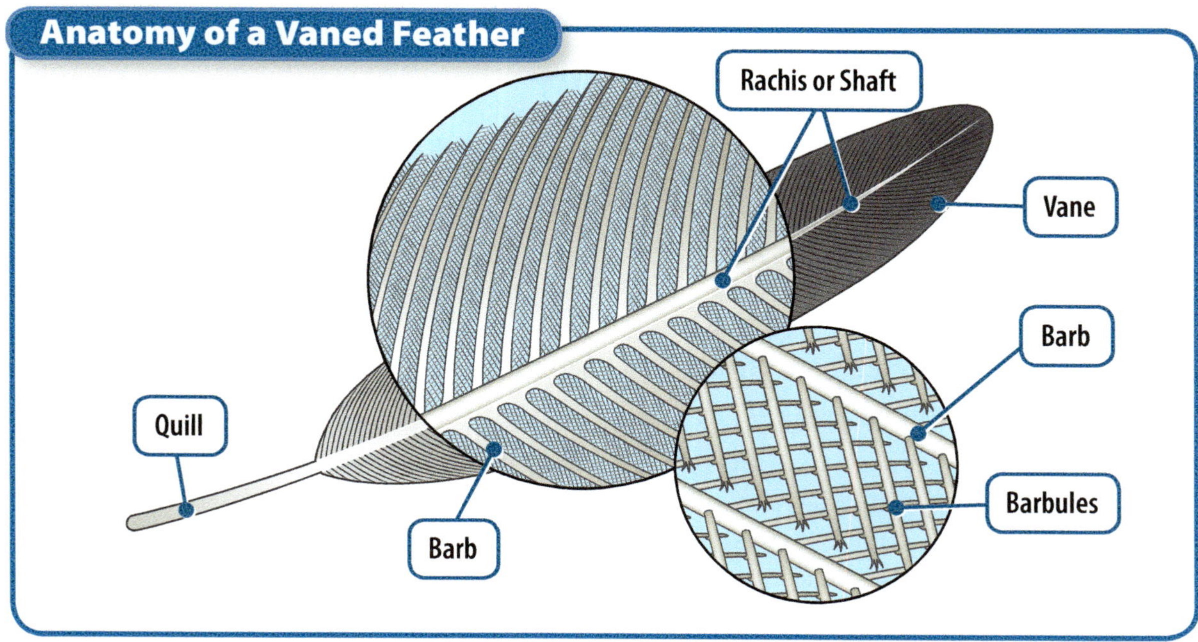

The Respiratory System of Birds Explain

Temperature control in birds takes considerable energy. So does flight. God designed birds with a unique respiratory system that provides the large amount of oxygen needed for both purposes. The system uses **air sacs**, air-filled cavities present in the bodies of birds. In most birds, there are nine air sacs in cavities in the bird's muscles and in the spaces in the hollow bones. The air-filled sacs make the bones lightweight.

How do air sacs give the bird a more efficient respiratory system? Follow the path of air through the diagram. This breathing system lets a bird process more air than it could if it had lungs only. What's more, the lungs are never filled with stale air that must be exhaled before fresh air can be taken in. Fresh air flows continuously through the bird's lungs. How does this compare with your respiratory system?

Bird Respiratory System

1. Breathed-in air enters the rear air sacs.
2. Compressing the rear air sacs moves air into the lungs. Oxygen and carbon dioxide are exchanged between the lungs and blood through cell membranes.
3. Air moves from the lungs to the front air sacs.
4. Air is exhaled.

air sacs

lungs

air sacs

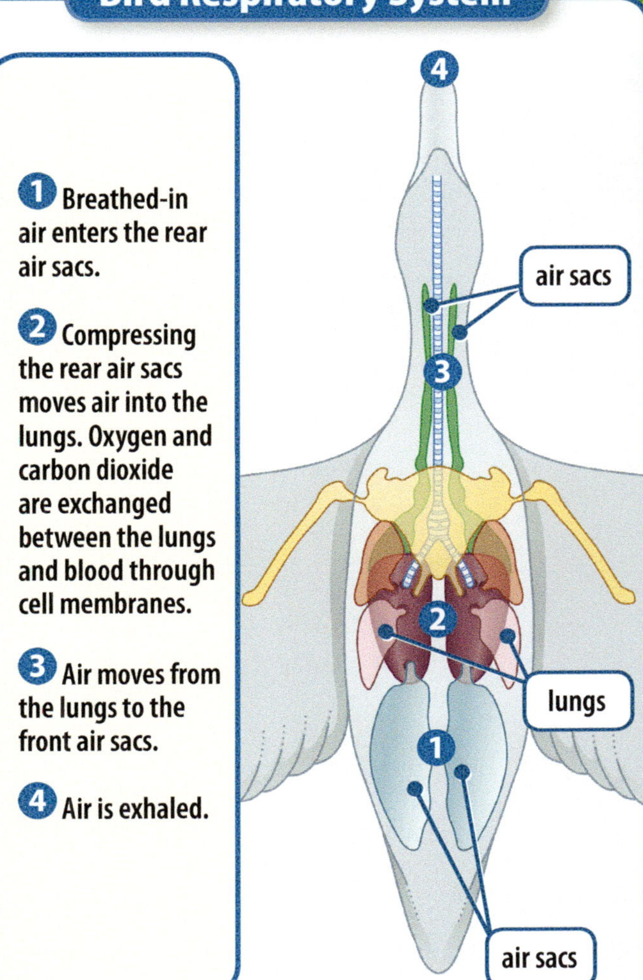

Explore-a-Lab

Guided Inquiry

How is the bone structure of birds related to their respiratory system?

Save the bones from a cooked chicken or turkey thigh. Remove any leftover meat (wear gloves) and soak the bones in a 1:10 solution of household bleach water for five minutes. Allow them to dry. Carefully cut or saw crosswise through the large, long bone of the thigh. **Observe** inside the bone. Draw what you see. What characteristics does the bone have that allow the bird to fly?

The Diversity of Birds Explain

Think about a bird that you have recently seen. Was it large or small? Was it hopping on the ground or on a branch, or was it flying? What was the main color of the bird? What other features do you remember about the bird? Birds are a very diverse group of vertebrates. They can be found in almost every habitat on the planet.

Study the table below. How does a condor compare with a mallard? How does a woodpecker compare with a heron? The table displays information about seven of the more than twenty-five orders of living birds. What are some types of birds that are not represented in the table?

Aves			
Order	**Characteristics**	**Habitat**	**Examples**
Falconiformes	• Sharply hooked beak • Strong legs and feet with claws to capture prey • May have broad wings, long narrow wings, or short stubby wings	Live in diverse habitats, including the desert, tundra, taiga, wetlands, and rainforests	Eagle, hawk, kite, vulture, condor, falcon
Galliformes	• Chicken-like in appearance • Small to large bodies • Short wings • Dark- to brightly colored plumage • Elaborate head and neck ornamentation	Live in diverse habitats, including the desert, tundra, taiga, forest, mountains, and rainforests	Grouse, pheasant, prairie chicken
Anseriformes	• Medium to large birds • Gray or brown to black and white plumage • Webbed feet	Distributed worldwide, except for Antarctica	Mallard, duck, goose, swan
Piciformes	• Two toes projecting forward and two toes projecting backward • Lack down feathers • Nest in natural cavities	Live in forests and woodlands	Woodpecker, barbet, toucan
Passerformes	• Three toes directed forward and one toe directed backward • Foot curls and becomes stiff when bird lands on a branch	Distributed widely; live in a variety of habitats	Wren, swallow, robin, finch, starling, warbler
Ciconiiformes	• Live near or on the water • Long legs, neck, and bill • Nonwebbed feet	Live in wetlands, tidal areas, swamps, marshes, damp meadows, and forest streams	Heron, stork, ibis, spoonbill
Charadriiformes	• Strong flight abilities • Long migrations • Breed in large colonies • Construct their nests on the ground	Live on beaches, rocky coasts, estuaries, riverbanks, lakeshores, and in a variety of other marine and freshwater wetlands	Sandpiper, murre, plover

Characteristics of Class Mammalia Explain

Mammals are endotherms with hair on their bodies at some time in their lives. How does this method of thermal regulation enable mammals to survive in habitats from the deep cold near the North Pole to the steaming heat of the tropical rainforest? What other ways do mammals regulate body temperature?

Female mammals produce milk in *mammary glands* to nurse their young through a period of growth and development. Is the composition of all mammal milk the same? Scientists have classified mammals into three subclasses, based on their embryonic and fetal development.

Scripture Spotlight

What mammals and their characteristics are mentioned in **Lamentations 4:3** and **Job 4:15**?

Subclass Prototheria (Monotremes)

Subclass *Prototheria* includes mostly animals that are extinct. The platypus and echida (spiny anteaters) are living members of this subclass. These animals are known as monotremes. Monotremes are mammals that lay eggs. There are only five species of monotremes living today. One is the duck-billed platypus, which lays its eggs in a burrow. The other four are spiny anteaters, also known as echidnas. Modern monotremes live only in Australia and New Guinea. Why might monotremes live only in this limited geographic region?

Subclass Metatheria (Marsupials)

Kangaroos, wombats, koala bears, and opossums are *marsupials*. They belong to the mammal subclass *Metatheria*. Marsupials get their name from the *marsupium*, a pouch of skin on the abdomen of the female. Marsupial mothers give birth to very underdeveloped young, which migrate to the pouch. There, the young attach to a nipple and suck milk from mammary glands as they develop for up to seven months. How does the pouch protect a newly born marsupial?

The echidna is native to Australia's rough scrublands. The baby echidna hatches from a leathery egg inside a fold of skin similar to a pouch.

What adaptations does this hardy survivor have that make it common in its habitat?

Lesson Activity

Mammals have many unique characteristics that distinguish them from other types of animals. Use books and online resources to research what other characteristics make mammals unique. How do mammals hear? How do the mouths of mammals differ from the mouths of other animals? What is different about a mammal's heart? Find answers to these questions and some of your own facts and present them in an illustrated poster.

What makes mammals unique?

Subclass Eutheria (Placental Mammals)

Placental mammals get their name from the placenta, an organ that grows along the wall of the uterus of the female during pregnancy. An umbilical cord connects the placenta to the offspring, which exchanges gases, food, and waste through the placenta with the mother. Why do you think the placental mammals are more common than other mammal groups?

Placental mammals develop for a relatively long period while attached to the placenta. How does this long developmental period benefit placental mammals?

Scientists have identified nine living orders of placental mammals. Study the information in the table. Compare a loris with a weasel. How are they similar, and how are they different?

Eutheria			
Order	**Characteristics**	**Habitat**	**Examples**
Lagomorpha	• Pair of incisors in each quadrant of the upper jaw • Teeth grow throughout the animal's life • Canines lacking	All regions except for Antarctica	Hare, pica, rabbit, cony
Primata	• Five digits on each appendage • Flexible and limber shoulders and hip joints • Stereoscopic vision • Nails instead of claws • Walks on two feet at least some of the time	Mostly tropical regions (other than humans)	Human, chimpanzee, lemur, baboon, gorilla, slow loris, capuchin monkey, colobus
Chiroptera	• Webbed wings formed from forelimbs • Habits that pollinate flowers and distribute seeds • Helpful in controlling insects • Highly developed ears	Tropical and temperate regions	Flying fox, fruit bat, free-tailed bat
Carnivora	• Strong, pointed incisors • Enlarged canines for stabbing prey • Powerful jaw muscles	Temperate, tropical, polar regions; terrestrial, saltwater, and freshwater	Tiger, hyena, wolf, lion, bear, cat, weasel
Artiodactyla	• Even number of toes • Hard hooves • Eyes positioned on sides of head • Long eyelashes • Rotating ears • Horns, antlers, or tusks	Temperate, tropical, polar regions; terrestrial	Sheep, goat, camel, pig, cow, deer, antelope

Eutheria *cont.*			
Order	**Characteristics**	**Habitat**	**Examples**
Cetacea	• Live entirely in water • Streamlined bodies • Horizontal tail • Nostrils located on top of the head • Forelimbs are flippers; no hind limbs • Teeth or baleen	Temperate, tropical, polar regions; saltwater and freshwater	Dolphin, whale
Rodentia	• Single pair of incisors in each jaw • Incisors grow throughout the animal's life • Mostly herbivorous • Many are social and live in groups	All regions, except for Antarctica	Beaver, muskrat, porcupine, woodchuck, chipmunk, squirrel, prairie dog, marmot
Perossodactyl	• Odd number of toes • Hard hooves • Eyes positioned on sides of head	All regions, except for Antarctica; primarily grasslands	Horse, zebra, donkey, tapir, rhinoceros
Insectivora	• Acute senses of smell and hearing • Most have a long, flexible snout with whiskers • Small eyes with poor vision	Temperate and tropical regions	Hedgehog, mole, shrew

Summary: What are birds and mammals? Birds and mammals are endotherms, and are able to maintain a nearly constant body temperature. Endotherms can be poikilotherms or homeotherms. Scientists have classified living bird species into more than twenty-five orders. The most familiar seven orders are Falconiformes, Galliformes, Anseriformes, Piciformes, Passerformes, Ciconiiformes, and Charadriiformes. Mammals have hair and feed their young with milk produced by their bodies. Based on the characteristics of embryonic/fetal development, scientists have classified mammals into three fundamental groups: monotremes, marsupials, and placental mammals. The nine most common orders of placental mammals are Lagomorpha, Primata, Chiroptera, Carnivora, Artiodactyla, Cetacea, Rodentia, Insectivora, and Perossodactyl.

1. How are endotherms different from ectotherms?
2. In what ways are birds well adapted for flying?
3. Compare and contrast the three modes of embryonic/fetal development found in mammals.
4. Select two mammals. Explain how their unique characteristics help each animal survive in its environment.

Lesson 4

Objectives

- Describe examples of behavior in vertebrates.
- Differentiate between innate and learned behavior.
- Identify examples of communication in vertebrates.
- Identify reasons for and modes of communication.

Vocabulary

habituation

sensitization

associative learning

courtship

Essential Question

How Do Vertebrates Behave and Communicate?

How do you perceive the world around you? How do you perceive changes in your own body, such as the sensation of hunger? A change in an animal's internal state or external environment is called a stimulus. Hunger is an internal stimulus. How might your response to stimuli differ from a bird's response or an insect's response? The roar of a predatory lion is an external stimulus. Animals respond to a stimulus with a behavior, or activity that can be observed. Animals behave in certain ways to obtain food, avoid predators, interact in social groups, and reproduce.

Behaviors can be as basic as blinking, eating, walking, running, or flying. They can also be as complex as hunting for food, gathering food, escaping from predators, finding a mate, building a nest, and fighting a rival or enemy. In general, all behaviors share the purpose of ensuring survival. What behaviors do you exhibit every day to ensure survival? What are some behaviors that a squirrel might use to survive?

Crested pigeons produce a whistling sound with their wings.

What stimulus might trigger sound production? What function might the whistling sound serve?

Two Types of Behavior Explain

Animals display a great variety of behaviors. Scientists classify these behaviors into two groups: innate and learned. Animals use a combination of innate and learned behaviors for survival. Many complex behaviors are really a series of smaller behaviors, some of which are innate and some of which are learned. Both innate and learned behaviors may require communication.

Innate Behavior

How does a baby let you know that she is hungry? How does the baby know how to behave this way? *Innate behaviors* are inborn. They are part of the makeup of a species, encoded in its DNA. God created animals to inherit innate behaviors just as they inherit physical traits. Innate behaviors are predictable. They occur reliably in response to a particular stimulus. The animal needs no experience with the stimulus to exhibit innate behaviors. The dance of the honey bees is an example of innate behavior in an invertebrate. Reflexes, such as the human newborn's sucking reflex, are also innate behavior. What other reflex behaviors can you identify?

Many innate behaviors are related to survival. For example, fish do not have to learn to swim. They hatch from their eggs with this ability and use it immediately. Salamanders exhibit a similar ability. Even when a newly hatched salamander is raised away from a body of water and does not observe any other salamanders swimming, it swims perfectly when it first enters the water. A newborn calf will stand up within minutes of being born. Kangaroo offspring will instinctively crawl toward the mother's pouch to feed. How does a reptile know how to gather food without any guidance from a parent? What are some other innate behaviors common in animals? Why might animals have more innate behaviors than humans?

Explore-a-Lab

Structured Inquiry

How does the snail become habituated to the cotton swab?

Place a large snail on a bed of moist leaves or newspaper. Use a stopwatch to time how long it takes for the snail to emerge fully from its shell. Record this data. Using a moistened cotton swab, gently touch the fully emerged snail between the eyestalks. Immediately start the stopwatch and measure the length of time between the touch and the snail emerging fully from its shell once again. Repeat the procedure for a total of 10 touches. Graph the time it takes the snail to reemerge after each touch.

Some species of vertebrates have innate behaviors that are related to a specific stimulus. For example, the male stickleback fish becomes aggressive when other male sticklebacks approach his territory. Researchers found that this aggressive behavior was stimulated by the red color of the male sticklebacks. When researchers placed other red objects into the tank with a red stickleback, they observed the same aggressive behavior. The female stickleback fish, which does not have the red coloring, does not provoke the aggressive behavior.

Migrating birds display innate behaviors that are related to migration and the direction of travel. Birds raised in captivity often behave restlessly during migration season and tend to move toward the direction to which they would naturally migrate.

Learned Behavior

How did you learn to ride a bike? How did you learn to read or write? What steps did you take to master any of these or other skills? *Learned behavior* is a behavior that results from experience or observation. God also created animals with this special ability. Learned behaviors are not inherited. They arise in response to repeated exposure to a stimulus. Learned behaviors can change over time. What are some learned behaviors that a kit will learn from the mother fox? Learned behavior does change in response to changing environmental conditions. This type of behavior typically does not appear in animals that grow up alone. It seems that animals learn new behaviors best from other animals. What behaviors have you learned from your parents or in school?

Scripture Spotlight

Read **Ecclesiastes 3:11**. What do you think it means that God has put eternity in our hearts? Does that sound innate or learned?

Learned Behaviors		
Learning	**Definition**	**Questions**
Habituation	A simple type of learning; if an animal is continually exposed to the same stimulus, it may stop responding to the stimulus.	Have you been exposed to a strong odor? Did you eventually stop noticing it? Did the odor vanish?
Sensitization	Becoming more responsive to a stimulus upon repeated exposure	How could you respond to stimuli if you could not see?
Associative learning	A more complex type of learning; learning that occurs when an animal learns to associate different stimuli and events	Would your experience riding a bike be habituation, sensitization, or associative learning?

Explore-a-Lab

Structured Inquiry

Did the animals you observed exhibit more learned or more innate behavior?

In small groups, go to some green space outside the school or a local park. Spend 10 to 15 minutes observing the behaviors of vertebrate animals in the environment. Record your observations, and then gather as a class to create a master list of observations. Classify the list of observations as innate or learned.

Animals also learn behaviors by observing other animals. One famous example of this was described in the 1950s. Japanese scientists studying a group of macaques started providing sweet potatoes for the animals to eat. A year after the feeding started, the scientists observed a young female macaque washing a soiled sweet potato in a stream before eating it. The young female, it seems, invented the behavior. Other macaques imitated the behavior. Before long, washing sweet potatoes was a widespread behavior in the population. Why do you think this behavior was adopted by the group? What other behaviors have animals developed?

Whether a behavior is learned or innate can depend on the species. The song of the dove, for example, is innate. Young doves raised alone and in silence sing the adult song perfectly the first time they try it. Most other birds, however, need exposure to the adult song during a critical period in their youth to learn the song.

Scripture Spotlight

What do **1 Peter 2:21** and **1 Timothy 4:12** say about learning by example?

Animals such as these macaques can respond in novel ways to new situations.

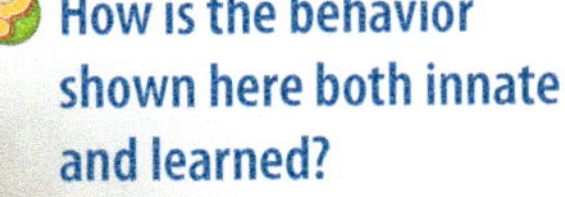

How is the behavior shown here both innate and learned?

Record your work for this inquiry. Your teacher may also assign the related Guided Inquiry.

Innate and Learned Behaviors

How do innate and learned behaviors compare?

Procedure

1. Work with a partner. **Observe** the pupils of your partner's eyes and note their size.
2. Turn off the lights and wait five minutes. Then turn the lights back on and begin the stopwatch. Observe your partner's eyes until the pupils contract to the size they were before you turned the lights off. **Record** how long it took the pupils to contract to the original size.
3. Repeat Step 2 two more times. Then switch roles, performing Steps 1 through 3 for your partner.
4. In the next challenge, record how long it takes your partner to complete a maze.
5. Flip the paper over so that the maze cannot be seen. Give your partner a second copy of the same maze. Record how long it takes your partner to complete the maze. Do this with the remaining two copies of the same maze.
6. Switch roles and complete Steps 4 and 5 with the second set of identical mazes, while your partner times you.

Materials

- clock with second hand, or stopwatch
- lights you can turn on and off
- 4 copies of a maze (for partner 1)
- 4 copies of a different maze (for partner 2)

Analyze Results

Graph the results of your pupil contraction investigation and the results of your maze investigation. Did either behavior (pupil change or ability to complete the maze) change with repetition?

Create Explanations

1. How do innate and learned behaviors compare?
2. Which behavior was innate? Which was learned? How can you tell the difference?
3. How does each type of behavior benefit animals?

Animal Communication Explain

Many animal behaviors are examples of communication. Think about the ways you communicate. Do you speak aloud, write letters, sing songs, make facial expressions, use gestures, or use a combination of verbal and nonverbal signals? Now consider the way a pet dog or cat communicates its need for food or a desire to go outside. How does communication help animals survive?

Courtship

Communication is used to attract or find a mate. **Courtship** behaviors help males attract a mate. Courtship may involve movements, sounds, and visual displays. Male peacocks, for example, spread their colorful tail feathers when courting a female. Prairie chickens do elaborate dances. They also make "booming" sounds when courting. Some bird songs are courtship songs. Have you ever heard a male bird singing to find a mate? The female bird judges the quality of the songs as a way to choose among possible mates. How do you let a fellow student know that you think he or she is interesting and that you want to hang out with him or her? Birds have songs for courting and calls for alerting others in the area of an approaching predator.

Check out your ***Science Journal*** for a Guided Inquiry that explores communication in fish. Discover

Protecting Territory

Animals also communicate about territory. Cats and dogs mark their territories with urine. Birds have specialized calls that warn off others that may be nearing their territory. Among social animals that live in groups, such as wolves and baboons, communication helps establish dominance. This allows the group to live peacefully. Think about your group of friends. Which friend is the leader of the group? What behaviors alert the group that he or she is the leader?

Lesson Activity

Work in groups of three or four. Invent a method of communication that does not use words. Does it rely on touch, sound, or visual cues, such as color? In a presentation to the class, explain your method and demonstrate your skill in using this form of communication.

Is your method of communication learned or innate?

Differences in When Animals Are Active

Different types of communication arise from the sensory abilities that animals have, as well as from differences in the times that they are active. Animals that are active in daylight hours most often use visual cues, such as posturing aggressively or displaying plumage. How do male deer communicate as they vie for mates in the fall?

Communicating by means of sound and scent can be effective during both the day and night, so many nocturnal species rely on these modes of communication. What do you think the purpose is of a cricket's night chirp? How do humans communicate that they care for one another? What sounds, other than words, do human use to communicate?

Pheromones

You have learned about the chemical senses of invertebrates. Remember how ants release *pheromones*? Pheromones affect the behavior of another individual of the same species. What information do ants communicate through pheromones? Like invertebrates, vertebrates also release pheromones and respond to them. Vertebrate pheromones may keep groups together, define a hunting territory, identify another individual of the same species, or attract a mate. For example, a female rabbit releases a scent that tells her young that she is ready to nurse them. This scent immediately causes the young rabbits to begin nursing. How else might animals communicate with pheromones?

A green anole flashes his red dewlap (pouch) during a courtship display.

Do you think this communication behavior is innate or learned? Explain.

Schooling

Many innate behaviors involve communication among individuals in a group. Fish form groups to enhance foraging, defend against predators, and find mates. They aggregate into tight groups, called schools. Fish in schools swim in synchronized patterns.

Schooling fish can change direction rapidly while staying together as a single unit. While in a school, fish maintain visual contact with their eyes. The fish can see what other members of the school are doing in relationship to themselves and respond accordingly. Fish with a lateral line use this feature to stay in a neat, orderly pattern in the school. How are schooling fish similar to a flock of birds? Like geese, herds of grazing wildebeest, and packs of wolves, fish are born knowing how to move together. Their cooperative behaviors help them survive. Prey animals find safety in numbers, and pack hunters may have an advantage over solitary predators. Do humans communicate in large groups? If so, how?

Explore-a-Lab

Structured Inquiry

How do birds communicate?

Find a location outside where you can observe a flock of birds. Why do birds travel in groups? Does there appear to be a leader for the flock? How do the birds know when to change directions? What type of behavior are they exhibiting? How would you describe the flock's movement? How is it different from how humans move? Record your observations. Also reflect on why you think God gave these birds such different abilities from what you possess.

Concept Check **Assess/Reflect**

Summary: How do vertebrates behave and communicate? Behaviors are actions performed in response to external or internal stimuli. Vertebrates exhibit innate, or inborn, behaviors, as well as learned behaviors. Communication is the sending and receiving of messages from one animal to another. Animals communicate for mating purposes, to maintain territory, to warn of danger, and to provide social structure (if the animals are social). Animals can communicate by visual means and by means of sound, scent, and touch.

1. Define behavior and give an example.
2. Explain the difference between innate and learned behaviors.
3. How do animals communicate with pheromones?
4. How are animal behavior and communication related?

People in Science

Extend

Called to Serve

Douglas Dewar was a highly trained scientist, lawyer, and public servant. He used his training and skills into his 70s and 80s, writing books supporting creationism.

Get to Know

Douglas Dewar

Douglas Dewar was born in 1875 in England. He studied natural science at Cambridge, a well-respected British university. He had a distinguished career in law, civil service, and politics in India with the title "barrister and Auditor General of the Indian Civil Service."

While in India, Dewar also was able to apply his natural science background and his love of birds. He wrote more than 20 books on the birds of India, Kashmir, and the Himalayas, many of which are still in print today.

Although Dewar was an evolutionist in his early years, at about age 50, Dewar changed. He became a strong voice for creationism. He personally founded the Evolution Protest Movement in London, England, in 1932 and led the movement for 12 years. This group is still active, but is now known by the name Creation Science Movement.

Dewar turned his writing energies away from ornithology to the support of antievolution. He published his well-known book, *The Transformist Illusion*, in 1957.

Dewar studied and wrote about white-breasted kingfishers (*Halcyon smyrnensis*)

Concept Check

1. In what way(s) did Douglas Dewar contribute to our understanding of ornithology, the study of birds?
2. How did Douglas Dewar express his belief in God and Creation?

Science and Technology

Extend

Using Biotechnology to Save Endangered Species

The area around Baja, California, at La Paz used to be a major feeding ground for hammerhead sharks and many other shark species each October. Shark biologists would use this opportunity to tag sharks for later monitoring. There were so many sharks in the area that divers would simply use a snorkel and mask to dive only a few meters below the surface to tag the sharks they were studying.

Today, there are few, if any, sharks at La Paz. Due to overfishing for the highly prized shark fin soup, sharks have all but disappeared from La Paz and many other locations in our oceans.

With many shark species and so many other species endangered and near extinction due to overfishing and habitat loss, how are scientists able to help? The answer may lie in biotechnology. Although captive breeding programs help, genetic diversity is small, a concern for the longevity of a species. However, "frozen zoos" may have the answer.

In San Diego, California, and other locations around the world, scientists have collected embryos, stem cells, DNA, sperm, and ova of more than 800 species of endangered animals using cryopreservation. Freezing cells under normal conditions causes ice crystals to form, killing the cells. Adding the chemical dimethyl sulfoxide to these types of cells can prevent the formation of ice crystals, making it possible to freeze cells without killing them. Dehydration can also cause cell death. A chemical called trehalose is added to help the cells survive dehydration. By using this valuable genetic material, scientists could save threatened, endangered, and even extinct species and protect Earth's biodiversity.

Concept Check

1. Why is protecting Earth's biodiversity important?
2. How can scientists help regenerate Earth's endangered species?

Chapter 3

Review

Assess/Reflect

Study Guide

Lesson 1

1. Fish are ectothermic vertebrates that use gills to obtain oxygen from the water environments they live in. Most fish are covered with scales. Their fins and streamlined shape help them move easily in water.
2. The fins of a fish are designed for different purposes. Some fins propel the fish forward, while other fins help the fish steer. The lateral line helps fish sense movement, pressure, and currents. The operculum controls the movement of water through the fish's gills. The exchange of oxygen and carbon dioxide take place in the gills.
3. Bony fish have true bones. The skeletons of cartilaginous fish are made of cartilage. A swim bladder allows bony fish to control their buoyancy. Cartilaginous fish lack a swim bladder and must swim continually to maintain buoyancy.
4. There are two main superclasses of fish: Gnathostomata, jawed fish, and Agnatha, jawless fish. Gnathostomata includes both bony fish, such as trout, and cartilaginous fish, such as sharks and rays. The superclass Agnatha includes hagfish and lampreys.

Lesson 2

1. Amphibians are ectothermic vertebrates that reproduce in or near water and lay their eggs in water. They have smooth, mucus-covered skin. Most adult amphibians have four legs.
2. Most amphibians go through metamorphosis. They begin life as an egg, hatch into a larval or juvenile form, and later become adults. Young amphibians live in water and have gills for gas exchange. Adult amphibians may have lungs, or may breathe through their skin or mouths.
3. Reptiles are ectothermic vertebrates. They have tough, dry, scaly skin that slows the loss of water from their bodies. Reptiles generally lay their eggs on land. Most reptiles have four legs, but some have no legs.
4. Snakes capture and kill their prey in different ways. Some snakes swallow their prey live. Others use constriction to suffocate their prey. Some produce venom, a poison, to paralyze or kill their prey before they eat it.

Lesson 3

1. Endotherms, such as birds and mammals, generate their own heat. Ectotherms regulate their body temperature using external sources.
2. Contour feathers cover the outside of a bird. Down feathers help the bird maintain its body temperature. Semiplume feathers provide insulation and form for the bird. Filoplume feathers help birds sense things in their environment.
3. Air in the respiratory system of a bird first goes to rear air sacs, then to the lungs, next to the front air sacs, and is finally released. This system allows fresh air to flow continuously through a bird's lungs.

4. Falconiformes, Calliformes, Anseriformes, Piciformes, Passerformre, Ciconiiformes, and Charadriiformes are a few of the orders of birds. See the chart on page 122 to identify the important characteristics of each order.

5. Monotremata, lagomorpha, primata, chiroptera, carnivora, artiodactyla, cetacea, and rodentia are a few of the living orders of mammals. See the chart on pages 124 and 125 to identify the important characteristics of each order.

Lesson 4

1. A behavior is an action that can be observed. Some examples of vertebrate behavior include running, blinking, building a nest, and gathering food.

2. Behaviors can be innate or learned. Innate behaviors are inherited from parents. Learned behaviors are a result of experience and observation.

3. Vertebrates communicate in many different ways. Communication can be by visual displays, through vocalization, by pheromones, by marking territory, and by touch.

Vocabulary Check

Fill in the blanks with the correct terms.

1. The bony protective covering over the gills in fish is the __________.
2. The protective shell of a reptile is made of bone covered by __________.
3. The study of reptiles and amphibians is __________.
4. The tiny hooks that hold the barbs of vaned feathers together are __________.
5. Becoming more responsive to a stimulus as a result of repeated exposure is known as __________.

Multiple Choice

Choose the best answer.

6. Which two vertebrate groups reproduce by releasing sperm and eggs into water?
   A. fish and amphibian
   B. amphibian and reptile
   C. reptile and bird
   D. bird and mammal

7. What is the scientific name for the juvenile stage in an amphibian's life cycle?
   A. spawn
   B. ectotherm
   C. larva
   D. metamorph

Check Point

Answer the following questions.

8. What structures adapt a fish for its life in the water, and how do they work?

9. What are three organs involved with respiration in amphibians?

10. Polar bears and arctic terns live near the North Pole. There are no snakes and frogs there. Suggest one reason why.

11. What are common characteristics of vertebrates?

12. What examples of God's Design do we see in vertebrates?

13. What types of learning take place in animals?

How Organisms Inherit Traits

Scripture Spotlight

God has created an amazingly complex and diverse world. Scientists are still making new discoveries regarding our genetic makeup. You will read the following passages in this chapter.

Psalm 139:13–14 (p. 146)
1 Peter 1:4 (p. 150)
Psalm 139:16 (p. 163)
Genesis 30:29–43 (p. 166)

The Big Idea

God designed organisms that inherit traits from their parents via genes, or segments of DNA. Patterns of inheritance can often be predicted if the genetic makeup of the parents is known.

Why don't these puppies look exactly like their mother?

The dog and its puppies share some traits, yet each animal is unique.

Inquiry Kick-Off

Engage

God designed animals to reproduce, or make more of their species. How does reproduction lead to differences among animals in the same species? In your ***Science Journal,*** you will compare traits of parents and offspring.

Lesson 1

Objectives

- Describe the structure of deoxyribonucleic acid (DNA).
- Explain how DNA carries the genetic code.
- Summarize the function of genes in cells.
- Distinguish between mitosis and meiosis.

Vocabulary

codon
genetics
mutation
karyotype

Essential Question

How Are DNA, Genes, and Cell Division Connected?

What characteristics have you noticed that parents and their offspring often share? For example, a girl may have her father's eye color and nose shape. A kitten may have its mother's long, brown hair. Now think about whether you have ever seen someone who does not look like either parent. For example, a boy may have blue eyes, while his parents both have brown eyes. A kitten may have long, black and white hair, but its parents both have short, gray hair. How can an offspring have features that are so different from its parents? To understand how offspring get their features such as eye color and fur length, we need to know how a trait is determined. It all starts with a substance you have learned about before, DNA, the "master molecule."

The Master Molecule Explain

Deoxyribonucleic acid, (DNA), is called the master molecule because it determines the characteristics of an organism. For example, DNA determines your eye color, blood type, and the curliness of your hair. How can DNA do all that?

Although DNA has a complex function in controlling cell operations, it has a rather simple structure. DNA consists of four different building blocks called *nucleotides*, and each nucleotide consists of three components: a sugar called *deoxyribose*, a phosphate group, and a nitrogenous base. There are four nitrogenous bases found in DNA, and each defines the identity of the nucleotide of which it is a part. These bases are *adenine, thymine, guanine,* and *cytosine*, abbreviated A, T, G, and C, respectively.

These four nucleotides are arranged in a structure called a *double helix*. A convenient way to think of a double helix is as a twisted ladder; imagine a flexible ladder twisted around, rather like a *spiral* staircase. If DNA were a ladder, its sides would be made of alternating deoxyribose and phosphate units bonded to each other, and each rung would be made of two bases, or a *base pair*.

Explore-a-Lab

Structured Inquiry

How can you model DNA?

Use black and red licorice strands that have a hollow center to construct the sides of a DNA molecule. Cut the licorice into 2 cm pieces. Cut two pieces of string, each 24 cm long. Tie a knot at the end of each piece of string. Thread 10 licorice pieces onto each string, alternating the black and red pieces. Use four different colored candies to represent the four nitrogenous bases. Use a toothpick to join two candies (bases) and then insert the toothpick into the red licorice pieces (deoxyribose sugars) on each string. Label your model using toothpick flags. What do the black licorice pieces represent? How does your model show the base pairs that form the *rungs* of a DNA molecule?

In DNA molecules, there are only four possible types of rungs, since A pairs only with T, and G pairs only with C. The four types of pairs are adenine-thymine, thymine-adenine, guanine-cytosine, and cytosine-guanine. The rungs also attach to the alternating deoxyribose-phosphate chain only at the deoxyribose units.

Earlier in your study of science, you learned that DNA codes for the synthesis of protein molecules. You will remember that *proteins* are large molecules that are the building blocks of cells. These long, chain-like molecules are made of smaller units called *amino acids*. Although there are thousands of different proteins in organisms, they are all made up of only 20 different amino acids.

Imagine a DNA molecule splitting into two, where each half consists of the alternating deoxyribose and phosphate units and one nitrogenous base (or half of each rung) attached to each deoxyribose unit. We can describe each strand of that divided DNA molecule by the sequence in which the adenine, thymine, guanine, and cytosine appear along its length (because the alternating deoxyribose and phosphate units are the same throughout the molecule). Even though there are only four bases, there are millions of possible ways to sequence the bases. Why do you think so many combinations are necessary?

The code that DNA carries lies in the order of the bases along a single strand.

When these DNA strands separate, what two code sequences will be revealed on the individual strands?

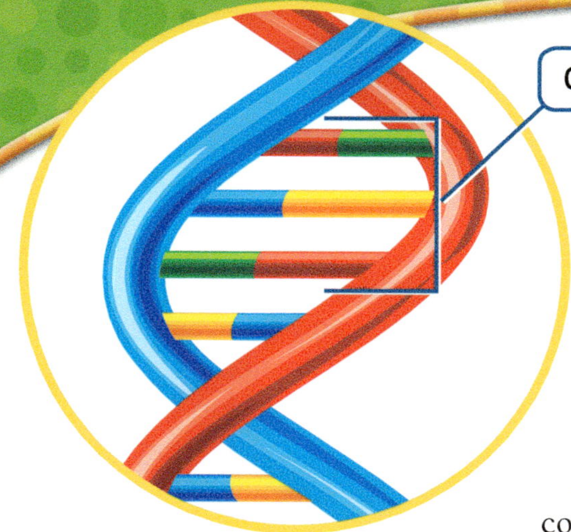

A codon provides instructions for making a specific amino acid.

Each sequence of three bases (such as ACG, CGT, CAC, etc.) in a DNA strand is called a **codon**. Each codon codes for a particular amino acid (although a number of these three-base sequences can code for the same amino acid). To see how the genetic code works, let's look at an example. The codon AGC stands for the amino acid serine. The codon TGG stands for a different amino acid, threonine. The codon CTG stands for aspartic acid. So the DNA sequence AGCTGGCTG codes for the amino acid sequence of serine-threonine-aspartic acid. What would the sequence of nucleotides be for a protein that contains the amino acid chain threonine-serine-aspartic acid-serine? How is the building of base sequences similar to using the alphabet to make words?

Genes

A segment of DNA that codes for a single protein is a *gene*. Recall that the DNA sequence AGCTGGCTG codes for a three-amino-acid protein. In reality, amino acid chains are much longer than that, and they also curve and twist themselves into fantastically complicated shapes that determine their function.

Genes make living things what they are—whether a rose, a sparrow, or a whale. The observable characteristics, or traits, of an organism arise from its genes. In humans, genes make eyes brown, hazel, or blue. They make hair straight or curly. Genes account for other characteristics that run in families, such as right- or left-handedness, hairlines, straight or curved thumbs, or a cleft in the chin. What are some genetic traits that you share with your family members?

Transmission of traits from one generation to the next is called *heredity*. **Genetics** is the study of how traits are inherited.

DNA Replication

As you know, traits are inherited because DNA passes from parent cells to daughter cells and from parent organisms to their offspring. The process of transmitting DNA from parent cells to daughter cells occurs (usually) without error. Why? It does so because DNA can *replicate*, or copy, itself. The process of replication is accurate because of the Creator's Design of the particular way that the nitrogenous bases pair with each other: A only with T, and G only with C.

DNA replicates during the process of cell division. During replication, the two strands that make up DNA separate, like a zipper unzipping. Each strand then serves as a template for a new DNA chain.

DNA Replication

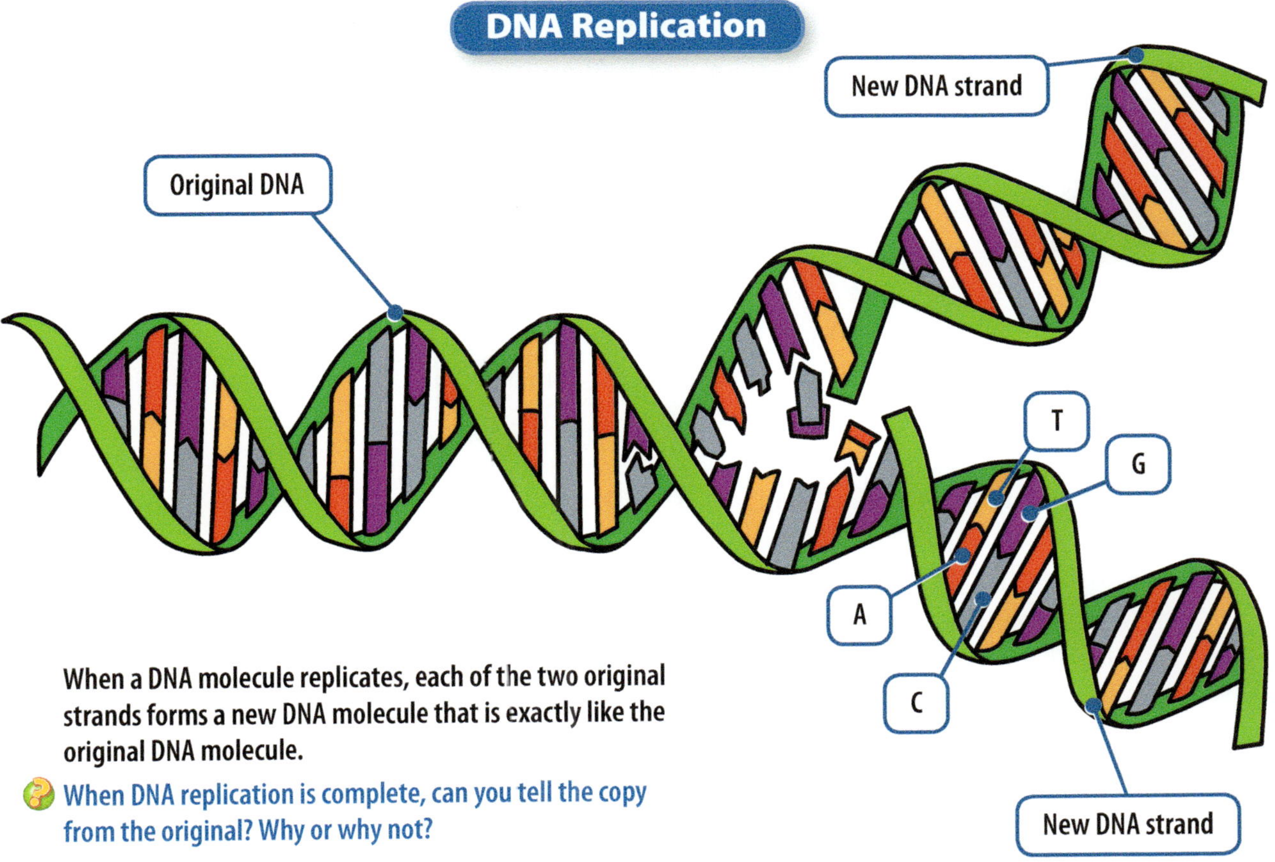

When a DNA molecule replicates, each of the two original strands forms a new DNA molecule that is exactly like the original DNA molecule.

When DNA replication is complete, can you tell the copy from the original? Why or why not?

Notice in the diagram above the place where the two strands of DNA are separating. Do you see the exposed bases that are attached to the deoxyribose/phosphate groups? Can you see the free bases near that area? These unattached bases will attach to the bases of the DNA chain, forming the base pairs that were in the original DNA molecule with an A always attached to a T and a C always attached to a G. Each DNA strand will again become a double helix. The result is two double-stranded DNA molecules that are exactly the same as the original one.

The process of DNA replication occurs without errors most of the time, but occasionally, DNA fails to replicate perfectly. For example, G may mistakenly pair with A. This can happen when these two bases do not align themselves in the proper positions. When something like that happens, the genetic code is changed. Any change in the genetic code is known as a **mutation**. A single error in sequence can have a serious effect on new cells. Fortunately, cells have ways to correct mistakes that occur during DNA replication. These mechanisms, however, do not correct all mistakes.

Record your work for this inquiry. Your teacher may also assign the related Guided Inquiry.

Extracting DNA from Living Cells

How can you extract DNA from cells?

Materials

- strawberry
- zip-top liter bag
- large plastic cup
- 100 mL graduated cylinder
- 10 mL graduated cylinder
- dishwashing detergent
- table salt
- funnel
- cheesecloth
- test tube
- cold 90% isopropyl (rubbing) alcohol
- glass stirring rod or plastic coffee stirrer

Procedure

1. Remove the top green leafy part of the strawberry and place the strawberry in the bag.
2. Seal the bag and use your fingers to gently smash the strawberry for two minutes.
3. In the plastic cup, mix about 100 mL of water, 10 mL of dish detergent, and about 1 mL of salt. Add 10 mL of this solution to the crushed strawberry.
4. Reseal the bag and gently mix the contents for one minute. Avoid making any soap bubbles.
5. Use the funnel and cheesecloth to filter the strawberry solution into the test tube. Fill the test tube about ¼th full of the solution.
6. Add cold alcohol equal to the volume of strawberry solution. Carefully pour the alcohol down the side of the test tube so that it forms a distinct layer above the strawberry solution.
7. Look closely at the two layers and record your observations.
8. Put the stirring rod in the test tube and slowly twirl or spin it in the solution so that the DNA wraps around it.
9. Remove the glass rod and observe the DNA that you extracted.

Analyze Results

Describe the appearance of the DNA you have collected.

Create Explanations

1. How can you extract DNA from cells?
2. Why did you crush the strawberry inside the plastic bag?
3. The procedure works with other fresh fruits and vegetables but not with cooked ones. Suggest a reason why.

Chromosomes

Recall that a gene consists of a sequence of DNA nucleotides that determines the order of amino acids in a protein. We find genes in the nucleus of a cell, within chromosomes. A gene is composed of DNA and protein. A chromosome may contain many thousands of genes.

Each kind of organism has its own number and type of chromosomes, which generally occur in pairs. The common fruit fly has 4 pairs of chromosomes. A petunia has 7 pairs. A chicken has 39 pairs. Humans have 23 pairs of chromosomes containing 20,000–25,000 genes.

Since chromosomes occur in pairs (except in sex cells), each gene on one chromosome has a corresponding gene on its corresponding chromosome. One gene in the pair comes from the mother. The other comes from the father. The two copies of each chromosome are called *homologous chromosomes*. Homologous chromosomes carry the genes for the same trait. For example, if one homologous chromosome has a gene for hair color, the other chromosome will also have a gene for hair color. Why do organisms have pairs of chromosomes?

If you look at a cell through a microscope, you will not see its chromosomes unless the cell is about to divide. You can take a picture of the nucleus just before the nucleus divides and you will see that the chromosomes are randomly arranged in the cell. You can cut out and arrange their images. A picture of chromosomes arranged by their size and shape is called a **karyotype**. In a karyotype, the chromosomes are matched to form pairs, with one exception in the case of human karyotypes. Human females have a total of 23 matched pairs. Males have 22 matched pairs and one unmatched pair. This unmatched pair is the sex chromosomes, represented as X and Y in males. The female sex chromosome is represented as an XX.

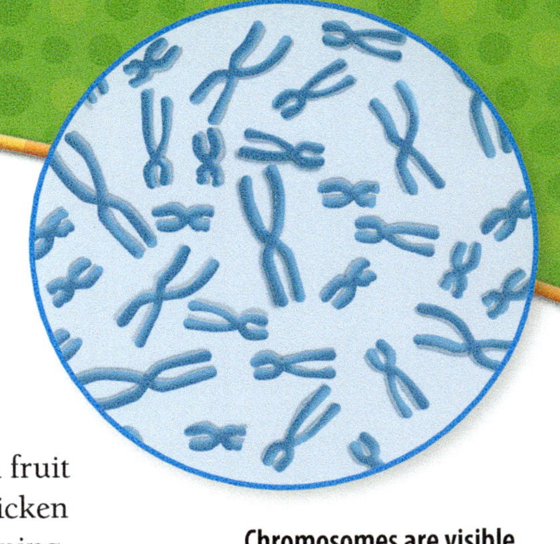

Chromosomes are visible in the nucleus of a human body cell when it divides. This illustration shows chromosomes that have already copied themselves.

Why is the process of replication so important?

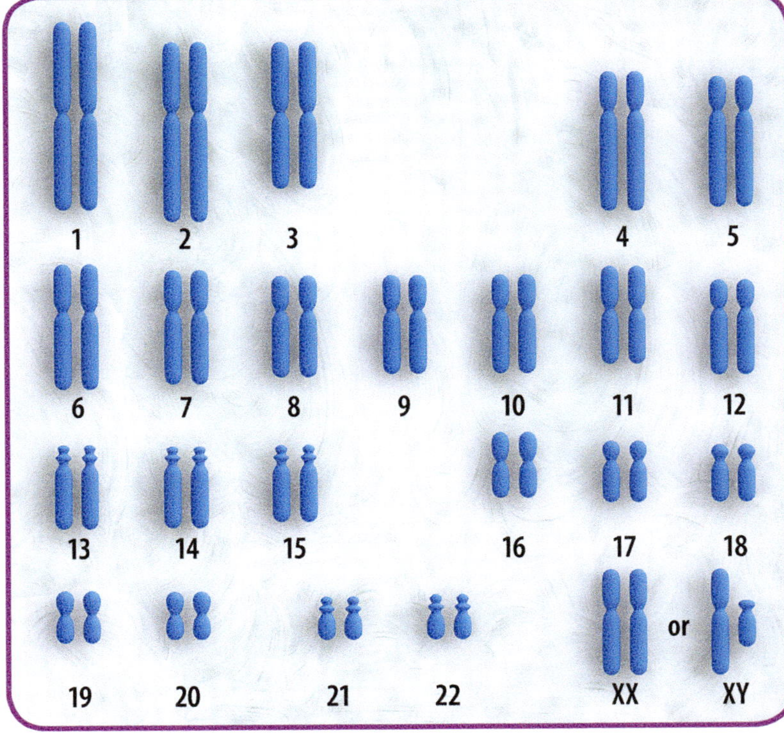

In this karyotype of human chromosomes, both the male and female chromosomes are shown.

How would the actual karyotype of a male differ from what is shown? How would the female karyotype differ?

Cell Division Explain

For species to persist, the genetic code in DNA must be passed on to future generations. Not only must the genetic code be transmitted from one generation to the next, it must be passed on intact. In other words, the entire genetic code must be passed from parents to offspring without any missing information or changes. This must happen every time a cell divides. There are two types of cell division.

Mitosis: the process that produces body cells, such as skin cells, that have a full set of chromosomes ($2n$). What are examples of other kinds of cells produced by mitosis?

Meiosis: the process that produces sex cells, which have one-half the complete number of chromosomes ($1n$). How many chromosomes would be in a human sex cell?

Scripture Spotlight

God knows us even before we are born. Read **Psalm 139:13–14.**

Mitosis

What happens to genes and chromosomes when cells divide depends on the kind of cell that is dividing. In *mitosis*, the cell's chromosomes are duplicated so that an exact copy of each one is made. The cell then has a full duplicate set of chromosomes. One set of chromosomes will pass into a daughter cell when the cell undergoes mitosis. The result is that each daughter cell receives an exact duplicate (assuming no mistakes were made) of all the chromosomes and genes that were contained in the parent cell. In humans, the body cell contains 46 chromosomes. So do the daughter cells after the cell divides.

Explore-a-Lab

Structured Inquiry

How can you model cell division?

Using two strands of multicolored beads that are the same length, you can demonstrate the relationship of genes, chromosomes, and chromosome pairs. If one strand of beads represents a single chromosome, what would each bead represent? Hold another strand of beads beside the first one. The second strand represents the second member of the chromosome pair. Where did the chromosomes in the pair come from? Hold the strands side by side and find the fifth bead on each string. Explain why these genes are a pair. Repeat with other pairs at other positions on the strands. Use the model to explain the chromosomes, genes, chromosome pairs, and gene pairs to a partner.

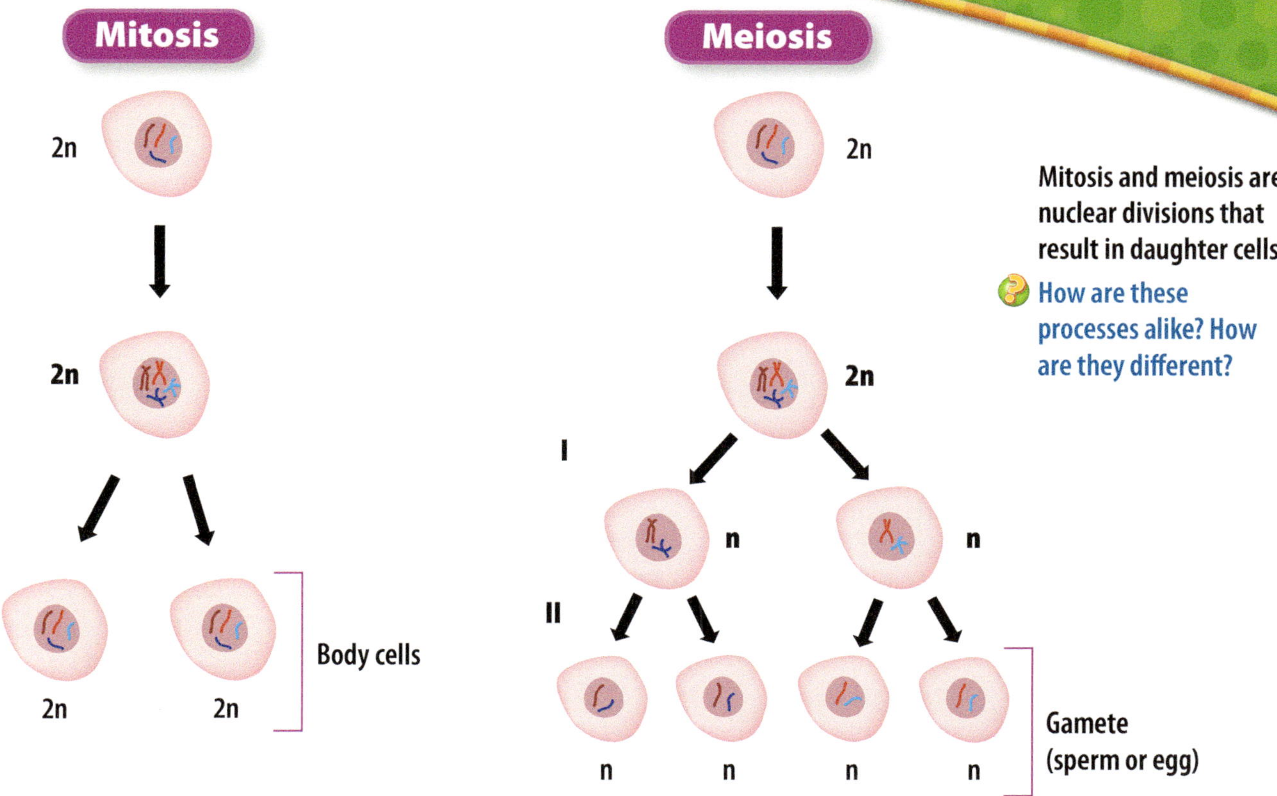

Mitosis and meiosis are nuclear divisions that result in daughter cells.

How are these processes alike? How are they different?

Meiosis

Meiosis results in reproductive cells, called *gametes*. Gametes are either eggs (from the female) or sperm (from the male). Gametes contain half as many chromosomes as body cells, one chromosome from each chromosome pair. Why do you think sex cells only have half the number of chromosomes? What are the processes involved in creating gametes?

Math in Science

A cell begins its process of division with 38 chromosomes.

1. Assume that this cell undergoes mitosis. How many daughter cells will it produce? How many chromosomes will each daughter cell have?
2. Now assume that this cell is a sex cell that undergoes meiosis. How many gametes will it produce? How many chromosomes will each gamete have?
3. After fertilization (union of gametes) in this organism, how many chromosomes will each cell contain?

When meiosis begins, chromosomes replicate and line up in matching pairs. The pairs separate into the nuclei of two new cells. Then a second division occurs, forming four cells. Because the chromosomes do not duplicate before the second division, the four cells each contain half as many chromosomes as the original cell did. Scientists call this the *haploid* number of chromosomes, or 1*n*. These cells are the gametes, sperm and egg. When gametes unite at fertilization, the full, or diploid, number (2*n*) of chromosomes is restored. Each chromosome from the father pairs with its partner from the mother. The fertilized cell that is the start of the new offspring contains a complete set of the chromosomes. In this way, meiosis ensures that offspring have the same numbers and pairs of chromosomes and genes as their parents. It also provides a way to rearrange genetic material, which increases the variability of the offspring. The offspring contain half their mothers' genes and half their fathers' genes. The offspring are like both parents, but not identical to either parent, and not identical to any sibling (other than an identical twin). What would happen if the egg and sperm each had the 2*n* number of chromosomes? How does this rearrangement of genetic material create variation?

Summary: How are DNA, genes, and cell division connected? Each cell contains DNA that tells the cell how to work. A gene is a segment of a DNA molecule that codes for a specific protein. Genes in turn make up chromosomes. Each organism has a particular number of chromosomes in its cells. During cell division, DNA (genes) is replicated and equally distributed to new cells. During reproduction, genes of the mother and the father combine and produce offspring that have a mix of traits from the mother and the father. Cell division can be either mitotic or meiotic. Mitosis, which occurs in body cells, results in two daughter cells that have the same genetic information as the parent cell. Meiosis, which occurs in sex cells, results in four gametes that have half the chromosome number as the original cell. The original chromosome number is restored when a male gamete and female gamete combine at fertilization.

1. How does the arrangement of genes determine an organism's characteristics?
2. What is a gene, and what is its function?
3. How are DNA, a gene, and a chromosome different?
4. Scientists predicted that a process other than mitosis must be involved when gametes are formed. Close observations led to the discovery of meiosis. Explain why scientists predicted that a process such as meiosis should exist even before they observed it.

Lesson 2

Essential Question

What Is Simple Inheritance?

Do people say that you look like your mother or father? What traits do you have that are similar to theirs? If you have brothers or sisters, do you share similar characteristics? Unless you have an identical twin, you probably notice differences between you and your siblings. Offspring inherit their physical characteristics from their parents. This process leads to variation. The characteristics of parents mix in new ways in their offspring. But how does inheritance work? In other words, how does an individual inherit certain characteristics but not others? The book of **Genesis** tells us that God created animals to reproduce after their kind. He established the laws of heredity at Creation. In this lesson, we will explore these laws and see how they are responsible for the patterns of heredity.

Genotypes and Phenotypes Explain

In humans, there are 20,000 to 25,000 different genes. Recall that a gene is a segment of DNA that codes for a particular protein. That protein may be an enzyme that breaks down starches in food, or it may be a protein that helps to determine eye color. Cells contain genes in pairs, one on each member of a chromosome pair. Thus, every body cell has two genes for the same trait—one gene from the mother and one gene from the father. Do you think that both genes will work together to determine the trait that the offspring will have? Or will the gene from one parent have more of an influence on what happens? Why?

Scientists know that skin freckles is a trait that is controlled by two forms of the gene for that trait. Each form of a gene for a particular trait is called an **allele**. In the case of freckles, one allele is for freckles, and the other allele is for no freckles. In this example, an individual with two alleles for freckles will have freckles. An individual with two alleles for no freckles will not have freckles. But what trait would a person with one allele for freckles and one allele for no freckles have?

Objectives

- Distinguish between dominant and recessive alleles.
- Distinguish between homozygous and heterozygous conditions.
- Distinguish between phenotype and genotype.
- Determine how dominant and recessive alleles affect phenotypes.
- Use a Punnett square to determine the phenotypes and genotypes that can appear in the offspring.

Vocabulary

allele
dominant allele
recessive allele
genotype
homozygous
heterozygous
phenotype
simple inheritance
complete dominance
Punnett square

Some alleles are dominant while other alleles are recessive. A **dominant allele** is one whose trait always shows up when the organism has the allele. A **recessive allele** is always hidden by the dominant allele. A trait for a recessive allele will only show up if the organism's alleles are both for the recessive trait. The combination of alleles that an organism has for a particular trait is known as a **genotype**.

Scientists have developed symbols and terms to describe alleles. Alleles can be represented by letters. The dominant allele is represented by a capital letter. The recessive allele is represented by a lowercase letter. In humans the allele for freckles is dominant and the allele for no freckles is recessive. The allele for freckles would be represented by *F*. The recessive allele for no freckles would be represented by a lowercase *f*. Because an individual possesses two alleles for the trait, three combinations are possible: two dominant alleles (*FF*), two recessive alleles (*ff*), and one dominant allele/one recessive allele (*Ff*). If the two alleles are the same (*FF* or *ff*), they are described as **homozygous**. If the two alleles are different (*Ff* or *fF*), they are described as **heterozygous**.

Scripture Spotlight

We inherit traits from our parents. What kind of everlasting inheritance do we receive from God, our Heavenly Father? Read **1 Peter 1:4**.

When you look at an organism or study how it functions, you cannot see its genotype, or the alleles that are in its cells. You can observe only the characteristics that are expressed. Expressed characteristics take many forms. You notice eye color and hair color in your family and friends. Together, these characteristics are the **phenotype** of the organism, or the characteristics that can be seen and measured. For example, the phenotype of an individual might be "widow's peak." The genotype that produces a widow's peak might be written as a pair of gene symbols—either *Ww* or *WW*. Why did we use those pairs? How do you suppose that scientists determined that the allele for widow's peak in humans is dominant?

The hair on some people's heads forms a pattern known as a widow's peak at the front of the scalp. A widow's peak is the expression of a dominant allele.

Why are there two possible genotypes for the widow's peak?

Record your work for this inquiry. Your teacher may also assign the related Guided Inquiry.

Determining Genotype

How can you use phenotypes to determine the genotype of an individual?

Procedure

1. The ability to roll the tongue is determined by a dominant allele. Check to see whether you can roll your tongue into a tube.
2. **Observe** the pedigree chart in your ***Science Journal***.

   A pedigree chart shows how a trait is inherited. A square represents a male, and a circle represents a female. A horizontal line connecting a square and a circle indicates that they are parents. A vertical line indicates their children. For example, the male (#1) and the female (#2) are the parents of three children—one son (#6) and two daughters (#7 and #8). The son (#6) and the female (#5) have two children—a son (#13) and a daughter (#14). Darkened circles and squares indicate individuals who possess a recessive trait.
3. Examine each individual on the pedigree chart and determine his or her genotype. Keep in mind that a person who can roll his or her tongue can have two possible genotypes. How could you eliminate one possibility?
4. Prepare another pedigree chart that includes the genotype for each individual.

Analyze Results

What is the genotype of a person who cannot roll his or her tongue?

Create Explanations

1. How can you use phenotypes to determine the genotype of an individual?
2. Would it be possible for individual 13 to roll his tongue if his mother (individual 5) could roll her tongue? Explain your answer.
3. If two people who are tongue rollers have children, is it possible for their children to be non-tongue rollers? Explain your answer.

Two recessive alleles must be present in a pair, or the characteristic they create in the phenotype does not develop. In humans, we know that the allele for no widow's peak is recessive. We use *w* to represent it. The genotype for a person without a widow's peak must be *ww*. No other genotype results in no widow's peak in the phenotype.

Simple Inheritance Explain

In humans, the allele that produces dimples shows complete dominance.

Which of these students must be homozygous for the recessive trait?

Knowing the phenotype of a particular trait that an individual has can be easy to determine, but how can you determine the genotype? How can you predict the likelihood of a certain phenotype being passed on to offspring of two parents? Let's consider dimples in humans, which is an example of simple inheritance. **Simple inheritance** is inheritance that involves one set of alleles that produce only two kinds of phenotypes. Having dimples results from a dominant allele, *D*. Only one dominant allele is needed for a person to have dimples, so the genotype may be *DD* or *Dd*. The absence of dimples requires two recessive alleles, or the genotype *dd*. We say that having dimples is a trait that has **complete dominance**.

Knowing that the allele that produces dimples shows complete dominance, you can use a table called a **Punnett square**, like those on the next page, to determine the likelihood of a certain phenotype being passed from parents to offspring. Suppose one parent is homozygous for dimples (*DD*). The other parent is homozygous for no dimples (*dd*). What is the likelihood that the offspring will have dimples? What are the possible genotypes and phenotypes of the offspring? Notice that in the top row of the Punnett square, the alleles for parent 1 (*DD*) are listed. The dominant allele (*D*) appears at the top of each column because this parent will produce gametes with only the D allele. Along the left side of the chart, the alleles for parent 2 (*dd*) are listed. This parent can produce gametes with only the recessive *d* allele. To determine the possible genotypes of the offspring, simply read across and down the chart as shown by the arrows. What are the possible genotypes of the offspring of these two parents? Will any of the offspring have no dimples? How do you know?

	D (allele in gamete, parent 1)	*D* (allele in gamete, parent 1)
d (allele in gamete, parent 2)	*Dd*	*Dd*
d (allele in gamete, parent 2)	*Dd*	*Dd*

Now let's consider another situation. One parent has dimples and is known to be heterozygous, with the genotype *Dd*. The other has no dimples. Use the Punnett square to determine the possible genotypes and phenotypes of their offspring.

	D (allele in gamete, parent 1)	*d* (allele in gamete, parent 1)
d (allele in gamete, parent 2)	*Dd*	*dd*
d (allele in gamete, parent 2)	*Dd*	*dd*

In this example, half the offspring ($\frac{2}{4} = \frac{1}{2}$) have the genotype *Dd* and the phenotype of dimples. The other half ($\frac{2}{4} = \frac{1}{2}$) have no dimples and the genotype *dd*. Under what circumstances would parents have an offspring with the homozygous dominant pair of alleles (*DD*)? Why?

Explore-a-Lab

Structured Inquiry

 What are the odds offspring will inherit recessive alleles?

Use four pennies to represent alleles found in the gametes of a male parent and a female parent. Assume that heads represents the dominant allele and tails represents the recessive allele. Arrange the pennies in a Punnett square to represent the gametes of both parents so that the offspring of these parents have a 50 percent chance of showing the dominant trait when the Punnett square is completed. Next, arrange the pennies so that the offspring have a 75 percent chance of displaying the dominant trait.

Gene Distribution during Meiosis

When you use a Punnett square, why do you write only one allele from each parent at a time? Recall that in meiosis, only one allele from each gene passes into a gamete. Because it is a matter of chance, a gamete is as likely to receive one allele as the other. Thus, each allele has a 50:50 chance of passing into a particular gamete.

Consider a cell that undergoes meiosis and has the genotype *Aa*. Half the gametes produced will have the *A* allele, while the other half will have the *a* allele. If 100 gametes were produced, about 50 should have the *A* allele, and the remaining half should have the *a* allele. All of the gametes, however, will have only one allele of this gene.

Ratios in Genotypes and Phenotypes

Now let's consider another situation. Suppose both parents have dimples. The genotype of each parent is *Dd*, or heterozygous for the trait. Use the Punnett square to determine the phenotypes and genotypes of their offspring.

	D **(allele in gamete, parent 1)**	***d*** **(allele in gamete, parent 1)**
D **(allele in gamete, parent 2)**	*DD*	*Dd*
d **(allele in gamete, parent 2)**	*Dd*	*dd*

Notice the ratio of genotypes: 1 *DD*: 2 *Dd*: 1 *dd*. So, the chance of a homozygous dominant (*DD*) offspring is $\frac{1}{4}$, or 25 percent. The chance for a homozygous recessive (*dd*) is also $\frac{1}{4}$, or 25 percent. The chance of a heterozygous condition is $\frac{1}{2}$ ($\frac{2}{4}$, or 50 percent). The phenotype ratio is different. Because both *DD* and *Dd* produce dimples in the phenotype, the phenotype ratio is 3 dimples to 1 no dimples, or $\frac{3}{4}$ (75 percent) dimples and $\frac{1}{4}$ (25 percent) no dimples.

Explore-a-Lab

Structured Inquiry

What will be the ratios of genotypes and phenotypes in the offspring of a pea-comb chicken and a normal-comb chicken?

In chickens, the pea-comb allele, *P*, is dominant over the normal-comb allele, *p*. Make a Punnett square to determine the genotype and phenotype ratios of the offspring of two heterozygous pea-comb chickens.

Pea-comb chicken, *PP* or *Pp*

Normal-comb chicken, *pp*

If this pair of chickens produced 80 chicks, predict the percentage of the offspring that would have each genotype. How many chicks would you expect to have the pea comb?

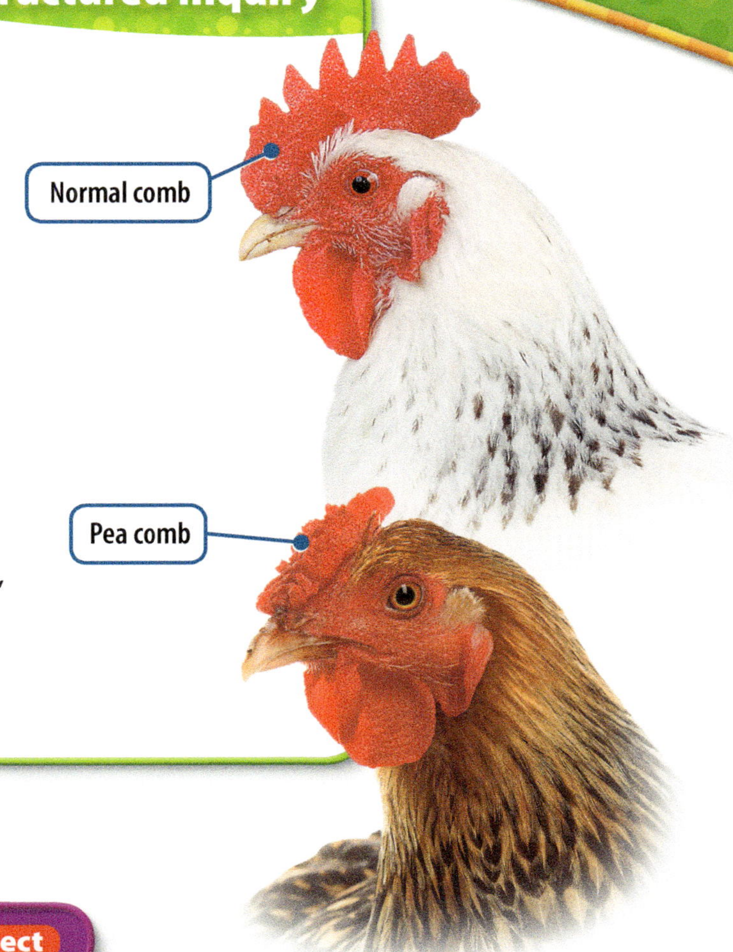

Concept Check Assess/Reflect

Summary: What is simple inheritance? Simple inheritance is inheritance that involves one set of alleles that produces only two kinds of phenotypes. The traits that offspring inherit from their parents depend on alleles, or different forms of particular genes. Alleles can be dominant or recessive. You can use uppercase letters to indicate a dominant allele and lowercase letters to represent a recessive allele. One or two dominant alleles must be present for a dominant trait to be present. Two recessive alleles must be present for a recessive trait to be present. You can use a Punnett square to determine the ratios of genotypes and phenotypes that you would expect in the offspring of matings of known genotypes.

1. Explain how alleles determine genotypes and phenotypes.
2. If you observe a 3:1 ratio of phenotypes among a large number of offspring from two parents, what can you infer about the genotypes of the parents?
3. Why are possible genotype and phenotype ratios among offspring of a particular set of parents sometimes the same and sometimes different?

Lesson 3

Objectives

- Differentiate incomplete dominance, codominance, and sex-linked inheritance.
- Predict patterns of inheritance that result from multiple alleles.
- Explain how environmental factors influence gene expression.

Vocabulary

complex inheritance
incomplete dominance
multiple alleles
codominance
sex-linked trait
hemizygous
carrier
polygenic

Essential Question

What Is Complex Inheritance?

Suppose you plant some red flowers and some white flowers in your garden. They look so great that you save some of the seeds that the flowers produce to plant in your garden next year. The following year you plant your seeds expecting red and white flowers. But mixed in with those red and white flowers are pink ones. What do you think happened? How can red and white flowers produce pink flowers? Will the pink flowers produce more pink flowers? What do you suppose is the genotype of the pink flower?

Incomplete Dominance Explain

Not all traits follow the patterns of simple inheritance. In fact, the inheritance patterns of most traits are more complicated. These patterns of inheritance that differ from simple patterns of inheritance are called complex inheritance. In **complex inheritance**, traits result from the influence of multiple factors.

For some traits, alleles are neither dominant nor recessive. Both are expressed. The result is a blend, or mix, of characteristics in the phenotype. This pattern of inheritance is called **incomplete dominance**. The color of snapdragon blossoms is one example. Neither red nor white is dominant. Thus, when homozygous red-flowered plants (genotype *RR*) are crossed with homozygous white-flowered plants (*WW*), the resulting offspring all have pink flowers (with the genotype *RW*). Both the *R* and the *W* are expressed, and the phenotype is a blend of both.

The red and white snapdragons would produce pink offspring.

How does the inheritance pattern for color in these snapdragons differ from the inheritance of dimples in humans?

With incomplete dominance, the chance of returning to the homozygous condition still remains. Suppose, for example, that two hybrid snapdragons (*RW*) are crossbred. The Punnett square reveals a familiar pattern.

	R **(allele in gamete, parent 1)**	***W*** **(allele in gamete, parent 1)**
R **(allele in gamete, parent 2)**	*RR*	*RW*
W **(allele in gamete, parent 2)**	*RW*	*WW*

Among the offspring, 25 percent ($\frac{1}{4}$) are red (*RR*), and 25 percent ($\frac{1}{4}$) are white (*WW*). The remaining 50 percent ($\frac{2}{4} = \frac{1}{2}$) are pink, with the hybrid genotype *RW*. Why is the reappearance of the two original phenotypes (in this case, red and white) always possible?

Multiple Alleles

Some genes have **multiple alleles**, or more than two allele forms. Traits that are the result of multiple alleles include blood type and eye color in humans, hair color in mice, and coat color in rabbits. In human eye color, for example, different alleles produce brown, hazel, and blue eyes. Multiple alleles often show a specific order of dominance. For example, brown eye color is dominant over both hazel and blue, but hazel is dominant only over blue. Blue eyes result from a homozygous recessive genotype. We can represent the hierarchy this way:

brown > **hazel** > **blue**

The ">" indicates that the trait on the left of this symbol is dominant over the trait on the right of the symbol. What can you state about blue eye color based on this hierarchy?

Let's look at coat color in rabbits to examine the pattern of multiple allele inheritance. The allele for coat color is represented by *C*. A superscript is used to indicate the different alleles for this trait. The table on the next page shows the four alleles that determine coat color in rabbits. Of the four alleles for rabbit coat color, only the allele for albino color does not use a superscript.

Coat Color in Rabbits			
Color	Allele	Color	Allele
Wild	C^{+}	Himalayan	C^{h}
Chinchilla	C^{ch}	Albino	c

The allele for the wild rabbit coat color (C^{+}) is dominant over all the others. The chinchilla allele (C^{ch}) is recessive to the wild type but dominant over Himalayan (C^{h}) and albino (c). The Himalayan (C^{h}) is recessive to wild type (C^{+}) and chinchilla (C^{ch}), but dominant over the albino. A homozygous recessive cc genotype is needed to produce an albino phenotype. The series of dominance relationships among these multiple alleles can be represented in this way:

$$C^{+} > C^{ch} > C^{h} > C$$

Which alleles are dominant over C^{h}? Although multiple alleles are involved, the genotypes of the rabbits can still be determined in the usual way because the alleles are present in pairs in a single individual. The table summarizes genotypes and phenotypes for coat color in rabbits.

Coat Color in Rabbits: Genotype and Phenotype	
Phenotype	**Possible Genotype**
Wild	$C^{+}C^{+}$ OR $C^{+}C^{ch}$ OR $C^{+}C^{h}$ OR $C^{+}c$
Chinchilla (C^{ch})	$C^{ch}C^{ch}$ OR $C^{ch}C^{h}$ OR $C^{ch}c$
Himalayan (C^{h})	$C^{h}C^{h}$ OR $C^{h}c$
Albino (c)	cc

Explore-a-Lab

Structured Inquiry

How can you obtain the color you want?

Using a Punnett square and the chart on the previous page, show how it is possible to get an albino rabbit from two parents that both have the wild-type coat color. Then use a Punnett square to show what the genotypes of the parents must be to produce offspring that are all homozygous for the Himalayan coat color.

Use three different colors of beads (one for each of the three alleles in this example) to show the results of all possible crosses between possible genotypes.

Although coat color is an example of multiple alleles, an individual rabbit can possess only two alleles for this trait. Explain why there are more possible genotypes for the wild-type coat color than all the other coat colors.

Codominance

Multiple alleles can exhibit a pattern of inheritance called **codominance**, which means that more than one dominant allele is expressed in the phenotype. The phenotype is not a blend as it is in incomplete dominance. Instead, both characteristics can be observed. An example of a codominant trait is coat color in horses. As you can see in the picture, both red coat color and white coat hairs can be seen in the horse. Two alleles determine this coat color, red (C^R) and white (C^W). Both alleles are expressed. What genotype is expressed in this horse?

The coat color of this horse is known as roan.

What would be the possible genotypes and phenotypes of the offspring of two roan horses?

The inheritance of the blood types A, B, AB, and O in humans is an example of an allele that involves both multiple alleles and codominance. Your blood type depends on whether a certain *antigen* (blood protein) occurs on your red blood cells. There are two antigens, A and B. If you have the A antigen only, your blood is type A. If you have the B antigen only, your blood is type B. People with type AB blood have both antigens. Type O blood means neither antigen is present. Knowing your blood type is important for medical reasons. For example, during a blood transfusion, mismatched blood can lead to blood clotting, which can result in death. Do you know your blood type? What can you infer about your parents' blood types?

The ABO Blood Group				
	Group A	**Group B**	**Group AB**	**Group O**
Red blood cell type				
Antibodies present	Anti-B	Anti-A	None	Anti-B and Anti-A
Antigens present	A antigens	B antigens	A and B antigens	None

Three alleles determine A, B, AB, and O blood types. They can be written as I^A—type A, I^B—type B, and i—type O. Both I^A and I^B are dominant. If either of those alleles is present, the antigen will be made—regardless of the other allele. The letter *i* indicates the recessive allele. A person with an *ii* genotype has a type O phenotype.

The table below summarizes the inheritance of A, B, AB, and O blood types in humans.

Phenotype	Genotype	Antigen in Red Blood Cells
Type A	I^AI^A or I^Ai	A
Type B	I^BI^B or I^Bi	B
Type AB	I^AI^B	A and B
Type O	ii	none

Check for Understanding

Why is a person's blood type important?

Notice that a person with type A or type B blood type can have either of two possible genotypes. You can determine which genotype is present by studying the family tree and looking at the phenotypes of parents and grandparents.

Explore-a-Lab

Structured Inquiry

How can blood type disclose who your parents are?

In the 1940s, a woman claimed that a famous Hollywood actor named Charlie Chaplin was the father of her baby. Blood tests revealed that Chaplin had type O blood, the woman had type A blood, and the baby had type B blood. However, blood types were not admissible as evidence at that time. Based on the testimony of witnesses, the court ruled in favor of the woman. Based on blood types, would you agree or disagree with the court's decision? Use Punnett squares to present your argument.

Sex-Linked Inheritance **Explain**

Chromosomes occur in pairs; therefore, alleles also occur in pairs. One allele of each pair is present on each chromosome. In male mammals, however, the sex chromosomes are not the same. One sex chromosome is called X. The other is called Y. Because the Y chromosome carries fewer genes than the X chromosome, males have only one allele for some traits. A trait that is determined by the alleles on a sex chromosome is a **sex-linked trait**.

Use your ***Science Journal*** to explore how sex-linked inheritance of eye color in fruit flies can help us understand genetics. **Extend**

Hemophilia

Perhaps the best-known example of a sex-linked trait in humans is hemophilia, a disease that prevents blood from clotting normally. In males, the phenotype for many sex-linked traits is determined by a single allele on the X chromosome. For males, why would a single allele determine traits such as hemophilia? To examine how sex-linked inheritance works, follow what types of gametes the parents produce and how they can combine to produce offspring.

Hemophilia Inheritance

Unaffected father: X Y × Carrier mother: X X

X X Unaffected daughter | X X Carrier daughter | X Y Unaffected son | X Y Affected son

The red on these chromosomes represents the allele for hemophilia.

Explain why only the son in this example could have hemophilia.

Record your work for this inquiry. Your teacher may also assign the related Guided Inquiry.

Sex-Linked Inheritance

How can you predict color blindness?

Materials

- chenille sticks (of one color)
- large-holed beads in two colors: red and white
- colored markers or pencils

Procedure

1. Use a straight chenille stick to model an X chromosome. Bend another chenille stick in half to model a Y chromosome.
2. Red beads represent the dominant allele for normal color vision, *C*. Thread a red bead on the straight chenille stick. Twist the end down so that the bead-allele stays attached to its chromosome. Do this for six X chromosomes.
3. White beads represent the recessive allele for color blindness, *c*. Make six X chromosomes with the recessive bead-allele attached.
4. Do not place any beads on the Y chromosomes.
5. Place side by side the chromosomes and attached alleles that would make up the genotype of a carrier female and a color-blind male. Draw and color this arrangement of chromosomes in your ***Science Journal***.
6. In your ***Science Journal***, draw a Punnett square. Place chromosome models in each of the cells to show the gametes and the results of a cross between a color-blind male and a carrier female. Use colored pencils to draw each chromosome and allele in its correct place.
7. Repeat Steps 5 through 6 for the following crosses:

   a) color-blind male × noncarrier female

   b) color-blind male × color-blind female

   c) non-color-blind male × carrier female

   d) non-color-blind male × noncarrier female

   e) non-color-blind male × color-blind female

Analyze Results

Describe the genotypes and phenotypes of the offspring. Which offspring will be color-blind?

Create Explanations

1. Why is color blindness considered a sex-linked trait?
2. Why is color blindness more common in males than in females?
3. Can two color-blind parents have children who are not color-blind? Why or why not?

Color Blindness

In discussing red-green color blindness in humans, we call the allele for normal color vision *C*. The allele for color blindness is *c*. The alleles are on the X chromosome. The Y chromosome does not carry them. When a parent organism carries only one allele for a certain trait, it is described as **hemizygous**. In the example of red-green color blindness, males would be hemizygous for color blindness because the Y chromosome does not carry a color blindness allele.

A female has two alleles for the characteristic. If she has both the *C* allele and the *c* allele, the normal allele is expressed and her color vision is normal. The genotype for a normal female who carries a recessive allele for color blindness on one of her X chromosomes is X^CX^c. A female with the genotype X^CX^c has normal color vision because she has the dominant allele, *C*. However, she also has the recessive allele, *c*. As a result, she is known as a **carrier** because she carries one recessive allele for this sex-linked trait. Why would a woman with the genotype X^CX^C be known as a noncarrier?

The genotype of a normal male is X^CY. The genotype of a color-blind male is X^cY. How can you use the information in the chart to determine the probabilities of genotypes and phenotypes in the offspring of a female carrier and a normal male?

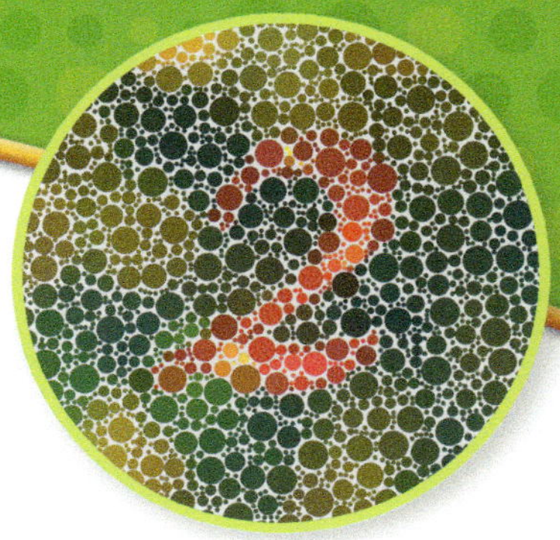

What number do you see in the circle? A person with normal vision will see the number 2. A person who is color-blind will not be able to see the number.

Color Blindness

Genotype	Phenotype
X^CX^c	Carrier female
X^cX^c	Color-blind female
X^CY	Normal male
X^cY	Color-blind male

Polygenic Inheritance Explain

Most inherited traits are more complicated than any of our examples. For one thing, most traits are **polygenic**, which means several genes are involved, and those genes may have multiple alleles. For example, consider a trait that is controlled by five genes. Assume that a person is heterozygous for all five genes. The genotype of this person could then be shown as *AaBbCcDdEe*, with each letter representing a heterozygous condition for each of the five genes. Height and skin color in humans are polygenic traits. At least three gene pairs are involved in height, and several pairs of alleles result in skin tones. In real life, the expression of traits also depends on environmental factors.

Many of the diseases that we worry most about today are both polygenic and environmentally influenced. High blood pressure (hypertension) is not the result of a single allele pair, although a tendency to have high blood pressure does run in families. The tendency is polygenic. The expression of this tendency depends on the environment, including the amount of salt in a person's diet. Some people who are genetically prone to high blood pressure develop it only if they consume a certain amount of salt. Practicing healthful habits can help reduce the risk of developing certain diseases.

Scripture Spotlight

Read **Psalm 139:16**. What does this passage have to do with your traits?

Explore-a-Lab

Structured Inquiry

How is a person's height determined?

Gather graph paper, plain paper, six pennies, and a metric ruler. On the plain paper create a table to track the number of heads and tails you flip during each of the 10 trials, and at the bottom make a row to record the height of 10 individuals. Flip all 6 coins. Each flip represents the genotype of an individual. In your table record the number of heads (dominant allele) and the number of tails (recessive allele) in the column for Flip #1. Repeat this procedure nine more times, recording the results for each set of flips. Use the chart below to determine the height of each individual and record the height in your table.

0 tails and 6 heads = 185 cm	4 tails and 2 heads = 165 cm
1 tail and 5 heads = 180 cm	5 tails and 1 head = 160 cm
2 tails and 4 heads = 175 cm	6 tails and 0 heads = 155 cm
3 tails and 3 heads = 170 cm	

Create a graph that displays your data. What was the average height for your sample? In this model of polygenic inheritance, how many genes determine height? How much genetic information does a parent give to his or her offspring?

Concept Check

Assess/Reflect

Summary: What is complex inheritance? Patterns of inheritance that differ from simple patterns of inheritance are called complex inheritance. Incomplete dominance, such as color in snapdragons, occurs when two alleles interact to produce a blend of the traits they determine. Codominance, such as coat color in horses, occurs when both alleles for a gene are fully expressed. Sex-linked inheritance, such as red-green color blindness, occurs when the allele is present only on the X chromosome. A gene can have more than two alleles, although an individual can possess only two of them. Complex inheritance is the result of multiple alleles. An example is blood types in humans.

1. Explain the difference between incomplete dominance and codominance.
2. Describe some environmental factors that can affect genetic traits.
3. Why do sex-linked traits usually affect male offspring rather than female offspring?
4. A fourth allele that contributes to coat color in rabbits is chinchilla, which can be represented as C^{ch}. This allele is recessive to the wild type allele but dominant to the alleles for Himalayan and albino colors. What would be the possible genotypes of a chinchilla rabbit?

Lesson 4

Essential Question
What Is Genetic Engineering?

God created creatures with many unique features. As you learned earlier, features such as color can even vary within a species. For thousands of years humans have selected certain traits to develop new varieties of plants and animals with desirable traits. What are some plants or animals that were developed in this way?

Today scientists manipulate the genetic material of organisms to create varieties that have new traits. Do you think it was God's intention for us to use the genetic material of living organisms to create new varieties of organisms? Why or why not?

DNA technology can be used to cure diseases and create better food crops.

Would you use a cure for a disease if it were created by manipulating the genes of an organism?

Objectives
- Distinguish between genetic engineering and selective breeding.
- Explain the process of genetic engineering.
- Evaluate the benefits and risks of genetic engineering.

Vocabulary
selective breeding
genetic engineering
genetically modified organism (GMO)

Genetic Engineering Explain

Even in biblical times, humans changed the genetic make-up of organisms through **selective breeding**. It works this way: People choose individual parent organisms that exhibit desirable traits. Examples of desirable traits are large yields in wheat plants and high milk production in cattle. The offspring of the chosen parents often express the same trait as their parents. In that way, new strains or breeds are produced. People decide which plants or animals will breed with each other and so manipulate the recombination of genetic material in the species being bred. No new genes are introduced into any species, and none are taken out. The practice of selective breeding has continued and has become more scientific. Can you think of some recent examples of selective breeding?

Scripture Spotlight

Read **Genesis 30:29–43**. What characteristic of sheep did Jacob use when selecting the animals for his herd? Why do you think he did this?

Transferring Genes

In the 1970s, a new science that went beyond selective breeding was born. **Genetic engineering** enables scientists to select a gene from one kind of organism and insert it into the genetic make-up of another. In this way, traits can be transferred from one species to another. Genetic engineering can also involve the insertion or deletion of genes within individuals of the same species. For example, genetic engineering has been used to try to cure diseases in humans by inserting a normal gene found in humans into a person with a particular genetic disease. This type of genetic engineering has been tried on individuals with a disease known as cystic fibrosis, which is caused by a recessive gene. Individuals with cystic fibrosis produce a thick mucus that clogs the lungs and breathing passageways.

The result of genetic engineering is new genetic types among bacteria, mice, crop plants, and farm animals. These new forms are now being used widely in medicine, scientific research, agriculture, and the drug industry. What fruits, vegetables, and grains do you eat that have been either selectively bred or genetically modified?

How Genetic Engineering Works

The purpose of genetic engineering is to combine DNA in new ways. The process usually begins with enzymes that break apart DNA molecules. Other enzymes then isolate and "snip out" a segment of DNA. A particular chemical environment and a series of temperature changes cause the snipped DNA to copy itself many times. The DNA is then inserted into the genetic make-up of another organism. Sometimes a harmless virus carries the new material into the target cell of the organism. Or the DNA may be injected directly into the cell nucleus.

Either way, when the modified cells divide and reproduce, they copy and carry the transplanted DNA with them. If the transplanted DNA works as it should, the **genetically modified organism (GMO)** expresses a trait received from another kind of living thing. How are GMOs used? What are some of the dangers and benefits of genetic engineering?

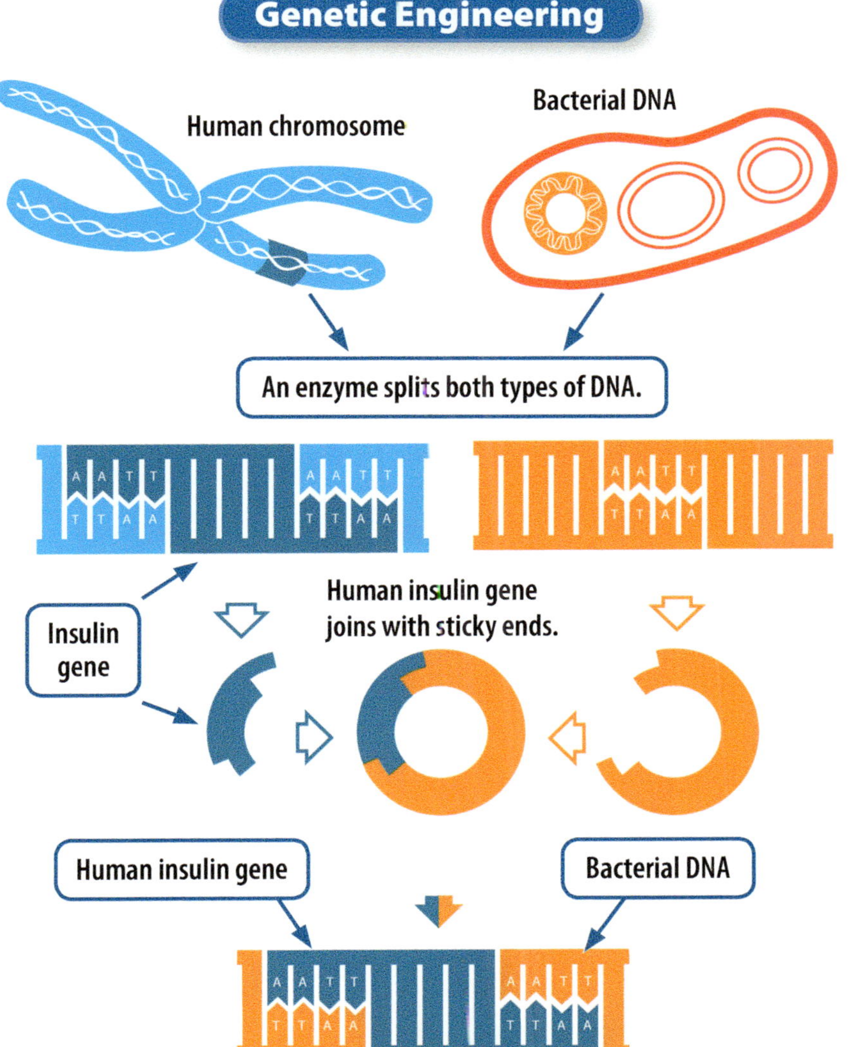

Enzymes can "cut and paste" a human gene into the chromosome of a bacterium to mass produce insulin to treat people with diabetes.

What kind of protein will this bacterium produce? Why?

An estimated 60–90 percent of food consumed in North America has some ingredient with a genetically modified basis to it, because many crops have been genetically modified. Consequently, it is very difficult to purchase foods that are GMO free. Only foods that are specifically labeled as GMO free can be presumed to contain no GMO protein or DNA.

Early Examples of Genetically Modified Organisms

The first GMOs were bacteria that scientists genetically engineered to produce the human hormone insulin in large quantities. Human insulin made by GM bacteria has been on the market since the early 1980s. Many people with diabetes depend on it.

Another early example of genetic engineering was the development in the 1970s of strains of mice for scientific research. In this case, however, instead of adding genes to the mouse DNA, scientists removed, or "knocked out," one of the genes. These "knockout mice" allowed researchers to study how particular genes work.

Faith Connection

God created humans with the power to think and acquire knowledge. However, He also created humans with the power of choice. Genetic engineering is an example where choices can lead to good or evil.

Math in Science

Study the graph below. Compute the percent change in acreage of genetically modified crops that occurred between 1998 and 2005. What does the percent change tell you about worldwide use of genetically modified crop plants? Why do you think this is so?

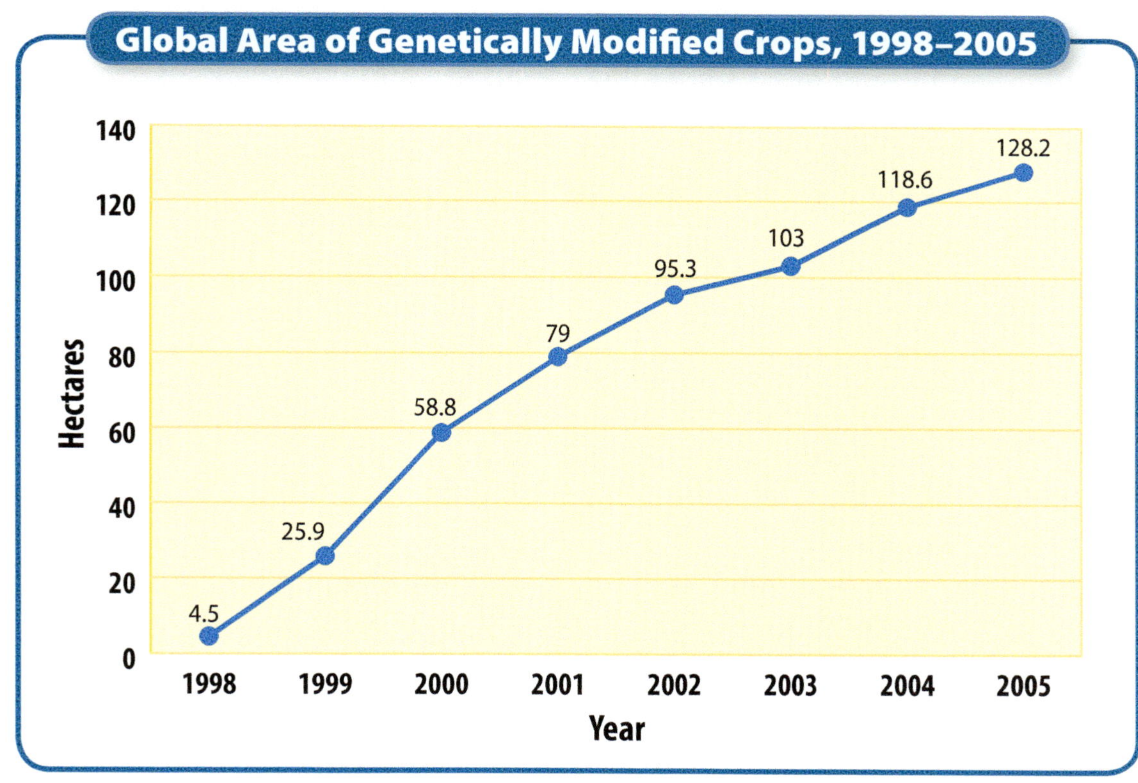

Record your work for this inquiry. Your teacher may also assign the related Guided Inquiry.

What's That You're Eating?

How common are GMOs in the foods you eat?

Procedure

1. Check the food products sold in a supermarket for items that are listed as being GMO free or that claim to use only GMO-free ingredients. Try to identify 10 such products.
2. Check for 10 similar items that do not claim to be GMO free.
3. Compile a class list of both types of food products, eliminating any duplicate finds.
4. Use the class list to check which of these products can be found in your home.
5. Mark all the products on the class list that were found in students' homes.

Analyze Results

What percentages of each type of food product (GMO and GMO free) were found in students' homes?

Create Explanations

1. How common are GMOs in the foods you eat?
2. Present your results to adults, including your parents. Have them identify which products they consume. Ask them for their opinion regarding the consumption of foods prepared with the use of GMOs.
3. Which food products do you think should not be genetically modified? Why?

Explore-a-Lab

Structured Inquiry

How does recombination affect genetic expression in a modified cell?

Cut two long strips of construction paper—one red, one blue. Each will represent a single strand of DNA. On the red strip, equally spaced apart, write the following sequence of bases: A T C C G A T T C A T G C A G G C C. On the blue strip, write the following sequence of bases: T C A T T C A T G G C A T T G C A G.

Assume that you apply an enzyme that cuts a strand of DNA after (to the right of) any A T G sequence. Use scissors to cut the strips apart as the enzyme would. Mark the cut ends. The cut ends of DNA are "sticky." Show how the cut sequences might recombine into one or more strips of genetically modified DNA. Make two new strips with the same sequence and show how a different enzyme might cut the strips in a different place to produce "sticky ends."

Genetic engineering is now contributing to the development of new food crops. Scientists have modified plants to tolerate certain herbicides. Farmers can then apply those herbicides to their crops, which will kill the weeds but not the crop plants. Crop plants have been genetically modified to resist insects, parasites, plant diseases, drought, cold, and salty soil. Some GM vegetables have greater nutritional value than non-GM varieties. The GMO crops grown in the United States include corn, soybeans, canola, squash, and papaya. In Canada, GMO food crops have been produced since the mid-1990s. Corn, sugar beets, soybeans, and canola are the GE crops that Canada has allowed to be produced and then used in conventional food. Certain crop plants have been engineered to produce edible vaccines and other drugs. Farm animals can be genetically modified, too. In 2013, the U.S. Food and Drug Administration approved the first GMO flu vaccine. This vaccine was developed by genetically modifying the genes of a caterpillar so that its cells produce large amounts of flu proteins.

Risks and Benefits

Genetic engineering is controversial. Some people promote its use, while others oppose it. Those who see benefits argue that GM food crops can increase food production while reducing the use of pesticides and fertilizers. They say that drugs produced in GM organisms are both safer and less expensive to produce than medicines manufactured in other ways.

They point out that the use of GM animals and plants in research is yielding possibly life-saving knowledge that could not be obtained any other way.

Opponents of genetic engineering raise concerns about safety. They believe that eating GM foods may cause health problems in humans. They also worry that transplanted genes may be introduced accidentally into other species with negative outcomes. They point out that the effects of GM crop plants on natural ecosystems are unknown. Those who oppose genetic engineering may also object on the grounds of religion, ethics, and morality. They ask, "Is it right for people to change organisms?" What do you think?

Now that you have studied genes, inheritance, and genetic engineering, think about how they are related. How can scientists apply the principles of inherited traits to selective breeding? In what ways can scientists use genetics to help people? What ways might be harmful? **Genesis** tells us that God created the animals. Do you think selective breeding and genetic engineering are consistent with biblical principles? Why or why not? You can research the topic and have a class debate about this question by having two sides take opposing viewpoints.

Summary: What is genetic engineering? Selective breeding is a method used to change the genetic make-up of organisms by selecting individuals to mate with each other. Genetic engineering is a method used to change the genetic make-up of organisms by transferring genes between species or by adding or removing genes from an individual. Organisms whose genetic make-up has been changed by means of genetic engineering are known as genetically modified organisms (GMOs). Such GMOs are used to produce new crops and medicines.

1. What is the difference between genetic engineering and selective breeding?
2. Explain how the DNA from two different kinds of organisms can be combined.
3. How does genetic engineering make the production of a human hormone safer, cheaper, and faster? Is genetic engineering always beneficial? Explain.
4. What information do you need to help you decide whether the benefits of genetic engineering outweigh the risks? Be specific.

People in Science

Extend

Get to Know
Gregor Mendel

Gregor Mendel, a monk, teacher, and biologist, is known as the "father of modern genetics." His now-famous experiments on pea plants were the basis for the development of the field of modern genetics.

Mendel was born in Austria and lived on his family's farm, where his fascination with fruit trees and gardening began. He excelled in his studies, especially physics and math. Upon completing studies at the Philosophical Institute of the University of Olmütz, he entered the monastery. During this time, Mendel began his famous pea plant experiments.

Before Mendel's findings, people believed traits were inherited equally from each parent. Through his experiments with pea plants, Mendel discovered the existence of dominant and recessive traits in all organisms. Mendel determined that dominant and recessive traits passed on randomly from parents to offspring based on the mixing of genes from both parents. This is known as the law of segregation.

While researching, Mendel recognized that traits are transmitted to offspring independently of one another. To explain this, he formulated the law of independent assortment. These traits are passed on independently and are not affected by other traits or the traits of the parents.

Today we are able to explain genetically linked disorders and diseases and even engineer plants due to Mendel's work. However, during Mendel's lifetime, much of his work was disregarded by other scholars. His work laid the foundation for genetics and gene therapy and is the basis for many other medical discoveries.

Concept Check

1. How did Mendel's experiments change people's belief of how traits are inherited?
2. How do Mendel's law of segregation and law of independent assortment differ?

Careers in Science

Extend

Genetic Counselor

Genetic counselors are health professionals who specialize in the study of genetics and counseling. Many people who enter the field of genetic counseling have backgrounds in biology, genetics, nursing, psychology, public health, and social work.

A genetic counselor is trained to help people with a wide range of situations that involve human genetics. Some areas they focus on are infertility, cardiovascular genetics, familial cancer-risk counseling, hematology, neurogenetics, pediatric counseling, and prenatal counseling. Genetic counselors also educate individuals and communities about genetic diseases, potential risks, and treatments. They often work closely with physicians and other medical professionals.

If you or someone you know has a disorder or disease that runs in your family, a genetic counselor can help you understand your risk or the risk for other family members. The genetic counselor can recommend testing, explain prevention strategies, refer families to community organizations, or provide research that may help your situation.

Many genetic counselors earn a master's degree in genetic counseling. As part of their training, they may take courses in anatomy, reproductive genetics, research skills, clinical studies, human genetics, chronic illness, molecular biology, health behavior, and interdisciplinary care. They must be well educated in a variety of topics to help people to the best of their ability.

Concept Check

1. Why might someone seek the help of a genetic counselor?
2. Why is it important for genetic counselors to work closely with physicians and other medical professionals?
3. What kinds of questions do you think a person with a family history of cancer might ask a genetic counselor?

Chapter 4

Review

Assess/Reflect

Study Guide

Lesson 1

1. DNA is a double helix composed of a sugar called deoxyribose, a phosphate group, and a nitrogenous base pair. The nitrogen bases are composed of adenine, thymine, guanine, and cytosine.
2. A segment of DNA is composed of a sequence of nucleotide pairs. It is this arrangement of the base pairs that provides a code of instructions.
3. A segment of DNA that codes for a single protein is a gene. The genes that an organism inherits from its parents determine the organism's characteristics.
4. Nuclear division of body cells is called mitosis and produces cells that are exact copies of the original cells. Meiosis is the nuclear division of reproductive cells, called gametes. During meiosis, two episodes of division occur, producing four cells. Each resulting cell contains only half of what the original cell contained.

Lesson 2

1. Alleles are different forms of a gene. Dominant alleles are represented by an uppercase letter. Recessive alleles are represented by a lowercase letter.
2. A homozygous condition consists of two dominant or two recessive allelles. A heterozygous condition consists of one dominant allele and one recessive allelle.
3. An organism's genotype is its genetic make-up for a particular gene. Its phenotype is its observable or measurable characteristics.
4. If one dominant allele for a given characteristic is inherited, the dominant phenotype will be expressed. Recessive alleles must be present in a pair to express the phenotype.
5. A Punnett square is a convenient tool for predicting the genotypes and phenotypes inherited by complete dominance. See examples of Punnett squares in the lesson.

Lesson 3

1. Incomplete dominance is a form of inheritance where neither allele for a specific trait is dominant over the other. This results in a third phenotype that is a blend of the two alleles. In codominance, more than one dominant allele is expressed in the phenotype. A sex-linked trait is determined by the alleles on a sex chromosome.
2. Multiple-allele inheritance involves more than two alleles.
3. The tendency toward a disease is inherited but the expression of the tendency depends on the environment.

Lesson 4

1. Selective breeding is the process by which individuals with specific traits are chosen to be parents of the next generation. Genetic engineering is the transfer of genetic material from one cell to another.

2. Enzymes break apart DNA molecules. Other enzymes then isolate and "snip out" a segment of DNA that contains a particular gene. The snipped DNA makes multiple copies of itself. The DNA copies are then inserted into the genes of another organism. When the modified cells divide and reproduce, they copy the inserted DNA. The genetically modified organism (GMO) expresses the gene it received from the other organism.
3. Genetic engineering offers specific benefits, including improved medical treatments and agriculture. Risks include safety concerns, unknown health problems, and the accidental introduction of new genes into ecosystems.

Vocabulary Check

Fill in the blanks.

1. __________ is the study of how traits are inherited.
2. A picture of chromosomes arranged by size and shape is a __________.
3. A(n) __________ is a trait that is always visible.
4. If the two alleles for a trait are different they are called __________.
5. You can determine a person's genotype by observing the __________.
6. Transferring genes from one species to another is __________.

Multiple Choice

Choose the best answer.

7. Which is an example of a codon?
   A. ACG
   B. ACGTGGCATG
   C. thymine-adenine
   D. deoxyribose-phosphate group
8. A gene is a segment of DNA that codes for the production of what substance?
   A. chromosome
   B. protein
   C. sugar-phosphate backbone
   D. genotype
9. What can you use to determine the odds of offspring inheriting a specific trait?
   A. Alleles
   B. Simple inheritance
   C. Selective breeding
   D. Punnett square

Check Point

Answer the following questions.

10. Suggest two important functions for enzymes in genetic engineering.
11. Decide whether you agree or disagree with the following statement and give two reasons to support your opinion. *Genetic engineering will someday produce crop plants that will end world hunger.*
12. A black cat mates with a tan cat, and all of their kittens have black and tan stripes. Does this result from incomplete dominance or codominance? Explain.
13. A father has type B blood, and a mother has type O blood. Their child has type O blood. What can you **infer** about the father's genotype? Explain.
14. Duchenne muscular dystrophy is a sex-linked disorder. If a father has a normal X chromosome and the mother is a carrier, what are the odds that a male child would have Duchenne muscular dystrophy? What are the odds for a female child to have the disease or be a carrier?

Unit 2 The Human Body

Unit Overview

Chapter 5 begins by describing the two main forms of reproduction. The lessons focus on how you grow and develop, starting even before your teenage years. As you grow, you begin to look more like an adult. You get better at solving your own problems. Your friendships change, and your emotions change quite often, as you may already know. You also become able to have children.

Being able to have children is a huge responsibility. You must decide whether to become sexually active and risk becoming a parent too soon. Chapter 5 will help you weigh the risks of sexual activity.

Chapter 6 focuses on how diseases spread from person to person. Your body helps you fight off most diseases. Staying disease-free is a problem in today's world. Travel can—and does—quickly spread infectious diseases from one part of the world to another. The guidelines in this chapter will help you stay healthy.

In Chapter 7, you will discover how the digestive and excretory systems function and why taking care of these systems is important.

How can the relationships between these young people affect them? How might they affect other people?

ADVENTISTS InStep for Life

Your teacher may assign an Open Inquiry lab and a Lifestyle Challenge activity. Use your ***Science Journal*** to record your work.

Science Journal

Human Development and Sexuality

Scripture Spotlight

Adolescence can be an exciting time in a young person's life. Throughout this chapter you can strengthen your faith by connecting what you are learning to your faith. You will read the following passages in this chapter.

Psalm 46:1–3 (p. 185)
Genesis 2:18–24 (p. 187)
Matthew 7:12 (p. 193)

The Big Idea

Becoming an adult involves physical changes in your body and changes in the way you interact with others.

Why do you think many people describe this period in life as one of the most challenging?

Inquiry Kick-Off Engage

You're in the stage of life when people move from childhood into adulthood. What worries you about this stage? Will your concerns be similar to your classmates' concerns? In your ***Science Journal,*** you will investigate some of these challenges.

Growth and development bring about many changes, along with learning and wisdom.

Objectives

- Describe the importance of asexual and sexual reproduction.
- Compare and contrast mitosis with meiosis.
- Explain the difference between adolescence and puberty.
- Describe what happens at puberty.

Vocabulary

budding
meiosis
zygote
testosterone
estrogen
primary sex characteristic
secondary sex characteristic

Essential Question

How Do Organisms Reproduce?

No organism is immortal. Reproduction is essential for the continuation of every species. Did you know that a pet mouse can reproduce quickly, breeding as often as every 20 days? A mouse, which has an average lifespan of 18 months, produces an average of 10 babies per litter. How many babies could an adult female mouse have in one year? How many children could a human woman have in her lifetime? What factors might limit the number of children that she can produce?

During most forms of reproduction, genes are exchanged to provide genetic diversity in the offspring. What genetic diversity can you observe in dogs? How does genetic diversity help an organism survive?

Asexual Reproduction Explain

Some organisms reproduce through *asexual reproduction*. The offspring are produced by *mitosis*, the process in which a single cell divides, resulting in two identical cells. How does mitosis occur? How could you use several colors of modeling clay to represent mitosis? During mitosis, no sex cells are involved. The new organism inherits genetic information from only one parent. As a result, the daughter cells that form are identical to the parent.

Have you ever seen a potato like the ones pictured here? New potato plants sprout from the "eyes" of the potato. These new sprouts are the result of asexual reproduction. For each sprout, one cell in the potato started the process by dividing. These new cells also divided. As the cells continue to divide, new stems, leaves, and roots develop. In time, the new plant will be able to grow on its own. Why

Each "eye" on this potato sprouts.

How can you use this information to grow more potatoes?

do you think God created some organisms to be able to reproduce on their own? Why do you think He didn't create all creatures with this ability?

Organisms can reproduce asexually in several ways.

- During **budding**, an offspring grows out of the body of the parent.
- Another form of asexual reproduction is *fragmentation*, in which the parent breaks into distinct pieces that become offspring.
- During *regeneration*, a damaged or lost piece of the organism can be regrown.

Asexual reproduction benefits the organism in many ways. Organisms that reproduce asexually do not need to expend energy to look for mates or produce offspring. What environments might be best suited to asexual reproduction? What organisms reproduce through asexual means? What do you think are some disadvantages of asexual reproduction? Asexual reproduction produces offspring that are genetically identical. Do you think this is an advantage or disadvantage to the organism. Why?

A sea star can grow a lost arm through regeneration.

Explore-a-Lab

Structured Inquiry

How do you know when budding has occurred?

Hydra are jellylike organisms slightly less than 1 cm long. Their entire reproductive process takes place over three to four days. Place a hydra on a depression slide with a few drops of water. Do not place a cover slip over the specimen. Using a microscope, observe the hydra and note the size of the bud. Observe several more budding hydra and compare the sizes of the buds. Record your observations in the form of drawings. Review the steps that follow. Try to determine which stage the budding is in for each hydra.

1. **Beginning of Bud**—This is the first step of the hydra's asexual reproduction.
2. **Tentacles Begin to Grow**—The tentacles and the mouth of the new hydra begin to develop.
3. **Beginning of Separation of New Hydra**—The new hydra is generally around one half the size of the parent.
4. **Breaking Off of New Hydra**—The bud will become a new hydra, identical to its parent.

Record your work for this inquiry. Your teacher may also assign the related Guided Inquiry.

Asexual Plant Growth

Can a new plant start from a cutting or stem?

Procedure

1. Label one plastic cup *A* and one plastic cup *B*.
2. Place half of the perlite into each plastic cup. Add about 20 mL of water to the perlite, making sure that all the perlite is evenly wet.
3. Place each plant cutting in a plastic cup with the water and perlite.
4. Place 5 mL of root hormone into plastic cup *B*.
5. Observe each plant every day for 2–3 weeks.
6. **Measure** the length of each plant cutting every other day for 10 observations. **Record** the data in the table in your ***Science Journal***.

Materials
- 2 plastic cups
- 0.5 L perlite
- 2 plant cuttings of household geraniums
- root hormone
- 10-mL graduated cylinder
- ruler

Observation	Length of Plant Cutting, Cup *A*	Length of Plant Cutting, Cup *B*
1.		
2.		
3.		
4.		
5.		
6.		
7.		
8.		
9.		
10.		

Analyze Results

Graph the data obtained for plant length for each plant during the period of observations. Use a different color for each plant.

Create Explanations

1. Can a new plant start from a cutting or stem?
2. What are the advantages of the root hormone in propagating plant cuttings?
3. What is the advantage of asexual reproduction for the plant investigated in this lab?

Sexual Reproduction Explain

Recall that many species, including humans, reproduce sexually. Sexual reproduction occurs when gametes—*eggs* and *sperm*—fuse to form new offspring. Gametes form during **meiosis**, a type of cell division that reduces the number of chromosomes to half the original number. Females produce eggs, and males produce sperm. When an egg and sperm unite, they form a **zygote**, the initial cell of a new organism produced during sexual reproduction.

Reproduction in Plants

Flowers contain the reproductive organs of most vascular plants. Recall that eggs form in the *ovary* of a flower's female reproductive structure, called a *pistil*. The male reproductive structures of a flower, called *stamens*, produce the sperm. Each of the tiny *pollen grains* made by a stamen contains a sperm cell. Before fertilization can occur, pollen grains must be transferred to a pistil. You might recall that this process is called *pollination*. What factors can aid pollination? During pollination, the sperm then move through the pistil and into its ovary, where they fertilize the eggs. After fertilization, cell division with mitosis causes the zygotes to develop into seeds. What happens to the flower after fertilization? Think about what conditions would need to be in the plant's environment for the seeds to grow into plants.

The flowers of some plants have both male and female parts and can fertilize themselves. Will the plants that result still have diversity? Other plants must be pollinated by another plant of the same type to form seeds. The flowers of a few plants, such as holly bushes, produce only one type of sex cell in each plant. Thus, a holly bush is either male or female. A male and female holly bush must grow near each other for the sperm of one bush to fertilize the eggs of the other bush.

This bee is covered in pollen.

How does the bee get this pollen on its body?

Explore-a-Lab

Guided Inquiry

How can you scientifically show asexual reproduction?

Working in groups, choose one of the following plants to demonstrate asexual reproduction: ivy (from stems), garlic (from bulbs), carrot (from root top), or potato. Work out a method to propagate your plant. Write out the steps of your plan and then conduct your experiment.

Prepare a presentation to explain your scientific approach and the results of your investigation. How is the method you used to grow new plants different from how plants are produced during sexual reproduction?

Human Reproduction

Humans also reproduce through sexual reproduction. A human zygote grows and develops into a baby. The mother contributes genetic information for the baby in the egg, and the father contributes genetic information in the sperm. Sexually produced offspring are never identical to either of their parents. Based on this information, what are the advantages of sexual reproduction over asexual reproduction?

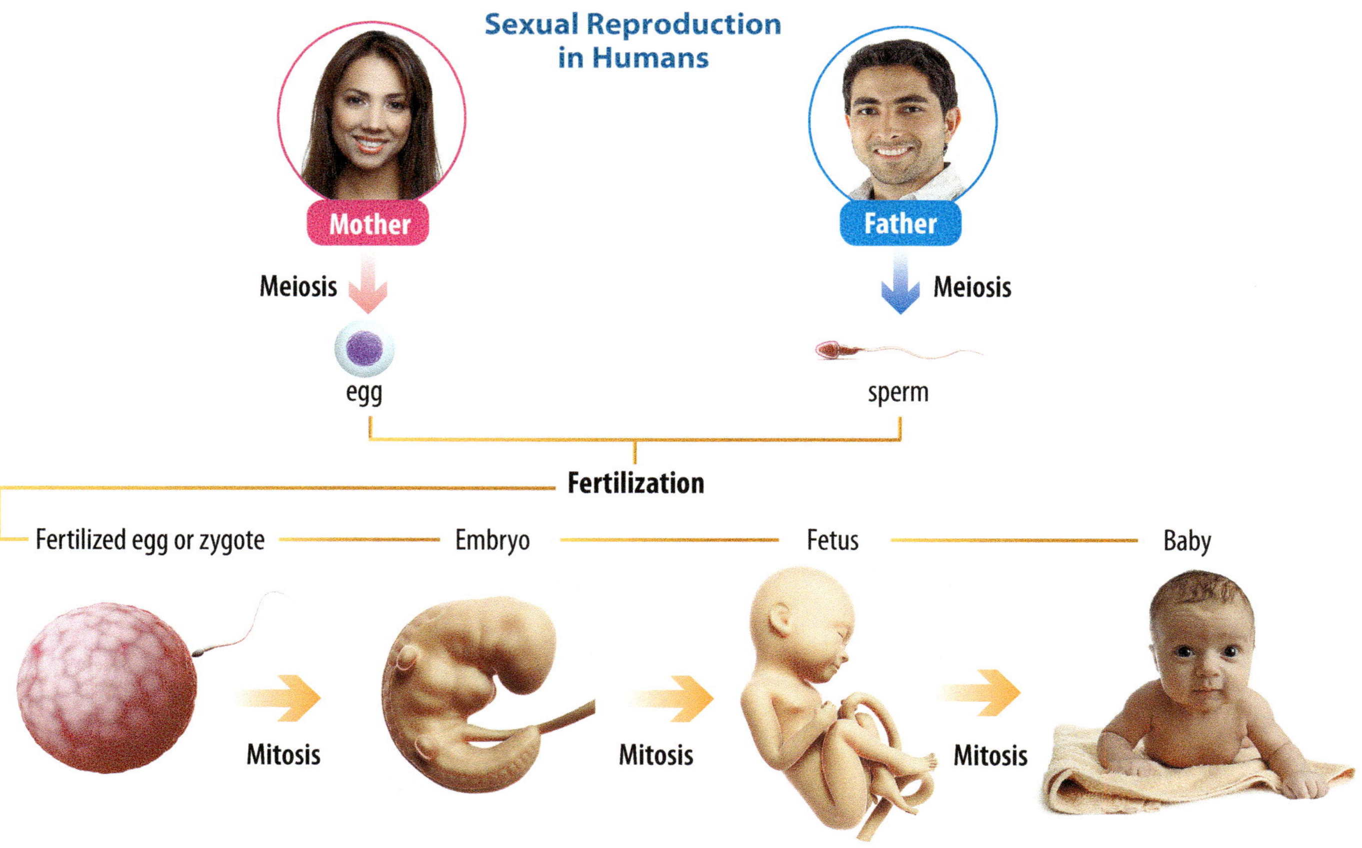

God gave each of us our own unique personality with the ability to think and choose for ourselves.

What is the benefit of diversity for humans?

Sexual Maturity

As you know, once you have completed the growth stages of infancy and childhood, you enter *adolescence*, the stage of development during which children become adults. Adolescence begins between the ages of 8 and 13 in girls and between the ages of 9 and 15 in boys. Adolescence ends when the body stops growing. This marks the transition to adulthood. During the transition through adolescence, how does the body change?

Reproduction is a gift from God, but the human body must mature before it is capable of reproduction. *Puberty* is the time in life when your body develops sexually, enabling reproduction. You will notice that at this time, your body grows faster than at any other time, except at infancy. Boys' bodies start producing sperm, while girls' bodies start the process that will produce eggs. You may sometimes feel as if no one else ever had to face these confusing changes before, but everyone does, including your parents and grandparents. Why are the changes of puberty necessary? Why might you feel you are living in a new body during puberty?

Changing from a child to an adult is a complex process. It is controlled by your endocrine system. Recall that the endocrine glands produce hormones. As the body's chemical messengers, hormones transfer information and instructions from one set of cells to another. Many different hormones move through the bloodstream, but each type of hormone is designed to affect only certain cells. How are hormones like traffic signals?

Puberty is initiated by the pituitary gland, a pea-shaped gland located at the bottom of the brain. This gland releases hormones, which affect growth, sexual development, and metabolism. How does the pituitary gland know to release hormones that initiate puberty? The hormones travel throughout the body in the bloodstream. In boys, the hormones target the *testes*, the egg-shaped glands in the scrotum. The hormones cause the testes to produce **testosterone**, a male sex hormone. Testosterone is responsible for the development of male sex organs and the production of sperm. In addition, testosterone produces secondary sex characteristics, such as voice deepening and growth of facial and body hair. What are additional secondary sex characteristics in males?

The hormones released by the pituitary gland also affect girls. These hormones target the *ovaries*, a pair of organs located on either side of the uterus. The ovaries produce **estrogen**, the female sex hormone. Estrogen is responsible for the development of female sex organs and the production of eggs. *Ovulation* occurs when a mature egg is released from the ovary, is pushed down the fallopian tube, and is available to be fertilized. The onset of *menstruation* is the shedding of the lining of the uterus. Menstrual periods last three to five days and are characterized by the flow of menstrual blood from the uterus through the cervix and out the vagina. In addition, estrogen causes secondary sex characteristics at puberty, such as widening of the hips and swelling of the breasts. What is the benefit of secondary sex characteristics in females? Why do a female's hips widen as she matures?

Scripture Spotlight

Psalm 46:1–3 tells us that no matter what is happening in our lives, God cares for us and is in control.

Check out your ***Student Journal*** for a Guided Inquiry that explores sexual and asexual reproduction using models.

Extend

Primary and Secondary Sex Characteristics Explain

Faith Connection

From creating us with an amazing reproductive system to numbering the hairs on our head, God cares for each of us personally. See **Matthew 10:29–31**.

A sex characteristic is any physical or psychological trait that is typically associated with sexual reproduction or sexual differentiation. Any organism—even a flower—that reproduces sexually has sex characteristics. What might be sex characteristics in a plant? Scientists separate sex characteristics into two categories, primary and secondary. Some characteristics are actively involved in reproduction, while the others are simply associated with a specific sex. Why do you think it may be important to have secondary sex characteristics that are not involved in reproduction?

Primary sex characteristics are associated with the production of gametes, which enable sexual reproduction. These characteristics can be external, such as the pistils of a flowering plant, or internal, such as the ovaries found in female mammals. What might be the primary sex characteristic of a cactus?

Secondary sex characteristics are not actively involved in reproduction. Rather, these characteristics differentiate between members of the opposite sex within a species or attract members of the opposite sex. Male peacocks, for example, use their elaborate feathers in this manner to attract female peacocks as partners. What are secondary sex characteristics of a platypus, a cow, or a silverback gorilla?

Male secondary sex characteristics in humans include facial hair and the Adam's apple. Men also typically have shoulders that are wider than the hips and bodies that are more muscular than the bodies of females. Secondary sex characteristics in human females include developed breasts (or mammary glands). Women also typically have wider hips than shoulders.

Lesson Activity

Working as a group, select from one of the following animals: mouse, cat, panda, or elephant. Use the Internet to research the different stages of sexual maturity for the selected animal. Compare and contrast the stages of maturity for the selected animal with human sexual development. Prepare a brief presentation to summarize your findings. After the presentations, gather as a class to discuss the information presented.

How do the stages of development of your animal compare to those of humans?

Your Changing Body

Have you noticed your friends and classmates suddenly getting taller? The rapid growth can seem to appear "almost overnight." For both boys and girls, a growth spurt can add 10 cm or more in height in one year. Because girls tend to enter puberty before boys, they are often taller than boys the same age, at least until the boys catch up. Some parts of the body, such as the hands and feet, grow quickly, but an adolescent's muscles may not be ready for all these changes. This can lead to awkwardness and clumsiness. What are some ways that you can help minimize awkwardness as your body shape changes? In time, the changes will end, and your body will have time to adjust to its adult size.

Scripture Spotlight

Genesis 2:18–24 tells us that God created male and female.

What Smells?

Hormones also cause your sweat glands to become more active, and now, unfortunately, the sweat smells more, too. Why might sweat odor increase during puberty? You can take care of this problem by showering more often, especially after playing sports. Daily showers are an essential part of your health as you mature and your body changes. Deodorant and antiperspirant can help relieve the odor and reduce the amount that you sweat.

Mood Swings

As you grow, your endocrine glands are releasing hormones at an unsteady rate. How do you think the fluctuation in hormones affects how you feel? Do you sometimes find it harder to control strong emotions? You might have noticed how your friends have quick mood changes. One minute they're happy, but the next minute they ignore you. Have you experienced similar fluctuations in your mood? What can you do to pull your emotions into order?

Skin Challenges

During puberty, oil glands in the skin become larger and produce more oil. When this oil combines with dead skin cells, it can clog the pores in your skin. You know the result: pimples called acne. Squeezing or picking at pimples can lead to infection. Keeping your skin clean is the best way to avoid acne, and medicated treatments can help stubborn cases.

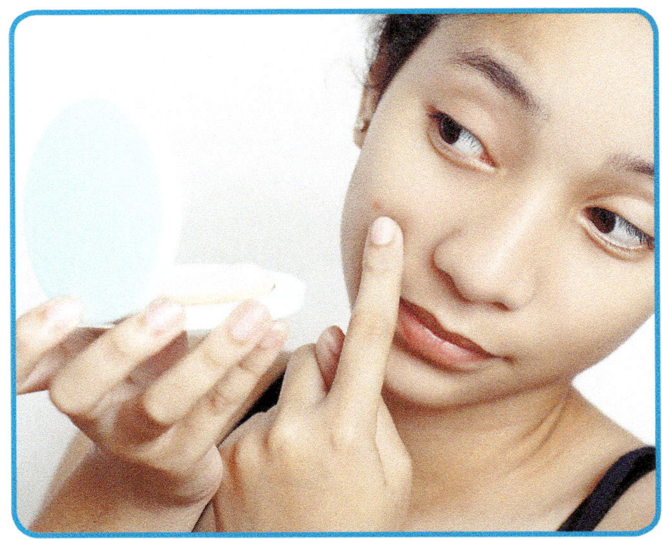

This girl is concerned about the skin changes caused by puberty.

What advice can you give her?

Math in Science

Most adults are within 5 cm of the average of their parents' heights. Use the equations below to predict how tall Chloe and Jacob will be as adults.

Chloe's dad is 1.85 m tall. Her mom is 1.68 m tall. Jacob's father is 1.78 m tall. His mom is 1.6 m tall.

Equation for girls:

$$\frac{(\text{Father's height} - 13\text{ cm}) + \text{Mother's height}}{2}$$

Equation for boys:

$$\frac{(\text{Father's height} + 13\text{ cm}) + \text{Mother's height}}{2}$$

Concept Check Assess/Reflect

Summary: How do organisms reproduce? All living organisms reproduce either asexually, sexually, or both. Asexual reproduction occurs when cells duplicate through budding, fragmentation, or regeneration. The offspring produced during asexual reproduction are genetically identical to the parent. Sexual reproduction involves the production of sex cells (eggs in females and sperm in males) through meiosis. A zygote forms when sex cells fuse. Sexual reproduction can occur in plants and animals. As humans mature, the body develops and gains the ability to reproduce. Puberty marks the period of development during adolescence that leads to adulthood. Hormones produced by the pituitary gland initiate this development. The gland targets the testes in males, which produce testosterone, and the ovaries in females, which produce estrogen. Sexual maturity also results in the development of sex characteristics, any physical or psychological trait that is typically associated with sexual reproduction or sexual differentiation. The changes initiated by hormones lead to growth spurts, mood swings, and changes in body chemistry that lead to body odor and skin problems.

1. How does ovulation relate to menstruation?
2. In what way is sexual reproduction different from asexual reproduction?
3. Explain the role of the pituitary gland in the episode of human development called puberty.
4. What are secondary sex characteristics in humans? What are their functions?

Essential Question

What Are Challenges of Sexual Maturity?

Some of the most noticeable changes that occur during puberty happen to your body, but these are not the only changes that you are experiencing. Have you also noticed changes in your interactions with others? Do people suddenly seem to behave differently than they did in the past? Do you sometimes feel that you want to be left alone? Do the activities that you once enjoyed with your friends now seem childish? During puberty, your relationships will change, too. What is causing these changes?

Objectives

- Describe the intellectual, emotional, and social changes that occur during puberty.
- Describe several types of abuse.
- Describe ways to avoid sexual violence.
- Suggest ways to cope with the changes that occur during puberty.

Vocabulary

cognition
risk taking
bullying
incest
sexual harassment
rape

Physiological Changes Explain

Puberty and adolescence are a time of change. Generally, puberty begins in girls between the ages of 8 and 12, and in boys between the ages of 9 and 14. The changes that take place during puberty are controlled by your endocrine system. The endocrine system helps to regulate mood, growth and development, tissue function, metabolism, sexual function, and reproductive processes.

Puberty produces more than just the physical changes in the body that are easily observed. During puberty, hormones increase an adolescent's sex drive. In addition, adolescents may become moody and more sensitive about their self-image and how they are perceived. What are some changes in yourself and your friends that you have noticed? How have these changes affected the way you interact with one another? You cannot control the changes or the rate at which the changes happen. You can control how you respond to them. How do changes in your body affect the way you think, feel, and interact with others?

Intense emotions and mood swings can be common during adolescence.

Can you relate with the emotions shown by this teenager's expression?

How the Brain Matures

Like your body, your brain also goes through changes as you mature. You gain in **cognition**, the ability to gain knowledge and comprehension. For example, when young children see someone cry, they assume that person is hurt or sad. Now that you are older, you realize that this person might actually feel very happy or proud. What are some other ways you see things differently now than you did as a child? Why are these changes happening?

Consider the changes in your brain as being similar to the physical changes in your body. As your body grows and develops, you may feel awkward and uncoordinated. At the same time, the changes in the brain do not necessarily mean an adolescent will suddenly begin making wise decisions. Adolescents need experience and practice to strengthen their decision-making skills. They might not even realize that they are not following their own standards. Have you ever felt unwelcome by friends at a sporting event? For example, some friends sit together at the game every week. Lately, they use cold looks to discourage others from sitting with them. At the same time, they complain about not being welcome with the "popular" kids. In time, this group might become aware of the mismatch between its actions and words. Do you think the "popular" kids are experiencing the same feelings of isolation and exclusion? Why or why not?

During adolescence, young adults begin to experience increased independence. These episodes of independence provide the opportunity to begin to make important decisions. The independence is not without peril. Decision making is a learned skill that takes practice to do well. There will be some mistakes during the practice.

During adolescence, young adults learn to make good decisions.

How do puberty and adolescence help prepare you to make wise decisions as an adult?

Record your work for this inquiry. Your teacher may also assign the related Guided Inquiry.

Investigating Bullying

How common is bullying on the playground?

Procedure

Materials
- Access to playground

1. With a partner, visit the playground at your school when younger students are playing.
2. **Observe** how young children interact with one another for at least 15 minutes if possible.
3. Look for examples of bullying behavior. For each bullying behavior you observe, describe what you see, classify the type of bullying behavior as either verbal bullying or physical bullying, and finally **record** whether the bully is an individual or a group of students. Record your observations and data in the chart provided in your ***Science Journal***.
4. Repeat Steps 1–3 two more times. Try to make your observations on different days.

Date	Bullying behavior observed	Type of bullying		Bully	
		Verbal	Physical	Individual	Group

Create Explanations

1. How common is bullying on the playground?
2. How did the students you observed interact?
3. What was the most common type of bullying behavior you observed?
4. What could you do to help eliminate bullying at your school?

Steps in Decision Making
1. Explore the alternatives.
2. Choose from among the alternatives.
3. Choose after considering the consequences.
4. Choose after considering what God says.
5. Consider what is most important to you.
6. Affirm your decision by speaking to others about it, when appropriate.
7. Act on your decision.

As young people begin to make more decisions, some of them take risks. **Risk taking** means acting in ways that may have unpleasant or dangerous results. In general, a risk is something that you may not be able to recover from. What are some risky things that you have seen or heard that other teens have done? Risky behaviors can result in expulsion from school, arrest, pregnancy, illness, injury, or even death. As young people mature, they learn to consider the consequences of risky behavior. How do you evaluate risks? Can you think of a situation where an everyday activity can increase in risk? How do you avoid those situations?

Young people who do not consider risks may make unwise decisions that can affect their entire lives. Staying true to biblical values will help you make responsible decisions and avoid these dangerous risks that could have lifelong consequences. What are risks that adolescents face? What are some risky behaviors that you have already encountered? How did you deal with these experiences? How will these experiences help you prepare for future scenarios when you are exposed to new risks?

Lesson Activity

The ability to measure risk is very important. Rotate a piece of notepaper to landscape. Write these five categories horizontally across the top of the page: Death, Dangerous, Risky, Problems, and Secure. Place the following activities under the appropriate category: skydiving, walking in a city park, underage driving, consuming alcohol, listening to music, eating high-fat foods, and sleeping eight hours.

Where did you rank each activity? Why?

Bullying Explain

In addition to the normal stress many young adults feel during adolescence, many adolescents worry about being bullied. What are some examples of bullying? How would you use these characteristics to define bullying? **Bullying** is an intentional act that causes harm to others and may involve verbal harassment, verbal or nonverbal threats, physical assault, stalking, or other methods of coercion, such as manipulation, blackmail, or extortion. A bully seeks power or attention by physically or verbally abusing others. For school-age children, bullying is unwanted, aggressive behavior that involves a real or perceived power imbalance. The behavior is repeated or has the potential to be repeated over time. Both the child who is the victim of bullying as well as the bully may suffer from serious, lasting problems. Where might bullying occur?

Bullying is more than an altercation in the school yard or hallway. Some bullies use the Internet or mobile devices to spread cruel gossip. Schools and parents are becoming more aware of the harm caused by bullying, and a growing number of schools have a "no tolerance" policy concerning it. Control of your emotions when being bullied is important. Don't be controlled by others. Learn to live above the bullying. Don't let it affect your self-respect. What else can you do to protect yourself and your classmates from bullying and abuse? Does your school have a person to whom you can report bullying incidents? Both at school and at home, who is a safe person to talk to about difficult topics like bullying?

How can you cope with the emotional roller coaster of adolescence? Remember that God loves you and understands what you are going through. Also realize that everyone your age is riding the same roller coaster. Don't forget that your parents and other trusted adults once rode this roller coaster, too. Talk to them about any feelings that trouble you, including facing bullies.

Scripture Spotlight

How does the Bible say we should treat other people? Read **Matthew 7:12**.

Math in Science

Researchers have determined that most instances of bullying occur between the ages of 9 and 13. About 70% of these instances are verbal, 50% are emotional, and 31% are physical. One large middle school had 136 reports of bullying last year.

- How many instances were likely to be verbal? Emotional? Physical?
- Which two percentages should add up to 100? Why don't they?
- Why don't all three percentages add up to 100?

Lesson Activity

Journaling is a way to express our emotions in a safe way. After putting your thoughts on paper, you can share your journal with a trusted adult to seek help or feel relief. Take a moment to write your thoughts to one of the three following prompts.

1. Were you ever bullied when you were younger? How did this make you feel? Explain the situation.
2. Did you ever act like a bully yourself? How did this make you feel? Explain the situation.
3. Have you seen a schoolmate being bullied? What did you do? Would you behave differently now?

Abuse

Bullying is not the only form of harassment that young adults might experience. Abuse is the misuse of power to control the behavior of another person. Abuse can be physical, psychological, sexual, verbal, or a combination of any or all of these. What environments or situations could allow an abusive relationship to continue? Abuse can also be neglect, which is when parents or guardians do not take care of the basic needs of the children who depend on them. Does your school have a policy on abuse? If so, what does it include?

Physical Abuse

Physical abuse can take many forms, such as pushing, shoving, slapping, kicking, punching, hitting, spitting, pinching, pulling hair, choking, throwing things, hitting victims with an object, and using or threatening to use a weapon. What steps can you take to end or help a friend end an abusive relationship? What can you do if you witness physical abuse at your school? What can you do if you witness physical abuse at a public space?

Psychological Abuse

Psychological abuse can cause anxiety and depression, which can lead the victim to withdraw from everyone. This type of abuse can include insulting your family or friends; ridiculing your beliefs, race, or religion; using constant put downs; threatening suicide if you leave; keeping you prisoner in your home; threatening to take the children if you leave; and threatening to have you deported. Do you think it is easier to spot signs of physical or psychological abuse? What can you do if you suspect a friend is a victim of psychological abuse?

Sexual Abuse

In sexual abuse, one person forces another person into sexual acts. Often the abuser is a family member or friend living with the family. It can even be a parent, teacher, or pastor. When the abuser is a family member, the sexual abuse is called **incest**. Usually the abuser wants to keep the sexual activity a secret and threatens to punish or even kill the victim if he or she tells. When the abuser is someone the child loves or trusts, the threat to withhold love may be enough to secure silence. This causes fear, hurt, and even despair to the abused. Often the abused is made to feel guilty, even though it is never the victim's fault. The abuser is the one who is guilty.

Warning Signs of Potential Sexual Abuse
• Emotional or verbal abuse
• Taking control, such as demanding that you stop seeing certain friends
• Excessive jealousy, suspicion, or mistrust
• Heavy drinking or drug use
• A history of childhood violence—having been abused by parents
• Inability to handle frustrations
• Cruelty to animals
Protective Steps against Sexual Abuse
• Avoid being alone with anyone who makes unwanted sexual advances.
• Do not make friends with those who use alcohol or drugs.
• Avoid dangerous areas.
• Be home at a reasonable time.
• Contact your local social services if you are the target of sexual abuse or harassment.
• Pray for God's protection.

Sexual Harassment

Sexual harassment is unwanted sexual advances. It may take many forms, including physical contact (grabbing, pinching, or unwanted kissing), sexual comments (such as name calling), sexual propositions, or unwanted communication (e-mails, text messages, or phone calls) that makes you feel uncomfortable. Are these scenarios common in your school? Have you experienced any of these scenarios? If so, you are not alone. Eighty-one percent of students will experience some form of sexual harassment at some time in school. What programs could you sponsor at your school to raise awareness of sexual harassment?

If you feel that you are being harassed, tell the perpetrator to stop immediately. If the person does not stop, it is imperative to speak to an adult whom you trust. *Sexual harassment* is a broad term that covers many behaviors. If the harassment continues, keep a journal to document each event. Share this information with an adult whom you trust. Although many of these behaviors might not constitute a crime, they may violate school rules. Follow the school's policies and guidelines to handle the situation to ensure that you are safe.

Sexual Violence

Sometimes abuse takes the form of sexual violence. Forcing sexual acts on individuals without their permission is called **rape**. Rape is an act of violence. The rates of rape are now so high in North America that one in three females will be raped in her lifetime. Date rape, or rape that occurs between two people who are friends or acquaintances, is more common than rape by a stranger. As many as 75% of all rapes are committed by acquaintances.

All types of sexual abuse are illegal and should be reported to a trusted adult. Only in this way can the one abused be protected and save others from being abused also. How do you think God feels about sexual abuse? Victims of sexual abuse are often filled with shame. Investigate what programs your community sponsors to help console victims of sexual abuse. The best way to reduce the risk of sexual violence is to be informed about prevention strategies. Look for informative classes in your school or local community.

Finding For-Real Friends Explain

As you reach adolescence, having friends—and belonging to a group—become very important. How can friends be a source of confusion, hurt feelings, and discouragement during this time? Why? As you change, your interests change, too, and you naturally look for people who share your new interests. This is normal. What are qualities that you seek in a friendship? Why might it be difficult to find these qualities in friends during adolescence?

During adolescence, friends can change from day to day. What if you come to realize you no longer have the same interests? That's part of the exploration that every young person needs to do. In time, you'll find friends with common interests who support you and enjoy being with you—and you with them.

Many young people yearn to be popular, and some take unhealthy risks to try to fit in with a popular group. Is being a part of a certain group or being popular an important goal to you? Belonging to a group can be comforting but not when it requires you to do things that do

Faith Connection

The Bible encourages us to listen to the advice of our parents. See **Proverbs 1:8–9**.

not match your values as a Christian. The importance of similar Christian values is an important aspect when choosing friends. What are some examples of positive friendships from the Bible? What characteristics of these friendships can you use in your life? What are some examples in the Bible of individuals who chose their friends wisely and those who made poor choices?

As friends become more important, it's natural for young people to start gaining more independence from their families to learn how to make decisions on their own. Sometimes, this need to pull away can lead to problems at home. Why is it a natural process to pull away from your family? What would happen if you didn't? How can you resolve conflict with your family in a positive way? Don't forget that your parents and grandparents were once teenagers themselves. You might be surprised to learn that your experiences sound a lot like what happened to them.

Balancing Life Explain

Becoming an adolescent is a time of huge change. How can you cope with these changes? Change requires adjustment, and that requires energy. As a teen it is easy to become overwhelmed with all the adjustments you must make physically, emotionally, and socially. What can you do to deal with these changes in a positive way? Think about simplifying your goals so that changes in your life are minimized. What is the best way for you to go about doing this?

Teens need to participate in activities that help them feel better. You may find some of these activities useful:

- **Relaxing activities**—Participate in relaxing activities, such as reading, listening to music, taking a walk, riding a bike, running, etc.
- **Diaries**—You may benefit from writing your thoughts and feelings in a diary.
- **Recreational activities**—Return to previous fun activities. Vigorous physical activities help reduce stress, help maintain personal fitness, and make people feel more alert, happy, and energetic.
- **Volunteer activities**—Volunteering helps you grow up caring, confident, and responsible. It will also help you deal with events in a positive way. Try calling the Volunteer Bureau in your community to find out how to help with services, such as helping elementary children with reading and homework, visiting senior citizen centers, or assisting at an animal shelter. Check with your church, your local conference youth department, and with nearby Pathfinders organizations.
- **Fundraising**—Organizing a fundraising event (for example, a car wash) to aid those in need is one way of directing your concern, compassion, and energy in a positive way.

Being part of a worthwhile activity can bring about personal satisfaction.

How might volunteering reduce the stress level of adolescence?

Explore-a-Lab

Structured Inquiry

Is it possible to use phones, apps, gadgets, and computers too much?

Keep a journal for seven days. In the journal, identify the time you spend in the following activities: text messaging, talking on the phone, video chatting, using social networks, and playing video games. Calculate the amount of time you spend in each activity per day. Graph your data collected for the week.

- **Peer-group activities**—Seek out activities where you can get together with friends to discuss things that happen, share your thoughts, and try to make sense of events.
- **Faith studies**—Spend time where you can connect with others in your peer group and learn more about what the Bible teaches about coping with the changes in your life. Finding time every day for personal Bible study and prayer to develop an active relationship with God can help ease the stress and worry of adolescence.

Name 10 technologies that you use every day. These modern conveniences provide hours of entertainment, but they can steal time to make memories with family and friends. How does technology change how you interact with your family and friends? Do you send text messages? One study suggests that U.S. teens send an average of 100 text messages a day. Can texting be harmful? Can it hurt someone? Poor judgment can get teens into trouble when texting. How might text messaging lead to stress in adolescence?

Concept Check

Assess/Reflect

Summary: What are challenges of sexual maturity? Puberty is more than a time of physical change; it is also a period of emotional and cognitive change. During adolescence, most teens make large leaps in the way they think, reason, and learn. Young teens may be able to think more like adults, but they still do not have the experience that is needed to act like adults. As a result, their behavior may be out of step with their ideas. Some teens become victims of bullying or harassment as they struggle to define who they are. It is important to deal with stress associated with adolescence in a positive way. Participation in activities that help others or that keep you active helps you feel better emotionally and socially. Talking things out with a friend or trusted adult helps you find a new way of dealing with your struggles.

1. Why are you better able to solve problems now than when you were a young child? Why are you less able to solve them now than you will be when you are older?
2. Are you likely to find a best friend for life in elementary school? Why or why not?
3. What would you do to help a friend who is being harassed or bullied?
4. Describe three ways to deal with new stresses in a positive way.

Lesson 3

Essential Question

Why Is Sexual Abstinence Important?

Intimacy is a special, close, familiar relationship that one shares with another person. It is important to have intimate relationships with friends and loved ones because it brings people closer. Who are the people in your life with whom you share this type of relationship? What are some of the benefits of intimate relationships? As people become close, separation becomes increasingly painful. How have these emotions been used in poetry and literature? During puberty, the desire to explore new intimate relationships becomes more important. What are some dangers with making poor decisions when engaging in new intimate relationships? To have a happy adulthood, it is important to make responsible choices.

Objectives

- Explain sexual feelings.
- Describe the consequences of premarital sex.
- Identify ways to support a decision of sexual abstinence.
- Analyze sexual issues in today's society in light of God's plan for sexuality.

Vocabulary

sexual feelings
sexual abstinence
unprotected sex
sexually transmitted disease (STD)
birth control
condom

Sexual Feelings Explain

It is normal for teenagers to feel attraction to people of the other sex. It is also normal for teenagers to experience sexual thoughts and feelings. They may be for the "crush" in your neighborhood this year, but these feelings do not last for long. **Sexual feelings** describe attraction to another person. Young adults often feel confused or embarrassed by these new feelings, but sexuality is a gift from God. Sexual maturity is a normal part of growing up. Learning how to deal with these new feelings is important. As boys and girls mature, they have difficult decisions to make about how to express their feelings.

Young adults must learn to handle new sexual feelings responsibly.

Lesson Activity

Interview 10 people about their first feelings of love. Write down at what age the person first experienced this feeling and who the object of the affection was. Return to the class and make a list of your findings. What are common trends in the information listed? Are these experiences different from what you and your classmates are experiencing today? Why are the experiences so similar despite generational differences?

Sex and intimacy are not synonymous. Sexual activity can have serious consequences. **Sexual abstinence** is a decision to avoid all kinds of intimate sexual behavior until marriage. Men and women should not engage in sex until marriage, because marriage is an important commitment where a man and woman pledge before God to remain together until death. Through this union, men and women share an increased intimacy and form a bond that is necessary to grow a strong, loving family. What are some things that can happen if you become sexually active before marriage?

Benefits of Choosing Sexual Abstinence
• You will never have to worry about getting pregnant—or getting someone else pregnant—before you are ready to care for a baby.
• You won't have to struggle with the worries, decisions, and responsibilities of being a parent.
• You won't become infected with a sexually transmitted disease.
• You can base relationships on trust and friendship, not on a temporary sexual attraction.
• You won't need to pressure someone you care about into an unwanted physical relationship.
• You will have an easier time maintaining a relationship with God because you respected His plan for you.
• You won't feel the guilt and regret caused by disappointing yourself and the adults who trusted you.
• You will have nothing to hide from your family, other adults who trust you, siblings who look up to you, and friends who admire you.
• You will not have limited your choices in life because you had to drop out of school to raise a child or marry too soon.
• You won't have to face extremely difficult decisions that you might regret for the rest of your life, such as giving up your first-born child for adoption.

Record your work for this inquiry. Your teacher may also assign the related Guided Inquiry.

Sexually Transmitted Diseases

How do you avoid sexually transmitted diseases?

Materials
- 3 in. × 5 in. note cards
- 4 pairs of gloves

Procedure

1. Take a note card provided by your teacher.
2. Write the numbers 1–4 down the side of the note card.
3. Go around the class and sign your name on the cards. Only collect four names on your card.
4. Identify the student with a red dot on his or her card. Write the student's name in the center of the board.
5. Ask this student to read the first name listed on his or her note card. Write that name on the board and connect the two names with a straight line. Repeat this for the remaining three names on the note card.
6. Ask the first student identified on the red-dotted card to tell the name of the second person on his or her card. Write the name on the board and connect the two with a line. Repeat this step with the third and fourth persons listed on the card.
7. Ask the first newly identified person to list the names of the third and fourth persons on his or her card. Write the names on the board and connect with a line.
8. Put an *X* through the name of any student listed on the board who is wearing gloves. Put an *X* through the name of any person subsequently connected to the person who is wearing gloves.

Analyze Results

Draw the web of people in the activity. Connect the web of students in the activity. Does overlap occur? What is the significance of the gloves in this activity? Why are names skipped in the list with each round of the activity?

Create Explanations

1. How do you avoid sexually transmitted diseases?
2. What risk does the student who abstained from the activity face?
3. As the activity progressed, how did you feel as more names were added to the web?

Explore-a-Lab

Structured Inquiry

Label one beaker *A* and one beaker *B*. Fill each beaker three-fourths full of water. Place two Alka-Seltzer tablets into beaker *A* and record your results. Place three mothballs into each beaker. Record your results.

 **Which beaker represents an elementary school student, and which beaker represents a junior high student? Why?**

Waiting Is the Right Choice Explain

Engaging in sexual activity can have a tremendous impact on your physical, emotional, social, and spiritual health. It can also influence your future financial success, your family happiness, and your feeling of fulfillment. Remember to refer to your value system when making decisions. What should you consider? Difficult decisions are easier to figure out when you consider important factors that guide your life, including the following:

- Being responsible and trustworthy
- Being honest and fair
- Respecting yourself and others
- Always doing your best
- Taking care of the body that God gave you
- Following the biblical principles you have learned at home, in school, and in church
- Wanting to establish a stable family of your own when you are ready

Faith Connection

God calls us to be sexually pure in our relationships. See **1 Thessalonians 4:2–5**.

Lesson Activity

Create a Futures Wheel and complete as instructed.

- What other consequences can you think of that are not listed?
- What consequences affect the baby?
- What consequences affect the mother?
- What consequences affect the father?
- What consequences affect friends of the teenage parents?
- What consequences affect the relatives of the teenage parents?
- What consequences affect the community?

 **What consequences could your decision cause?**

Lesson Activity

Use the Internet to learn the age restrictions for the following activities in your local state or province and the surrounding states or provinces: driving, gambling, purchasing cigarettes, purchasing alcohol, voting, and getting married. As a class, discuss what age you think is appropriate for each activity.

Why is age important?

Physical Health

When making a decision about sexual abstinence, remember that premarital sex can affect your *physical health*. You must be aware that the possible effects on your physical health can include having to cope with an unplanned pregnancy. You might also have to ask for help in treating a disease that can affect your health for the rest of your life and your ability to have children in the future.

Emotional Health

What about the possible effects on your *emotional health*? Emotions are difficult to deal with during adolescence. Imagine how hurt you would feel if your sexual partner started spending time with someone else. How would that affect your emotional health? This is a time in your life when you must actively try to reduce extra stress. How might a sexual relationship impact your daily stress level?

The decisions you make now can affect the rest of your life.

What kinds of struggles do you think this young mother faces?

Social Health

How might premarital sex affect your *social health*? Adolescence should be a time for trying out new friendships. People in a sexually active relationship tend to avoid forming friendships with others. How can this affect your ability to relate with your friends? How might this isolate you from your peer group?

Spiritual Health

Going against your spiritual values can also have unwanted consequences. While God can and does forgive any sin including premarital sex, the consequences are often life changing and lifelong. You must be prepared to make an informed decision about your sexual life during adolescence. Living in harmony with God's plan and in close relationship to Him ensures your spiritual health. How can you work with your parents, pastor, and church to regain your spiritual health?

Maintaining Abstinence

You are already trying to cope with all the changes and challenges of adolescence. In the middle of this turmoil, how can you follow through with your decision to abstain? It's never too late to abstain, by the way. Even if you have been sexually active, you can choose to abstain from now on. If you haven't yet faced the problems already discussed, you are guaranteed to avoid them if you decide to abstain now. This chart offers some actions and attitudes that will help you abstain.

Actions for Maintaining Abstinence

What to Do	How to Do It
Remember your values.	Think about what's important to you. Don't let a long-term goal, such as graduating from college, be wiped out by a few minutes of sexual activity.
Set limits on how you express affection, and share them with your partner.	Remembering your values, decide how far is too far. Keep your brain in charge. Share your limits and the reasons for them with your partner. A caring partner will respect your limits. If your partner keeps pushing them, this relationship may not be as caring as it seemed.
Avoid situations where you will be pressured to have sex.	Date with another couple or in a group. Avoid situations and parties where no adults are present.
Do not use alcohol or other drugs. Don't date someone who does use alcohol or other drugs.	Drugs and alcohol make it harder to think clearly. When people use drugs and alcohol, they are more likely to take risks. Some people are more likely to pressure their partners to take risks.
Choose friends who have also decided on abstinence.	You can support each other and find fun things to do together.
If you have questions or doubts, talk with a trusted adult.	A parent or other trusted adult can help you sort out and manage your feelings during this time.

Respectful Relationships Explain

Young adults begin to experience strong emotional feelings of attraction toward the opposite sex. These are natural feelings. These feelings are initiated by hormones, which begin during the preteen and early teen years. The changes to your body that you have learned about now make it possible for you to reproduce.

Unfortunately, the media can manipulate these feelings in a way that might lead you astray from your spiritual path. It is important to understand that the sexual drive that draws you toward a member of the opposite sex needs to be controlled. Sex is a gift from God. It should be treated with respect, just as you treat your friends and family with respect. However, Satan has set out to sabotage this gift. From the time sin entered this world, people have practiced immoral sexual relations. What are some examples of Satan's sabotage in Bible times? Throughout human history, Satan has brought misery to something that God intended for happiness. How does Satan continue to spoil this important gift from God?

How can you tell someone whom you find attractive that you are abstaining from sex? If you are attracted to someone, it is possible to do things together and enjoy each other's company without having sex. The foundation to any strong relationship is respect. Although sexual feelings may be intense, these feelings are temporary. Respect and friendship grow over time. People who really care for you respect your values and decisions. They care when you do well and share your feelings when things do not go well. People who respect you will not use you to satisfy their own desires. They will not encourage you to do something that is morally wrong. They will not ask you to break any of God's laws.

One way to control feelings of attraction is to keep social interactions to group settings.

If someone you are attracted to indicates that they want to be alone with you, what can you say to nicely and respectfully decline?

Lesson Activity

Take five minutes to flip through newspapers and magazines to find five examples of Satan's sabotage of sex. Create a collage of your findings and prepare a two-minute presentation to explain your work to the class.

Why should you be aware of the way sex is portrayed today? A great many forms of product advertising and entertainment may try to convey the impression that it is okay to be sexually active. Many of these ads and forms of entertainment are market driven. They strive to make money from people who either purchase the products advertised or watch them. This marketing ploy has little or no concern for the personal or emotional life-altering effects it can have upon your life values. What kinds of costs can be involved if you are sexually active?

Guidelines for sexual behavior include choosing to follow God's plan and making friends among those who share your moral values and goals in life. A Christian school and church are places where you can make such friendships. Teenagers can safely express their sexual feelings by choosing activities that demonstrate caring, yet set specific limits. What are some activities that you can engage in that are moral and follow God's plan?

Sexual Issues Explain

Many people today ignore God's plan for a variety of reasons, both right and wrong. These differences of opinion about sexual behavior in society become sexual issues. None of today's sexual issues is new. Most have been debated since ancient times, although people's ideas about them have changed from one generation to another. As you consider these issues, think about how each sabotages God's original plan and what a Christian's position should be.

Lesson Activity

Magazines and television shows often feature "Top 10" lists. Now you will make a Top 10 list of reasons to avoid becoming sexually active before marriage. First jot down the reasons that mean the most to you. They might include the consequences of sexual activity, the benefits of being sexually abstinent, and your spiritual commitment to live according to the principles found in the Bible. Then put your reasons in order from most important (#1) to least important (#10). List your reasons on a sheet of paper, putting #10 at the top. Compare your list with your classmates' lists. What is similar about the lists that the members of your class made?

What are my top 10 reasons to abstain?

Premarital Sex

Even though an adolescent girl is physically able to become pregnant, she is not ready to become a mother or to raise a baby. A boy of the same age is not ready to handle the responsibilities of being a father and supporting a family. Sexual intercourse belongs only in marriage. It is a very special activity by which God made human beings to become as one. God's instruction is very clear: sexual intercourse is part of God's plan for marriage, but not before. Why do you think God wants people to wait until marriage before engaging in sexual intercourse?

Some people think that having only unprotected sex can spread disease. During **unprotected sex**, partners do not use a method of protection. But both unprotected and protected sex can spread diseases. A **sexually transmitted disease (STD)** is an infection that can spread from person to person through sexual contact. Some STDs cause painful sores or warts, while others can make men and women unable to have children later in life. Still other STDs can be fatal.

When a person chooses to become sexually active, pregnancy is what the body is intended to do. When a married or unmarried couple has sex and is not ready to raise a child, the couple should select a safe method of birth control. **Birth control** is a way to reduce the chance of pregnancy. A **condom** is a barrier method of birth control. It is a thin sheath (usually made of latex, a type of rubber) that is worn on the penis. Condoms provide only some protection against STDs and pregnancy. The only 100% effective way of preventing STDs and pregnancy is by sexual abstinence. Also, having sex, even protected sex, without being married is not right in God's eyes. It is important to remember that it is never too late to return to God's plan. If you make a bad choice, you can always make a commitment to strive to make good choices in the future. God always forgives.

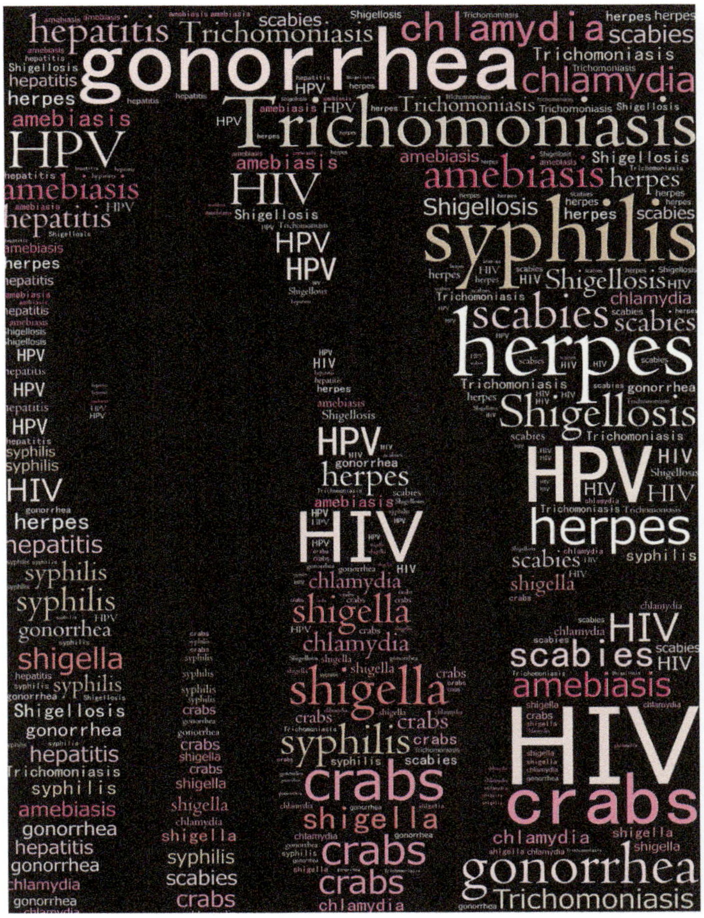

There are several types of STDs.

What kinds of risks do you face if you do not choose sexual abstinence?

Homosexuality

God created men and women to be attracted to each other. This is *heterosexual orientation*, which leads ideally to marriage, homes, and children. Men who are attracted to other men and women to other women have a *homosexual orientation*. A homosexual woman is sometimes called a lesbian. A homosexual man may be called gay. Scientists are not in agreement as to why some people have a homosexual orientation. Christians believe that this is part of Satan's effort to sabotage God's plan for men and women. A person with a homosexual orientation, however, need not practice homosexual behavior. Homosexual behaviors were not part of God's plan.

Most young people have close friends of the same sex. This does not mean that they are homosexual. These friendships provide an opportunity to practice thoughtfulness and selflessness, which are necessary to have a close friendship with any person.

Prostitution

Prostitution describes selling one's body for sexual activity. Prostitutes appear in several Bible stories, and the practice is still common today. Prostitution is promoted by poverty, lack of education, and the use of drugs. People who see no other way to earn a living or are young and vulnerable may become prostitutes. Prostitution is not the answer. What can happen to women who enter into prostitution?

In most places prostitution is illegal, and both the prostitute and the person buying the sexual services can be arrested. Do penalties fall harder on the prostitute or the customer? Why? But arrest is not the worst outcome. Loss of self-respect, high risk of sexually transmitted diseases, suffering, and sometimes murder are common dangers for both the prostitute and the customer. Prostitution and its accompanying dangers are certainly not part of God's plan.

Abortion

Abortion is one of the most controversial issues today. It is the termination of a woman's—oftentimes a teenager's—pregnancy, which also causes the death of the unborn child. Biblically, human life at any stage is sacred and is to be protected (**Psa 139:13–16; Exo 21:22–24; Lev 20:2**). God relates to the unborn as a person (**Isa 44:2; Psa 139:13–16; Luke 1:31, 36, 41, 44**), made in God's image (**Gen 9:6**). By joining His divinity to human nature in fetal form, Christ affirmed the sacred value of human life in the womb. Pregnancy places the mother in a unique role and relationship to the new human life within her, making her a steward of a life not her own.

Typically, pregnancy is a case of joy and celebration. The desire of a married couple to have a child has come true. But sometimes a woman considers terminating pregnancy. This may be triggered by a variety of reasons. For example, the pregnancy may have been caused by rape or incest. The woman may be unmarried (for example, a teenager) and fears a burden, social stigma, or financial hardship. The unborn child may have been diagnosed with a serious birth defect. Additionally, some pregnancies may pose a significant threat to the life of the mother.

Such circumstances produce great stress. However, making a decision about abortion is traumatic. Unmarried teenagers who are pregnant need the help of parents and the counsel, care, and support of Christian adults. They need to know that in such situations, options other than abortion are available. Therefore, personal decisions about abortion should always be made after much prayer, a study of the will of God regarding life and death in Scripture, and a reflection on God's purpose for the unborn child **(Jer 1:5; 29:11; Psa 139:16)** and the legitimate interests of the unborn child's father in the life of that child. Abortion is not part of God's ideal plan.

Summary: Why is sexual abstinence important? Puberty is the onset of physical maturation as well as sexual feelings. These feelings draw people together in close intimate relationships. Young adults must treat these new feelings responsibly. One effective way of making good decisions is to understand the benefits of sexual abstinence and to refrain from participating in sexual activity. Even though you may have very strong feelings about someone at this time in your life, realize that these feelings are the result of the increased hormones in your body. If you make the decision to become sexually active, you put yourself at risk for many different problems, one of which is a sexually transmitted disease, or STD. There are methods of preventing pregnancy and the transmission of disease, such as the use of a condom. As you grow into an adult, you will become aware of many sexual issues in society, including homosexuality, prostitution, and abortion. These issues are not part of God's plan for sexuality and bring misery and sadness to many.

1. What are three benefits of abstaining from sexual activity?
2. What are three ways that young people could support their decision to remain sexually abstinent?
3. Describe one sexual issue and how it deviates from God's plan.

People in Science

Extend

Get to Know Rosalind Franklin

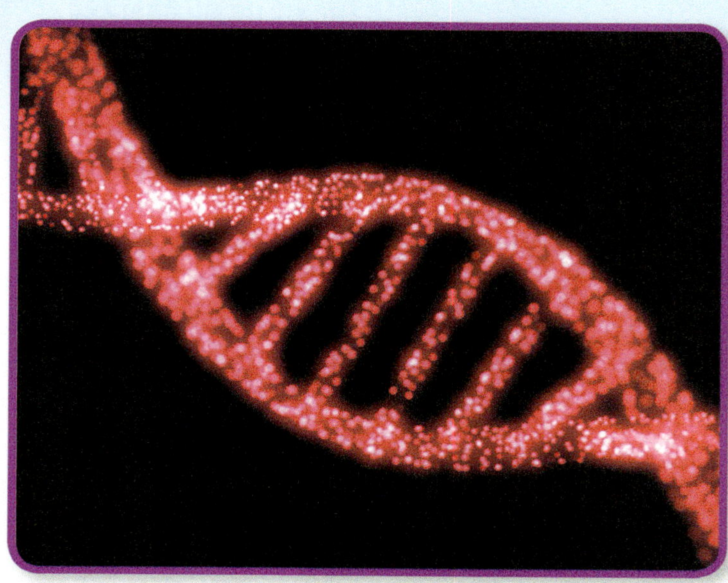

Pioneer molecular biologist Rosalind Franklin is best known for her role in the discovery of DNA structure and for her groundbreaking use of X-ray diffraction. She was born in London, England, in 1920 and died there in 1958. Even at a young age, she exhibited exceptional intelligence and decided at the age of 15 that she wanted to be a scientist. Her father was against his daughter becoming a scientist, since it was very difficult at the time for women to have that type of career. Eventually, he gave in to her wishes. She earned a doctorate in physical chemistry in 1945.

Franklin spent the next three years in Paris, where she learned X-ray diffraction techniques. She returned to England as a research associate at King's College, London, in the biophysics unit, where her expertise and X-ray diffraction techniques were used in the study of DNA.

Studying DNA structure with X-ray diffraction, Franklin made an amazing discovery—that there were two forms of DNA. One of her pictures, titled Photo 51, became famous as significant evidence for identifying the structure of DNA. The photo was completed through 100 hours of X-ray exposure from a machine Franklin had improved.

Even though Franklin had an excellent work ethic and took great pains in her work, she did not get along well with her colleague Maurice Wilkins. In 1953, Wilkins changed the course of DNA history by showing Franklin's Photo 51 to a competing scientist. Using the evidence from Franklin's photo, two scientists used what they saw in Photo 51 as the foundation for their famous model of DNA, which they published. They received a Nobel Prize award in 1962 for their work. Franklin had come very close to solving the DNA structure, and the debate about the amount of credit due to her for the model of DNA continues.

Rosalind Franklin's DNA research has provided a basis for understanding genetics and how reproduction contributes to genetic variation. This knowledge is important in understanding human growth and development. She was a scientist held in high esteem. It is believed that she never knew that the two scientists based their famous DNA model on the research she had completed.

Concept Check

1. What techniques did Franklin learn before she began her scientific research?
2. Why was Photo 51 so significant?

Obstetrician

An obstetrician is a specialized physician who supervises the stages of pregnancy, labor, and the weeks following childbirth. A woman's gynecologist will confirm that she is pregnant and will usually refer the expectant mother to an obstetrician. Obstetricians and gynecologists, also known as OB/GYNs, are medical doctors who treat the health of women. A gynecologist mainly specializes in caring for the reproductive system of women, which means diagnosing and treating issues such as cervical cancer and hormonal imbalances. They perform regular checkups on women to make sure that their reproductive health is in good standing.

Obstetricians treat pregnancy and pregnancy-related medical conditions, and gynecologists treat non-pregnancy-related medical conditions. Both physicians provide treatment for sexually transmitted diseases and other personal matters. Therefore, this type of work requires objectivity and an understanding of the patient's stress level. A bachelor's degree, a doctorate in medicine or osteopathy, and a four- to seven-year residency is required for both obstetricians and gynecologists.

Obstetricians provide prenatal care (care before the baby is born), pregnancy-related testing, and education about childbirth to patients. Prenatal care includes monitoring the growth and health of the fetus through ultrasounds and checking the fetal heart rate. In some cases, especially high-risk pregnancies, obstetricians offer genetic testing and testing of the amniotic fluid. Obstetricians also check the mother's health by monitoring blood work, blood pressure, heart rate, and weight gain. It is important that the woman chooses an obstetrician whom she is comfortable with and who respects her wishes, as pregnancy and childbirth can be an anxious time.

Obstetricians deliver babies and perform other birth-related duties as needed, such as inducing labor and performing cesarean-section births and emergency surgery when needed. They monitor the mother and the baby throughout childbirth and oversee the administration of medications. Obstetricians must be attentive in order to take appropriate action if the mother is bleeding uncontrollably, if the baby is not positioned correctly for birth, or if the baby is not breathing at birth.

Concept Check

1. How is the work of a gynecologist different from that of an obstetrician?
2. Why is it important that an expectant mother go to an obstetrician?

Chapter 5 Review

Assess/Reflect

Study Guide

Lesson 1

1. Asexual reproduction results from mitosis. Sexual reproduction occurs when gametes—eggs and sperm—produced during meiosis fuse to form new offspring.
2. Mitosis is the process where a single cell divides, resulting in two identical cells. Meiosis is a process of two cell divisions. The process results in four cells. Each cell has half the genetic material as the original parent cell.
3. Adolescence is the stage of development after childhood during which children become adults. Puberty is the time when the body develops sexually, enabling reproduction.
4. During puberty, the body develops sexually and begins to show the secondary sex characteristics of the male or female, enabling reproduction. Puberty is controlled by hormones produced by the endocrine system.

Lesson 2

1. During puberty, the brain gains cognition, the ability to gain knowledge and comprehension. Young adults begin to experience increased independence and the opportunity to begin to make important decisions.
2. Abuse can be physical, psychological, sexual, verbal, or a combination of any or all of these. All types of sexual abuse are illegal and should be reported to a trusted adult.
3. Sexual violence can be avoided by respecting yourself and others. Take care of the body that God gave you. Follow the biblical principles you have learned at home and in church.
4. Spend time where you can connect with others in your peer group and learn more about what the Bible teaches about coping with the changes in your life.

Lesson 3

1. Sexual feelings describe attraction to another person. Young adults often feel confused or embarrassed by these new feelings, but sexuality is a gift from God.
2. Consequences of premarital sex include pregnancy, increased responsibilities of parenthood, sexually transmitted disease, decreased trust, lack of respect, guilt, shame, financial difficulty and difficult choices.
3. Ways to support a decision of sexual abstinence are to remember your values, set limits on how you express affection, avoid situations where you will be pressured to have sex, do not use drugs or alcohol or date someone who does, choose friends who have also decided on abstinence, and talk to a trusted adult about any questions and doubts.
4. Homosexuality, prostitution, and its accompanying dangers are not part of God's plan. Personal decisions about abortion should always be made according to a study of the scriptures and the laws of God.

Vocabulary Check

Explain how each pair of terms is related.

1. endocrine system—reproductive system
2. puberty—cognition
3. birth control—STD
4. rape—abortion
5. asexual reproduction—sexual reproduction

Multiple Choice

Choose the best answer.

6. Why is reproduction essential?
   A. for the continuation of species
   B. to prolong the life of individuals
   C. to provide genetic material
   D. to form gametes
7. What is unwanted sexual behavior?
   A. bullying
   B. sexual harassment
   C. sexual abuse
   D. rape
8. What is true of sexual reproduction?
   A. It produces offspring exactly like the parents.
   B. It requires only one parent.
   C. It requires sperm and eggs.
   D. It does not require meiosis.
9. Which of these would most likely help a young person avoid sexual activity?
   A. friends who drink alcohol
   B. a boyfriend or girlfriend who has a car
   C. a need to be accepted by others who push limits
   D. friends who like to have fun together
10. Which hormone do the ovaries produce?
    A. pituitary
    B. testosterone
    C. estrogen
    D. endocrine

Check Point

Answer the following questions.

11. How is your reproductive system similar to that of a plant with flowers?
12. **Infer** how the changes of puberty are affecting this person. Explain.

13. What are two **variables** that could affect the changes that a young person experiences during puberty?
14. Why might bullying become a bigger problem as young people enter puberty?
15. What are three reasons why some young people become sexually active?
16. Why would it be important to decide what your spiritual values are and what your physical limits are before entering a dating relationship?
17. Name three reasons why it would be a good idea to develop friendships with people who share your spiritual values and who have also decided to abstain from sexual activity until marriage.

Infectious Diseases

Scripture Spotlight

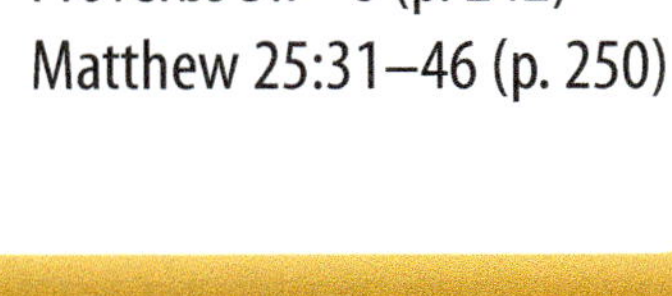

God created our bodies to resist disease in amazing ways. Learning about the immune system is one way to see how God cares for us. You will read the following passages in this chapter.

Psalm 139:14 (p. 217)
Matthew 15:29–31 (p. 228)
Psalm 103:3 (p. 233)
Proverbs 3:7–8 (p. 242)
Matthew 25:31–46 (p. 250)

The Big Idea

Knowing how disease is spread and how the immune system protects you from disease can help you stay healthy.

Does avoiding the risk of disease mean that you must avoid being close to other people? Explain your answer.

Some diseases spread easily from person to person. Knowing how diseases spread and what steps to take to prevent them can reduce your risk of infection by these diseases and infecting others.

Inquiry Kick-Off Engage

Why is it important to understand how diseases spread? Does your family know how diseases spread? How can all of you become educated on this topic? In your *Science Journal* you will be asked to conduct a survey to find out what your classmates and family members know about diseases and how they are spread.

Science Journal

Lesson 1

Essential Question

How Does the Immune System Protect the Body?

Objectives

- Describe the immune system and its function.
- Identify the body's lines of defense against infection.
- Compare and contrast the different kinds of immunity.
- Explain how the immune system can harm the body.

Vocabulary

inflammatory response
macrophage
immune response
antigen
antibody
autoimmune disease

In the beginning, God created the human body perfectly—no disease, no death. When sin came into the world, things changed. Even with the effects of sin, your body is still amazing. It works even when it is not taken care of as well as it should be. At times, the body can become infected, or an illness can take hold. A simple virus can cause a cold or the flu. Bacteria may cause an earache or sore throat. What other ways can infection or illness set in? How many invaders do you think your body fights off in an average day? What things can you do to keep your immune system strong? When might there be times when your body needs help fighting off invaders?

The Immune System Explain

Every surface you touch, even the air you breathe, has the potential to make you sick. *Pathogens*, organisms that can cause disease, lurk all around us. They are bacteria, viruses, fungi, and protists. What illnesses do you know of that might be caused by pathogens? Fortunately, our bodies have several defenses that protect us against all types of pathogens. The initial defenses are known as first-line defenses. Your skin, your respiratory system, and your digestive system all have first-line defenses that form obstacles to harmful pathogens.

Sometimes, however, harmful pathogens manage to get past these first-line defenses. Then your *immune system*, a complex group of your body's defenses against disease, works against specific pathogens. The immune system is your strongest weapon in your body's defense against specific harmful pathogens. There are three lines of defense that the immune system has at its disposal.

These bacteria, called streptococcus, can cause strep throat.

 In what ways might this pathogen enter the body?

First-Line Defenses: The Barriers to Infection Explain

The first-line defenses against pathogens are made up of physical and chemical barriers. These barriers can trap or kill pathogens before they can attack you. Your skin is the largest physical barrier against pathogens. The outer layer of skin is made up of tough cells that are constantly shedding, making it difficult for pathogens to enter. Cilia, or tiny hairs, line the nose and respiratory tract. Pathogens are trapped by cilia and can be expelled by sneezing, coughing, or blowing your nose.

Chemical barriers, such as mucus, tears, saliva, and sweat, trap or wash away pathogens. Your saliva and tears contain an enzyme called *lysozyme* that protects you from bacterial invasion. Acid and other types of disease-fighting enzymes line your stomach and intestines. These first-line defenses are illustrated in the diagram. What physical and chemical barriers might act on a pathogen that enters the body through your nose? What about the pathogens that enter through a cut on your hand?

Scripture Spotlight

Psalm 139:14 tells us that our bodies are "fearfully and wonderfully made."

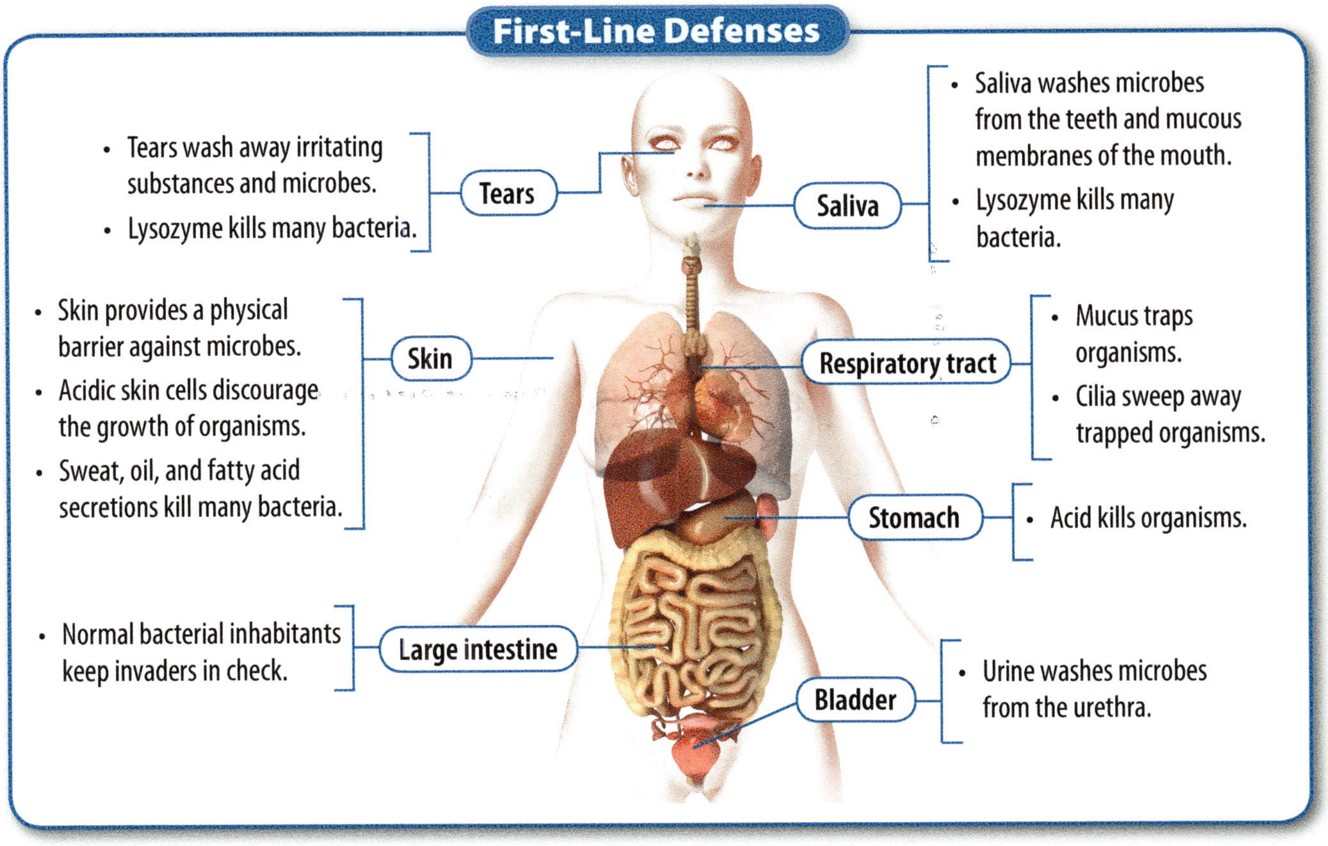

First-Line Defenses

Tears
- Tears wash away irritating substances and microbes.
- Lysozyme kills many bacteria.

Saliva
- Saliva washes microbes from the teeth and mucous membranes of the mouth.
- Lysozyme kills many bacteria.

Skin
- Skin provides a physical barrier against microbes.
- Acidic skin cells discourage the growth of organisms.
- Sweat, oil, and fatty acid secretions kill many bacteria.

Respiratory tract
- Mucus traps organisms.
- Cilia sweep away trapped organisms.

Stomach
- Acid kills organisms.

Large intestine
- Normal bacterial inhabitants keep invaders in check.

Bladder
- Urine washes microbes from the urethra.

Which of the body's defenses are physical barriers? Which ones are chemical defenses?

Record your work for this inquiry. Your teacher may also assign the related Guided Inquiry.

Observing White Blood Cells

How do white blood cells compare with each other?

Materials

- 2 prepared slides of white blood cells
- microscope

Procedure

1. Mount a prepared slide of blood cells on the microscope and focus on low power. Turn to high power and look for white blood cells.
2. Find a neutrophil, monocyte, eosinophil, and lymphocyte. You may see a basophil, although it is rare. Refer to the photos in your ***Science Journal*** to identify these cells.
3. Count the total number of cells in the viewing area and **record** how many of each type of white blood cell you see.
4. **Calculate** the percentage of each type of white blood cell that is present in the sample. Record the percentages. Draw a diagram of each kind of cell.
5. Repeat Steps 1–4 using a different prepared slide.

Analyze Results

Construct a graph that shows a comparison (percentage) between slide 1 and 2 in reference to the number of each type of white blood cell that is present.

Create Explanations

1. How do white blood cells compare with each other?
2. What type of white blood cell was most common?
3. What role might white blood cells play in preventing illness?
4. Are white blood cells a first-line defense? Explain.

Second-Line Defense: The Inflammatory Response

Explain

Sometimes a pathogen manages to get past your body's physical and chemical barriers. For example, you might get a splinter in your foot. Pathogens from the splinter can invade your body. Or maybe you are really tired or stressed, have not been eating well, or are recovering from another illness. If so, your body's physical and chemical barriers might not be able to fight off pathogens. They can enter your body and begin to damage your cells. What are other examples of how pathogens might be able to get past the body's first-line defenses?

If pathogens invade your body, the attackers are met by a second line of defense. The **inflammatory response** takes over. If a splinter has entered your skin, chemicals from the damaged cells cause the walls of nearby blood vessels to expand. This allows more blood to flow into the area under attack. This extra blood flow can also cause the area around the splinter to become red, swollen, and painful.

Other chemicals from the damaged cells attract white blood cells, known as leukocytes. There are two main types of leukocytes, phagocytes and lymphocytes. *Phagocytes* are white blood cells that engulf foreign particles, bacteria, and other harmful pathogens. One type of phagocyte, a monocyte, can migrate out of blood vessels and into tissues. There, the monocyte develops into a **macrophage**. Macrophages work slowly, taking hours to surround and break down pathogens. They consume large quantities of materials and release some undigested particles back into the bloodstream to be devoured by other phagocytes.

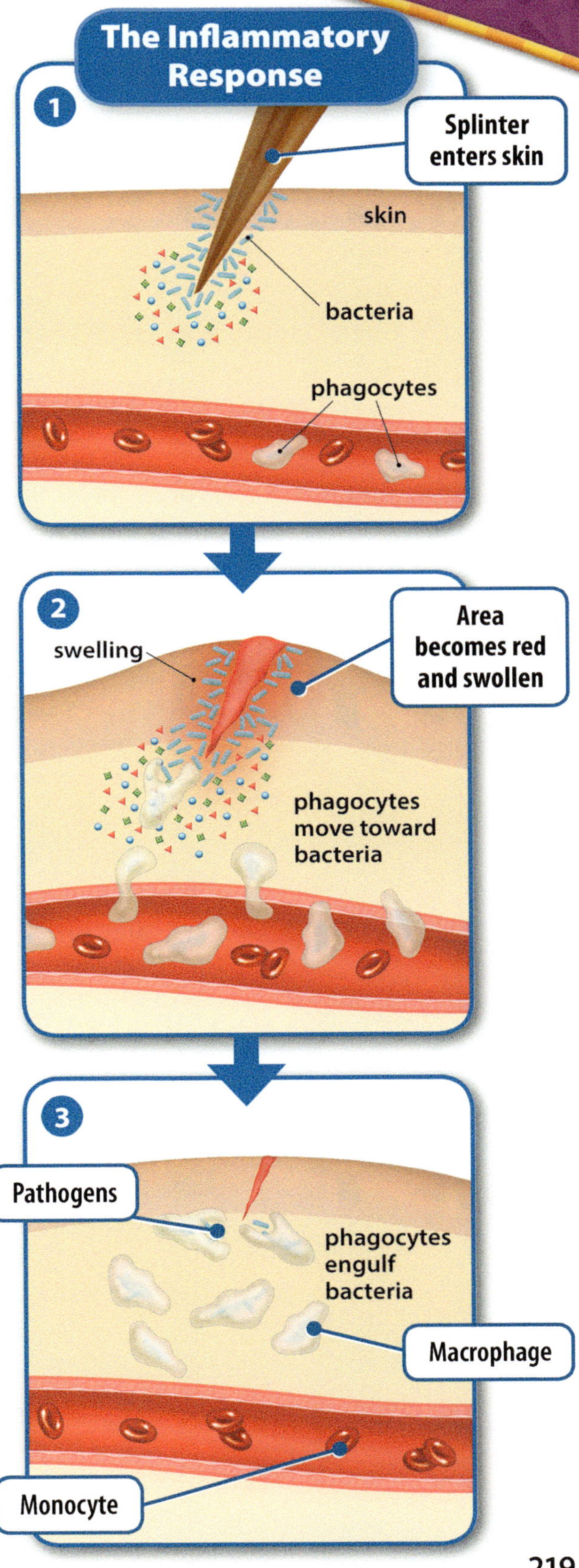

Pathogens can easily enter the body through breaks in the skin.

You accidentally cut yourself. Why are pain and swelling a good sign?

You have probably seen the pus that sometimes forms around an infection. What do you suppose causes this pus to form? Sometimes the inflammatory response also causes a fever. A fever may make you feel uncomfortable, but it is actually helping your body. Higher body temperatures slow the reproduction of some pathogens. Thus, a fever helps fight off infection. The inflammatory response also helps your body deal with injuries, burns, and other damage to tissues.

Faith Connection

We believe that our bodies are a temple of the Holy Spirit (see **1 Corinthians 6:19–20**), so we want to do everything possible to keep our systems in good working order.

Third-Line Defense: The Immune Response

The first and second lines of defense are nonspecific responses. That is, the chemical and physical barriers and the inflammatory response do not distinguish one invader from another. Sometimes a pathogen gets by these first two lines of defense. That is when a third line of defense steps in. A specific **immune response** occurs against specific pathogens.

How does the immune response work? Your immune system contains tens of millions of *T cells*, which are special white blood cells known as lymphocytes. T-helper cells turn on the immune system. Killer T cells identify specific pathogens. That is because each pathogen contains a specific marker called an **antigen** that only one type of killer T cell can recognize. Thousands of kinds of pathogens exist, but you have so many different T cells that they can recognize nearly all of them.

Think about how you can look at a large crowd of students and immediately identify your classmates. In a similar way, T cells can distinguish one pathogen from another. If a T cell recognizes the antigen of a pathogen on a body cell, it destroys that cell.

Your immune system also contains another kind of white blood cell, a lymphocyte. B lymphocyte cells, also known as *B cells*, produce proteins that help destroy pathogens. Each kind of B cell produces just one kind of protein. Each protein, called an **antibody**, has a structure that fits onto the molecules from a certain pathogen. The antibody and the pathogen fit together like puzzle pieces. The antibody sticks to the pathogen, marking it for destruction. What might happen if all the T-helper cells in your body were destroyed?

After an infection is gone, memory B cells and memory T cells remain in your body. They will help your immune system respond faster if you are exposed to the same kind of pathogen again. They also help you to resist some illnesses by giving you *immunity*, the body's ability to fight off pathogens before they can cause illness.

The Immune Response

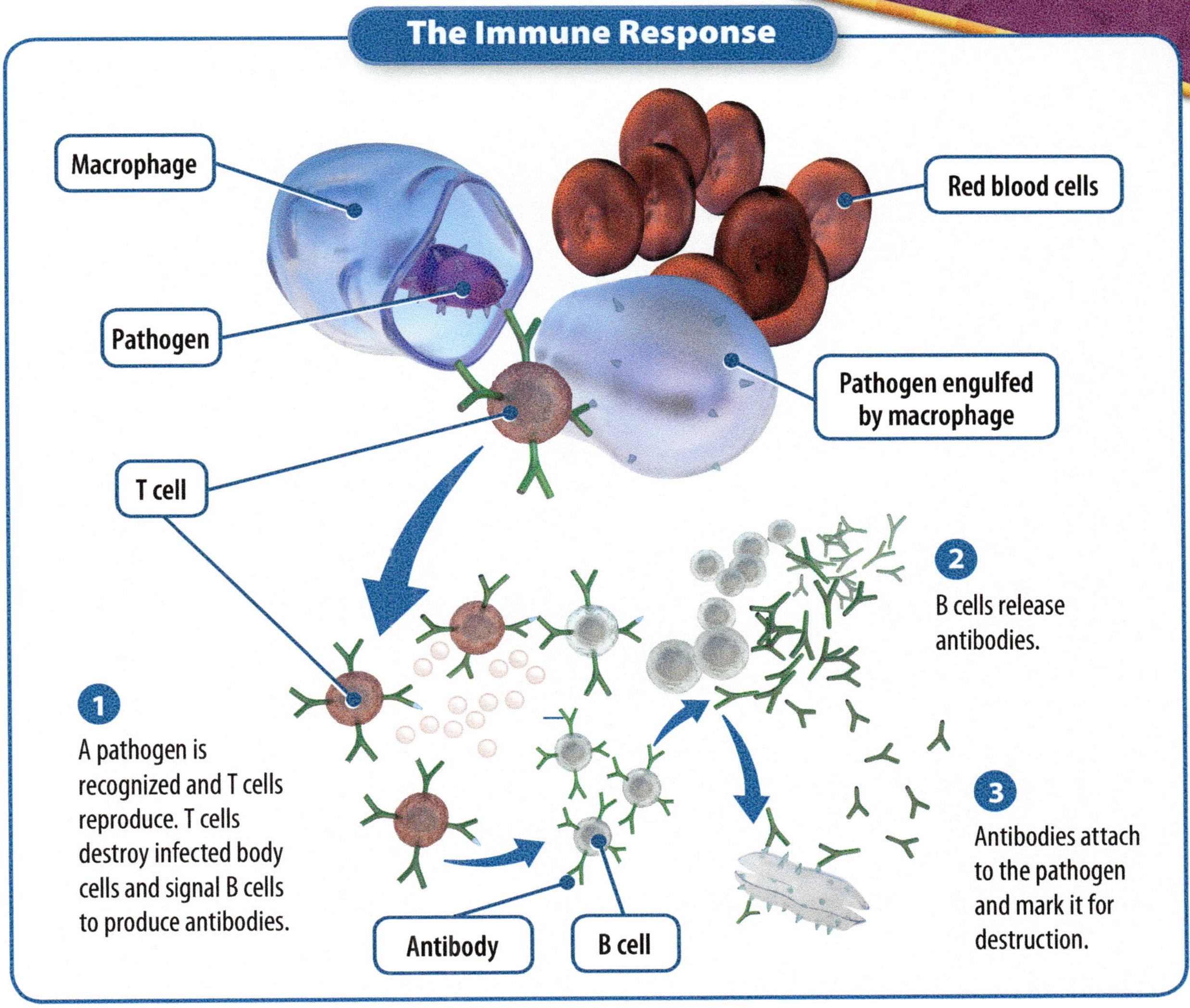

Macrophage

Red blood cells

Pathogen

Pathogen engulfed by macrophage

T cell

1 A pathogen is recognized and T cells reproduce. T cells destroy infected body cells and signal B cells to produce antibodies.

2 B cells release antibodies.

3 Antibodies attach to the pathogen and mark it for destruction.

Antibody

B cell

T cells and B cells in the immune system work together to fight off an infection. This infection might cause the flu, chicken pox, or another disease.

Are all antibodies the same shape as the ones in this diagram? Why or why not?

Explore-a-Lab

Structured Inquiry

How do antibodies stop pathogens?

Your teacher will distribute a cutout of a half circle with a jagged pattern on the cut end. The circle represents a pathogen that can cause an illness, such as strep throat or influenza. Matching halves to your circle will be scattered about the room. On your teacher's signal, find the half circle that matches your cutout. When you find the matching half, return to your seat. If time is called before you find your circle, stand at the front of the room. What do the two halves of the circle represent? How does this activity model how antibodies stop pathogens?

Your parents or grandparents might remember getting chicken pox as children. After people have been infected with chicken pox, their immune systems produce memory T cells and memory B cells. These memory cells remain in their bodies, ready to fight the chicken pox virus if it attacks their bodies again. The antibodies in their blood quickly destroy the chicken pox virus before it can make them sick. Because of the memory cells, many people who have had chicken pox are immune to the disease. What other childhood diseases might a person have immunity to because of memory cells?

Suppose the red objects are chicken pox viruses. Describe what is happening in this diagram.

Types of Immunity **Explain**

The complexity of your immune system goes beyond the three lines of defense built into your body. Immunity also takes three different forms: innate immunity, adaptive immunity, and passive immunity.

Innate Immunity

The innate immune system is made up of your first and second lines of defense. But the innate immune response is not specific to a particular pathogen. As soon as a body is infected with a pathogen, the innate immune system is activated. The quick response in the first hours of exposure to a new pathogen allows this system to protect the body. Why is this particularly helpful when a single invading bacterium can multiply into millions of bacteria in a matter of hours?

Adaptive Immunity

The adaptive immune system takes a longer period to react to a new invading pathogen. It may take several days or a week for the response to be effective. Unlike the innate immune system, adaptive immunity is antigen specific. The immune response is tailored to specific invaders. The adaptive immune system also "remembers" when it comes into contact with a certain pathogen. If the pathogen invades the body again, the immune response is quicker. The innate immune system, however, does not remember an encounter with a pathogen.

Passive Immunity

Passive immunity is a bit different. Although it also is specific to a certain antigen, it is acquired passively rather than formed by the immune system in response to the presence of the antigen. One type of passive immunity is that given from the mother to a newborn baby. A baby has not yet had the time to develop antibodies. Passive immunity protects the baby against most antigens for its first few months of life, until it develops the ability to produce its own antibodies. Why do you think the baby has not developed its own antibodies by the time it is born? What would be the consequences for the baby if it did not acquire passive immunity from the mother?

Another type of passive immunity is given through special types of vaccines that work by giving the body antibodies. The protection given by these vaccines does not last as long as that from traditional vaccines. What do you think is the reason for this difference? Often the two types of vaccines are used together to prevent immediate infection and to help the body develop long-term protection. One example of this is in the prevention of rabies after a possible exposure to the rabies virus. Several vaccines are administered over time, the first being an antibody-based vaccine given for short-term immediate immunity while the others are given for longer protection.

Antibodies produce immunity as long as they remain in the body. Passive immunity may last for several months or years before the body destroys the antibodies. You will learn more about vaccines in the next lesson.

Lesson Activity

The lines of defense in your body can be compared to the idea of a war being waged. There are good guys and bad guys. There are ways that you, as a person, can take steps to fortify your lines of defense. Work with a group to develop a Body War Game in which the object is to protect the body from an attack of enemy pathogens. Use your knowledge of infectious disease and the body's immune system to design your game board and game pieces. Develop a game board that simulates the battlefield of the body. Design your game pieces with different strengths and weaknesses. There might be warriors, enemies, and lines of defense. For example, a general can be effective in combating a battalion of pathogens. A scout can warn of an approaching invader. Once you have developed the rules of the game and your game board and pieces, share your game with classmates. Were you able to win the war in combating invading pathogens?

How can you simulate the immune response in your body?

Problems with the Immune System Explain

Our immune system protects us from thousands of diseases. This protection usually allows us to stay healthy. However, sometimes an immune system can harm a person's health instead of protecting it.

Autoimmune Disease

An **autoimmune disease** is a condition in which the immune system targets a person's cells, tissues, or organs by mistake. This person's immune system cannot tell the difference between pathogens and the body's own parts. It may attack normal tissues, such as red blood cells, blood vessels, endocrine glands, joints, muscles, or skin.

The more than 80 types of autoimmune diseases include multiple sclerosis (MS), rheumatoid arthritis, lupus, and type 1 diabetes. In MS, the nerves of the brain and spinal cord are mistakenly attacked by the person's own immune system. The result is loss of muscle control, vision, balance, and feeling. A person can have more than one autoimmune disease. The treatment for these diseases often includes medicines to reduce the body's immune response. What challenges do you think researchers face in developing medicines to prevent or control autoimmune diseases?

Allergies

An *allergy* is an immune system response to a foreign substance that an average person would not be affected by. That substance might be pollen, foods, animal hair, dust, or mold. It could also be chemicals, insect venom, or medicines. A substance that triggers an allergic reaction is called an *allergen*. Do you know anyone that has allergies? What kind of allergies are they? How do they affect the person?

Lesson Activity

Some foods contain ingredients that can be problematic for people with food allergies, including gluten, lactose, or peanuts. Federal law requires that food labels list whether a food product contains certain food allergens. These allergens are not always easy to spot on food labels. A food label may state that a certain food either "contains" or "may contain" a certain allergen.

Examine 5–10 packaged food labels, such as granola bars, cereals, soups, and frozen meals, for ingredients that may be problematic for people with food allergies. Identify which ingredients may create an allergy problem. List the ingredient under the allergy it may cause. Were you surprised to find the ingredients in those foods? Why or why not?

 **How can "hidden" ingredients affect someone with food allergies?**

When the immune system detects an allergen, it forms antibodies. The body reacts to the antibodies by releasing chemicals called *histamines*. Histamines cause uncomfortable reactions. They can include red, swollen tissues or itchy bumps.

Being aware of one's allergies is an effective way to prevent such potential health problems. Precautions such as avoiding the things that cause allergies and taking medications can help reduce or relieve symptoms. Allergy medicines usually contain antihistamines. Based on its name, what do you think an antihistamine does?

Many allergic reactions result in sneezing, itching, or other mild reactions. However, some allergic reactions can be severe and even life threatening. A severe allergic reaction is called *anaphylaxis*. Some symptoms of anaphylaxis include difficulty breathing (caused by the swelling of breathing passages), nausea, vomiting, feeling light-headed, skin rash, and a weak pulse. Certain foods, such as peanuts, or insect bites, such as bee stings, can result in death if not treated immediately. How do you think people with these types of allergies can take extra care to guard and take personal responsibility for their health?

Itching and pain are closely related because the same nerves transmit impulses to the brain for both of these sensations.

Concept Check Assess/Reflect

Summary: How does the immune system protect the body? Your body has three lines of defense that can trap or kill pathogens that invade the body. If a pathogen gets past the physical and chemical barriers of the first line of defense, the inflammatory response takes over. During this second line of defense, white blood cells called macrophages surround the pathogens and break them down. The third line of defense is the immune response, in which a complex group of your body's defenses works against specific pathogens. Immunity can be innate, allowing a quick response to pathogens; adaptive, which is tailored to specific invaders; or passive, which is found in newborn babies or acquired through vaccines. Certain problems of the immune system can result in allergies or autoimmune diseases in which the immune system mistakenly attacks normal tissues.

1. What is the difference between an immune response and an inflammatory response?
2. Our bodies have an immune system. So why do we need physical and chemical barriers to pathogens?
3. How would your health change if your T cells could not recognize specific pathogens?
4. How does the immune system work with other body systems to protect the body from pathogens?

Lesson 2

Essential Question

What Are Infectious Diseases?

Objectives

- Distinguish between infectious and communicable diseases.
- Describe common infectious diseases.
- Distinguish between diseases caused by direct contact and those caused by indirect contact.
- Explain how vaccines work to protect against disease.
- Describe guidelines for the proper use of antibiotics.

Vocabulary

noninfectious disease
infectious disease
communicable disease
virus
direct contact
indirect contact
pasteurization
antibiotic

For thousands of years, doctors did not understand what caused infection or how to prevent it. Many used leeches to remove "poisons" from their patients. When the leeches removed the "poisons," what else were they removing from the body? Surgery during this period often led to death by infection. Why do you think this happened?

Scientists first saw bacteria after the microscope was invented in the early 1700s. However, it was another hundred years before they began to understand how bacteria cause disease. Now we know the causes of many diseases and how disease is spread. What are some factors that put you at risk for diseases caused by bacteria?

Identifying Causes Explain

A disease is a condition that can damage the structures or function of your body. Diseases are often classified as noninfectious diseases and infectious diseases. **Noninfectious diseases** are not caused by disease-causing organisms. They include genetic diseases, such as Down syndrome and hemophilia. They also include diseases related to lifestyle or environment, such as heart disease and skin cancer. **Infectious diseases**, on the other hand, are caused by pathogens that get into the body and cause problems. Infectious diseases include malaria and tetanus, as well as many others. Some infectious diseases are also considered **communicable diseases** because they can spread from person to person. Colds, flu, chicken pox, and sexually transmitted diseases are communicable diseases. Knowing how common communicable diseases spread can help you avoid infection and stay well.

This patient is having blood drawn so the doctor can test for diseases.

Record your work for this inquiry. Your teacher may also assign the related Guided Inquiry.

Detecting the Source of Infection

How do diseases spread?

SAFETY: Wear goggles while working on this lab. Do not drink the liquid.

Materials
- goggles
- clear plastic cups with liquid
- medicine droppers

Procedure

1. To simulate how disease spreads, obtain a numbered cup from your teacher.
2. During the next five minutes, talk with several of your classmates. For example, you might discuss your plans for after school or this weekend's basketball game. As you talk with each person, remove a half dropper full of your liquid and put it in the other person's cup. The other person will do the same.
3. **Record** whom you talked to and the numbers of the cups you came in contact with.
4. After five minutes, return to your desk or table. Your teacher will add a drop of pH indicator to your cup. **Observe** what happens to the liquid.

Analyze Results

See whether you can **identify** the source of the disease that has spread through the class. **Display your data** in a flow chart, web, or graphic of your choice to trace the source of infection. Your graphic should show how students are connected to the source of the disease. How many students ended up with pink water? What might that indicate? Did you have trouble identifying the source of this disease? Why or why not?

Create Explanations

1. How do diseases spread?
2. What did you model in this activity? What do the results show?
3. If you wanted to monitor the spread of a disease, what should you do differently?
4. Is the disease modeled here an infectious or noninfectious disease? Explain.

The main pathogens that cause infectious diseases are viruses, bacteria, protists, and fungi. What infectious diseases do you know of that are caused by viruses? By bacteria?

Check out your *Science Journal* to learn more about how pathogens spread. **Extend**

Viruses

Viruses are pieces of genetic material coated with protein. Viruses are neither dead nor alive. To reproduce, a virus requires a living host cell, but without a host cell, the virus is inactive. Why do you think it needs a living cell? What happens to the host cell?

Some diseases, such as food poisoning and pneumonia, can be caused by either bacteria or viruses. Viral pneumonia is more like the flu. It tends to last a shorter time than bacterial pneumonia. In fact, some people have viral pneumonia without even realizing it.

Diseases caused by a virus include polio, measles, mumps, chicken pox, rabies, West Nile virus, and AIDS, a disease that affects the immune system. You will learn more about AIDS in Lesson 3. Some viral diseases such as influenza—more commonly known as the flu—can be protected against through vaccination. Below are two other common viral diseases.

Common Cold

At least 200 different viruses cause the symptoms of a cold. The cold virus causes irritation and swelling in the respiratory tract. Preschool and elementary school students can have 3–12 colds a year. As you get older, you are likely to catch a cold less often. Why do you think this happens? Although some people think you catch a cold from being cold, you actually get a cold through direct contact with someone who has a cold. Why do you think more people get colds in cold weather? How can you protect yourself from catching a cold? How can you decrease the chance of spreading a cold you have to others?

Scripture Spotlight

Read **Matthew 15:29–31** to see how Jesus responded to those with feared communicable diseases.

Hepatitis

This disease has several forms and causes. Viral forms of hepatitis are spread through contact with fluids, especially blood, from an infected person. Hepatitis can also be spread through contaminated food or water. Some forms of hepatitis cause severe liver damage. A liver transplant might then be needed. People can prevent being infected with certain types of hepatitis by getting a vaccination. What are other ways to protect yourself from getting hepatitis?

Explore-a-Lab

Structured Inquiry

How long does it take for bacteria to grow?

Break three dried beans in half. Put them in a beaker and add about 10 mL of distilled water. Place the beaker in a place where it will not be disturbed. Observe how many days it takes for the water to become cloudy and develop an unpleasant odor. Use the methylene blue to dye a drop of water from the beaker and observe it under a microscope. Make a sketch of what you see through the microscope. What did you observe on the slide that would make the water cloudy and give it the unpleasant odor?

E. coli

Bacteria

Did you know that scientists have learned that you have 10 times more bacteria cells in your body than human cells? That means that 90 percent of the cells that make up your body are nonhuman! *Bacteria* are single-celled microorganisms that live everywhere on Earth. Most bacteria are harmless. Some are even helpful. For example, the bacteria in your intestines help you digest food. Bacteria in the environment cause dead plants and animals to decay. How do you think they do this? Some bacteria produce a poison, or *toxin*, that can damage or kill body cells. On the following pages are some common diseases caused by bacteria. As you study the chart, think about the Body War Game you created in Lesson 1. What other pieces might you include in your game to fortify your body's defenses against invading pathogens? As you read through the rest of this lesson, take time to work with your group, adding game pieces and revising the rules of your game.

Bacteria called *E. coli* can live on some raw foods. If you eat these foods without washing them thoroughly, you could end up with food poisoning.

If food looks clean and fresh, is it free from harmful bacteria? Why or why not?

Focus on Health

Studies have shown that Seventh-day Adventists live longer. A focus on healthful eating, abstinence from alcohol and tobacco, and exercise all contribute to longevity. What effect would healthful living have on any of the diseases listed in the chart?

Bacterial Diseases			
Disease	**Description**	**How It Spreads**	**Prevention**
Food Poisoning	• Caused by microorganisms or toxins present in food • Typically results in vomiting and diarrhea • Symptoms take anywhere from a few hours to a few days to develop.	• Spread by undercooked meat and eggs, improper food preparation and preservation, contaminated utensils and surfaces, and improper hand washing	• Properly prepare and store fresh food. • Wash food-preparation surfaces and utensils with hot, soapy water. • Wash hands for 20 seconds with soap and warm water before and after food preparation.
Tetanus	• Nervous system disease caused by bacteria that live in soil • Results in muscle tightening and spasms • Can make breathing difficult, resulting in death	• Caused by a deep wound that allows bacteria to enter the body • Bacteria produce a toxin that damages the nervous system.	• Get updated tetanus vaccinations. • Clean wounds with soap and warm water. • Seek medical attention for serious, deep wounds.
Strep Throat	• Caused by *Streptococcus* bacteria attacking cells in the throat • Symptoms include sore throat, fever, and swollen lymph nodes under the jaw.	• Spread by airborne droplets from infected sneezes or coughs • Also spread by sharing food and drink utensils and by contact with contaminated objects and surfaces	• Cover mouth when sneezing or coughing. • Wash hands regularly with soap and warm water. • Wash dishes, glasses, and eating utensils in hot, soapy water. • Avoid sharing food or drink.
Pneumonia	• Caused by *Pneumococcus* bacteria attacking cells in the lungs. • Symptoms include cough, fever, fatigue, nausea, shortness of breath, and increased heart rate.	• Spread by airborne droplets from infected sneezes or coughs • Also spread by sharing food and drink utensils and by contact with contaminated objects and surfaces	• Maintain a healthful lifestyle. • Get up-to-date immunizations. • Do not share drinking glasses or eating utensils. • Wash hands regularly with soap and warm water.

Bacterial Diseases			
Disease	**Description**	**How It Spreads**	**Prevention**
Staph Infection	• Caused by *Staphylococcus* bacteria and results in boils, which may fill with pus • Carried by the bloodstream and can infect organs • Can lead to pneumonia and heart disease and may result in death	• Spread through skin-to-skin contact with an infected area • Spread through contact with contaminated objects	• Keep clean and practice good hygiene. • Wash hands regularly with soap and warm water. • Keep skin injuries (cuts, scrapes, etc.) clean and covered. • Avoid sharing towels, sheets, or clothing unless they have been washed properly.
Lyme Disease	• Symptoms include rash, fever, headache, and fatigue. • If untreated, can affect the joints, heart, and nervous system	• Spread by the bite of certain kinds of ticks • Caused by bacteria that multiply in the ticks' bodies and spread to the animals or people the ticks bite	• Avoid tick-infested areas. • Wear long pants and long sleeves when in the woods. • Use an insect repellent with 20% or higher concentration of DEET. • Check yourself and pets for ticks after spending time in wooded or grassy areas.

Lesson Activity

Traveling Diseases Many infectious diseases are spread as people travel within a nation or around the world. These diseases include plague, measles, malaria, HIV/AIDS, smallpox, tuberculosis, swine flu, hepatitis, and avian flu. Choose a disease that has been spread by travelers. Describe where it seemed to have started and how far and how fast the disease spread. Is it still affecting people? How can it be prevented or treated? Create a graphic organizer to share what you learn with the class.

How do travelers spread disease?

Deer ticks spread Lyme disease.

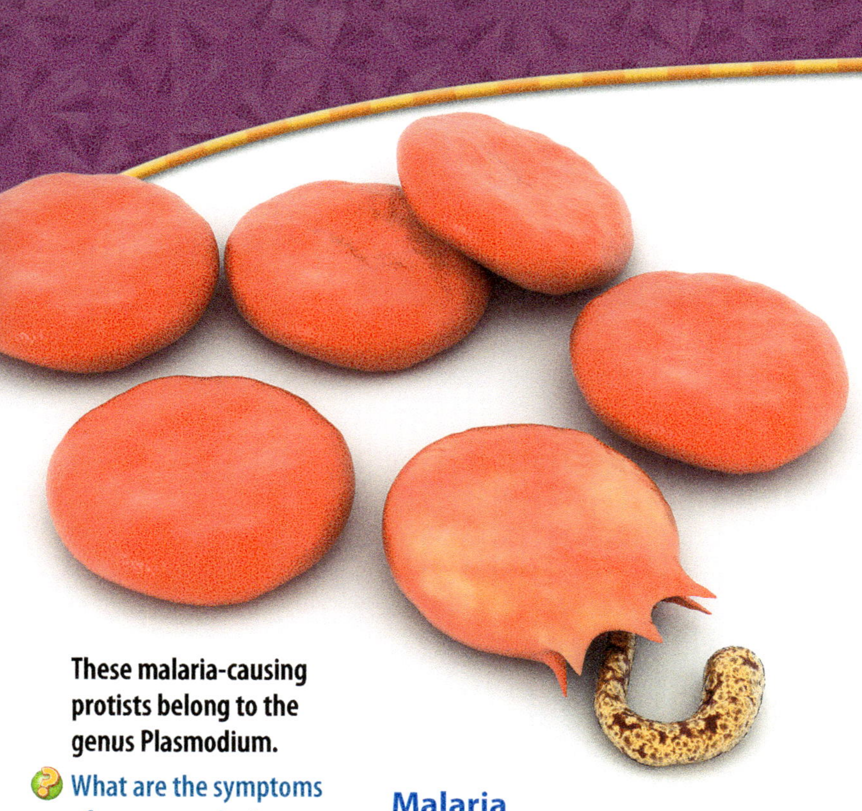

These malaria-causing protists belong to the genus Plasmodium.

What are the symptoms of someone who is infected with malaria?

Protists

Protists are the most diverse group of living organisms. They vary in structure and function more than any other kingdom of living organisms. Protists are typically unicellular organisms that live in soil, oceans, lakes, or streams. Most protist microorganisms are harmless. However, some protists are a major cause of disease in the world. Malaria and amoebic dysentery are two such diseases.

Malaria

Protists that cause malaria are carried by mosquitoes and infect people's blood through the mosquito's bite. The word *malaria* means "bad air" in Italian. However, in 1880 scientists discovered that the disease was caused by protists in the blood, not bad air. Still, they continued to call it "malaria." Today over 200 million people are infected with malaria, mostly in Africa, Asia, and Central and South America. Each year, hundreds of thousands of people die from this disease. Estimates are that about 90% of malaria deaths occur in Africa. Travelers going to countries where malaria is prevalent can take precautions by taking preventive medications. Imagine you are on safari in Africa. What additional precautions could you take to prevent yourself from becoming infected with malaria?

Amoebic dysentery

A certain type of amoeba is the culprit for amoebic dysentery, a disease that causes painful abdominal cramps, fever, and diarrhea. People can become infected with the protist that causes this disease by eating contaminated food or drinking contaminated water. Amoebic dysentery is common in developing countries. Why do you think this is so?

Fungi

Fungi are organisms that include molds, yeasts, and mushrooms. Unlike plants or animals, fungi get their energy from living and once-living things. They secrete enzymes into a food source, such as decaying plant matter or animal waste. Then the food is broken down and absorbed. Fungi typically grow in dark, warm, moist places, including your skin or mucous membranes. Two common fungal diseases are athlete's foot and ringworm.

Athlete's Foot

Damp socks and plastic shoes can keep your feet moist, which encourages the growth of the fungus that causes athlete's foot. The fungus infects the top layer of skin, causing the next layer to produce extra skin cells. These cells push to the surface, making the skin thick and scaly. You can catch athlete's foot by direct contact with someone else's infected foot or by sharing shoes or socks. You can also pick up this disease from floors, showers, swimming pools, and pets.

Ringworm

Ringworm is named for the itchy, red, circular sores it causes. However, unlike its name suggests, ringworm has nothing to do with worms. This disease is caused by a fungus that lives in the outer layer of the skin. Ringworm spreads easily in the same ways as athlete's foot. What animals other than humans might get ringworm?

Scripture Spotlight

In **Psalm 103:3** the Bible says that the Lord is the one who heals all our diseases.

How Diseases Are Spread Explain

You have probably heard of many ways that you can prevent spreading or catching illnesses. Cover your mouth when you cough or sneeze. Wash your hands after using the bathroom. These are just some ways you can prevent the transmission of disease.

Pathogens can be transmitted in three ways: by direct contact, by indirect contact, and by animals. The Centers for Disease Control and Prevention continually conducts studies to identify factors contributing to the spread of disease. Monitoring diseases is an important tool in the fight against communicable diseases. What are some factors that might affect how the spread of disease can be slowed or stopped?

How Diseases Are Spread

Cause	Description
Direct Contact	**Direct contact** occurs when a pathogen from one person is carried directly to another person. Touching, kissing, and sexual intercourse are examples of direct contact. Diseases transmitted by sexual activity, such as gonorrhea, syphilis, and the HIV virus, are transmitted by direct contact. Medical technologists and those who draw blood wear gloves, masks, and/or face shields to prevent the transmission of pathogens by direct contact.
Indirect Contact	**Indirect contact** means that pathogens move from an infected person to some object, perhaps a doorknob, and then to the next person who touches that object. Drug abusers who share needles may infect each other through indirect contact. Diseases such as chicken pox and the flu can be transmitted by indirect contact.
Animals	Animals become infected with pathogens through their food, water, or the bites of other animals. In many cases, an animal itself is not affected by the pathogen but carries the pathogen in its body. Rabies, a viral disease, is transmitted by the bite of an infected dog, skunk, or other mammal. Lyme disease is carried by the deer tick. Insects carry many other diseases. Mosquitoes spread malaria and yellow fever. The fleas that live on some rodents can transmit bubonic plague, which killed millions of Europeans during the fourteenth century. Since the animals that spread these diseases are all around us, how might people control the infectious diseases that are spread by them?

Faith Connection

Most pathogens are not visible to the naked eye, and spiritual enemies may also be unseen. As we defend ourselves from communicable diseases, we must also defend ourselves against the spiritual forces of evil.

Explore-a-Lab

Structured Inquiry

What places are at a high risk for contamination?

Bacteria are present everywhere. What surfaces in your school do you think are most likely to be contaminated with bacteria? Work with a partner to formulate a hypothesis. Your teacher will give your team a Petri dish filled with nutrient agar and a cotton swab. Obtain a sample of the surface you chose with your cotton swab and rub it across the agar. Label the Petri dish and place it with other teams' dishes. After several days, observe and record your observations. Do not open the dish! Which surface had the most growth of bacteria? Why do you think so? How is this related to how diseases are spread?

Fighting Off Pathogens Explain

Few of us manage to avoid all pathogens. Some can still get past our physical and chemical barriers. You already know that the easiest way to ward off potential invaders is to wash your hands often. But there are other effective ways to control and fight off diseases. Vaccinations, antibiotics, and pasteurization are just a few modern medical technologies that can help control disease. You have probably had a vaccination at some point in your life or know someone who has. How do you suppose vaccinations fight off infectious disease?

The red objects in this image are bacteria that cause gonorrhea.

What is the most effective way to prevent contracting an STD?

Vaccinations

Soon after you were born, you may have begun to get vaccinations. They might have been an injection, a liquid you swallowed, or a nasal spray. They safely protected you from serious diseases such as measles, mumps, whooping cough, and tetanus. As a result, there has been a marked drop in deaths from diseases that can be prevented by these vaccinations. For example, in the first half of the twentieth century, polio was a serious infectious disease that paralyzed thousands. Today, the number of polio cases worldwide has dramatically decreased. How has the near eradication of polio been accomplished through mass vaccinations? What other infectious diseases have shown a marked decline or near eradication due to vaccination?

Some—but not all—of these vaccines protect you for life. For example, you will need another tetanus vaccination every 10 years. This booster shot will make sure you are still immune to this deadly disease. The diagram to the right shows how a vaccination works.

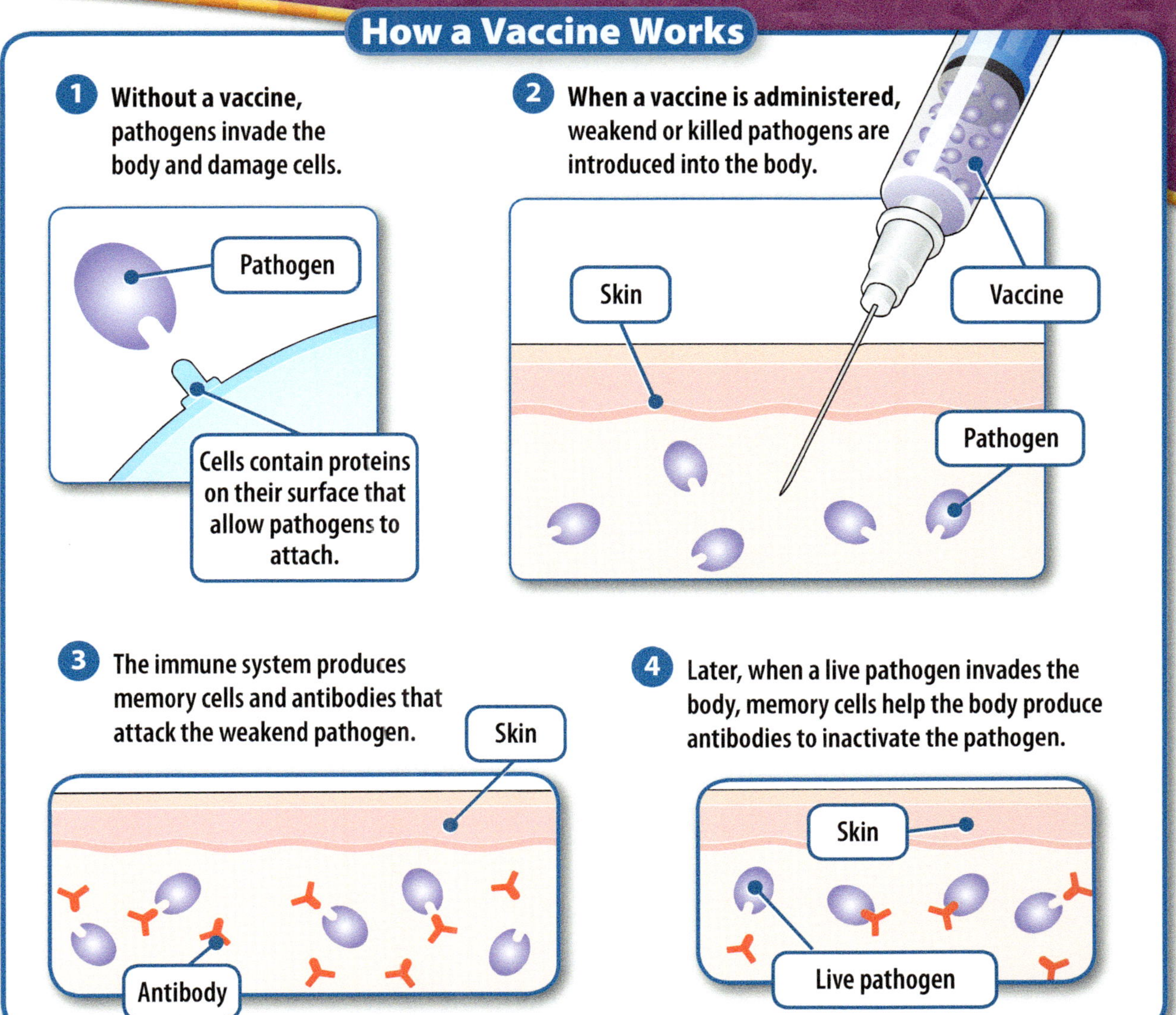

Lesson Activity

Many infants in the United States receive vaccinations. Most school systems require children to be vaccinated against several diseases before they can start school. Each state has its own laws, so the requirements vary. Even though vaccines prevent disease, some people question their safety. For example, most babies are vaccinated when they are two to four months old. This is also the peak time when some babies die suddenly, with no clear cause, from a syndrome known as *sudden infant death syndrome*, or *SIDS*. Some parents worry about a link between vaccines and *autism*. Autism is a condition that causes a person to have problems in communicating and relating to others. With a partner, choose one of the conditions mentioned in the text about vaccinations. Use the Internet to research whether there is any evidence of a connection between the vaccine and the condition. Do the risk factors of vaccinations outweigh the benefits? Support your conclusions with the latest data and scientific research.

What are the risks of vaccines?

Math in Science

A few people believe that a reduction in disease is not due to vaccines. They think that less crowded living conditions, new medicines, and better health care have reduced the number of cases.

Does this graph support the value of vaccines? Why or why not?

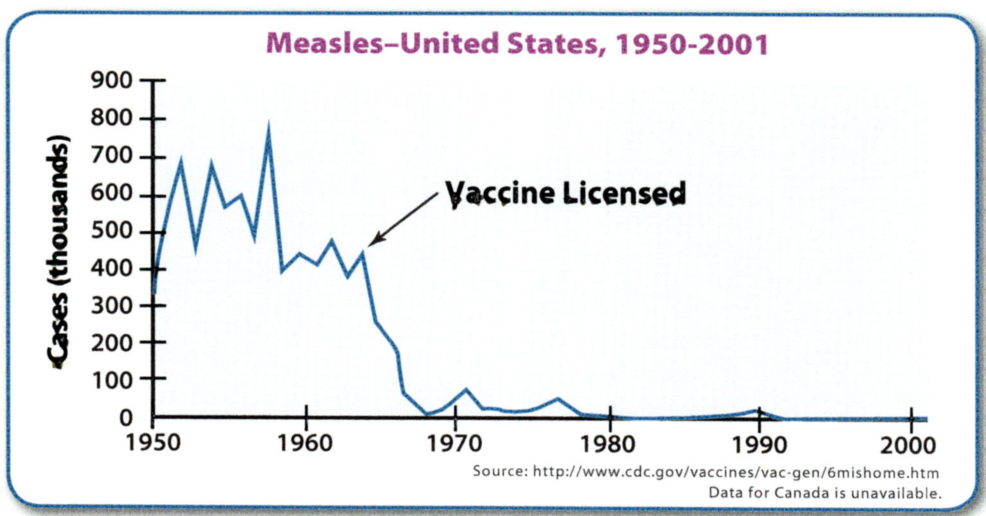

Use the graph above to answer these questions:

1. In which year did measles cases peak? What was the highest number of cases?
2. How many fewer measles cases were there within five years after the vaccine was licensed?
3. Based on this graph, what might cause the number of measles cases to rise again?

Pasteurization

In the 1800s, Louis Pasteur was one of the first to understand that microorganisms can cause disease. In his experiments, Pasteur prevented milk from spoiling by heating it and then cooling it quickly. The heat killed most of the bacteria in the milk. However, some harmless bacteria remained. They still caused the milk to spoil if the milk were left at room temperature. Named for Pasteur, **pasteurization** is the process of heating a substance to destroy or slow the growth of microorganisms in it.

Pasteur wondered whether the microorganisms that cause milk to spoil would also cause disease. Through experiments, he demonstrated that microorganisms cause many diseases in humans, other animals, and plants. In time, Pasteur was able to identify virus particles. He experimented with weakened viruses. Pasteur himself produced a vaccine against rabies. Through the process of pasteurization and his discovery that bacteria cause disease, this scientist has saved millions of lives. Other scientists have built on his work. For example, Joseph Lister used Pasteur's discoveries to greatly reduce the rate of infection in hospital patients after surgery.

Pasteurization has lowered the cost of food and has made food less likely to carry harmful bacteria. How does this improve our economy? How do you think this technology has impacted trade between countries? Many products today, such as juice, cheese, and other dairy products, are pasteurized before we bring them into our homes. Read the labels of foods you eat. How many of them have been pasteurized?

Other scientists repeated Pasteur's experiment. However, sometimes the broth spoiled in flasks with long necks. How could this happen?

Evidence of Microorganisms in the Air

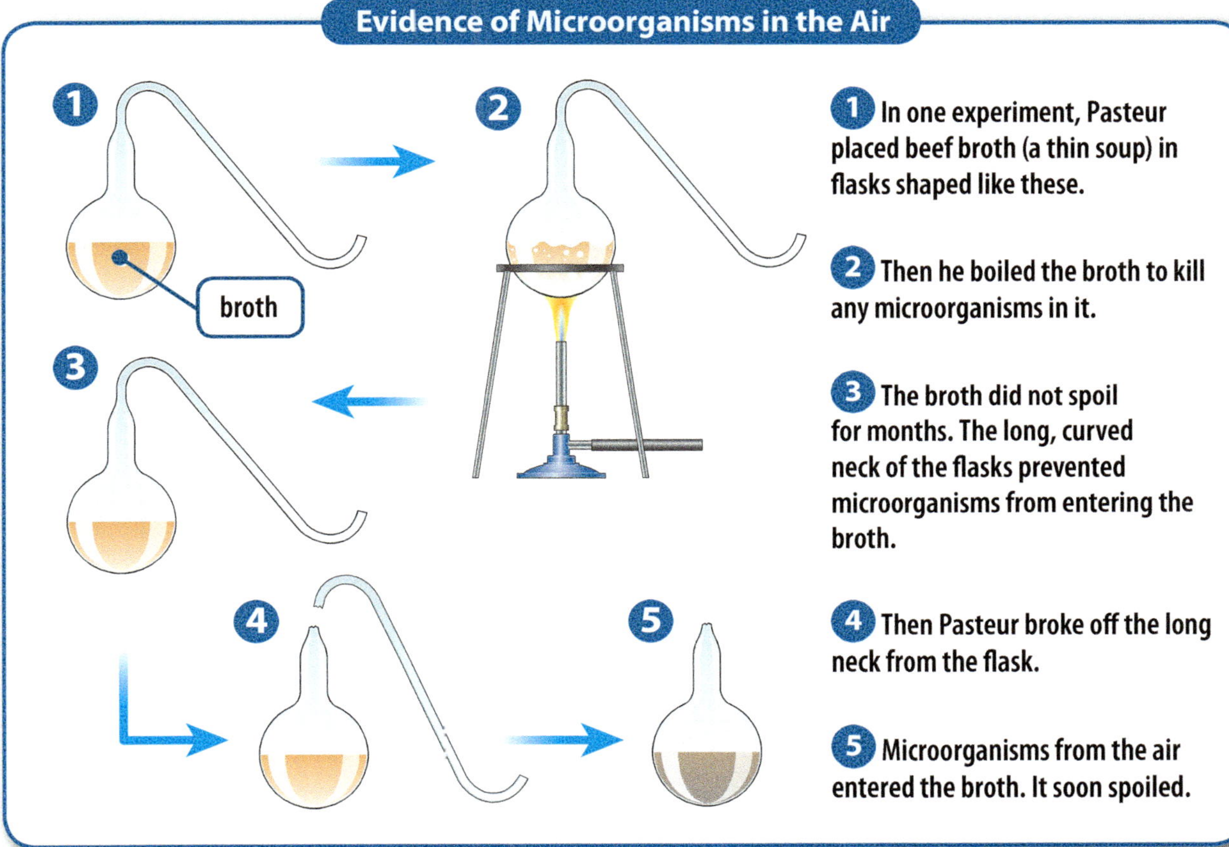

1 In one experiment, Pasteur placed beef broth (a thin soup) in flasks shaped like these.

2 Then he boiled the broth to kill any microorganisms in it.

3 The broth did not spoil for months. The long, curved neck of the flasks prevented microorganisms from entering the broth.

4 Then Pasteur broke off the long neck from the flask.

5 Microorganisms from the air entered the broth. It soon spoiled.

Explore-a-Lab

Structured Inquiry

How can you observe the benefits of pasteurization?

Obtain two containers of raw milk from your teacher. Heat one container to 63°C and keep it at that temperature for 30 minutes. Cool the milk to room temperature. Do not heat the second container. Observe the milk every day for three days. Each day, obtain a sample of milk from each container and place each sample in a separate test tube. Place a few drops of methylene blue into each sample. In the presence of bacteria, the methylene blue will turn colorless and the milk will change from blue to white. Record your observations. Discuss with your classmates how this activity demonstrates the process of pasteurization.

Focus on Health

During Bible times, the communicable disease leprosy was a major public health issue. Today, the diagnosis and treatment of leprosy are quite easy. However, in some remote areas of the world, leprosy is still a problem. Health organizations are trying to reach these areas with multidrug therapy.

Antibiotics

Not every disease has a vaccine to prevent it. When you do get sick, certain medicines might help you fight off the pathogens. An **antibiotic** is a chemical that kills bacteria or prevents them from reproducing. Antibiotics work only on bacteria, not viruses.

Antibiotics have limitations. If antibiotics are overused, they may not be as effective. The antibiotic may kill the weakest harmful bacteria in your body, but the strongest bacteria will survive. The antibiotic can also cause those bacteria to change their genetic structure. They become immune to that antibiotic. The next time those bacteria make you sick and you take that antibiotic, it may not work! The bacteria have become resistant to that antibiotic. How can this create huge health problems? What happens if you take an antibiotic to treat a viral disease?

Antibiotics can be misused in other ways, too. If you have a prescription for an antibiotic and you stop taking the medicine as soon as you feel better, you may get sick again. You did not take enough antibiotic to kill all of the bacteria. The bacteria may infect your body again. This second infection may be worse than the first infection because the bacteria may have become resistant to the antibiotic.

These guidelines can help you use antibiotics safely:

1. Take antibiotics only if your doctor prescribes them for your condition.
2. Follow the directions on the label.
3. Take the medicine as long as the doctor says to, even if you start feeling better.
4. Tell your family whether you have any side effects, such as an upset stomach or a rash.
5. Never take a prescription drug prescribed for someone else.

Antibiotics most frequently are prescribed in pill form. They also come as liquids and ointments.

Why is an antibiotic not an effective treatment for the common cold?

Fighting Pathogens at Home

Has the water supply in your neighborhood ever been contaminated? If so, what did you do to protect yourself from pathogens in the water? People can be diligent at home about protecting themselves from infection by taking everyday precautions. Household disinfectants can be used to wipe down counters or cutting boards that may be contaminated with salmonella, a bacterium found on raw chicken. Thoroughly cooking food reduces the risk of becoming infected with harmful bacteria that may be found on foods. Good hygiene is important, too. What steps can you take in your own personal hygiene to protect yourself from potential pathogens?

Keeping surfaces clean, especially in the kitchen, can help reduce the spread of diseases caused by pathogens.

How do disinfectants fight pathogens?

Summary: What are infectious diseases? Infectious diseases are caused by pathogens that have entered the body. A communicable disease is a disease that can spread from person to person. Viruses, bacteria, fungi, or protists can cause infectious diseases. The flu, common colds, malaria, strep throat, and Lyme disease are just a few infectious diseases. Some infectious diseases are transmitted by direct contact with other people, while others are transmitted by indirect contact with an object. Animals also carry pathogens that can cause infectious diseases in humans through animal bites. Vaccinations, pasteurization, and antibiotics are ways to prevent and control the spread of infectious diseases.

1. What are two reasons why so many people get communicable diseases?
2. Are you more likely to get a disease caused by bacteria or fungi? Explain your answer.
3. How does vaccinating children protect others from disease?
4. In your opinion, what was Louis Pasteur's most important discovery? Explain your answer.

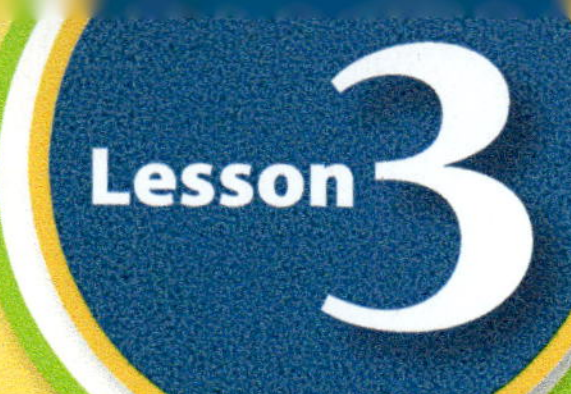

Essential Question

How Are STDs Prevented?

Objectives

- Identify common sexually transmitted diseases.
- Explain how HIV/AIDS affects the body.
- List and describe the ways in which HIV can be transmitted from one person to another.
- Identify effective ways to reduce and prevent the risk of STD infection.
- Identify ways of obtaining current information about STDs.

Vocabulary

AIDS
HIV
opportunistic infection
retrovirus

As you learned in the previous lesson, communicable diseases can spread from person to person. Some of these diseases are specifically spread through sexual contact with another person. Why do you think it is important to know about sexually transmitted diseases? How does God's plan for sexuality help protect you from contracting these diseases?

Sexually Transmitted Diseases Explain

When the U.S. population was in excess of 3 million people and growing, the Centers for Disease Control and Prevention estimated that about 19 million people in the United States became infected with an STD every year. Do you think that this trend continues today? Do you think that the number of people infected with STDs has increased or decreased since the last population census? Explain. Sexually transmitted diseases (STDs) are spread through sexual contact with an infected person. Like other communicable diseases, STDs are caused by viruses, bacteria, fungi, and protists.

No one is immune to sexually transmitted diseases. People of all ages can become infected with an STD. Unborn babies may become infected from their mothers. It is even possible to become infected with more than one STD at the same time.

In recent years, the number of STD infections occurring in the teenage population has risen. This increase is due to several factors: teenagers are becoming more sexually active, they often have a greater number of sexual partners, their use of drugs and alcohol is increasing, and they often do not take protective measures.

What other factors do you think might account for a rise in the incidence of STDs? What do you think is the current trend? Explain.

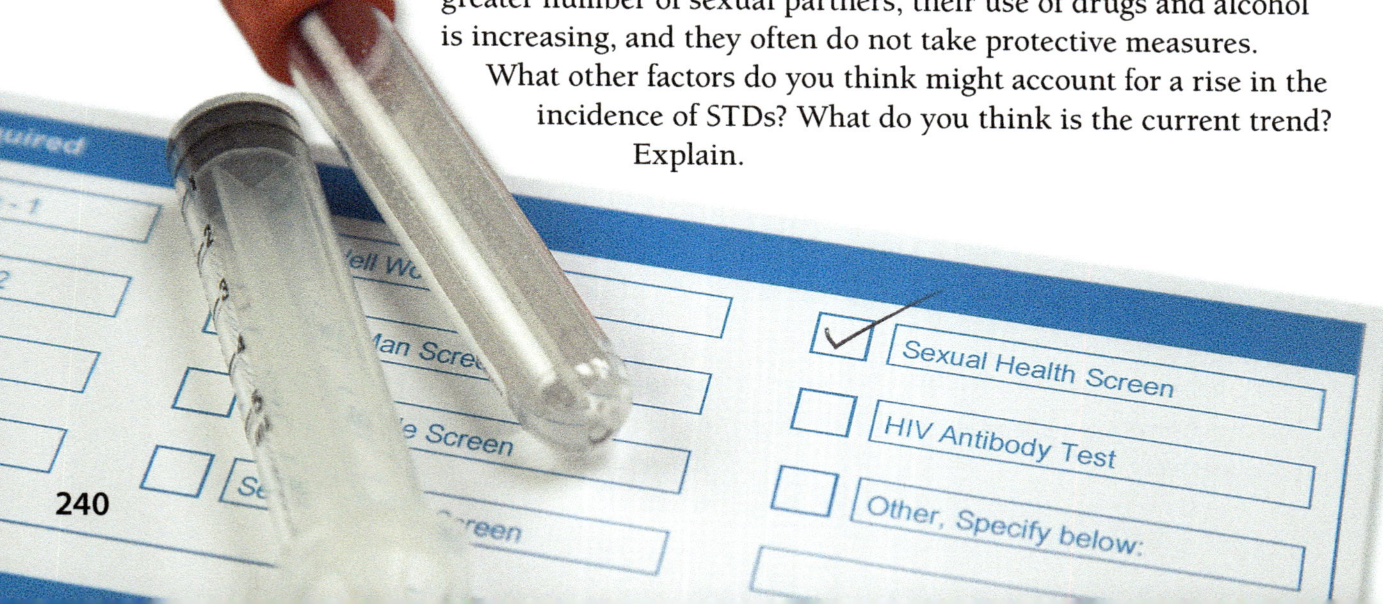

Record your work for this inquiry. Your teacher may also assign the related Guided Inquiry.

Hey, Wanna Trade?

How is HIV spread?

Materials

- colored construction paper (5 cm x 5 cm)

Procedure

1. Obtain a square of colored construction paper.
2. When given the signal, you may trade your square of construction paper with anyone else in exchange for his or hers, or you may choose not to trade. If you choose to trade squares, **record** the color you traded for in the data table.
3. Continue trading as many times as you want during the time allotted. Be sure to record how many times you traded and the color each time.
4. When time is called, stop trading, record the color of paper you ended up with, and complete the rest of the table.

Trade Number	Red	Orange	Yellow	Green	Blue	Purple
1.						
2.						
3.						
4.						
5.						
6.						
7.						
8.						
9.						
10.						

Analyze Results

Calculate the average number of trades students in your class made, the average number for students without HIV, and those with HIV. **Compare** the numbers.

Create Explanations

1. How is HIV spread?
2. How did the number of trades a person made affect their odds of getting the disease?
3. How would following God's plan reduce the spread of HIV?

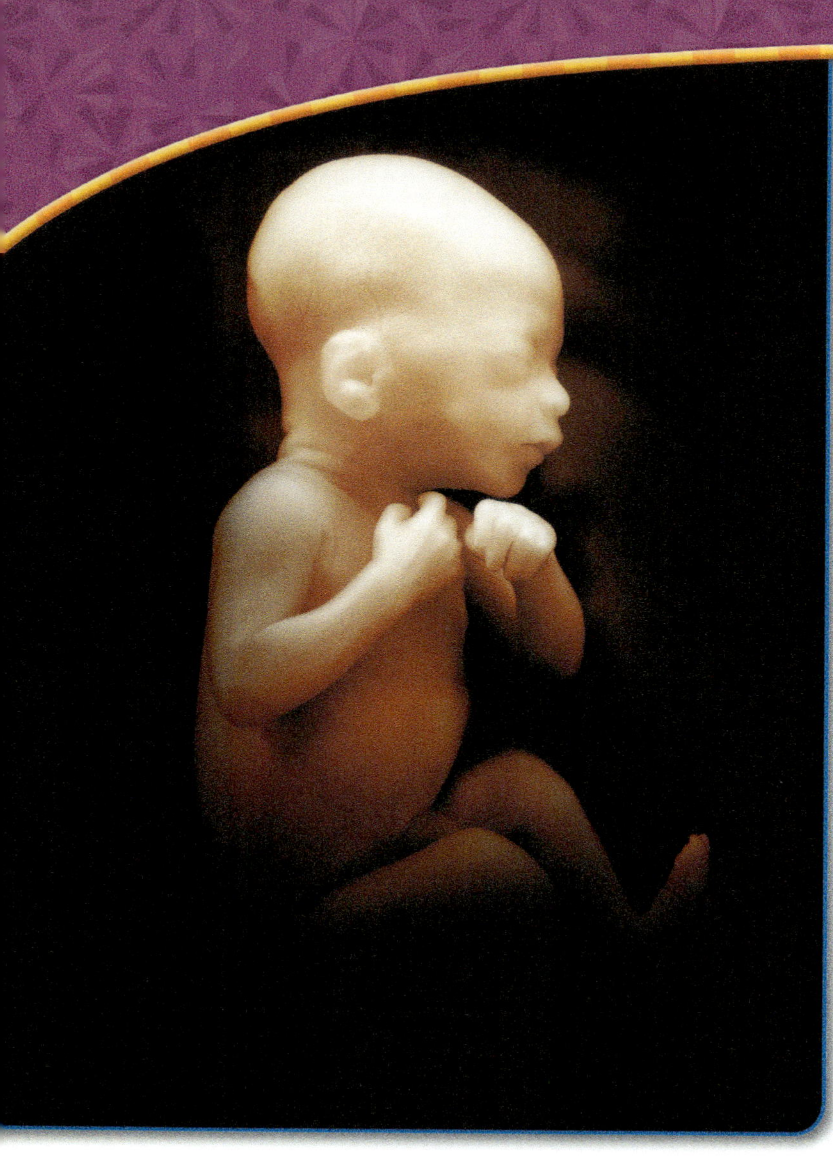

STDs can make both men and women infertile, or unable to have a baby. They can also cause organ and nerve damage. STDs affect more than just the people involved in sexual activity. If a woman has an STD that goes untreated and she becomes pregnant, the STD can also harm the baby. An example is syphilis, which can pass from mother to unborn baby during pregnancy. Other STDs such as gonorrhea, chlamydia, and herpes can infect the baby as the baby passes through the birth canal. Without preventive medications, HIV can be passed from mother to fetus. STDs passed to babies can lead to death, low birth weight, blindness, brain damage, and many other health problems. Why is it important to know what these STDs are? What are ways that people can protect themselves from becoming infected?

Unborn babies become victims of STDs when their mothers are infected.

How do you think the STD infection travels from the mother's body to the unborn child?

Common Sexually Transmitted Diseases

The most common sexually transmitted diseases are chlamydia, genital herpes, genital warts, gonorrhea, syphilis, vaginitis, and AIDS. Read the following information about the causes, symptoms, and treatment of STDs.

Chlamydia

- Chlamydia, the fastest spreading STD in the United States, is caused by bacterial infection.
- In males, it produces inflammation of the urethra, causing a thin mucous discharge and a burning sensation when urinating. Or, there may be no symptoms.
- In females, no symptoms occur until the disease is in its late stages.
- If untreated, chlamydia may cause damage to the reproductive organs or sterility.
- An infected mother can pass the disease to her baby, and it can cause blindness.
- Chlamydia is treated and cured with antibiotics.

Scripture Spotlight

Read **Proverbs 3:7–8**. How can making good choices benefit us both physically and spiritually?

Genital Herpes

- Genital herpes is caused by the herpes simplex virus, which also causes cold sores.
- Once infected, a person is infected for life. The disease disappears and reappears throughout life.
- Symptoms include red, tender skin in the genital area. One or more blisters or bumps form, which may burn or itch and may cause flu-like symptoms.
- An infected mother can pass the disease to her baby. It can cause severe brain fever, which can be fatal.
- No cure exists, but herpes outbreaks can be treated with antiviral drugs.

Genital Warts

- Genital warts are caused by *Human papillomavirus*, which also causes cancer of the cervix.
- Warts on other parts of the body are caused by different viruses. Genital warts are growths on the skin of the genital area and around the anus.
- These warts can be treated with cryotherapy (freezing the wart with liquid nitrogen) or with lasers, electrocautery, or antiviral drugs.
- In recent years, a vaccine for *Human papillomavirus* has been developed for teens to prevent infection from certain forms of this virus.

Gonorrhea

- Gonorrhea is caused by bacterial infection.
- In males, gonorrhea causes discharge from the penis and painful urination.
- In females, gonorrhea may cause vaginal discharge, abdominal pain, fever, and pelvic inflammatory disease.
- Symptoms of gonorrhea may not appear at first. If left untreated, it can cause sterility in woman.
- Gonorrhea is treated and cured with antibiotics. However, antibiotic-resistant strains of gonorrhea have developed, making treatment of this disease more difficult.

These bacteria, *Neisseria gonorrhoeae,* cause gonorrhea. They have progressively developed resistance to the antibiotics used to treat the disease.

Syphilis

- Syphilis is caused by a type of bacterium called a spirochete.
- This disease has three stages. In stage 1, a painless sore forms on the genitals or mouth. Stage 2 symptoms include rash, fever, headache, and weight loss. If left untreated, Stage 3 of syphilis can result in damage to the cardiovascular and nervous systems.
- Symptoms may disappear, but the disease does not unless treated.
- Syphilis is treated and cured with antibiotics in the early stages.

Vaginitis

- Vaginitis, an infection of the vagina, occurs only in women.
- Not all forms of vaginitis are sexually transmitted.
- Trichomoniasis is a type of vaginitis caused by a protist. Symptoms include a yellow discharge, itching, and burning.
- Candidiasis is another form of vaginitis. It is a fungal infection caused by yeast. Symptoms include a cottage-cheese-like discharge and intense itching.
- Gardnerella is a vaginitis caused by bacteria. It produces a grayish, watery, strong-smelling discharge.
- Different forms of vaginitis can be treated with appropriate medications and specific drugs. Good personal hygiene and cotton undergarments ensure air circulation and reduce the risk of vaginitis.

Lesson Activity

With a partner, draw a chart like the one below. Determine the cause of each disease. Place an *X* under *V* for virus, *B* for bacteria, or *O* for other. Determine who is most affected by the disease. Place an *X* under *M* for males or *F* for females. Briefly describe the symptoms and treatment for each disease listed. Finally, rate the seriousness of the disease, using this rating system: 1 = extremely serious (life threatening), 2 = serious, 3 = moderately serious, 4 = not serious. What can you conclude from this activity? How will your conclusions help you make healthful choices?

Disease	V	B	O	M	F	Symptoms	Treatment	Rating
Chlamydia								
Genital herpes								
Genital warts								
Gonorrhea								
Syphilis								
Vaginitis								

 **Which STDs have the most serious impact?**

HIV and AIDS Explain

In 1981, a new disease was discovered. When young men in several parts of the United States began to die of a rare type of cancer called Kaposi's sarcoma, physicians suspected that something was damaging their immune system. As scientists found out more about the disease, it was named acquired immune deficiency syndrome. The disease is often referred to by its acronym, AIDS. **AIDS** is a sexually transmitted disease in which the immune system can no longer protect the body against invasion by pathogens or opportunistic infections. In 1983, the virus that causes the disease was identified, and in 1985 it was named the human immunodeficiency virus, or **HIV**.

As epidemiologists began to trace the number of persons with symptoms of AIDS, they discovered cases in Europe, Africa, North and South America, and the Caribbean. Within five years, AIDS was found in every country in the world. Many people become infected with HIV during their teenage years. Why do you think it is valuable for teenagers to know the risks associated with this disease?

HIV virus attacking cell

HIV and Its Action in the Body

Opportunistic infections are caused by microorganisms commonly present in or on the human body or found in the environment. Normally, these microbes do not cause illness in a person with a healthy immune system. However, if the body's immune system is damaged, these microbes gain a foothold and cause infections. AIDS allows such opportunistic infections to occur.

The name of the virus, human immunodeficiency virus, describes the action of the virus; it causes the immune system to become deficient. Recall that viruses are capable of reproducing only in living cells. The common cold virus attacks only cells in the respiratory tract, the hepatitis virus attacks only liver cells, and the rabies virus uses nerve cells for its growth and reproduction. Viruses also have a host cell preference. HIV prefers invading the cells that play a major role in defending the body against infection—the white blood cells. Why do you think that a person who is infected with HIV might not be aware that he or she is infected?

Retroviruses carry their genetic information in RNA.

How else is HIV different from other viruses?

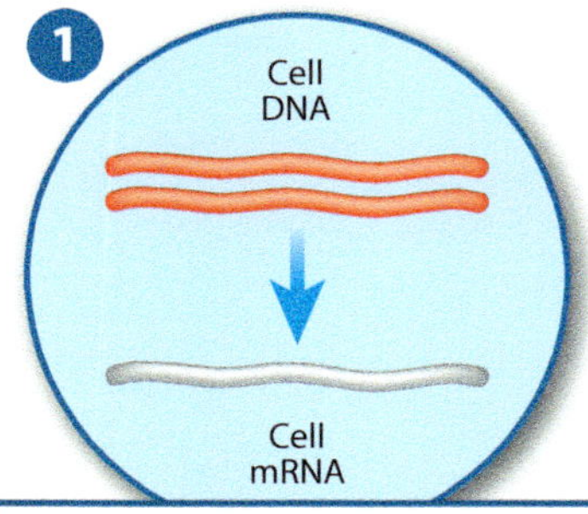

Normally, in a viral infection, the flow of genetic information is from the DNA of the virus to the RNA of the host cell. HIV reverses this flow.

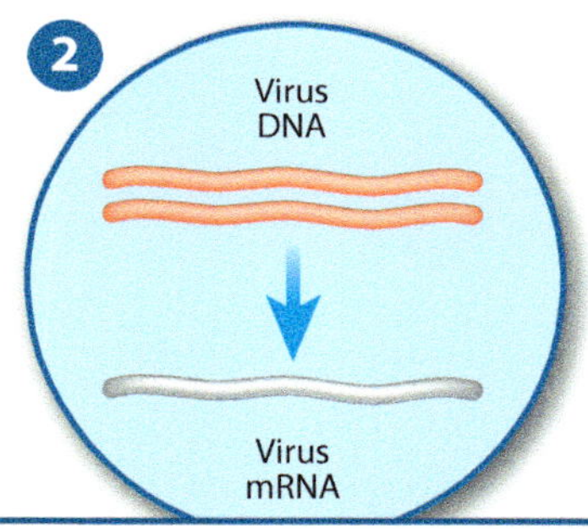

The genetic material of HIV is encoded in RNA.

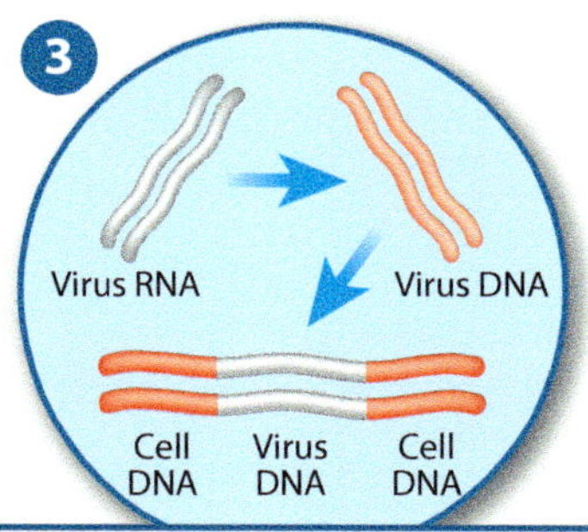

When HIV attacks a cell, the genetic information of the virus passes from its RNA to the host cell's DNA. Because of this reverse flow of genetic information, HIV and similar viruses are called retroviruses.

When HIV takes over the cells of the human immune system, it enters the genetic material of the cells. It acts like a pirate, taking over the cells of the immune system by inserting itself into the DNA of the host cell. HIV grows quietly in the lymph system, gradually destroying white blood cells called T-helper cells. When enough of these cells are destroyed, the infected person has more difficulty fighting off other diseases. The infected person then begins to have symptoms of AIDS.

Symptoms of AIDS are flu-like and can include fever, chills, rash, muscle aches, sore throat, fatigue, swollen lymph nodes, and mouth ulcers. In the early stages of AIDS, a person may experience symptoms as early as two weeks or up to three months. Some people do not experience any symptoms. As the disease progresses, symptoms may disappear. Without treatment, it can take up to 10 years before a person moves into full-blown AIDS. By then the immune system has been severely damaged. Why do you think that the symptoms of AIDS can take such a long time to develop?

Human T-Helper Lymphocyte

HIV attacks the body by attacking the T-helper cells, causing them to produce copies of HIV virus.

When HIV infects a T-helper cell, the T-helper cell can no longer signal the other cells of the immune system, making the system unable to protect the body from disease.

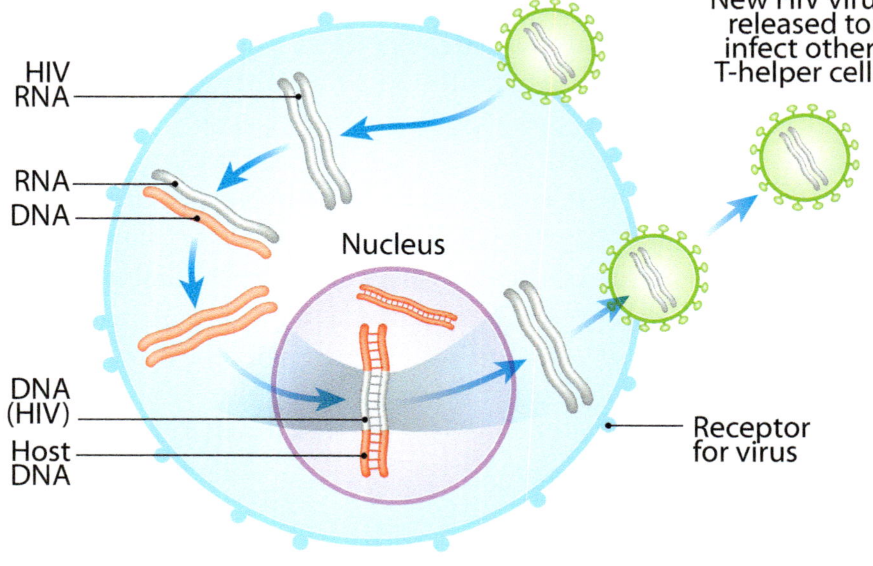

How HIV Is Transmitted

HIV cannot multiply outside of living cells. It is unable to survive outside the body. Infected persons can transmit HIV to other persons by direct contact, through blood and other body fluids, and between a mother and her unborn child. Not all body fluids transmit HIV. The chart below summarizes the ways that HIV can and cannot be transmitted. How does the information in the chart clear up any misconceptions you may have about how HIV is transmitted?

Transmission of HIV	
Yes	**No**
Sexual intercourse	Mosquitoes
Sexual activity	Touching hands
Blood transfusion	Hugging
Organ transplant	Food
IV drug use	Dishes
Mother to child	Clothing
Exposure to body fluids: blood, semen, vaginal fluids, pus	Toilet seat
Occupational exposure	Sneezing
Open sores or bleeding gums in the mouth	Coughing
	Saliva
	Tears

Treatment for HIV

Over the years, treatments for HIV have been developed that are effective in prolonging the onset of AIDS. In 1987, a drug called AZT was approved as the first treatment for HIV. Since then, more than 30 drugs have been developed for the treatment of HIV and AIDS. These drugs, called antiretrovirals, are often referred to as "cocktails." That is because many people take a combination of different drugs. The drugs work to keep HIV from growing and multiplying in the body. Thanks to antiretroviral drugs, people can now live in good health with HIV for decades. These drugs are not a cure and can't stop an HIV-positive person from infecting others. Scientists continue to work for a real cure.

Although the drugs work to control HIV in the body, they often have unpleasant side effects. These include diarrhea, dizziness, fatigue, headaches, nausea, vomiting, rash, pain, and nerve problems. What other kinds of issues do you think a person living with HIV has to deal with? How does HIV affect other areas of a person's life?

Preventing STDs Explain

You have learned that an STD infection can have serious consequences. Most STDs can be treated. However, early detection can be difficult. A person may have the disease without knowing it. In these cases, the treatment may come too late to prevent the serious effects of the disease. Often, an STD comes back to haunt a person years later. Married couples find out that the disease they had as teenagers makes it impossible for them to have children. Or a woman discovers she has a disease that has infected her unborn baby. What other consequences of STD infection may be unknown until some time in the future?

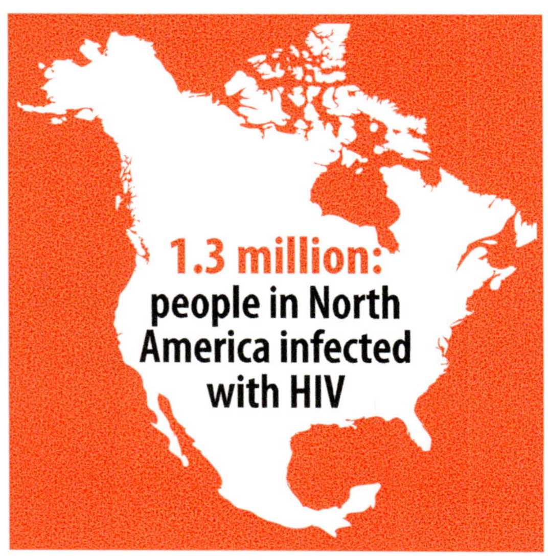

1 in 6:
Do not know they are infected

Estimated children living with HIV: 1400-2600
Most of them do not know they are infected, are not getting treated, and can unknowingly pass the virus on to others.

Responsible Sexual Behavior

The best way to deal with STDs is to prevent them. You have learned that sexual abstinence is refraining from all sexual activity (not just intercourse) until marriage. This is God's plan for all of us, and it is the responsible choice for teenagers. Choosing not to have sexual contact outside of marriage reduces the risk of having sexual contact with an infected person. Later in life when you are ready to marry, choose a faithful marriage partner who is not infected. Although it is true that barriers such as *condoms*—latex devices used to avoid pregnancy and reduce the risk of STD infection—may prevent the transmission of some STDs, they are not 100% effective. When you are ready to marry, learn all you can about your prospective partner. Both of you should have a physical examination, including tests for STDs.

If an individual has been infected with an STD, early treatment is the best means of preventing further damage or transmission to another person. People who have had several sexual partners and are diagnosed with an STD should take the responsibility to tell their sexual partners, who may have become infected without knowing it. In most states and provinces, public health workers trace such contact to be sure everyone has been treated. Pregnant women should be checked for STD infection. In most cases, early treatment can protect the unborn baby. In HIV/AIDS, however, protecting the unborn baby is difficult. Why do you think this is so?

Obtaining Current Information

You can obtain current scientific information about HIV and AIDS and other STDs by reading reliable magazines and newspapers and by watching educational television programs. Websites sponsored by organizations such as the Centers for Disease Control and Prevention (CDC), Communicable Disease Control Division (CDCD) of Canada, and the World Health Organization (WHO) provide up-to-date information based on research and data. Becoming informed about HIV can help control the rate of infection for a disease that has grown into a worldwide epidemic.

Lesson Activity

With a partner, find out the latest statistics for the incidence of STDs, including HIV infection. Choose at least three STDs discussed in the lesson along with HIV. Consult reliable websites such as those from the Centers for Disease Control and Prevention and the World Health Organization. Prepare a list of questions you would like to research. For example, how many people are estimated to contract a particular STD each year? What are the trends for infections? What part of the population is the most at risk for infection? What new technologies and medications have been developed for treatment and cure? Prepare your data in a report using charts, tables, and illustrations.

 **Why is being informed about STDs important?**

Concept Check Assess/Reflect

Summary: How are STDs prevented? Sexually transmitted diseases are transmitted by sexual intercourse and other sexual activity. Many STDs do not have immediate symptoms and can go undetected. Early diagnosis and treatment can cure some STDs, such as chlamydia, gonorrhea, and syphilis. Other treatments may reduce STD symptoms or prevent problems later in life. Once a person is infected with HIV, it develops into the life-threatening disease, AIDS. This disease slowly destroys the immune system, and the infected person becomes at risk for various diseases and opportunistic infections. The only reliable way to prevent STDs is to practice sexual abstinence.

1. Name two sexually transmitted diseases and how they could affect your body.
2. Describe some ways that HIV is transmitted.
3. How is HIV different from other viral infections? How is it the same?
4. What are some of the consequences of becoming infected with an STD that would affect your everyday life?

Lesson 4

Objectives

- Explain the role organizations have in controlling infectious disease.
- Describe situations that interfere with the use of health-care resources for infectious disease.
- Describe how nations around the world work to combat infectious disease.

Vocabulary

Centers for Disease Control and Prevention (CDC)

Communicable Disease Control Division (CDCD)

pandemic

vector

Scripture Spotlight

Read the special promise in **Matthew 25:31–46** for those who show mercy even to "the least of these."

Essential Question

How Does the World Combat Disease?

Your family probably uses technology to stay informed of disease threats in your area. Who tracks that information to share with your family? Do you think other countries have similar systems? What about a Third World country, where television and computers are not as readily available? In North America, health care is readily available, but that is not the case everywhere in the world. What health problems might exist in other countries that are not issues locally? Why should you be concerned about the health of people in other parts of the world?

Keeping Track of Disease Explain

How do we know when a disease threatens us? The **Centers for Disease Control and Prevention (CDC)** and the **Communicable Disease Control Division (CDCD)** of Canada monitor many kinds of diseases nationwide. These agencies watch for serious new diseases that might be brought into the United States or Canada. Their aim is to reduce the incidence and spread of infectious disease and to improve health through effective disease prevention. They determine whether diseases already active are becoming a threat.

The World Health Organization (WHO) performs the same functions as the CDC worldwide. Both the CDC and WHO watch for *epidemics*, disease outbreaks that affect a large population or region. They also watch for **pandemics**, disease outbreaks that affect a very large group of people spread over a wide geographic area.

Pandemics have occurred throughout history. The bubonic plague in Europe during the fourteenth century wiped out nearly 60% of the population. This plague, known as the Black Death, was transmitted to humans through fleas that had bitten infected rodents. In 1918 a flu pandemic known as the Spanish flu killed an estimated 30 to 50 million people worldwide. Pandemics still occur, but advanced health care and drugs such as antibiotics and vaccinations have cut down drastically on mortality rates. What epidemic do you know of that is going on now? Do you think the pathogen causing it will be destroyed? If so, how can it be destroyed?

One of the greatest challenges for organizations such as the CDC and WHO is finding ways to control emerging and epidemic diseases before they develop into pandemics. For example, new flu virus strains emerge each year. The potential for influenza outbreaks has increased over the years due to people living closer together, the ease of travel between countries, and the methods by which food is packaged and traded. WHO collects and analyzes data about these flu strains from countries all over the world. How do you think WHO uses the data to make recommendations for vaccines and treatment?

Called to Serve

The Adventist Development and Relief Agency (ADRA) was started by the Seventh-day Adventist church as a way to follow Christ's example of caring for those in need. ADRA seeks to identify and address social distortions and deprivation in developing countries. ADRA has been actively developing water resources for more than 20 years in developing countries. ADRA wants to put the words of **Isaiah 41:17–18** into action.

Treating Disease in Developing Nations Explain

Some nations cannot provide basic health care for their people. These nations are said to be "developing." Their citizens have very low incomes and often have an uncertain food supply. Small villages might—or might not—have a doctor, nurse, or other health-care worker to help villagers recover from injuries or illness. Poverty and lack of transportation often keep villagers from traveling to larger towns and cities where health care might be more available. Diseases such as dysentery, AIDS, and malaria are more common in developing nations. Why do you think this is so?

Countries all over the world are at risk for natural disasters, such as hurricanes, typhoons, and earthquakes. Many nations with unstable governments find themselves in the midst of political upheaval or war. Families uprooted by hurricanes, earthquakes, or war seek any available shelter. After fleeing violence or natural disasters, refugees often end up living in crowded tent cities in developing nations. In these conditions, proper sanitation and waste disposal is often unavailable. Human and/or animal wastes may easily contaminate the water sources near these people. Still, the people must drink water to survive. How might this affect the level of infectious disease? What kinds of diseases might people become infected with from contaminated water?

In 2008, this CDC researcher was checking for signs of growth in a flu virus. That virus had already killed 61 people in Mexico and was spreading into the United States.

Why is this worker dressed like this?

Record your work for this inquiry. Your teacher may also assign the related Guided Inquiry.

Disease Defense

How is hand washing related to the spread of infectious disease?

Procedure

Materials
- cooking oil
- cinnamon
- soap
- sink with running water
- towels or paper towels

1. Work with a partner to develop a quantitative rating system to rate the effectiveness of hand washing. Your rating system should consider how infectious diseases are spread through direct contact. **Record** your rating system in your ***Science Journal***.
2. Dirty your hands by covering them with oil and then rubbing them in cinnamon.
3. Wash your hands in cool water with no soap for 10 seconds and blot them dry. Then shake hands with your partner. Record your observations.
4. Both you and your partner are to wash and dry your hands thoroughly.
5. Repeat Steps 2–4, five more times, but each time wash your hands as described in rows 2–6 in the table.

Description	Cleanliness Rating
1. Cold water, no soap (10 s)	
2. Warm water, no soap (10 s)	
3. Warm water and soap (10 s)	
4. Cold water, no soap (30 s)	
5. Warm water, no soap (30 s)	
6. Warm water and soap (30 s)	

Analyze Results

Which method was the least effective? Which method was the most effective? Why do you think this is so?

Create Explanations

1. How is hand washing related to the spread of infectious disease? Use the results of your investigation to support your answer.
2. Why must health workers follow established standards for hand washing?
3. What are some ways that the pathogens from dirty hands can get into your body?

At times, distrust keeps people from accessing health care. For example, older family members may warn younger family members that going to a hospital—or even to a doctor—will lead to their death. In hospitals without proper cleanliness, they could be right. Why do you think this fear might exist in some cultures?

Methods of Controlling Disease Explain

You know that many infectious diseases are spread through the bites of insects or other infected animals. Animals that carry disease-causing pathogens and spread them to humans are called **vectors**. Vector control is any method that is used to eradicate animals such as birds, insects, or mammals to reduce or eliminate the disease pathogens they carry. Think about your own community. Is there a population of deer that might carry the ticks that cause Lyme disease? What methods does your community use to control the incidence of the ticks?

Lesson Activity

Many challenges exist in developing countries for disease control. These challenges often inspire questions for scientific research. For example, what technologies can be developed for improved waste management? What questions can you think of that might be objectives for such research? Do you think that social priorities in developing nations can influence research priorities? If so, how? Should social priorities take precedence over funding for research? What examples can you think of to support your point of view?

Which is more important, social priorities or scientific research?

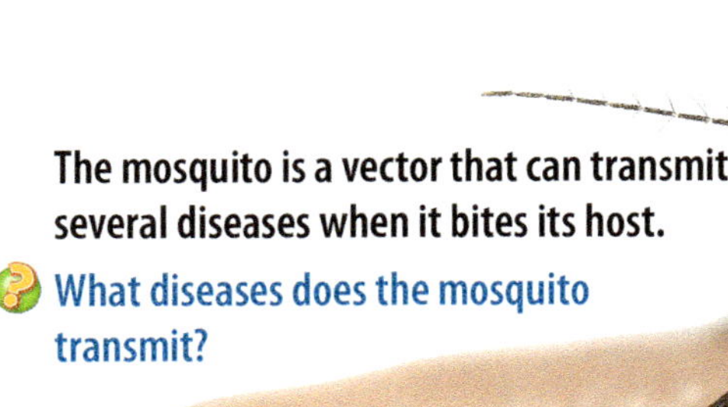

The mosquito is a vector that can transmit several diseases when it bites its host.

What diseases does the mosquito transmit?

Vector Control of Mosquitoes

The most widely used type of vector control is the control of mosquitoes. Malaria, West Nile virus, and dengue fever are infectious diseases that are spread by mosquitoes. Malaria is found in North America, and the incidence of dengue fever there is rare. West Nile virus is a potential threat.

Many developing countries are at a high risk for malaria infection. These countries are located in tropical and subtropical climates where the warmer temperatures allow the *Anopheles* mosquito to breed. Many countries in Africa lack resources to develop effective malaria control programs. However, several methods are effective in the control of these infectious mosquitoes. These are summarized in the chart below. What kinds of roadblocks do you think might exist in developing countries that could prevent or inhibit the use of such methods?

The use of mosquito nets has greatly reduced the incidence of infection in many countries in Africa.

What other personal precautions might a person take to reduce the risk of malarial infection?

Mosquito Control	
Method	**Description**
Indoor residual spraying	This method reduces the transmission of malaria by spraying insecticides around a house or where people sleep.
Long-lasting insecticidal nets	Mosquito nets that provide a physical barrier are typically used around beds at night to prevent bites while people are sleeping. The nets are treated with insecticides that are safe for humans.
Larval source management	This method of control involves applying insecticides to the places where mosquitoes breed, reducing the population. The insecticides kill the mosquito larvae before they hatch.
Habitat control	In many areas, stagnant water can be a source of mosquito breeding. Removal of stagnant water can reduce the spread of mosquitoes.
Biological control	In some places, control can be managed through the use of animals that are natural predators to mosquitoes. Fish that eat mosquitoes, for example, can be effective in controlling populations.

Lesson Activity

Working with a group, use what you have learned in this chapter to plan a campaign to control the spread of a communicable disease. You might choose strep throat, pneumonia, flu, hepatitis, malaria, rubella, or another disease. Review what causes the disease so you can plan ways to prevent it. Prepare a poster that details the plans of your campaign. Be sure to include a statement that summarizes the purpose of your campaign. Include diagrams, charts, photos, and illustrations to add content to your presentation. Describe whether you feel the government should fund research in your campaign and why or why not. Be prepared to share your campaign with the class.

 **How can you stop the spread of a disease?**

Summary: How does the world combat disease? The Centers for Disease Control and Prevention (CDC) monitors many kinds of diseases nationwide. The CDC and the CDCD watches for serious new diseases that might be brought into the United States and Canada. There is a need for precautions for people who travel easily around the world, potentially spreading disease. The CDC also collects data to determine whether diseases already active in our nation are becoming a threat. The World Health Organization (WHO) performs the same functions as the CDC, but it does so worldwide. Both the CDC and WHO watch for epidemics and pandemics. Vector control is the eradication of vectors—animals that carry pathogens and transmit them to humans. Vector control has been effective in reducing the spread of malaria in developing countries.

1. How do we know when a new disease is beginning to threaten American lives?
2. What prevents citizens from getting good health care in developing nations?
3. How are vectors controlled? Give two examples.
4. What risks are involved in the use of insecticides or other pesticides in the control of vectors?

People in Science

Extend

Get to Know

Bruce Beutler, Jules Hoffman, and Ralph Steinman

In 2011, three scientists won the Nobel Prize in Medicine for their work on the immune system and disease treatment and prevention. The Nobel committee stated that the three scientists had "revolutionized our understanding of the immune system by discovering key principles for its activation."

In 1973, Steinman discovered immune system dendritic cells. These specialized cells are responsible for triggering the immune system's secondary immune response, called adaptive immunity. In this immune response, B cells are activated, producing antibodies that mark a specific pathogen for killer T cells to seek out and destroy. The immune system "remembers" the pathogen and prevents you from getting the disease again.

In 1996, Beutler and Hoffman discovered protein receptors that trigger the immune system's first response, innate immunity.

With the combined knowledge of these scientists' discoveries, drug companies can create better vaccines, improve prevention of diseases, and create better and more effective treatments for a host of diseases, such as cancer, rheumatoid arthritis, type I diabetes, multiple sclerosis, and chronic inflammatory diseases.

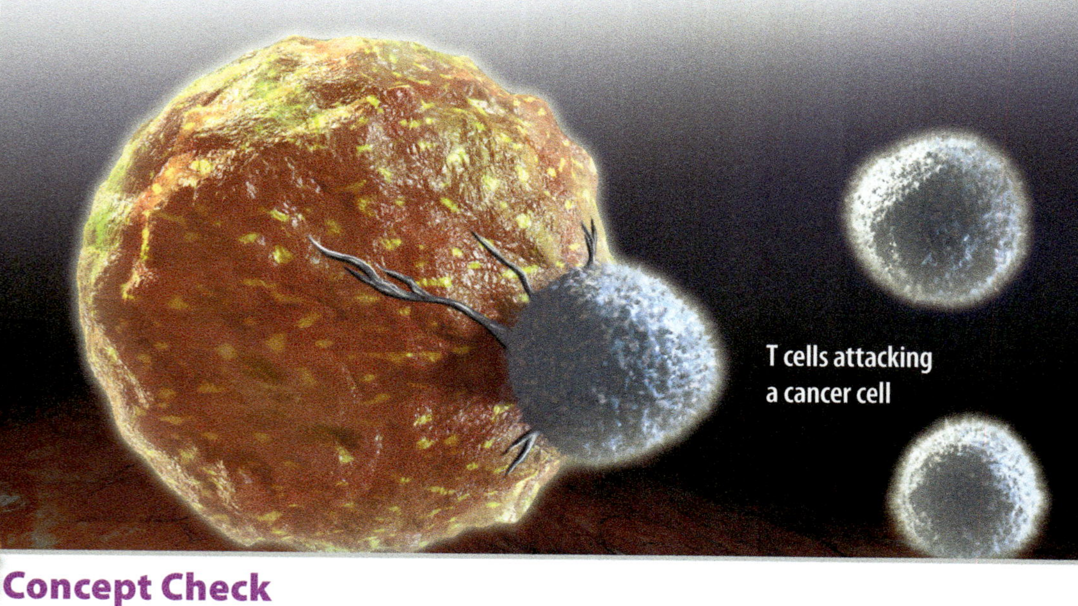

T cells attacking a cancer cell

Concept Check

1. How can knowledge of the immune system help prevent and treat diseases?
2. How are these scientists doing God's work?

Infectious Disease Specialist

Careers in Science

Extend

Do you like to solve problems? Does the idea of being a "medical detective" sound interesting? Then being an infectious disease specialist is a career you should consider. Infectious disease specialists diagnose, treat, and work to prevent infectious diseases. Patients go to infectious disease specialists for a variety of reasons. They may have an infection that other doctors are having difficulty diagnosing, may not be responding to current treatments for a disease, or may have plans to travel to a foreign country where the risk of infection is greater, for example.

Infectious diseases are the second leading cause of death in the world. More than half of the deaths caused by infectious diseases are children under the age of five. You might think that with our increasing scientific knowledge of diseases, the number of cases of infectious diseases would be decreasing. However, in the United States, the number of cases has doubled since the early 1980s. Today, the topic of infectious disease is among the hottest in medical research.

You can prepare now to become an infectious disease specialist by taking science classes, including biology and chemistry. You will need a bachelor's degree and a doctorate in medicine, as well as special training in this area of expertise.

This special training can last two or three years. Infectious disease specialists are usually hospital-based, so they can see patients and also maintain close contact with public health officials in the case of possible epidemics.

Infectious disease specialists may choose to do research and work in a laboratory studying diseases, the body's immune system responses, and potential treatments and preventions. They may specialize further, studying one particular disease and how to find a cure.

Concept Check

1. What are some reasons an infectious disease specialist needs so much training?
2. How does the work of an infectious disease specialist help save people's lives?
3. Why is it important to know how you can get an infectious disease?

Chapter 6 Review Assess/Reflect

Study Guide

Lesson 1

1. The immune system is a complex group of your body's defenses that protect against disease and specific pathogens.
2. The first lines of defense against pathogens are the physical and chemical barriers of the body. The second line of defense is the inflammatory response. The third line of defense is the immune response.
3. Innate immunity is the first and second lines of defense. It occurs quickly and is nonspecific. Adaptive immunity is specific to individual antigens and takes time to develop. Passive immunity acts quickly and is antigen-specific, but it does not last long.
4. The immune system can harm the body in the presence of an autoimmune disease and allergies. Autoimmune disease is a condition in which the immune system targets and attacks a person's cells.

Lesson 2

1. Infectious diseases are caused by pathogens that get into the body. Communicable diseases are infectious diseases that can spread from person to person.
2. Infectious diseases include malaria, tetanus, colds, flu, chicken pox, and STDs.
3. Some diseases are spread by direct contact, which occurs when a pathogen passes directly from one person to another. In diseases spread by indirect contact, pathogens move from the infected person to some object and then to another person.
4. Vaccines protect against disease by introducing an antigen to the body to trigger the production of antibodies.
5. There are a number of guidelines for proper use of antibiotics. These are listed in the box on page 238.

Lesson 3

1. Common sexually transmitted diseases include chlamydia, genital herpes, genital warts, gonorrhea, syphilis, and vaginitis.
2. HIV takes over the cells of the host's immune system and destroys the T-helper cells. When enough of these cells are destroyed, the infected person has more difficulty fighting off other diseases. This leads to AIDS. Individuals with AIDS have weakened immune systems and are subject to a variety of serious diseases.
3. HIV can be transmitted from one person to another through direct contact of blood or body fluids, sexual activity, blood transfusion, IV drug use, and organ donation. A mother also can transfer the infection to her unborn child.
4. Using barriers such as condoms can prevent the transmission of some STDs. However, they are not 100% effective. Practicing sexual abstinence outside of marriage will prevent contact with an infected person.
5. Current, reliable information regarding STDs is available through local, national, and international health departments and from doctors.

Lesson 4

1. Organizations try to reduce the spread of infectious disease through education and disease prevention measures. They find ways to control emerging and epidemic diseases before they develop into pandemics.
2. Poverty, poor sanitation, and lack of education, as well as natural disasters, war, and unstable governments all interfere with the distribution of health-care resources.

3. A vector is an animal or insect that spreads disease to humans. Vector control is any method that is used to eradicate animals such as birds, insects, or mammals in order to reduce or eliminate the disease pathogens they carry.

Vocabulary Check

Fill in the blanks with the correct terms.

1. A protein that has a structure that fits onto the molecules from a certain pathogen is a(n) __________.
2. __________ is a disease outbreak that affects a very large group of people over a wide geographic area.
3. The __________ response occurs when chemicals from damaged cells cause the walls of nearby blood vessels to expand.
4. The disease __________ is caused by the human immunodeficiency virus, or __________.
5. A chemical that kills bacteria and keeps them from spreading is called a(n) __________.
6. __________ disease is caused by pathogens that get into the body and cause problems.
7. A(n) __________ disease is a condition in which the immune system targets a person's cells, tissues, or organs by mistake.

Multiple Choice

Choose the best answer.

8. Food poisoning is infectious but not communicable. Why?
   A. It is caused by a virus.
   B. It is caused by bacteria.
   C. It can be spread from person to person.
   D. It cannot be spread from person to person.
9. A vaccine provides immunity by
   A. causing your body to produce antibodies.
   B. destroying pathogens before they enter your body.
   C. injecting macrophages into your body.
   D. producing live pathogens in your body.
10. Why are people in developing countries more likely to be at risk for malarial infections?
   A. A greater number of mosquitoes breed in these countries.
   B. These countries lack resources for effective malaria control programs.
   C. Stagnant water sources are a breeding ground for mosquitoes.
   D. Proper sanitation and waste development is often not available.

Check Point

Answer the following questions.

11. Tyler's doctor prescribed an antibiotic for a staph infection. Tyler is taking the medicine as directed on the label, but his infection is not getting better. What could be the problem?
12. Jessica says that people can become infected by HIV from indirect contact. Is her statement correct? Explain why or why not.
13. Why is it difficult to gather and record data to determine the source of an epidemic?
14. Why is bacterial resistance to antibiotics considered a significant and growing health problem?

Chapter 7 Other Systems of the Human Body

Scripture Spotlight

God created our bodies with the ability to obtain energy from the world around us. By learning about our bodies' systems we can better see how God cares for us. In this chapter you will read the following passages.

1 Corinthians 10:31 (p. 263)
Matthew 15:17 (p. 279)

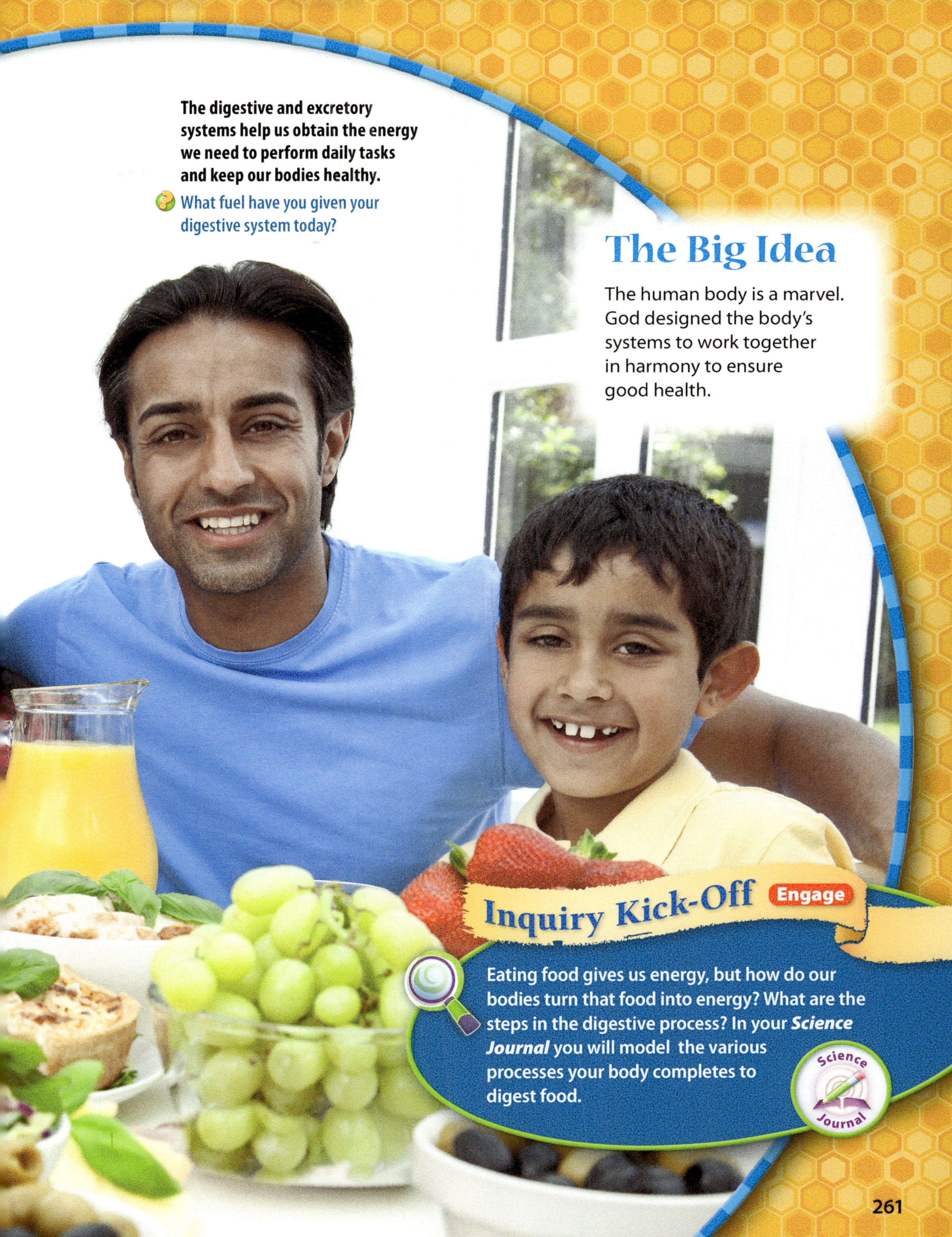

The digestive and excretory systems help us obtain the energy we need to perform daily tasks and keep our bodies healthy.

What fuel have you given your digestive system today?

The Big Idea

The human body is a marvel. God designed the body's systems to work together in harmony to ensure good health.

Inquiry Kick-Off

Engage

Eating food gives us energy, but how do our bodies turn that food into energy? What are the steps in the digestive process? In your ***Science Journal*** you will model the various processes your body completes to digest food.

Lesson 1

Objectives

- Describe the structure and function of the digestive system.
- Identify disorders or diseases of the digestive system.
- Explain how to care for the digestive system.

Vocabulary

digestive system
digestion
mechanical digestion
chemical digestion
enzyme
saliva
salivary gland
chyme
villi
pancreas
liver
bile
gallbladder
rectum
anus

Essential Question

How Does the Digestive System Work?

Each day, you use many things that need energy in order to work. For example, you might use a mobile device, a smartphone, or other electronic device. Where does the device get the energy it needs to function? What functions is the device able to do with the energy? What do you do when there is not enough energy to power the device? Consider how you get the energy you need to think and move. How do you know when your supply of energy is getting to low levels? How do you add more energy to your system? Is it possible to add too much? Does it make a difference what source you get your energy from?

God gave us many types of food to eat. Why do you think that was important in His plan? You know that taking care of your body is one way of honoring God. What can you do to keep your body working properly to process the food you eat?

How does eating healthful foods provide energy to fuel your body and honor God's creation?

Digestion Explain

Think about the walking or running you have done today. Was the energy that allowed you to do this from yesterday's lunch, last night's dinner, or this morning's breakfast? Suppose you eat a dinner of pasta and salad. These foods provide you with nutrients you need for energy, growth, and repair of cells. But these foods are too large to go directly into your cells.

Recall that *cells*, which are the basic units of life, are organized into *tissues*, groups of similar cells working together to perform a specific job. Tissues, in turn, are organized into *organs*, or groups of tissues that work together to perform certain tasks. *Organ systems* are groups of organs that work together to perform a series of related functions. All of the body's organ systems work together to keep the body healthy.

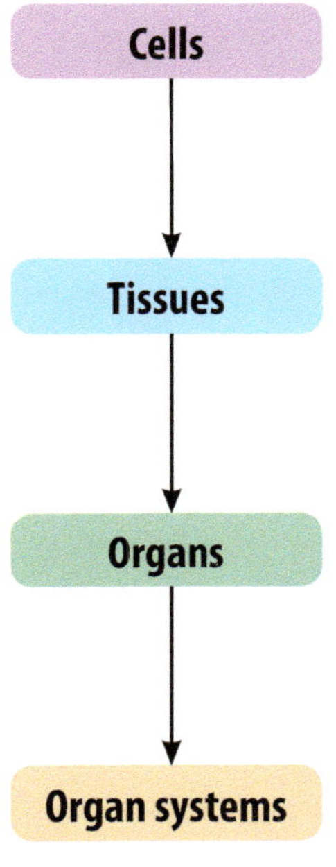

The group of organs that works together to process food for use by the body is the **digestive system**. The organs in this system break down foods into individual molecules so that the foods can be absorbed into your bloodstream and transported to your cells, a process called **digestion**.

Digestion is both a mechanical and a chemical process. **Mechanical digestion** is the physical tearing and grinding of food into smaller pieces. Teeth perform this function. The process is similar to that of a blender grinding up fruit for a smoothie. Can you think of another example of larger pieces being broken into smaller pieces? **Chemical digestion**, on the other hand, is done by digestive juices. These juices contain a combination of acids and digestive **enzymes**, which are substances that speed up chemical reactions in the body. The digestive juices help break food particles into molecules the body can use.

Digestion occurs in the digestive tract, a tube-like passageway that begins at the mouth and continues through the esophagus, the stomach, the small intestine, the large intestine, the rectum, and the anus. A garden hose makes a good model of the digestive tract. What length hose would make an accurate model? Talk with a friend about this. How does your idea compare with the ideas of others in your class? Do some research and find out the length of the digestive tract. Is it longer or shorter than you thought? Other organs that are part of the digestive system are the liver, the pancreas, and the gallbladder. How do you think these organs help with digestion? On the next few pages, you will find descriptions of the parts of the digestive system.

Scripture Spotlight

You can cooperate with God's design of your digestive system by being selective about what you eat, how much you eat, and even when you eat. How does **1 Corinthians 10:31** relate to this lesson?

Digestive System

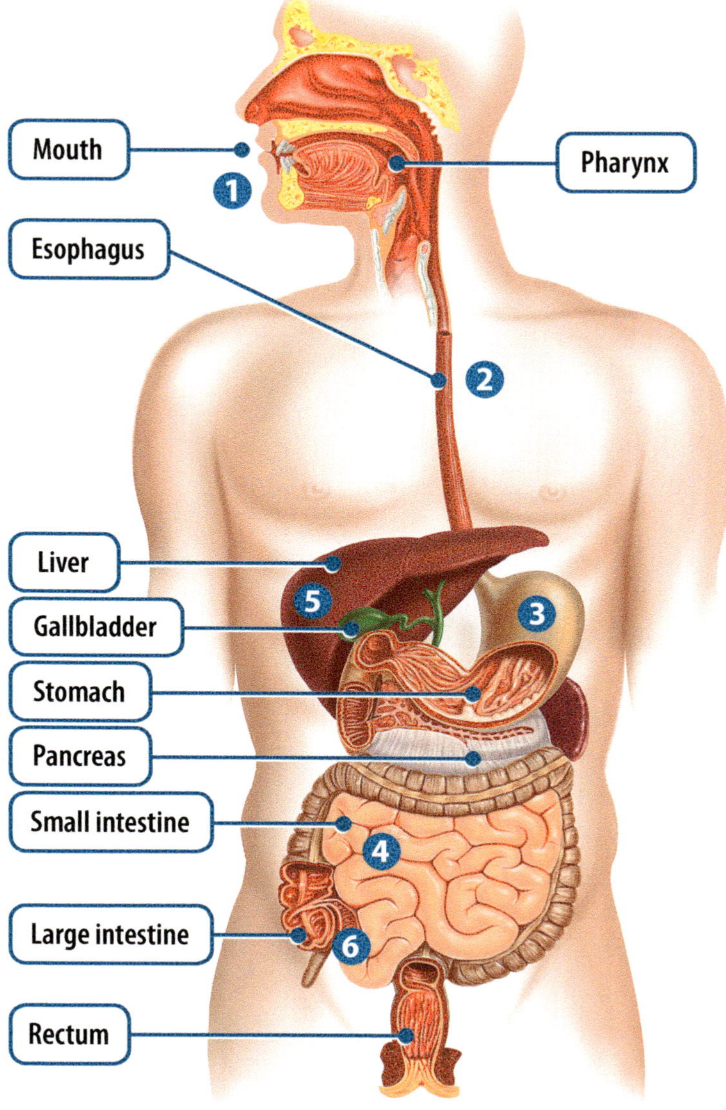

The digestive system contains multiple organs that process food into molecules the body uses for energy, growth, and cell repair. As you read about the organs in the digestive system, look back at this diagram.

What is the job of the mouth in the digestive system?

Organs	Process
1 Mouth (tongue, teeth, salivary glands, saliva)	Food is chewed. Saliva begins to break it down and moisten it. The tongue forms it into a ball for swallowing.
2 Pharynx and esophagus	A swallow pushes the food into the pharynx. Muscle contractions move the food through the pharynx and into the esophagus.
3 Stomach	Contractions move the food into the stomach, where it is churned and broken down by enzymes (mechanical and chemical digestion).
4 Small intestine (villi)	Partially digested food called *chyme* is released into the small intestine, where most chemical digestion and absorption of nutrients takes place through villi.
5 Pancreas, liver, and gallbladder	The pancreas releases an acid-neutralizer and digestive enzymes that help break down protein in the small intestine. The liver produces bile, which helps break down fats in food. The gallbladder stores the bile and releases it into the small intestine.
6 Large intestine (rectum, anus)	Muscle contractions push remaining material into the large intestine, where the remaining water is absorbed and waste is formed into feces. Contractions push the feces into the rectum and then out of the body through the anus.

The Mouth

Mechanical and chemical digestion begin in the mouth, the first step on food's journey through the digestive tract. But even before the food enters your mouth, the body starts preparations for digestion. When you smell, see, or think about appetizing food, the excretion of saliva (which is always present in your mouth) increases. **Saliva** is a fluid that contains digestive enzymes that begin the process of breaking down food into smaller pieces. It is released by the **salivary glands** in your mouth. Try it now; think about a food you like. What happens to the amount of saliva in your mouth? God could have chosen to let us fuel our bodies by some system like we fuel cars, where one substance provides everything that is needed. Instead, we have many choices regarding refueling. Thus, taste, texture, smell, and other attributes of food have an impact on how we view those choices. Since humankind has the choice and taste for refueling, what is the responsibility that comes with that choice?

Human Mouth

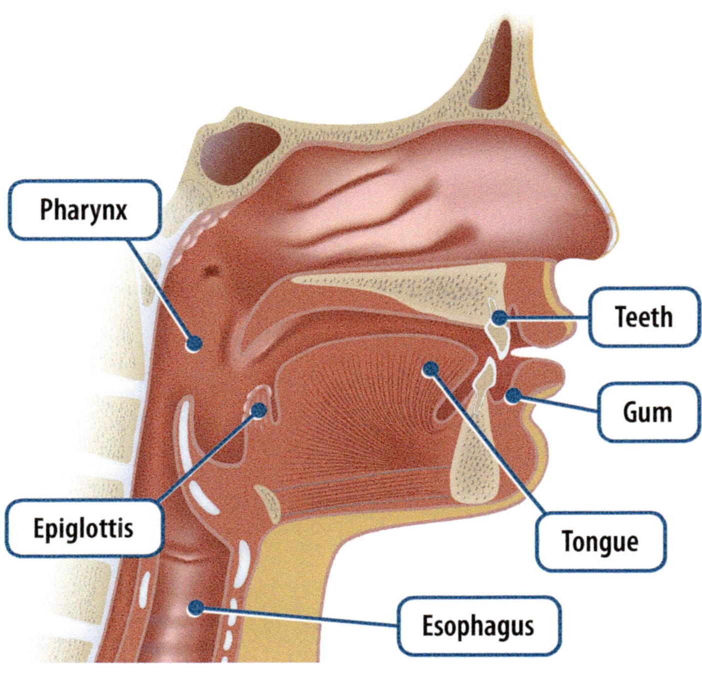

When you swallow, the epiglottis closes off the passage to the larynx so food does not enter the windpipe.

Why is this process necessary?

When food enters your mouth, your tongue and cheeks move it between your teeth, which grind the food into smaller bits. Why do you think it is important to always chew your food thoroughly? As you finish chewing, your tongue makes the food into a ball so it can be swallowed.

The Esophagus

When you swallow, the meal of pasta and salad enters the pharynx, a muscular passageway that connects the nasal cavity and mouth to the esophagus. The esophagus is a narrow tube that connects the pharynx to the stomach. When this happens, the *epiglottis*, a flap of cartilage, covers the entrance to the larynx to keep food and drink from entering the lungs. Food is pushed through the pharynx and the esophagus by a series of contractions. Do you have to consciously push the food through your esophagus, or does it just happen?

Explore-a-Lab

Structured Inquiry

How does food move through the esophagus?

Wet thoroughly a piece of rubber tubing (40 cm by 1 cm). Put a marble in one end of the tubing. Move the marble through the tubing by squeezing the tubing just behind it. Have your partner use a stopwatch to time how long it takes. Record the data. Then add 1 mL of vegetable oil to the inside of the tubing. Flatten the tubing to cover the entire inner surface with oil. Again move the marble through the tube while your partner uses the stopwatch to time you. Record your data. With your partner, compare the two times. Discuss the reason for any difference between the times.

The Stomach

Muscle contractions move the meal of pasta and salad from your esophagus into your *stomach*. Then the muscles of your stomach contract, mixing food with digestive acids and enzymes secreted from glands in the stomach walls. This combination of mechanical and chemical digestion breaks down food into even smaller pieces. To understand this process, think of how a washing machine mixes clothes with water and laundry detergent. The clothes are not broken down, but they are changed (cleaned) by the action of the machine and the chemicals in the detergent. What other illustrations of the digestive process can you think of? A layer of mucus, a thick fluid, covers your stomach walls. It protects the stomach from being damaged by the digestive acids and enzymes, much as a plastic sheet protects a floor from paint spatters when a room is being painted. Most food stays in the stomach from two to five hours. What might happen if the food stays for less time in the stomach?

Special muscles called sphincters control the flow of food into and out of the stomach. The lower sphincter is at the end of the esophagus.

What happens if the lower sphincter fails?

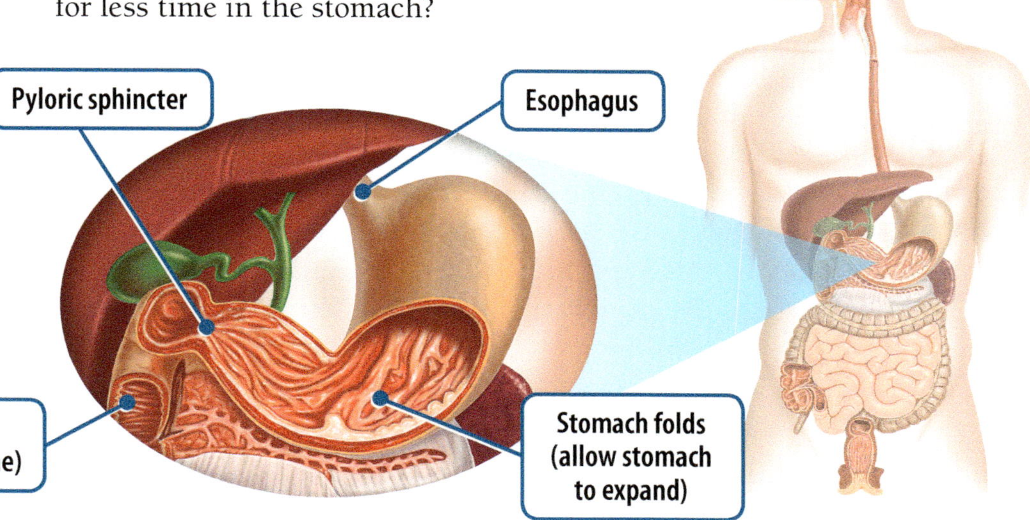

The Small Intestine

Food broken down and mixed with acids and enzymes in the stomach is called **chyme**. When the chyme is the right consistency, your stomach slowly releases it into the duodenum, the first part of the small intestine. Most chemical digestion and absorption occur in the small intestine. Enzymes produced by glands in the small intestine aid digestion, and chyme is moved through the organ by a series of muscle contractions. The lining of your small intestine has fingerlike projections called **villi**, where absorption occurs. The chyme can flow around and over the villi. There are 10 to 40 villi per square millimeter in the small intestine, with more toward the beginning of the organ and less toward the end. The villi increase the small intestine's surface area, allowing the organ to maximize the nutrients your body absorbs. Absorption is further increased by the undulating movements of the villi, which also contract to help chyme move through the small intestine. What would be the effect on the body if the small intestine did not have villi?

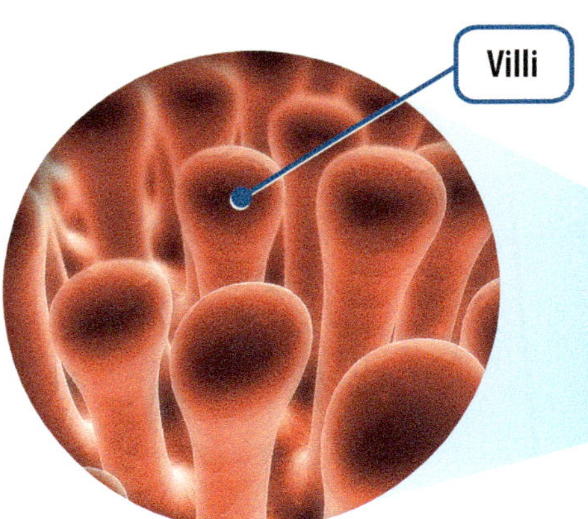

Villi are finger-like projections that extend from the interior lining of the small intestine.

What is the benefit of increased surface area?

Math in Science

The human small intestine is between 6.7 m and 7.6 m long.

Your teacher will assign each student a different mammal. Research the length of the small intestine in your assigned mammal. Compare the length of a human small intestine with the length of the small intestine in your assigned mammal. Then create a ratio. Share your results with the class.

Record your work for this inquiry. Your teacher may also assign the related Guided Inquiry.

Absorbing Nutrients

How do villi improve nutrient absorption?

Materials

- 7 square sponges (8.3 cm × 14 cm, all made of the same material)
- stapler
- scissors
- metric ruler
- pan deep enough to submerge the sponges completely
- water
- stopwatch
- large plastic cup
- graduated cylinder (100 mL)
- funnel

Procedure

1. Cut one of the sponges into 12 fingerlike strips. Each strip should be about 4.0 cm long and 2.3 cm wide. Staple 4 strips to each of 3 of the other sponges so that they project outward like the bristles of a hairbrush.

2. Submerge 1 (unaltered) sponge in the pan of water for 30 seconds. Remove the sponge from the water. Place it in your palm and squeeze the water into the large plastic cup. Each time you squeeze a sponge, use the same amount of force. Put the funnel in the top of the graduated cylinder. Pour the water from the cup into the funnel. **Record** the volume of the water. Pour the water from the graduated cylinder back into the pan of water. Repeat with the 2 remaining (unaltered) dry sponges.

3. Place 1 altered sponge in the pan of water so that the sponge and all of its finger-like strips are fully submerged in the water. Remove the sponge after 30 seconds. Use the same procedure as in Step 2 to remove the water from the sponge. Repeat with the 2 remaining dry, altered sponges.

4. **Calculate** the average volume collected from the 3 trials with the unaltered sponges and from the 3 trials with the altered sponges.

5. In your ***Science Journal***, make a bar **graph** to show the data.

Analyze Results

Look at the graph. Was there a difference between the volumes of water absorbed by the unaltered sponges and the volumes of water absorbed by the altered sponges? **Use numbers** to explain.

Create Explanations

1. How do villi improve nutrient absorption?

2. How did the volume of water absorbed by the three altered sponges vary? What do you think accounts for these differences?

3. Do you think the results would be different if the "fingers" of the altered sponges were longer? What if they were shorter? Explain.

The Pancreas, Liver, and Gallbladder

When chyme enters your small intestine, it is time for the **pancreas** to play its part. The pancreas is a gland located behind the stomach. It releases a solution into the small intestine that neutralizes the acids in the chyme so that the acids do not eat into the walls of the small intestine. The pancreas also releases a combination of digestive enzymes that help further break down the chyme. *Proteases* break down proteins, *pancreatic lipase* breaks down fats, and *amylase* breaks down starches.

The **liver** is a large, glandular organ in your body, second only to the skin in size. It is located under the rib cage in the right part of the abdomen. The main function of the liver is to filter harmful substances from the blood before passing it to the rest of the body. The liver has other functions in the body, including its role in the digestive process. Without the liver, body cells would die from lack of energy and nutrients. The liver secretes **bile**, a fluid that helps break large fat particles into smaller pieces.

Bile that is not immediately needed for digestion passes from the liver to be stored in the **gallbladder**. When food that contains fats enters the small intestine from the stomach, the bile stored in the gallbladder is released into the duodenum. There the bile breaks up large chunks of fat. Breaking up the large pieces of fat creates more surface area, increasing the area where digestive enzymes in the small intestine can break down the fats.

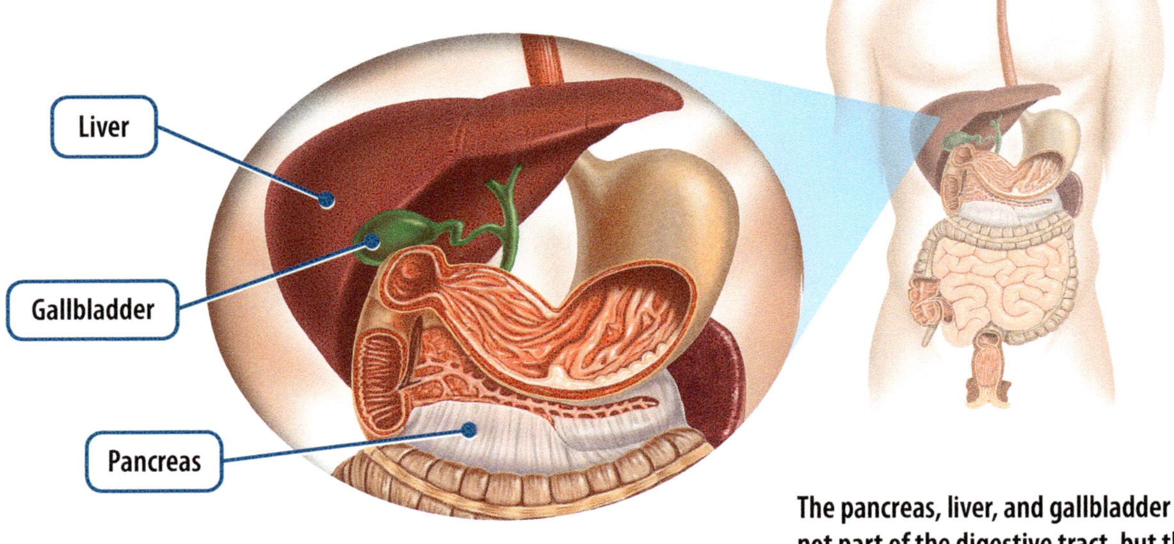

The pancreas, liver, and gallbladder are not part of the digestive tract, but they are important to digestion.

To what systems do these organs belong?

The Large Intestine

Any material your small intestine does not absorb is pushed into your large intestine by muscle contractions. The large intestine's main task is to absorb water and prepare undigested materials for expulsion from the body. Unlike the small intestine, the large intestine has no villi. However, it does use enzymes from the small intestine to finish digestion, and it absorbs any remaining nutrients. Bacteria in the large intestine produce vitamin B and vitamin K, which are absorbed through the walls of the intestine. However, the large intestine's main job is to absorb any remaining water and to form the remaining material into feces (solid waste). The feces are pushed by muscle contractions into the **rectum**, a lower part of the large intestine, and out through the **anus**, the lowest part of the large intestine, which is an opening to the outside of the body.

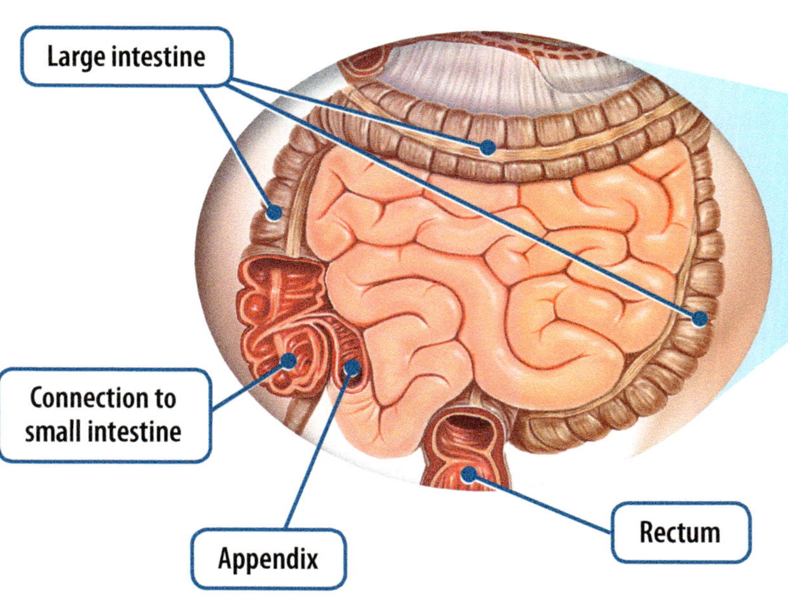

The large intestine is wider and shorter than the small intestine. The large intestine is about 1.5 m in length, compared with 6.7 to 7.6 m for the small intestine.

What is the advantage of a longer small intestine?

Explore-a-Lab

Structured Inquiry

Is chewing important to digestion?

Use a sugar cube, a crushed sugar cube, water, and a stopwatch. Fill a glass with water. Add a sugar cube to it. Note the time. Then record how long it takes to dissolve. Repeat for the crushed sugar. Compare. Which dissolved faster?

Your meal of pasta and salad is no longer recognizable. Through digestion, your body has received all the nutrients your meal has to offer. The cells of your body will use these nutrients for energy, growth, and repair. And tomorrow, the digestive process will begin all over again when you eat breakfast!

Eating plenty of fruits and vegetables will help your digestive system function properly.

What foods other than fruits and vegetables should you eat daily?

Called to Serve

You may have helped with a project in your community to distribute food to people who need it. That is similar to one function of your digestive system—getting nutrients to locations where they can be used. Compare **Psalm 119:103** with **Proverbs 25:11**. What else could you help to spread around your classroom and your community?

Explore-a-Lab

Structured Inquiry

Where does starch digestion begin?

Label three test tubes *A, B,* and *C*. Crumble a soda cracker. Put ¼ of it in test tube *A*. Add 3 mL of water. Mix the cracker and water by rotating the test tube between the palms of your hands. Add 3 or 4 drops of iodine solution to the test tube. Observe and record the results. Then place another ¼ of the crumbled soda cracker in test tube *B*. Add 5 mL of Benedict's solution. Fill a beaker about half full of water and place test tube *B* in the beaker. Then chew a cracker for about 1 minute. Place the chewed cracker in test tube *C*. Add 5 mL of Benedict's solution. Place test tube *C* in the beaker along with test tube *B*. Then heat the water in the beaker for about 5 to 7 minutes. Observe and record the results. Discuss what happened when you added the iodine solution to test tube *A* and what happened to the substances in test tubes *B* and *C* when they were heated. What happened to the starch in the chewed cracker?

Disorders and Diseases of the Digestive System Explain

There are many disorders and diseases associated with the digestive tract. A few of the more common ones are shown in the table below.

	Causes	Description
Celiac Disease	Celiac disease is hereditary.	This disease causes the immune system to attack and damage the lining and villi of the small intestine when a sufferer eats glutens, which are proteins found in grains such as wheat and rye. Because the villi are damaged, they cannot absorb nutrients properly. Symptoms include abdominal pain, bloating, diarrhea, fatigue, weight loss, and delayed growth. Celiac disease can cause anemia, fertility problems, osteoporosis, diabetes, and cancer.
Gastroesophageal Reflux Disease (GERD)	The cause of GERD is sometimes unknown. However, contributing factors include smoking, obesity, pregnancy, certain types of medications, and some types of hernias.	This condition occurs when the contents of the stomach, including acids, back up into the throat because a muscle at the end of the esophagus does not close completely. GERD causes heartburn and other problems.
Inflammatory Bowel Disease (IBD)	The cause of IBD is unknown, although it may be hereditary. An abnormal reaction of the immune system to food and bacteria may be the cause of Crohn's disease (a form of IBD). Immune system irregularities and genetic abnormalities may be contributing factors to other diseases in this group.	This is a group of diseases that affect the intestines. *Ulcerative colitis*, one type of IBD, is a chronic condition that causes inflammation and sores (ulcers) in the large intestine and rectum. Symptoms include abdominal pain, diarrhea that contains blood and pus, rectal bleeding, loss of appetite, weight loss, and fatigue. *Crohn's disease* is another type of IBD that most often affects the *ileum* (the end of the small intestine) but that can affect any part of the digestive system. The affected part experiences swelling and the buildup of scar tissue. Symptoms include abdominal pain, rectal bleeding, diarrhea, fever, weight loss, and anemia.
Lactose Intolerance	Sometimes lactose intolerance is caused by damage to the intestines caused by diseases, such as celiac disease. However, often the cause is unknown.	This condition is caused by the small intestine's lack of the enzyme *lactase*, which prevents the body from digesting *lactose*, a sugar found in dairy products. Symptoms include bloating, cramps, gas, and diarrhea.
Peptic Ulcers	Peptic ulcers can be caused by bacteria or the long-term use of medications such as ibuprofen and aspirin.	These sores form in the lining of the stomach or in the duodenum (the first part of the small intestine) when bacteria or the long-term use of some medications cause digestive stomach acids to eat into surrounding tissues. Symptoms include a burning feeling in the stomach, vomiting, nausea, and GERD.

Treatment	Further Inquiry
People who have celiac disease cannot eat foods that contain gluten.	What resources exist today to help people with celiac disease?
Medication or even surgery might be needed.	How does smoking cause GERD?
Drugs can control symptoms of ulcerative colitis, but sometimes the large intestine must be surgically removed. There is no cure for Crohn's disease, but medications, supplements, and surgery may help relieve symptoms.	What medications can help relieve symptoms of Crohn's disease, and what do they do?
People who have lactose intolerance can manage the symptoms by eating fewer dairy products, by eating lactose-free dairy products, or by taking over-the-counter medications that contain lactase when they eat dairy products. People who are lactose intolerant can get the calcium they need by eating other foods rich in calcium, including leafy green vegetables such as kale and broccoli. Other foods high in calcium include chickpeas, whole-grain flour, oranges, figs, and almonds.	How can people determine whether they have lactose intolerance?
Antibiotics can kill bacteria, and other medications can reduce stomach acids. Surgery might be needed to resolve ulcers.	What steps can a peptic ulcer sufferer take to avoid flare-ups?

Cirrhosis and Diabetes

Two of the most damaging diseases that can affect the parts of the digestive system are cirrhosis and diabetes. Cirrhosis is a replacing of normal liver tissue with scar tissue. Because the liver is scarred, it is less able to perform its normal functions. How do its scars interfere with its functioning? Cirrhosis can lead to bloating and swelling of your legs and abdomen, nosebleeds, bruising, kidney failure, and liver cancer. People can guard against cirrhosis by avoiding alcohol and by taking steps to keep from contracting hepatitis, a disease that causes inflammation of the liver.

Diabetes is also a very serious disease that causes the level of glucose in your blood to be too high, which is bad for your health. Glucose levels are controlled by the hormone insulin, which is produced by the pancreas. Insulin helps cells absorb glucose so they can use it for energy. The glucose levels of people who have diabetes are so high because their bodies either cannot produce enough insulin or are unable to respond to the insulin that is produced. As a result, too much glucose stays in their bloodstream. Imagine a freeway where the exits are all blocked by accidents. How does this situation correspond to that of a person with diabetes? Diabetes can lead to heart disease, stroke, blindness, and kidney disease. People can prevent diabetes by exercising regularly and maintaining a healthful weight.

Keeping the Digestive System Healthy

In many cases, it is not difficult to keep your digestive system healthy, but you must make good choices to do so. The table below shows several steps you and your family can take to help avoid diseases and disorders of the digestive system.

Maintaining Digestive Health		
Healthful Practices	**Effects**	**Further Inquiry**
Exercise portion control and eat small meals.	Can prevent or relieve symptoms of GERD, ulcerative colitis, and heartburn	What steps can you take to control your portions?
Eat a diet high in fiber, which can be found in vegetables, whole grains, seeds, nuts, and legumes.	Can help reduce constipation, gas, and diarrhea; also helps prevent some digestive disorders and treat their symptoms; may reduce the risk of developing colon cancer	What are your favorite foods that contain fiber? How can you work them into your diet more frequently?
Exercise regularly and drink plenty of water.	Can help keep food moving efficiently through the digestive system, reducing constipation	What other systems of the body are benefitted by exercise and drinking plenty of water?
Don't drink alcohol.	Can reduce the risk of GERD, peptic ulcers, cirrhosis, and the risks associated with intoxication.	What are other reasons to abstain from alcohol?

Maintaining Digestive Health		
Healthful Practices	**Effects**	**Further Inquiry**
Avoid spicy foods.	Can prevent or relieve symptoms of GERD, peptic ulcers, ulcerative colitis, and heartburn	If you are a fan of spicy foods, how can you cut down on the potential that they will negatively affect your digestive system?
Avoid eating meat and fatty foods.	Can prevent or relieve symptoms of GERD, peptic ulcers, ulcerative colitis, heartburn, and constipation	How can you substitute lean foods for fatty meats in your diet? How would a vegetarian diet help eliminate this problem?
Don't smoke.	Can help reduce peptic ulcer flare-ups and the risk of developing Crohn's disease	What are some other reasons to avoid smoking?
Avoid stress.	Can help reduce peptic ulcer flare-ups and prevent worsening of ulcerative colitis	What causes stress in your life? How can you reduce that stress?
Avoid carbonated drinks.	Can help prevent worsening of ulcerative colitis and heartburn; reduces tooth decay	Do you drink most frequently water or other types of liquids (such as carbonated beverages) during the day? How can you remind yourself to cut down on carbonated beverages and drink water more frequently?
Reduce coffee consumption and don't eat close to bedtime.	Can help prevent heartburn	How can you avoid feeling hungry around bedtime?
Maintain a healthful weight.	Can help prevent GERD and heartburn	What actions can you take to maintain a healthful weight?

Summary: How does the digestive system work? The digestive system breaks down food into molecules that can be used by the body. It includes the mouth, the esophagus, the stomach, the small intestine, and the large intestine, as well as the pancreas, the liver, and the gallbladder. There are many disorders and diseases of the digestive system, including celiac disease, GERD, IBD, lactose intolerance, and peptic ulcers. Cirrhosis and diabetes are two very serious diseases that can affect the digestive system. There are many actions you can take to maintain the health of your digestive system, including exercising, drinking water, eating a diet high in fiber and low in fat, and avoiding spicy foods, coffee, and alcohol.

1. Name three ways the body breaks down food during digestion.
2. How do the pancreas, liver, and gallbladder help in the digestive process?
3. What is the purpose of villi?
4. Your grandfather complains of an upset stomach. What might be the problem?
5. Make a list of the parts of the digestive system. What evidence of God's Design can be seen in the structure and function of the digestive system? What do we need to do to keep it working well?

Lesson 2

Objectives

- Describe the structure and function of the excretory system.
- Explain how the digestive, excretory, circulatory, and respiratory systems work together.
- Identify disorders or diseases of the excretory system.
- Explain how to care for the excretory system.

Vocabulary

excretion
excretory system
kidney
nephron
urea
ureter
urinary bladder
urethra
dialysis

Essential Question

How Does the Excretory System Work?

Has your family ever forgotten to put out the trash on garbage collection day? How did that affect the running of your household during the following week? Imagine that the trash collectors in a large city went on strike. Suppose that at the same time, the sewage system broke down. What would happen in the city? Who might be affected first by this breakdown? Who might be the last to feel the effects of the breakdown? Just as a city needs a waste-removal system, your body must be able to remove the waste it creates every day. What happens to the body if the wastes are not removed from the body?

The Excretory System Explain

After your body digests food, it must remove the wastes it produces to stay healthy. These wastes include carbon dioxide, nitrogen, and excess minerals and salt. The process by which wastes are removed from the body is **excretion**. The group of organs responsible for removing wastes from the body is the **excretory system**.

Several organs work together to remove wastes from the body. These organs are the kidneys, ureters, bladder, and urethra. What are the sources of body wastes? The skin and lungs also play a part in the excretory system.

Lungs and Skin

What systems do your lungs and skin belong to? Did you include the excretory system on your list? When your skin sweats, it excretes water, salt, other minerals, and fats. Bacteria on your skin break down the fats, which causes an unpleasant odor. Why is it important to take a daily shower? How do you think this promotes overall health? What wastes does your respiratory system get rid of?

Kidneys

Place your palms on your back just below your ribs. This is the location of your two **kidneys**. The purpose of the kidneys, which are the main organs of the excretory system, is to remove wastes and excess fluids from the body and to maintain the body's water balance. Your blood passes continuously through your kidneys. It enters through the renal artery and leaves through the renal vein. *Renal* means "having to do with the kidneys." What do you think would happen if God did not include kidneys in your body's design?

Each kidney has about one million filtering units called **nephrons** surrounded by masses of capillaries. Liquid from the blood flows through tubes in the nephron, which are surrounded by capillaries. As the fluid travels through the nephron, it is filtered and purified. The cleaned fluid is reabsorbed by the blood. The wastes and excess water form urine, a yellow liquid composed of water, minerals, salt, and **urea**, a nitrogen waste. What do you think might happen if additional water were not reabsorbed while in the kidney?

Ureters, Urinary Bladder, and Urethra

In the center of each kidney, the nephrons connect together in a cavity. The cavity is attached to a narrow tube called the **ureter**. The ureter from each kidney carries urine out of the kidney to a large sac of smooth muscle known as the **urinary bladder**. Here urine is stored until it can be removed from the body. Because the bladder is muscular and can contract, the urine can be forced out. Urine is removed from the body through another tube called the **urethra**.Why do you think a separate organ is needed to store the urine instead of keeping it in the kidneys? Do you think the movement of the bladder muscles is voluntary or involuntary?

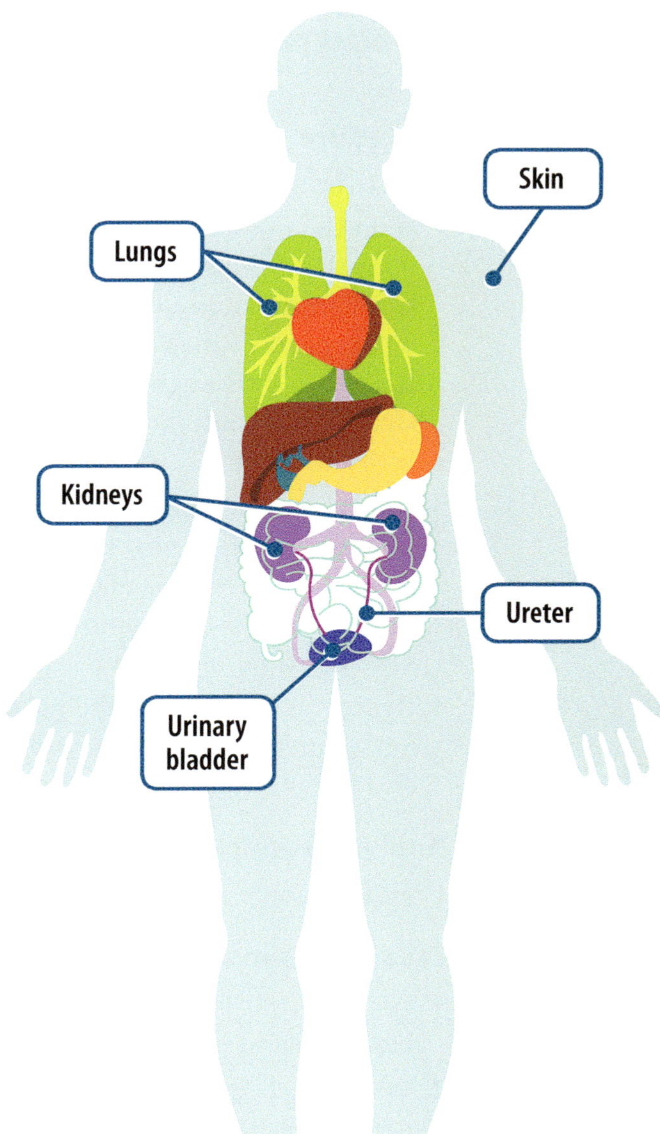

The excretory system has the job of removing waste from your body.

What do you think might happen if the excretory system did not function properly?

Record your work for this inquiry. Your teacher may also assign the related Guided Inquiry.

Filtering Waste

How do your kidneys remove wastes from your bloodstream?

Procedure

1. Mix the sand, pebbles, and water in a bowl.
2. Hold the poultry netting over the empty bowl. Pour the mixture of sand, pebbles, and water through the netting into the empty bowl. Examine and **record** your **observation** of the materials that passed through the netting. **Measure** the sizes of any material that did not pass through the netting.
3. Hold the hardware cloth with $\frac{1}{2}$-inch mesh over the first bowl (which now should be empty). Pour the mixture from the second bowl through the mesh into the first bowl. Again, measure the sizes of any material that did not pass through the netting.
4. Repeat these steps for the hardware cloth with $\frac{1}{4}$-inch mesh, the hardware cloth with $\frac{1}{8}$-inch mesh, and the plastic window screening.

Materials

- 250 mL sand
- 250 mL aquarium gravel
- 250 mL pea gravel
- 250 mL large landscape rock
- 250 mL water
- 2 large bowls
- 1 or 2 sheets of newspaper
- large funnel (must have large opening in neck)
- metric ruler

Each group also needs one 6-inch square piece of the following materials:

- window screening
- hardware cloth ($\frac{1}{2}$-inch mesh)
- hardware cloth ($\frac{1}{4}$-inch mesh)
- hardware cloth ($\frac{1}{8}$-inch mesh)
- poultry netting (1-inch holes)
- duct tape (to cover edges)

Analyze Results

Compare your results with those of other groups.

Create Explanations

1. How do your kidneys remove wastes from your bloodstream?
2. Which materials passed through the poultry netting? Which materials passed through the different types of hardware cloth? The plastic window screening?
3. How is this lab an accurate model of the way the kidneys work?
4. How is this lab an imperfect model of the way the kidneys work?

How Do Body Systems Work Together?

You know that during digestion, your body produces enzymes to break down food into molecules that can be used by the body. In the small intestine, nutrients are absorbed and passed into the bloodstream, which is part of the circulatory system. As the blood flows, it transports nutrients to every cell in the body, including the cells that make up the organs of the excretory system. Blood also transports oxygen from the lungs, which are part of the respiratory system, to every cell in the body. How do you think cells recover the oxygen and nutrients from the circulatory system?

Blood also carries waste products away from cells. These waste products include carbon dioxide. The bloodstream carries carbon dioxide to the lungs, where the gas stays until you exhale it through your nose or mouth. What would happen if the bloodstream stopped picking up waste? Other waste products, such as salts, ammonia, and urea, are filtered from the blood by the kidneys and are later expelled from the body as urine. After food passes through the large intestine, solid wastes are expelled through the rectum and anus.

Scripture Spotlight

Does the Bible talk about excretion? Check out **Matthew 15:17**. What do you think Jesus was trying to say?

How Body Systems Work Together

	Circulatory System	Digestive System	Excretory System	Respiratory System
Circulatory System		Moves digested nutrients to cells	Moves blood into kidneys so wastes can be filtered	Transports blood containing carbon dioxide waste to lungs for exhaling
Digestive System	Provides heart with nutrients so it can beat		Provides kidneys with nutrients so they can filter blood	Provides lungs with nutrients so they can inhale and exhale
Excretory System	Clears blood of wastes from food and respiration (carbon dioxide)	Clears blood of wastes from food		Clears blood of wastes from respiration (carbon dioxide)
Respiratory System	Provides oxygen for blood to circulate and removes carbon dioxide waste	Provides oxygen so cells can function	Provides oxygen so cells can function	

Explore-a-Lab

Structured Inquiry

 **Do people expel waste when they exhale?**

With a partner, label a test tube *A* and another test tube *B*. Fill test tube *A* about $\frac{1}{4}$ full of limewater and place a straw in it. Wait 1 minute. Then gently blow through the straw. Have your partner record the number of exhalations you must make through the straw in order for the limewater to turn cloudy. Make sure that you inhale through your nose and exhale through your mouth. Do NOT inhale through the straw! Now run in place for 3 minutes. While you are running, have your partner fill test tube *B* about $\frac{1}{4}$ full of limewater and place a straw in it. After 3 minutes have passed, blow through the straw into the limewater in test tube *B*. Have your partner record the number of exhalations required to turn the limewater cloudy. Last, discuss with your partner any difference between the two numbers and why the difference occurred.

Disorders and Diseases of the Excretory System

What do you think would happen if one of the organs in the excretory system stopped working? Would the body be able to compensate for the loss of that function? Many factors can cause kidney disease, including injury and infection. However, diabetes and high blood pressure are the usual culprits. People who have diabetes and high blood pressure often develop chronic kidney disease (CKD). Their kidneys cannot efficiently filter waste and excess fluids from their blood, so the wastes build up. This situation can cause anemia, heart disease, bone disease, and, eventually, kidney failure, which is also known as renal failure.

To survive, people who have renal failure need **dialysis**, which is the repeated removal of wastes and excess salt and water from the blood. The most common type of dialysis uses a machine to clean the blood. People who lose 85 to 90 percent of their kidney function require dialysis. People whose kidneys are failing must receive dialysis several times per week for the rest of their lives, unless they choose to have a kidney transplant—an operation to place a healthy donor kidney in their body. They also

Called to Serve

Undergoing dialysis can be physically and emotionally difficult for patients. Contact a local health-care facility and ask what items dialysis patients might appreciate in a care package (possibly magazines, books, and games). If you can't collect these items, just sending a card offering your support can help lift a patient's spirits.

must make changes to their diet, limit their fluid intake, and take certain medications. What changes to the diet do you think would help a person with renal failure? What can an individual do to reduce the risk of these problems in the first place?

There are other disorders and diseases associated with the excretory system. A few of the more common ones are shown in the table below.

	Excretory System Disorders		
	Cystitis	**Gout**	**Kidney Stones**
Causes	Cystitis is caused by bacteria or other factors.	Gout is caused by high levels of uric acid in the blood that result from the consumption of certain foods, including organ meats and anchovies, that contain substances called purines.	Certain foods and medications may make some people more susceptible to kidney stones, as does not drinking enough water. A family history of kidney stones and a personal history of gout, kidney disorders and diseases, and surgery on the gastrointestinal tract also make people more susceptible.
Description	This condition is an inflammation of the bladder. When bacteria that have entered through the urethra cause this condition, it is called a urinary tract infection (UTI). Symptoms include frequent urination, a burning sensation when urinating, blood in the urine, and discomfort in the pelvic area.	This condition occurs when urate crystals build up in a joint, usually the big toe but sometimes elsewhere in feet, ankles, hands, wrists, or knees. With gout, the kidneys fail to remove enough uric acid from the blood, causing intense joint pain and inflammation.	These "stones," composed of mineral and acid salts in urine, form in the kidneys and cause great pain as they pass out of the body through the urethra. If a stone becomes stuck in the urethra, though, it can cause severe pain, vomiting, fever, and bleeding.
Treatment	Antibiotics are the usual treatment for UTIs.	There is no cure for gout, but medications that inhibit the production of uric acids can help prevent flare-ups. Other medications treat the pain and inflammation of gout attacks.	Usually, kidney stones simply pass on their own. Your doctor might direct you to take painkillers and drink water to help a stone pass. If a stone becomes stuck in the urethra, though, it can be broken up and removed with the use of sound waves, a scope, or surgery.
Further inquiry	How can you protect yourself from UTIs?	We know what foods cause gout, but what are the foods that help prevent it?	Which foods and medications make people more susceptible to kidney stones?

Keeping the Excretory System Healthy

You can take several simple steps to maintain the health of your excretory system. To protect your kidneys, keep your blood pressure and cholesterol levels within a healthful range. Check with your doctor to find out what a healthful range is for you. Get plenty of physical activity, eat healthful foods, and reduce your salt intake. How does salt affect your kidneys? Maintain a healthy weight so that you do not develop type II diabetes, which often leads to kidney disease. If you do have diabetes, keep your blood glucose level within the acceptable range indicated by your doctor. What changes can you make in your life to help protect your kidneys?

To prevent cystitis, women should wipe front to back to avoid getting bacteria from the anus near the urethra. Take showers instead of baths and avoid perfumed soaps. These steps will prevent irritation of delicate tissues. Drink plenty of water and urinate frequently. Some people find that drinking cranberry juice helps prevent urinary tract infections (UTIs). Why do you think cranberry juice is useful in warding off cystitis?

People are more likely to develop gout if they have high blood pressure, diabetes, high levels of cholesterol and fat in the blood, and narrowed arteries. Take steps to prevent these conditions. Drink plenty of water, and abstain from alcohol and sweetened drinks. A healthful vegetarian diet can also help. Can you think of any fad diets that might contribute to people developing gout?

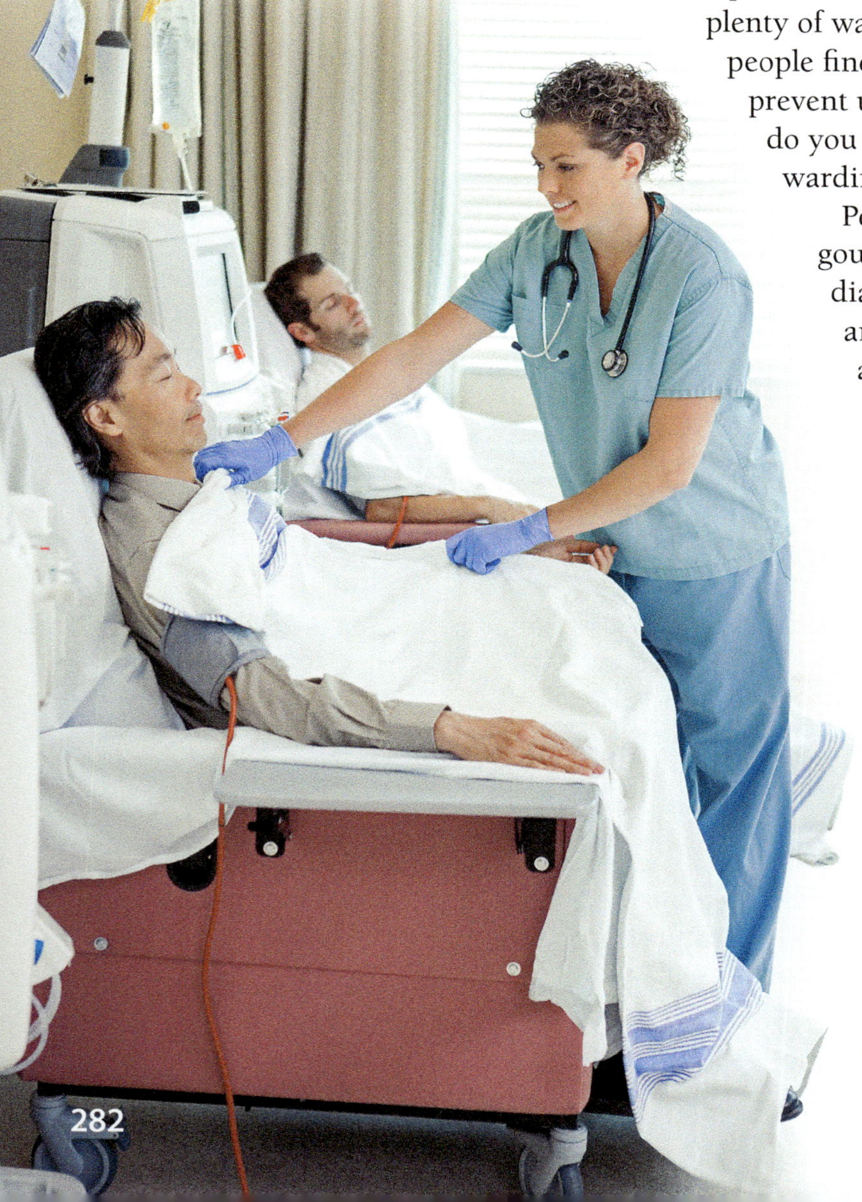

Keeping your body healthy not only makes you feel better, but it also saves you money. Dialysis does not cure kidney disease, and it is very expensive.

You can help prevent kidney stones by drinking plenty of water, maintaining a healthful weight, and making sure your diet isn't too high in animal proteins, salt, or sugar. How can you increase the amount of water you drink every day?

Explore-a-Lab

Guided Inquiry

How does kidney dialysis work?

You will design an investigation to model how the kidneys filter certain molecules out of the blood. In this case you will be trying to filter iodine out of a water-iodine solution, using a plastic sandwich bag, plastic cups, and some starch solution Begin by writing a hypothesis. Then design a procedure to collect and record data, using the materials provided. Get your teacher's approval, and then carry out your investigation. After you have gathered all your data, make a conclusion regarding how the kidneys filter toxins and other compounds out of the blood.

Concept Check

Assess/Reflect

Summary: How does the excretory system work? The excretory system, which removes wastes from the body, includes the kidneys, the ureters, the urinary bladder, and the urethra. The skin (part of the integumentary system) and lungs (part of the respiratory system) also remove wastes from the body. The digestive, excretory, circulatory, and respiratory systems work together to carry nutrients and oxygen to cells, carry wastes away from cells, and expel wastes from the body. There are many disorders and diseases of the excretory system, including chronic kidney disease (CKD) and renal failure. There are many actions you can take to maintain the health of your excretory system, including drinking plenty of water, exercising, and choosing your diet carefully.

1. Describe how the excretory system functions.
2. You drink a glass of water. After the water reaches your stomach, it is absorbed through the walls of your stomach into your bloodstream. Describe the path water molecules take after this point.
3. Make a list of the parts of the excretory system. How has God designed this system to make our lives better? What do we need to do to keep it working well?
4. Which disease of the excretory system do you think is most likely to affect people your age? Explain your answer.

People in Science

Extend

Get to Know
Neil Nedley, MD

In his private practice in Ardmore, Oklahoma, Neil Nedley, MD, cares for patients with a wide variety of ailments. He realized that many of his patients' illnesses could have been prevented or controlled by their lifestyle choices.

Dr. Nedley has written several books, including *Proof Positive: How to Reliably Combat Disease and Achieve Optimal Health Through Nutrition and Lifestyle*, in an effort to help people understand how their choices regarding food, physical activity, smoking, and drug and alcohol use effect their bodies. In his book, he explains that people take care of their personal belongs, such as maintaining their cars to keep them running as long as possible, but many say God will decide how long each person's life is. Dr. Nedley teaches that people can somewhat control how long their lives are by making healthy choices. They can also control the quality of their lives.

For example, in his book, Dr. Nedley explains how smoking affects the digestive system by causing ulcers and acid reflux. Dr. Nedley also cites research that supports Ellen White's teachings to abstain from alcohol in order to live longer.

Focusing on the digestive system, Dr. Nedley recommends that everyone eat breakfast because studies show that people who eat a good breakfast live longer. In addition, he says a smaller evening meal can help a person sleep better because the stomach is not working all night. He adds that the smaller evening meal will make you hungry in the morning for a good breakfast.

In addition to numerous scientific studies, Dr. Nedley uses Bible verses in his books to support his teachings because he is a devout Christian. *Proof Positive* includes an entire chapter on how faith in God enhances a person's health. Research shows that believing in God and attending church increase a person's lifespan, as does having close, reliable friends and family. Dr. Nedley explains that these strong bonds increase a person's social and emotional health, which have an impact on physical health.

In addition to his practice in Oklahoma, Dr. Nedley also runs a residential treatment program for patients with treatment-resistant depression and anxiety near Sacramento, California. Dr. Nedley is also a pilot. He and his wife Erica have four boys, and the entire family enjoys flying, biking, hiking, skiing, and many other endeavors that keep them active.

Concept Check

1. How do lifestyle choices affect your health?
2. How does a relationship with God make you healthier?

Careers in Science

Extend

Food Scientists and Technologists

Think about how food production has changed. How did your ancestors get their food 200 years ago? They didn't have grocery stores with aisles of food waiting for them. They also didn't have refrigerators to help keep foods fresh.

Do you know how many scientists may have had a role in getting food to your plate? Scientists start work on your food long before it even enters your home. Food scientists and technologists study the physical, microbiological, and chemical make-up of food to develop the safe and nutritious food you eat every day. Some food scientists focus on growing healthy crops of fruits, vegetables, and grains. Others study the most efficient way to get fresh food to consumers at the peak of ripeness. Those scientists might work on figuring out the best time to pick produce or on packaging that keeps the food fresh longer. One type of food technologist develops new products for a quick meal that might be easier for you to prepare at home.

Another type of food technologist focuses on developing foods free of the top eight allergens—eggs, dairy, soy, gluten, peanuts, tree nuts, fish, and shellfish. People with certain food allergies cannot eat the standard breads, cookies, and cakes that you buy in the store. Food scientists try to create new products that do not contain these allergens but still taste good. These scientists might be responsible for creating new recipes, figuring out how to mass produce the new product, and determining how long it will safely keep on the grocery store shelf. Why might mass production be harder than making a new product in the lab?

Depending on their area of expertise, food scientists may have a variety of educational backgrounds, including degrees in biology, chemistry, chemical engineering, and food science. As with many careers, food scientists can choose to specialize in many areas, and they often obtain more training for that specialty.

Concept Check

1. How have food scientists improved food quality and availability?
2. Do you think food science is a growing industry? Why or why not?

Chapter 7

Review

Assess/Reflect

Study Guide

Lesson 1

1. The organs of the digestive system work together to break down food for use by the body. Organs of the digestive system include the mouth, pharynx, esophagus, stomach, small intestine, pancreas, liver, gallbladder, and large intestine. Refer to the chart on page 264 for information about the function of these organs.
2. Disorders and diseases of the digestive system include celiac disease, gastroesophageal reflux disease, inflammatory bowel disease, lactose intolerance, peptic ulcers, cirrhosis, and diabetes.
3. Behaviors that promote a healthy digestive system include drinking plenty of water, exercising regularly, and choosing a healthful diet that includes plenty of fruits and vegetables.

Lesson 2

1. The excretory system removes wastes from the body. The major organs of the excretory system include the lungs, skin, kidneys, and urinary bladder. The lungs remove carbon dioxide. The skin excretes excess salt and other wastes present in the body. The kidneys filter and remove wastes from the blood. The urinary bladder stores urine until it is released from the body.
2. The digestive, excretory, circulatory, and respiratory systems work together to bring nutrients and oxygen to cells and to take wastes away from cells and expel them from the body. Refer to the chart on 285 for information on the specific ways these systems work together.
3. The disorders and diseases of the excretory system include chronic kidney disease, renal failure, cystitis, gout, and kidney stones.
4. You can keep your excretory system healthy by keeping your blood pressure, cholesterol levels, and glucose level in a healthful range. Get enough physical activity, eat properly, reduce your salt intake, and drink plenty of water. Avoid alcohol and sweetened drinks. Maintain a healthful weight.

Vocabulary Check

Fill in the blank with the correct terms.

1. To get energy, the body relies on the __________ to break food down into molecules that can be used by cells.
2. The fingerlike projections in the small intestine where absorption occurs are the __________.
3. __________ are substances that speed up chemical reactions in the body.
4. Urine is formed in the __________.
5. The lungs and skin help remove wastes from the body, so they are part of the __________.
6. The most common type of __________ uses a special machine to clean the blood.

Multiple Choice

Choose the best answer.

7. Which of these are the filtering units in the kidneys?
   A. arteries
   B. ureters
   C. nephrons
   D. capillaries

8. In which part of the body does chemical digestion take place?
   A. mouth
   B. stomach
   C. small intestine
   D. all of the above

9. Which statement about celiac disease is true?
   A. People who have celiac disease cannot eat dairy products.
   B. People who have celiac disease cannot eat glutens.
   C. It causes the immune system to attack the stomach lining.
   D. It causes the immune system to attack the esophagus.

10. How many villi are in a square millimeter of the small intestine?
    A. 10 to 40
    B. 20 to 60
    C. 30 to 70
    D. 40 to 80

11. What causes cystitis?
    A. alcohol
    B. cigarettes
    C. bacteria
    D. viruses

12. For people who have renal failure, what is the alternative to dialysis?
    A. regular exercise
    B. kidney transplant
    C. daily medication
    D. restricted diet

13. What moves food through the digestive system?
    A. gravity
    B. enzymes
    C. chemical digestion
    D. muscle contractions

14. What is the role of the gallbladder in digestion?
    A. aids the kidney in filtering waste
    B. important in absorbing certain nutrients
    C. stores bile that helps to digest fats
    D. helps to digest sugars and proteins

Check Point

Answer the following questions.

15. The digestive system and the excretory system both produce wastes that are expelled from the body. How are they alike, and how do they differ?

16. How does drinking water keep both the digestive and the excretory systems healthy?

17. What diet guidelines should you follow to protect against diseases and disorders of the digestive system and the excretory system?

18. How are the small intestine and the large intestine alike? How are they different?

19. Why is it important to take steps to avoid developing diabetes?

Earth and Space Science

Unit Overview

People have long been fascinated by the Sun, Moon, stars, and other objects in the sky. They gaze into the Heavens and are inspired by the vast Universe that God created. What makes the Sun shine? Why does the Moon seem to change shape each month? What are the stars made of? Is Earth the only planet with life?

To answer these and other questions, ancient peoples built observatories to study the skies. Today, we use spacecraft and other technology to take images of exploding stars and distant galaxies. We land robotic rovers on other planets to scoop up soil and rocks for study back on Earth. Space science has come a long way. In this unit, you'll learn about the tools scientists use to study the Universe. You'll take a journey through the Solar System and then outward to distant stars and galaxies.

Scientists think that Stonehenge in England is an ancient observatory. Built about 4500 years ago, it may have been used to study the changing of the seasons and the movements of objects in the night sky. Here, the Sun is rising over Stonehenge on the first day of summer.

ADVENTISTS InStep for Life

Your teacher may assign an Open Inquiry lab and a Lifestyle Challenge activity. Use your ***Science Journal*** to record your work.

Science Journal

Objects in Our Solar System

Scripture Spotlight

The care that God took when designing Earth as our home is evident when you study the other objects in the Solar System. You can see God's masterful Design as you study the stars and planets. You will read the following passages in this chapter.

Job 26:7 (p. 297)
Genesis 1:16 (p. 297)
Matthew 24:29 (p. 307)
Psalm 8:3–9 (p. 319)

The "tail" of a comet results from the effects of solar radiation and solar wind upon the core of the comet.

The Big Idea

When God created the Sun and the planets and their moons, He also created the many other objects that make up our Solar System.

How do you think objects in the Solar System differ from one another?

Inquiry Kick-Off

Engage

You have learned that your weight is the measure of gravity acting on your body. If you traveled to the Moon or another body in the Solar System, would your weight change? If so, would you weigh more or less? Would your mass change? The Inquiry Kick-Off in your ***Science Journal*** will help you answer these questions.

Science Journal

Lesson 1

Essential Question

What Are the Planets in Our Solar System?

Objectives

- Identify the objects that make up the Solar System.
- Describe Kepler's laws of planetary motion.
- Describe Newton's law of universal gravitation.
- Compare and contrast terrestrial planets with gas giants.
- Describe evidence in the Solar System that shows God's Design and plan.

Vocabulary

ellipse
perihelion
aphelion
foci
astronomical unit (AU)
retrograde

When you look up in the sky at night, how many planets can you see? Why isn't it possible to see all of the planets at once when you look into the night sky? As you know, Earth is one of eight planets in our Solar System. You have learned that the Solar System is highly organized, which is evidence of God's Design. Isaac Newton, the great physicist who explained the motions of the planets and who was the first to determine some of their masses, believed the Solar System was designed. He said, "When I look at the Solar System, I see the Earth at the right distance from the Sun to receive the proper amounts of heat and light. This did not happen by chance. The motions of the planets require a Divine arm to impress them." What is some other evidence of Design that you can see in the Solar System?

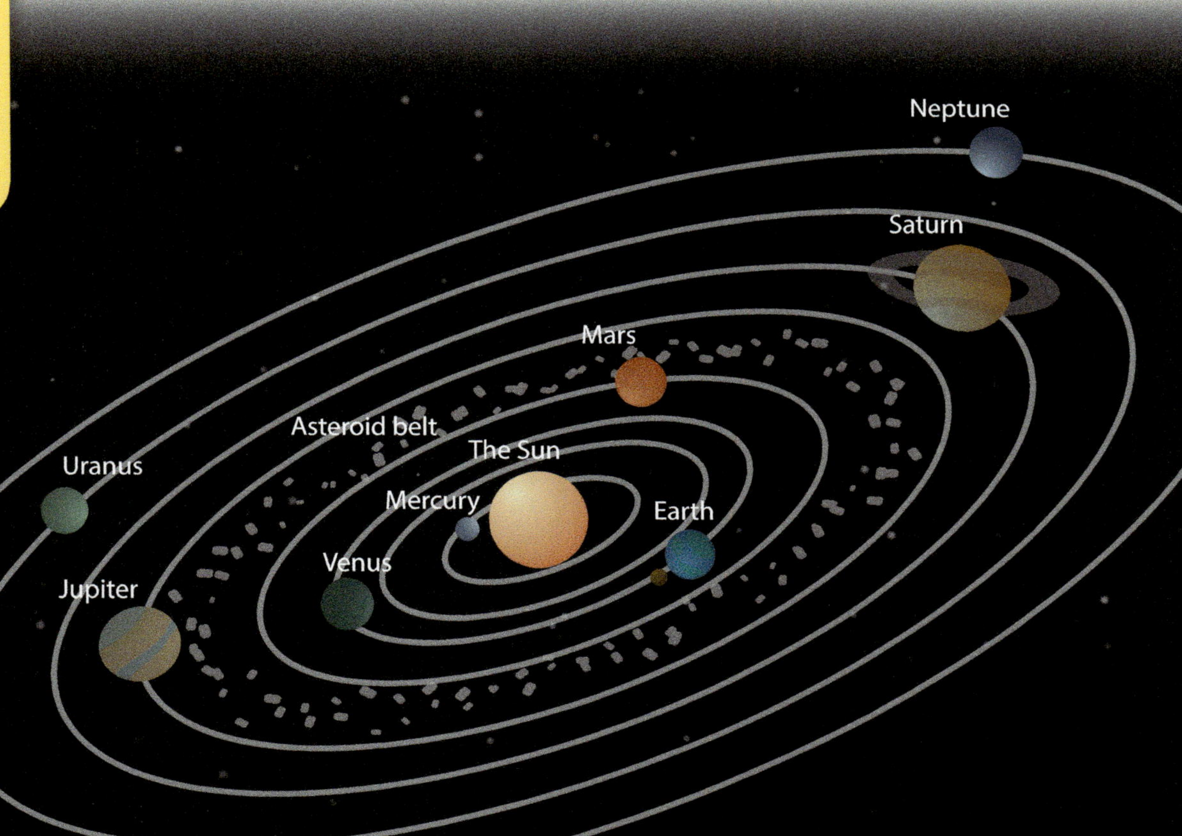

Our Solar System includes the Sun, the eight planets and their moons, and numerous smaller objects that orbit the Sun.

If our Solar System had two suns, how do you think this would affect planetary orbits?

The Solar System Explain

You may have heard many times before that Earth is the third planet from the Sun. But what exactly is a planet? Scientists define a *planet* as

- an object that orbits a star
- an object that has a round shape because of its gravity
- an object that does not have asteroids, meteors, or other space debris orbiting around it

The Solar System is made up of the Sun, eight planets, and the numerous smaller objects that orbit the Sun. The Sun is by far the largest object in the Solar System. It makes up more than 99% of the Solar System's mass. Given its huge mass, the gravitational pull of the Sun is immense. Gravity is the reason why objects in the Solar System orbit the Sun.

The Sun's mass is just right to provide a gravitational force that holds the Solar System together. What would happen to the planets if the Sun had greater mass? What would happen if the Sun had less mass?

Planetary Movements

The planets do not travel in circular orbits around the Sun. Rather, they travel in elliptical orbits. An easy way to think of an **ellipse** is as a flattened circle, as shown below.

Notice that a planet does not remain at the same distance from the Sun throughout its orbit. For example, Earth is closer to the Sun in winter than it is in summer. **Perihelion** is the point where Earth is closest to the Sun. **Aphelion** is the point when Earth is farthest from the Sun. Why do you suppose that it is warmer in North America during summer when Earth is actually farther from the Sun than it is during winter?

A balance between Earth's momentum and gravity keeps Earth in orbit around the Sun.

At what position in its orbit is Earth today?

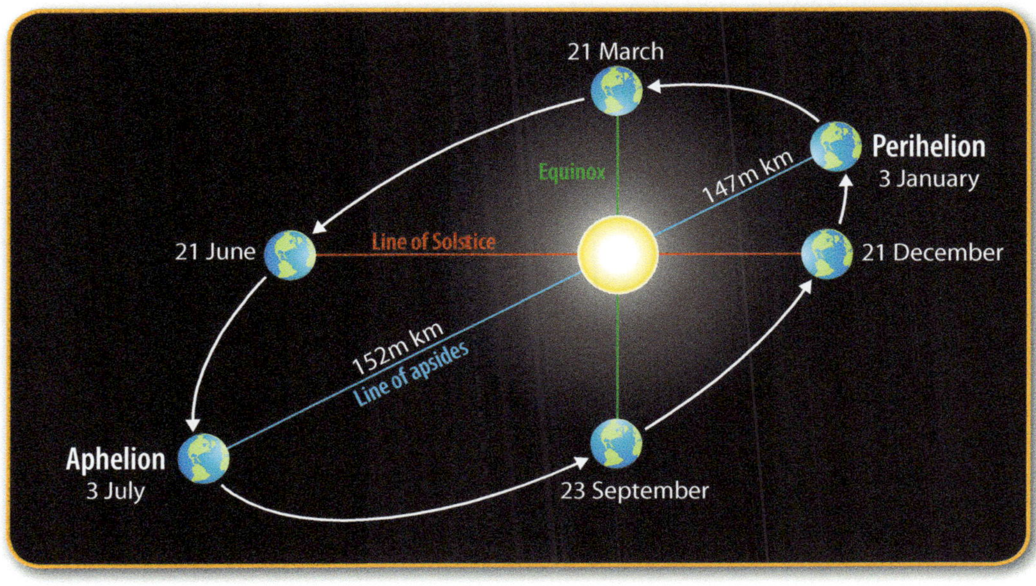

Kepler's Laws of Planetary Motion

A scientist named Johannes Kepler first proposed that planets travel in elliptical orbits. His observations of planetary motion led to the development of three laws that still describe the behavior of the planets.

Kepler's First Law

As you can see in the diagram below, an ellipse has two points, known as the **foci** (singular: focus) of the ellipse. Kepler's first law states that *the orbits of the planets are ellipses with the Sun as one of the foci.*

The sum of the distances between the planet and the two foci of its elliptical orbit is always the same regardless of its position.

At what positions in its orbit are the distances between the planet and the two foci equal? At what positions are these distances the largest and the smallest?

An Elliptical Orbit of a Planet

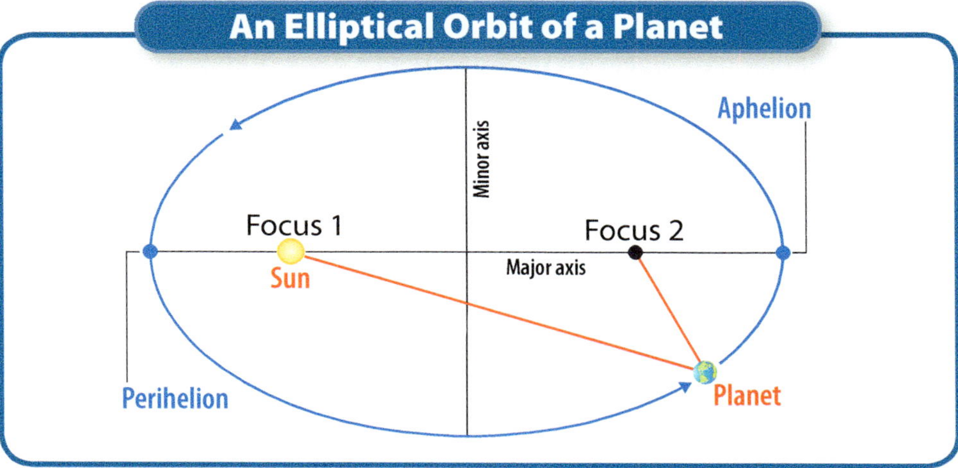

Kepler's Second Law

As each planet travels in an elliptical orbit, the distance between each planet and the Sun decreases, the distance between the planet and the other focus of its orbit increases, and vice versa. According to Kepler's second law, *an imaginary line from the Sun to a planet sweeps out equal areas in equal time intervals.* In the diagram below, the planet takes the same amount of time to go from point A1 to B1 as it does to go from point A to B. The two shaded areas have the same area. The planets move fastest when closest to the Sun and slowest when they are farthest away.

Kepler did not number the laws himself. It was the discovery of the "second" law that led to establishing the "first" law.

Equal Areas in Equal Time

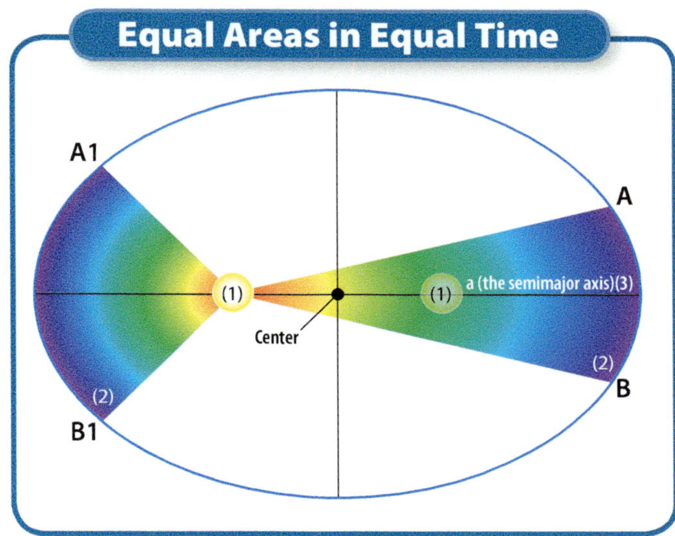

Kepler's Third Law

The farther a planet is from the Sun, the longer it takes to complete one revolution. Kepler's third law states that *the ratio of the squares of the periods of any two planets is equal to the ratio of the cubes of their average distance from the Sun.*

What major difference can you see between the first and second laws and the third law? How might the third law be used?

Universal Gravitation

Kepler's three laws explain the motion of the planets in astonishing detail. But Kepler could never answer the question: Why do planets travel faster when they are closer to the Sun? The answer had to wait until another astronomer considered this question. The astronomer was Isaac Newton, and his answer stems from Newton's law of universal gravitation.

Gravity is an attractive force that exists between any two objects, such as the Sun and Earth, or a mountain and a person. According to the law of universal gravitation, *the force of gravity between two objects is the product of the masses of two objects divided by the square of the distance between them.* This formula means that if the distance between two objects increases by a factor of two, then the force of gravity between them decreases by a factor of four. How will the force of gravity change if the distance between two objects increases by a factor of 10?

Gravity is the force that keeps planets moving in their orbits. What would happen to the planets if the gravity between them and the Sun did not exist? According to Newton's law of universal gravitation, the closer two objects are, the stronger the force of gravity. Therefore, the closer a planet is to the Sun, the stronger the force of gravity is between that planet and the Sun, and the planet will therefore move faster. How does this explain Kepler's second law of planetary motion?

Check out your ***Science Journal*** for a Structured Inquiry that explores the shape of planetary orbits.

Extend

Lesson Activity

Become an expert on a planet in our Solar System. Select a planet that you would like to learn more about. Using the Internet or other resources, record facts, such as its distance from the Sun, size, composition, the date of its discovery, number of moons (if any), unique features, geologic features, and so on. Choose a method to communicate your information and share what you learned with your class. A computer or visual presentation, an animation, or physical models are good options.

What features make the planet you chose special?

Record your work for this inquiry. Your teacher may also assign the related Guided Inquiry.

Solar System Distances

How can you show relative distances of the planets from the Sun?

Materials

- adding machine tape
- meter tape
- pencil
- colored pencils

Procedure

1. Use a scale of 1 AU = 1 m. **Calculate** the number of meters you need to represent each distance given in the chart. **Record** the calculations in the chart.

Planet	Distance from Sun (AU)	Distance of Machine Tape (m)
Mercury	0.38	
Venus	0.72	
Earth	1.00	
Mars	1.52	
Jupiter	5.20	
Saturn	9.53	
Uranus	19.28	
Neptune	30.06	

2. Draw a line across the end of the roll of adding machine tape. Label it *Sun*.
3. Unroll the adding machine tape and measure from the Sun line to the number of meters you calculated for Mercury in column 3 of the chart. Place an easily seen dot at that point and label it *Mercury*.
4. Repeat Step 3 for the remaining planets.

Analyze Results

What explains the relatively close distances of the inner planets to the Sun and the relatively far distances of the outer planets from the Sun?

Create Explanations

1. How can you show relative distances of the planets from the Sun?
2. Why is it difficult to make a model of the Solar System that is correct with reference to both planetary diameter and distance?
3. Alpha Centauri is the closest star to the Sun. Astronomers estimate that Alpha Centauri is about 277,600 AU from the Sun. Based on the scale you used for this model, how many meters of adding machine tape would be needed to show its distance from the Sun?

How the Planets Differ **Explain**

The table below shows all the planets in our Solar System. In order from the Sun, the planets are Mercury, Venus, Earth, Mars, Jupiter, Saturn, Uranus, and Neptune. Distances within our Solar System are often given in **astronomical units** (AU) based upon the average distance between Earth and the Sun. One astronomical unit is equal to 150,000,000 km. The table below shows the distance of each planet from the Sun in astronomical units.

Planetary Distances from the Sun	
Planet	**Distance (AU)**
Mercury	0.38
Venus	0.72
Earth	1.00
Mars	1.52
Jupiter	5.20
Saturn	9.53
Uranus	19.28
Neptune	30.06

Scripture Spotlight

Read **Job 26:7** and compare it with **Genesis 1:16.** How are the two passages similar?

The Terrestrial Planets

The eight planets in our Solar System can be classified into two major groups, the terrestrial planets and the gas giants. The terrestrial planet include Mercury, Venus, Earth, and Mars. When God created the planets of the Solar System, He created the solid inner planets with mostly silicate rocks and an inner iron core. These planets are small in comparison to the gas giants. Besides having a solid surface, terrestrial planets have a variety of surface features such as craters and mountains and relatively few moons.

Terrestrial Planet Characteristics					
Planet	**Distance from Sun (km)**	**Diameter (km)**	**Orbital Velocity (km/s)**	**Atmosphere**	**Number of Moons**
Mercury	58,000,000	4878	48	Oxygen, sodium, hydrogen	0
Venus	108,000,000	12,104	35	Carbon dioxide, nitrogen	0
Earth	150,000,000	12,756	30	Nitrogen, oxygen	1
Mars	228,000,000	6787	24	Carbon dioxide, nitrogen, argon	2

Mercury

Mercury is the planet closest to the Sun, and it is the smallest planet in our Solar System. Its surface is heavily cratered from collisions with small, rocky objects. Why do you think Mercury's surface is so heavily cratered? Long cliffs also cover parts of its surface.

Mercury has the greatest daily temperature extremes of any planet. Because it is so close to the Sun, daytime temperatures soar to 430°C. Mercury has virtually no atmosphere to trap this heat, so at night, temperatures plummet to around –180°C.

Mercury is the smallest planet in our Solar System.

What challenges can you foresee for landing any probes on Mercury?

Venus

Venus is the second planet from the Sun. Its surface is hidden by thick clouds of mostly carbon dioxide. Its dense atmosphere lets very little sunlight penetrate to the surface. But the sunlight that does penetrate the clouds cannot escape back into space. This results in surface temperatures on Venus that reach a scorching 470°C. Probes that land on Venus are destroyed by the heat within hours. They have been able, however, to send back stark images of sand dunes and ancient lava flows.

Venus is often referred to as Earth's twin planet. They are similar in size, differing in diameter by only 500 km. This means their mass and surface gravity are also very similar. The major similarities, however, end there. Venus's rotation is **retrograde**. This means that its daily rotation on its axis is the exact opposite of the rotation of Earth and of all the other planets in our Solar System. Thus, seen from Venus, the Sun would appear to rise in the west and set in the east. In addition, Venus's day is longer than its year! Its daily rotation takes 243 Earth days, yet it takes only 225 days to orbit the Sun. Why do you think it takes Venus so long to complete a single rotation?

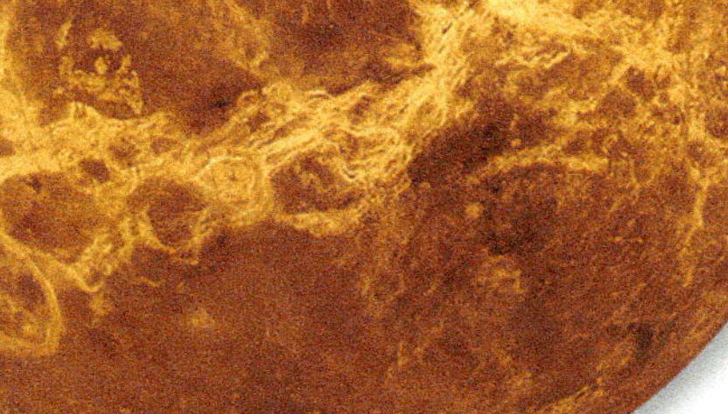

Venus, Earth's twin planet

Why do we know so little about the surface of Venus?

Faith Connection

Venus spins on its axis, or rotates, in a direction opposite from the rotation of all the other planets. This is called retrograde rotation. How does Venus's retrograde rotation pose a problem for the Big Bang model?

Explore-a-Lab — Guided Inquiry

How could you create a retrograde motion?

Start spinning a ball so that it rotates like Earth. Should it spin clockwise or counterclockwise? While the ball is spinning, try to reverse the spin without holding and stopping the ball first. What technique did you use to create this retrograde motion? Venus has retrograde motion. What do you think may have caused Venus's retrograde motion?

Earth

While Earth shares characteristics with the other inner planets, it is different in an important way. As far as we know, life in our Solar System exists only on Earth. Earth's features and the conditions that allow life on Earth to exist and thrive provide strong evidence of an intelligent Creator.

Earth looks blue from space because about 70% of its surface is covered by water. Earth is the only planet that we know of where water exists as a solid, a liquid, and a gas.

God placed Earth to be the proper distance from the Sun to support life, as you can see in this table that shows what would happen if Earth were closer or farther from the Sun.

Earth's Distance from the Sun	Amount of Heat Received
Twice as far	One-quarter of what it now receives
Twice as close	Four times what it now receives

Earth has the right type of atmosphere to sustain life. Our atmosphere contains a proportion of nitrogen, oxygen, carbon dioxide, and other gases. The atmosphere protects Earth's surface from being bombarded by dangerous radiation. The carbon dioxide in Earth's atmosphere also traps heat and moisture, preventing their escape into outer space. Earth is just the right size to support life. If Earth were smaller, its gravity could not hold our atmosphere or water.

Earth is the only planet known to contain life.

Astronomers search other solar systems for planets in the "Goldilocks Zone." What do you think that term refers to?

Earth has a Moon. It is often just called the Moon, but its official name is Luna. The Moon is the only object in space, other than Earth, that has been visited by humans. How do you think life on Earth would be different if there were no Moon? Could life on Earth exist without the Moon? Why or why not?

If we look closely at the features of our Earth, we see evidence that these features did not happen by random chance. The large number of critically connected features shows that an Intelligent Creator planned, created, controls, and supports our Earth.

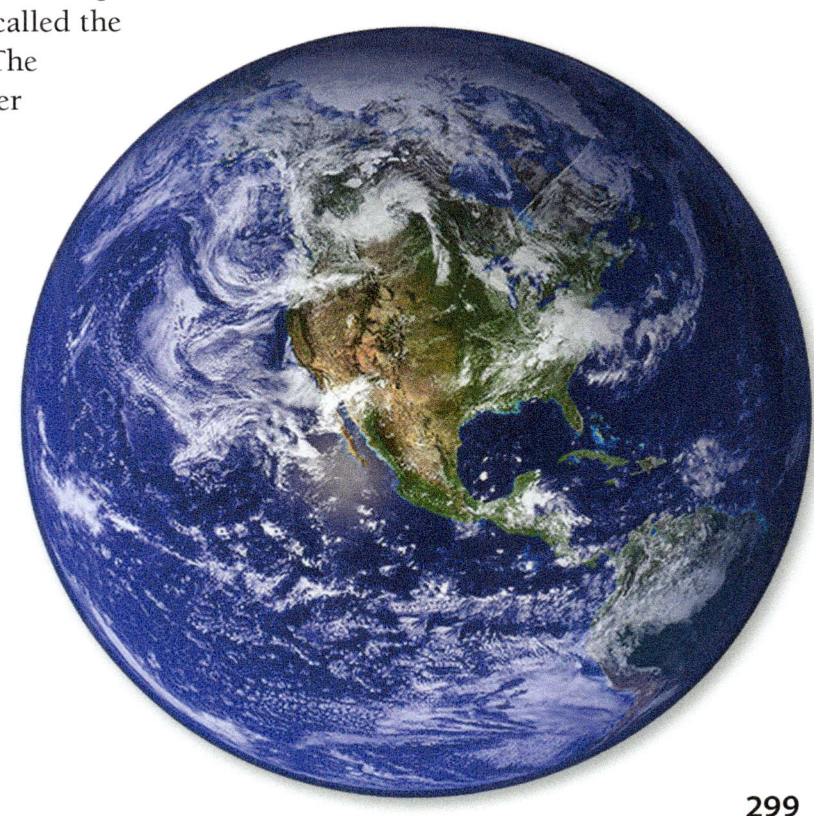

Mars has approximately 43,000 impact craters on its surface

How would you explain this high number?

Mars

Look at an image of Mars, the fourth planet from the Sun, and you can see why it is often called the "red planet." The soil of Mars contains iron oxides, giving the planet a reddish color. The polar areas of Mars are visibly white because Mars has polar ice caps made of frozen carbon dioxide layered over frozen water. The ice caps enlarge during the Martian winter and shrink during the Martian summer. Why are scientists so interested in the red planet?

Mars has a thin atmosphere and cold temperatures that range from –87°C to –5°C. Surface temperatures might seem too cold for liquid water to flow on Mars. But data gathered by spacecraft and space rovers indicates that Mars may have some flowing water. Data also indicates that the planet once had plentiful liquid water on its surface. The water most likely carved the long channels visible on Mars's surface. Do you think it is possible that there was or still is life on Mars? What evidence supports your answer?

In addition to channels on its surface, Mars has deep rift valleys and boasts the largest volcano in the Solar System, Olympus Mons. This giant volcano is almost three times taller than Mount Everest, the tallest mountain above sea level on Earth. If you were standing on the top of Olympus Mons, you could not see its base because it would extend beyond the horizon. Despite its massive size, Olympus Mons is one of the youngest of the large volcanoes on Mars. Mars also has two small, irregularly shaped moons: Phobos and Deimos.

Explore-a-Lab

Structured Inquiry

What does the Curiosity Rover look like?

Use information available on the Internet and other sources to build a three-dimensional model of the Curiosity Rover. Try to include as much detail as possible. Use inexpensive materials or materials you have on hand. Your model needs to be small enough to fit in a shoebox. What are important features on your model? What is the function of each?

Gas Giant Characteristics

Planet	Distance from Sun (km)	Diameter (km)	Orbital Velocity (km/s)	Atmosphere	Number of Known Moons
Jupiter	778,400,000	142,984	13.0	Hydrogen, helium	68
Saturn	1,427,000,000	120,536	9.7	Hydrogen, helium	62
Uranus	2,871,000,000	51,118	6.8	Hydrogen, helium, methane	27
Neptune	4,498,000,000	49,528	5.4	Hydrogen, helium, methane	13

Gas Giants

The second group of planets in our Solar System is known as the gas giants, which includes Jupiter, Saturn, Uranus, and Neptune. God made the outer planets, the gas giants, from lighter elements, such as hydrogen and helium, and the gaseous molecule methane. How do the atmospheres of the outer planets compare with those of the inner planets? What do you think is the reason for the difference?

Notice that all the gas giants have many more moons than the terrestrial planets. What might be a reason for this? Of the 68 moons that orbit Jupiter, only 50 have been officially recognized. The remaining 18 are awaiting "official confirmation." How do you think scientists officially confirm this information? Do you think more moons will be found orbiting these giant planets?

Jupiter

Jupiter, the fifth planet from the Sun, is the giant of the gas giants. It makes up about 70% of the mass of all the planets in the Solar System. Jupiter's atmosphere is mainly hydrogen and helium. Closer to the planet's surface, the hydrogen gas is a liquid. What might have caused this hydrogen gas to turn into a liquid? Deep within Jupiter may be a rocky core as big as Earth.

Jupiter is the largest planet in the Solar System.

Why do you think Jupiter has so many moons?

Jupiter has a banded appearance. Those bands are clouds of different colors that blow by strong winds in different directions at different latitudes. Powerful storms are common in the atmosphere. One such storm, called the Great Red Spot, has been visible since the 1800s.

In addition to its many small moons, Jupiter has four large moons, one of which is bigger than the planet Mercury. The four large moons—Io, Europa, Ganymede, and Callisto—are called Galilean moons because in 1610 the Italian astronomer Galileo Galilei discovered them. Io has more active volcanoes than any other object in the Solar System. Europa is covered with ice, and a vast ocean might be present beneath the ice. For this reason, astronomers are particularly eager to explore Europa.

Saturn is known for its spectacular rings.

Why do gas giants have rings, but rocky planets do not?

Saturn

Each gas giant has a ring system. But none can rival the rings of Saturn, the sixth planet from the Sun. Saturn has thousands of rings, each of which orbits the planet at a different speed. Why does each ring orbit the planet at a different speed? The rings are made of ice and rock. Some are less than a kilometer wide. Others have widths of more than 280,000 km. Where do you think the material for the rings of Saturn came from? One theory is the material for the rings most likely came from moons and other small, rocky objects in space that broke up as they neared the planet.

Like Jupiter, Saturn has an atmosphere of mostly hydrogen and helium. It, too, might have a rocky core. Saturn's numerous moons include Titan, the second-largest planetary moon. Unlike most other moons, Titan has an atmosphere. This atmosphere, which extends hundreds of kilometers into space, is rich in nitrogen.

Uranus

Uranus, the seventh planet from the Sun, is a lovely blue-green color. The color comes from the methane in its atmosphere. Other gases in the atmosphere include hydrogen and helium. Astronomers do not believe that Uranus has a rocky core. Instead, they think that its core is an icy mixture of water, methane, and ammonia. How would you determine what the composition of Uranus's core is like?

Like the other gas giants, Uranus has moons and a ring system. But the planet also has a unique axis, or imaginary line around which the planet rotates. Instead of having north and south poles as Earth does, Uranus's axis is extremely tilted, with the axis nearly at a right angle to the radius of orbit so that one of its poles is always oriented toward the Sun during its orbit. Scientists think that a collision with an Earth-sized object may have knocked Uranus on its side.

Uranus has a unique axis of rotation.

Neptune

Neptune, the eighth planet from the Sun, is the most distant of the gas giants, a whopping 4.5 billion km from the Sun. It takes Neptune nearly 165 Earth years to complete one orbit. How would you describe the irregular nature of Neptune's orbit? Given its distance from the Sun, the planet is extremely cold. Temperatures are about –214°C. Its largest moon, Triton, is one of the coldest places in the Solar System.

Neptune is the farthest planet from the Sun.

Like Uranus, Neptune has hydrogen, helium, and methane in its atmosphere. The methane gives the planet its distinct, bright blue color. It also has rings, a rocky core, and more than a dozen moons. The moon Triton has volcanoes that discharge gases and liquids. These freeze almost instantly and fall back to the surface as snow. Triton orbits its planet in a direction opposite to that of all of Neptune's other moons. Neptune's gravity is slowing Triton's movement, causing it to inch closer to the planet. Eventually, Triton will drop too close to Neptune, and the planet's gravity will break it into pieces. These pieces may form a ring around the planet in the distant future.

Concept Check Assess/Reflect

Summary: What are the planets in our Solar System? Earth is one of eight planets in our Solar System. A planet is an object that orbits a Sun and, because of its gravity, has a nearly round shape and doesn't have an accompanying field or band of asteroids or other space debris. All planets in our Solar System follow an elliptical orbit around the Sun. Kepler's three laws of planetary motion describe these elliptical orbits. Newton's law of universal gravitation explains why the planets remain in orbit. The planets are classified into two groups. The terrestrial planets include Mercury, Venus, Earth, and Mars. These planets are made of heavy elements such as iron and silicate rocks. The giant gas planets are located farther from the Sun and include Jupiter, Saturn, Uranus, and Neptune. These planets are made of lighter elements, such as hydrogen and helium.

1. What role does gravity play in the Solar System?
2. How do the terrestrial planets compare with the gas giants?
3. Jupiter's moon Europa has twice as much water as Earth's oceans and rivers. How is Europa's water different from that found on Earth?
4. Like the orbits of all the planets, Earth's elliptical revolution around the Sun obeys Kepler's three laws of planetary motion. But why is Earth's orbit not as elliptical as Neptune's orbit?
5. What examples of God's Design do you see in the Solar System?

Lesson 2

Essential Question

What Other Objects Are in Our Solar System?

Objectives

- Compare planets with dwarf planets.
- Describe the composition and movement of comets.
- Identify the location of the asteroid belt.
- Distinguish between asteroids and meteoroids.

Vocabulary

Kuiper Belt
asteroid belt
meteorite

Look up at the sky on a cloudless day and you will see the Sun. Look up at the sky on a clear night and you will see a countless number of stars and perhaps a planet or two. You may see small objects traversing the sky. Some of these are natural objects; others are human-made. How can you tell natural objects in the night sky from human-made objects? Which move faster? Which are brighter? Are there places on Earth where you can see more of these objects than others? Why or why not?

Dwarf Planets Explain

Some objects in space that orbit the Sun have a round shape but have debris in their path. Such objects are known as *dwarf planets*. Space debris typically consists of rocky and metallic materials that have broken off from celestial objects such as planets, moons, and stars. Why do you think dwarf planets have debris surrounding them while larger planets do not? Some scientists have also used the term *protoplanet* to refer to these smaller, planet-like objects. Which term do you think best fits the definition of these objects?

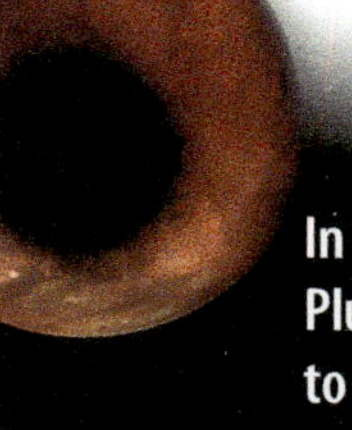

In 2006, astronomers changed Pluto's classification from a planet to a dwarf planet. It has five known moons. You can see one of those moons, Charon, in the picture.

What would you expect to see in the area surrounding Pluto's orbit?

Pluto

From 1930, when Pluto was discovered, until 2006, scientists considered it to be a planet. But Pluto was a misfit among planets. At more than 39 AU from the Sun, Pluto is far more distant than its nearest planetary neighbor, Neptune, which is about 29 AU from the Sun. Pluto's orbit also crosses that of Neptune. Unlike the outer gas giant planets, Pluto is rocky. It also is very small, only a bit larger than Earth's Moon, and it has its own moons. Astronomers discovered Pluto's moon Charon, about half the size of Pluto itself, in 1978. In 2005, they discovered and named two more moons, Nix and Hydro. More recently, they found two more moons, which have yet to receive names. Why did it take so long for scientists to discover these moons?

Eris

The question of Pluto's planetary status came to a head in 2005, when astronomers discovered Eris and its moon Dysnomia. Eris is nearly 97 AU from the Sun and was originally thought to be larger than Pluto. This caused scientists to ask, "Should Eris be considered a planet, or should we create a new classification?" For years, scientists had argued Pluto's status, but in 2006, the International Astronomical Union decided that Pluto, Eris, and similar objects should be classified as dwarf planets.

Eris is nearly three times farther from the Sun than Pluto. With a surface temperature of about –240°C, it would not be a nice place to visit. Astronomers consider Eris to be a Kuiper Belt object (KBO). The **Kuiper Belt** is a disk-shaped region beyond Neptune, from about 30 to 55 AU from the Sun, composed of millions of icy and rocky objects. Scientists also refer to these objects as transneptunian objects (TNO). Why do you think these objects are called "transneptunian"?

Dr. Michael E. Brown of the California Institute of Technology discovered the dwarf planet Eris. This dwarf planet was named for the goddess of strife and discord. According to Greek myths, she caused the Trojan War.

Think about what happened when Eris was discovered. Do you think the dwarf planet's name is appropriate? Why or why not?

Other Dwarf Planets

One family of KBOs is Haumea and its two small moons, Hi'aka and Namaka. Haumea is thought to have collided with another object about half its size, which caused it to spin end over end and assume the approximate shape of a football.

The farthest dwarf planet detected in our Solar System is Sedna. Orbiting the Sun between 76 and 1000 AU from the Sun, Sedna takes more than 10,000 years to complete one orbit. Some scientists believe that Sedna may have come from a region referred to as the Oort Cloud, which is even more distant from the Sun than the Kuiper Belt.

A comet is sometimes called a dirty snowball in space.

Comets Explain

One of the most dramatic sights in the night sky is a comet. A *comet* is a small, icy object that develops a glowing tail as it nears the Sun. As you can see in the diagram on the next page, a comet has several parts, including a nucleus, or rocky core. The nucleus may be only a few kilometers wide and is surrounded by a layer of ice and frozen gases.

Many comets have orbits that take them to the farthest reaches of the Solar System. When a comet nears the Sun, solar heat vaporizes the ice in the comet, causing a coma, or bright head, of dust and gas to form around the nucleus. Solar wind and the comet's motion push on the dust and gas, forming one or two glowing tails. The dust tail's glow is reflected sunlight, while hot gases cause the ion tail to appear. Why do you think comets sometimes have one or two tails?

Comets orbit the Sun at regular intervals that vary greatly in length. Some, such as Halley's Comet, complete an orbit in less than 200 years. People have noted their repeated appearance, and so we can predict when they will return. Astronomers believe these short-period comets to be Kuiper Belt objects that were disturbed in some way, causing them to plummet toward the inner Solar System. Other comets orbit in the Oort Cloud, a region some 100,000 AU from the Sun. According to current understandings, these comets will not complete one orbit of the Sun for 30 million years!

Long ago, many people considered comets to be harbingers of doom. Why do you think people were frightened by comets?

The orbits of comets can cross the orbits of planets. Sometimes the two objects collide—some craters on planets and moons are the result of collisions with comets. Most often, however, comets pass safely by the planets. Such a "fly by" occurred in 1997, when Comet Hale-Bopp came within 1.3 AU of Earth. Comet Hale-Bopp is named after the persons who discovered it in 1995. It was one of the brightest objects in the night sky during the winter and spring of 1997.

Scripture Spotlight

Matthew 24:29 describes signs of the end times. What does Scripture say will happen to objects in the Solar System?

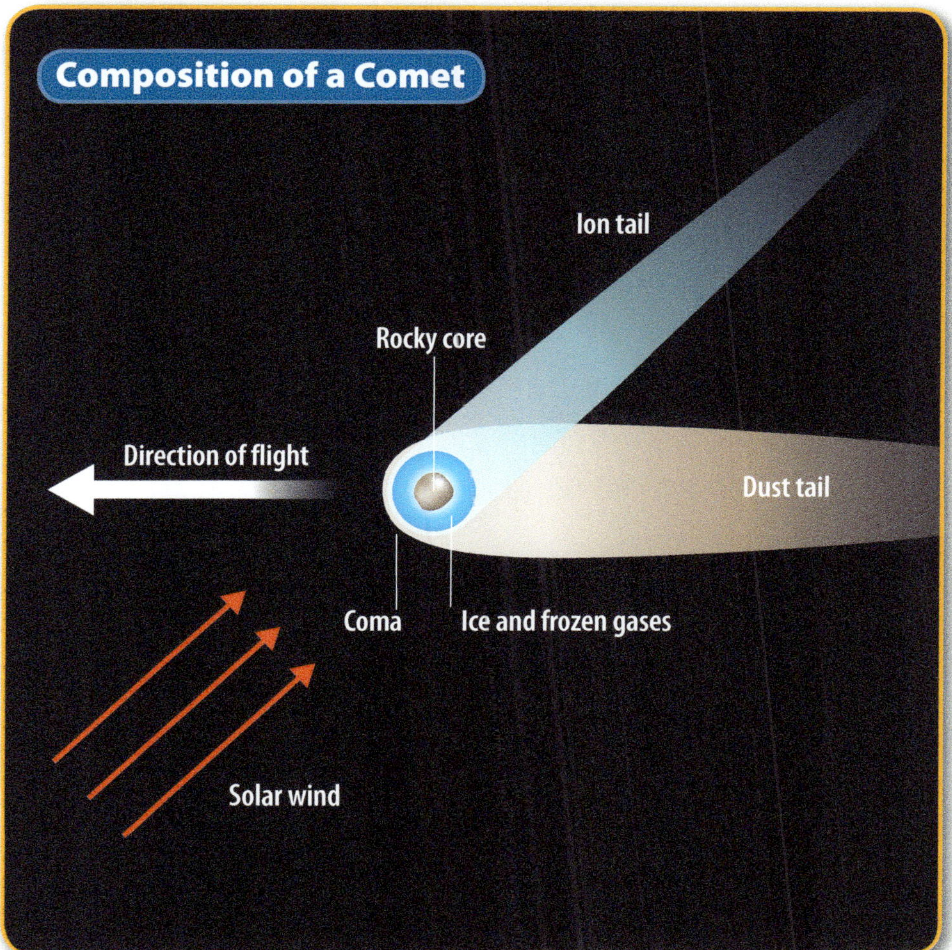

Explore-a-Lab

Guided Inquiry

 **How can you model a comet?**

Use a plate, an ice cream scoop, vanilla ice cream, chocolate cookies, and a whipped topping to make a model of a comet. Identify what each ingredient in your model represents in a comet. Scientific models have limitations. What might be a limitation of the model you created?

Record your work for this inquiry. Your teacher may also assign the related Guided Inquiry.

Making Dents

What can be learned by studying the craters on planets?

Procedure

1. Fill the aluminum pan with about 4 cm of flour.
2. Gently shake the pan back and forth to make an even layer. Sprinkle a thin layer of paprika over the surface of the flour.
3. Hold the fishing weight 10 cm above the surface of the flour and drop it into the pan. **Record** the diameter (mm) and depth (mm) of the crater formed. You may need to use a paperclip straightened out or a toothpick as a gauge for measuring the depth.
4. Use tweezers to carefully remove the fishing weight so as to not damage the crater.
5. Repeat Steps 3 and 4, dropping the fishing weight two more times, once from 20 cm and once from 40 cm. Be sure to record the diameter and depth of the craters formed.
6. Repeat Steps 3–5 with the marble.
7. Use the clamp, ring stand, and ring to adjust the marble ramp to a 15° angle. Set the end of the ramp at 10 cm above the surface. Roll the marble down the ramp and record the diameter and depth of the crater formed.
8. Repeat Step 7, but set the angle of the ramp to 45°.
9. Repeat Step 7 once more, but set the angle of the ramp to 75°.
10. Repeat Steps 3–5 and 7–9 with the wood bead.

Materials

- aluminum pan 20 cm × 20 cm
- burette clamp
- small fishing weight
- marble
- marble ramp
- metric ruler
- paprika
- protractor
- ring stand and ring
- sifted flour
- tweezers
- wood bead

Analyze Results

Construct a **graph** of your data.

Create Explanations

1. What can be learned by studying the craters on planets?
2. What factors affected the size and depth of the craters formed?
3. What characteristics of a crater allow you to determine a meteorite's direction and angle of impact?
4. Which would you expect to find more craters on: Mercury or Venus? Why?

Asteroids Explain

The dwarf planet Ceres was discovered in 1801. Since then, it has been reclassified several times. Originally, astronomers considered Ceres to be a comet. Then in 1802 they classified it as a planet, and later they thought it to be an asteroid. *Asteroids* are small, rocky objects that orbit the Sun mostly between the orbits of Mars and Jupiter.

Ceres's orbit, 2.9 AU from the Sun, is between the orbits of Mars and Jupiter. Astronomers refer to this region of the Solar System as the **asteroid belt**, which, like the Kuiper Belt, is a collection of millions of small rocky objects orbiting the Sun. Ceres contains one third of the entire mass of the asteroid belt. NASA launched a probe called *Dawn* to study Ceres and a nearby asteroid called Vesta.

Many scientists believe that the asteroid belt consists of leftovers from the Solar System's formation—pieces that were not able to collect and form into a planet. Others believe that the asteroids are what were left over when God created the gas giant Jupiter, reasoning that the planet's immense gravity caused several dwarf planets to collide and break into the irregularly shaped asteroids. No one knows for sure how the asteroids came to be. The asteroid belt consists of millions of asteroids. Some are tiny, and others are the size of small moons. Some even have moons that orbit around them. There are over 100 asteroids with moons. In 1993, scientists discovered that asteroid Ida had a moon, which they named Dactyl. What scientific methods do you think scientists used to determine Ida had a moon?

Like comets, asteroids can collide with planets and moons and leave craters. Scientists are particularly interested in studying asteroids that reach Earth's surface. Their composition and structure can tell us interesting things about the history of the Solar System.

Many asteroids look like giant potatoes. This asteroid is called 243 Ida.

What might happen when an asteroid collides with another body in space?

Lesson Activity

Work in small groups. Formulate definitions in your own words for each of the objects you studied. Write the definitions on cards. Collect images from the Internet of dwarf planets, comets, meteoroids, and asteroids. Place the pictures you found on cards. Exchange these images with another group. Use your definitions to classify the images you are given. Be sure to justify your classifications. Use a table or chart to identify the characteristics of the objects you classified. Reveal the classification of the objects you found to the other group and share your results with the class. Does this exercise help you better understand the task the International Astronomical Union faced when dwarf planets were discovered?

How can you classify the other objects found in our Solar System?

Meteoroids Explain

There are small bits of rock and dust in space whose paths cross Earth's orbit. You have learned that these objects, which may have come from asteroids or comets, are called *meteoroids*. Have you ever seen a bright streak of light flare and then fade across the night sky? These lights are *meteors,* which are meteoroids that fall through Earth's atmosphere. Friction with the atmosphere heats the meteor to extremely high temperatures, causing the bright streak of light we see. Many meteors burn up completely, but some make it all the way to Earth's surface. We call those meteors **meteorites**. How would you know if a rock you discovered is a meteorite?

During certain times of the year, Earth passes through interplanetary material, causing a meteor shower.

Scientists are very interested in meteorites because the rocky objects from space offer clues about the composition of the Solar System or about comets. Some even come from other planets and the Moon. Scientists have found many meteorites in Antarctica that they believe came from Mars. Why do you think so many of these meteorites are found in Antarctica? How do you think scientists came to the conclusion that these meteorites were from Mars?

This meteorite, which was found in Arizona, is made of iron.

 **What evidence do you see here that suggests this is a meteorite?**

Concept Check Assess/Reflect

Summary: What other objects are in our Solar System? In addition to the Sun, planets, and moons, our Solar System contains a variety of other objects, including dwarf planets. A dwarf planet is an object that orbits the Sun, has a round shape, but does not clear other space objects in its orbital path. Pluto, once classified as a planet, is now considered a dwarf planet. Eris, another dwarf planet, is found in the Kuiper Belt, a region in space beyond Neptune that is composed of millions of icy and rocky objects. In addition to dwarf planets, our Solar System also contains comets, which are small, icy objects that develop a tail as they approach the Sun while traveling in their orbits. The orbit of a comet can cross that of a planet. Still another object found in our Solar System is an asteroid, which is a small, rocky object that orbits the Sun. Millions of these objects are found in the asteroid belt, a region of space between the orbits of Mars and Jupiter. Meteoroids are also found in our Solar System. These are small bits of rocks and dust whose paths cross Earth's orbit. When they fall through Earth's atmosphere, they are called meteors. If they land on Earth, they are known as meteorites.

1. How are dwarf planets and planets similar? How are they different?
2. Why are scientists interested in studying asteroids and meteoroids?
3. How does the structure of a comet change as it approaches the Sun?
4. Where in the Solar System would you expect to find the highest risk of collisions with an asteroid? Explain your answer.

Objectives

- Compare and contrast the geocentric model with the heliocentric model.
- Describe how technology impacts our knowledge of space.
- Compare and contrast different types of telescopes.
- Compare and contrast different spacecraft.

Vocabulary

geocentric theory
heliocentric theory
optical telescope
refracting telescope
reflecting telescope
nonoptical telescope
space probe

Essential Question

How Do We Study Space?

What do you think lies beyond our Solar System? For a long time, traveling deep into space seemed more like science fiction than science. But fiction eventually became reality. In September 2013, scientists confirmed that a spacecraft named *Voyager 1* became the first human-made object to travel outside our Solar System. In going beyond our Solar System, *Voyager 1* was more than 19 billion km from the Sun. What objects did *Voyager 1* pass on its journey? The spacecraft had taken 35 years to get there. Why did it take so long? Although *Voyager 1* is unique in having traveled outside our Solar System, scientists have used many other tools to study space, including our own Solar System.

Early Ideas about the Solar System Explain

In ancient times, people who thought about the structure of the Universe believed that all objects in space moved around Earth. This model, known as the **geocentric theory**, however, did not adequately explain the movements of the planets. People accepted the geocentric theory for a long time for a number of reasons. One was related to the belief that humans were God's most perfect Creation on Earth, and so God placed this most perfect Creation at the center of the Universe. Another was also related to a belief in the perfection of God's Creation. Circles were assumed to be the most perfect geometric idea, and so God must have given the Sun, the stars, and the planets circular paths to traverse. How would you build a scale model of our Solar System based on the geocentric theory?

In 1543, Polish astronomer Nicolaus Copernicus proposed that the Sun was the center of the Universe and that the planets orbited the Sun in circular orbits. Why would Copernicus have proposed a different model of our Solar System? His view of the Solar System, called the **heliocentric theory**, helped explain scientific observations. The theory was controversial.

Lesson Activity

Copernicus is just one of the many scientists who contributed to our knowledge of our Solar System. Develop a time line that shows his contributions and those of nine other scientists. Include artwork and illustrations in your time line. Of the 10 scientists included in your time line, who do you think made the greatest contribution to our understanding of our Solar System? Defend your choice.

Which scientist contributed most to our knowledge of the Solar System?

Some religious leaders denounced the Copernican view. Why might religious leaders have found this theory threatening?

Early religious leaders felt that science and religion were in conflict. Today, we view science as a way to study God's Creation. The consistency of natural laws, the high level of organization in the Universe, and the beauty and design seen in the objects that make up space are evidence of God's Design and of His Creation.

These nonoptical telescopes are part of the Very Large Array (VLA) in New Mexico. This collection of 27 dishes, each 25 m in diameter, collects radio waves from space. Because radio telescopes do not depend on visible light, they can gather images 24 hours per day.

Telescopes Explain

Telescopes played a major role in the development of the heliocentric model. The Italian astronomer Galileo Galilei built one of the first telescopes in the early 1600s.

Optical Telescopes

The telescope that Galileo built is known as an optical telescope. An **optical telescope** uses visible light to magnify distant objects. Galileo used his telescope to learn about our Solar System.

With the help of his telescope, Galileo first observed Saturn with its rings and Jupiter along with four of its moons. Galileo first observed these four moons in 1610. Why did Galileo see only four of Jupiter's more than 60 moons? Galileo also looked closer to home, observing that Earth's Moon was dotted with craters and mountains. What do you think Galileo could deduce from this observation? How did Galileo's observations confirm the ideas of Copernicus?

Specifically, Galileo's optical telescope is known as a refracting telescope. A **refracting telescope** uses glass lenses to collect and focus light onto an eyepiece lens. In the late 1600s, Sir Isaac Newton invented a different type of optical telescope called a reflecting telescope. A **reflecting telescope** uses curved and flat mirrors to collect and focus light onto an eyepiece lens. A reflecting telescope avoids the problems that arise when light waves are bent as they pass through a lens. The bending of light waves causes each color of light to be focused at a different point. How would this affect the object that is being viewed through a refracting telescope?

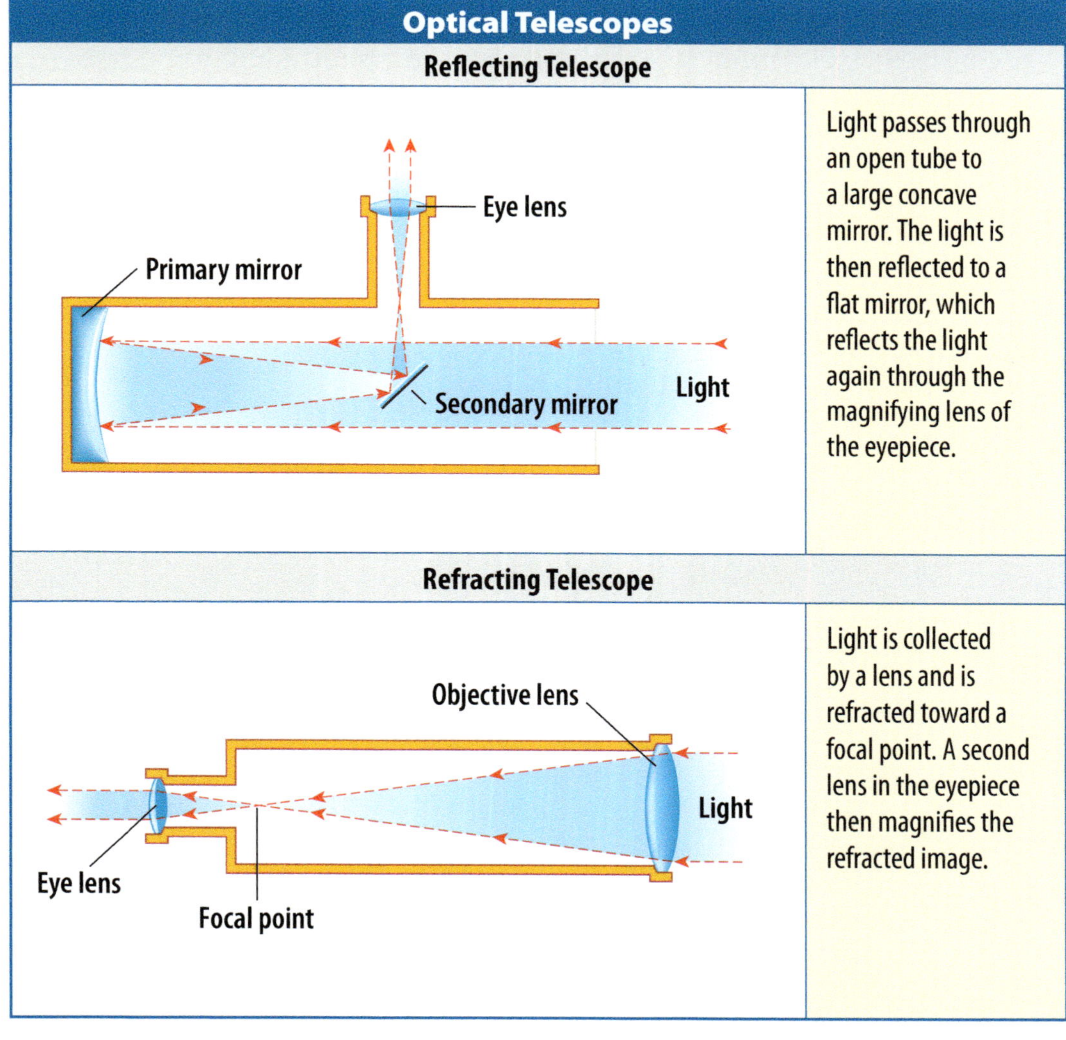

Record your work for this inquiry. Your teacher may also assign the related Guided Inquiry.

Build a Model Telescope

How do telescopes make distant objects seem closer?

Procedure

1. Look out the window through the small lens. Hold the larger lens farther away. Adjust the distance between the lenses until you can see a distant object while looking through both lenses.
2. Have a partner **measure** the distance between the lenses.
3. Place the small lens in one end of the smaller tube. Place the large lens in one end of the larger tube. Use tape to secure the lenses.
4. Slide the smaller tube inside the larger tube. Use modeling clay to secure them together. The length of the tubes should equal the measurement you obtained in Step 2.
5. Test your telescope outside. Adjust the lengths of the tubes, if necessary. (Do not look at the Sun.)

Materials

- large convex lens
- small concave lens
- metric ruler
- two cardboard tubes of different sizes
- tape
- modeling clay

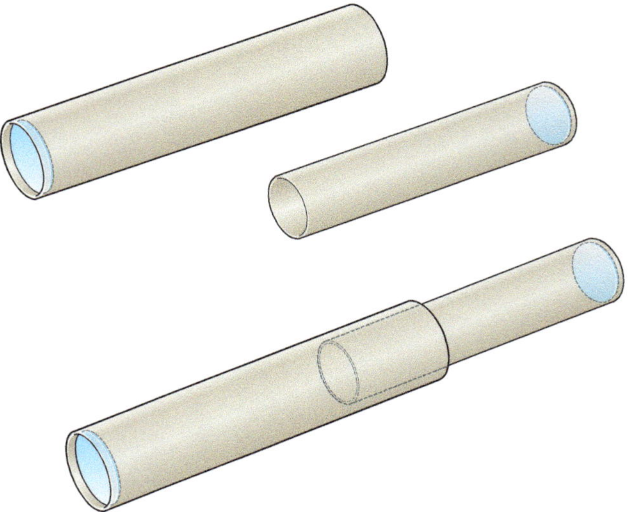

Analyze Results

Draw your telescope. Show how light entered it and reached your eye.

Create Explanations

1. How do telescopes make distant objects seem closer?
2. Do you think you could see the planets and their moons using your telescope? Explain.
3. What could you do to improve your design?
4. What kind of telescope did you make?

Nonoptical Telescopes

Some telescopes do not use visible light to create an image of an object. Known as **nonoptical telescopes**, they use invisible radiation to observe distant objects. Examples of nonoptical telescopes include radio telescopes, infrared telescopes, UV telescopes, and X-ray telescopes. Astronomers used the first X-ray telescope in 1963 to observe the Sun. In 1999, NASA launched the most sophisticated X-ray telescope it had ever built. Why is an X-ray telescope used to study objects in space, such as black holes and dark matter?

Space Telescopes

Some of the most powerful telescopes are not located on Earth. Instead, they travel in space. Because these telescopes orbit above Earth's atmosphere, their images are not affected by the dust, gas, other materials, and heat in the atmosphere that can distort our view of space. One of the first space telescopes is the Hubble Space Telescope. Using light measurements taken by Hubble, scientists who believe the big bang theory have estimated the age of the Universe to be about 13.7 billion years.

Space telescopes are spacecraft that make observations as they orbit Earth. In the future, other spacecraft may transport astronauts to the Moon, and robotic spacecraft may explore distant planets.

Space telescope over Earth

Lesson Activity

Nonoptical telescopes use invisible radiation, including radio waves, infrared waves, UV waves, and X-rays. Prepare a table that illustrates the differences between these types of radiation, including the frequency, wavelength, and energy of each type of radiation.

How can you differentiate types of radiation?

Missions with Astronauts

In the 1960s, the United States engaged in a "space race" with the Soviet Union. Both countries were determined to be the first to put a human on the Moon. This race involved launching successively more complex missions, most of which included human astronauts.

In 1957, the Soviet Union launched *Sputnik* into orbit. It was the first artificial satellite. Soon after, in 1961, the Soviet Union launched a human into space. This marked the beginning of manned space travel. The United States had a slower start, but it won the race when astronaut Neil Armstrong walked on the Moon in 1969. As part of the *Apollo* program, 12 U.S. astronauts walked on the Moon between 1969 and 1972. They conducted experiments and gathered crucial information about the Moon's structure and composition. What kinds of information do you think astronauts learned from visiting the Moon? Do you think this spirit of competition helped or harmed space exploration? Explain your answer.

Explore-a-Lab

Guided Inquiry

How are astronauts protected in space?

Your astronaut's mission is to repair part of the International Space Station (ISS). Design a space suit that will protect your astronaut from small space objects, such as meteoroids. The suit must use a maximum of four layers of different materials and be as lightweight and flexible as possible. You will test your space suit's ability to protect a potato from damage during impact. How will you test your space suit? Which materials will work best?

Astronauts used a "moon buggy" to explore the surface of the Moon.

Explore-a-Lab

Guided Inquiry

How did the winning spacecraft design accomplish the greatest payload?

Your next mission is to design and construct a spacecraft that will deliver the most payload to the International Space Station. Work in teams by using the materials provided by your teacher. Then launch the spacecraft and record the payload delivered. The team with the greatest payload wins!

The space shuttles were reusable spacecraft.

What advantages do reusable spacecraft have?

Space Shuttle

As NASA was wrapping up the *Apollo* program, it went forward with plans to build a reusable spacecraft. In 1981, it launched the first space shuttle into space. When the shuttle's mission was complete, it glided back to Earth and touched down like a plane.

Engineers designed and built the space shuttle fleet to make multiple missions into space. Onboard the shuttles, astronauts performed various experiments. They also repaired satellites, launched robotic spacecraft, and carried equipment and crew to the International Space Station, which was launched in 1998. The various space shuttles made more than 130 missions into space and had many successes and two devastating disasters. In 1986 and again in 2003, space shuttles exploded, killing all crew members onboard. What action do you think NASA took following each of these disasters? NASA ended the space shuttle program and retired the space shuttle fleet in 2011. What might have been a reason why NASA retired the space shuttle program?

The solar arrays of the ISS cover a surface area of 2500 square meters. These panels convert sunlight into electricity at a rate sufficient to power 10 average-sized homes.

Space Stations

The space shuttle program helped scientists learn about how space travel affects the human body. The missions, however, were short and did not allow time for extended study. Space stations do not have that limitation. A space station is a spacecraft that humans can live in for months at a time. How do you think living in a spacecraft for months at a time might affect a person?

The first space station was launched by the Soviet Union in 1971. The crew followed in a spacecraft but could not open the hatch and had to return to Earth. A second attempt succeeded; the crew stayed aboard the space station for 22 days. But unfortunately their spacecraft malfunctioned on the return trip, and the crew died. The Soviet Union had greater success with later space stations, including *Mir*, which was launched in 1986 and remained functional for 15 years, completing 86,000 orbits around Earth. Why did the Russians have greater success with later space stations such as *Mir*?

The United States launched its first space station, *Skylab*, in 1973. It was functional for only one year and fell out of orbit within five years. It actually killed a cow when it crash-landed in Australia. In 1998, the United States and 15 other countries joined resources to launch and build the International Space Station. The first crew arrived in 2000. As part of their mission, astronauts conducted experiments on weightlessness, loss of bone density, high levels of radiation, and other conditions that affect human health in space.

Scripture Spotlight

Read **Psalm 8:3–9.** How does this poetic verse relate to this lesson?

Robotic Space Missions Explain

One concern about long-term space travel is its effect on human health. What are some health issues that may develop as a result of long-term space travel? In addition to health concerns, humans need fresh water, food, air, and other resources that are difficult to provide during long journeys in space. Thus, all wholly robotic space missions that have ventured beyond the Moon use remotely controlled technology rather than humans to gather data. For example, **space probes** are spacecraft that travel beyond Earth's orbit to explore the Solar System. They gather data and transmit it back to Earth. Space probes can land on planets, moons, and even asteroids. They may carry rovers, or robotic vehicles that can travel along the surface of a planet, collecting samples of soil and rock.

Space probes explore the Solar System and transmit data back to Earth.

How does a space probe differ from a satellite?

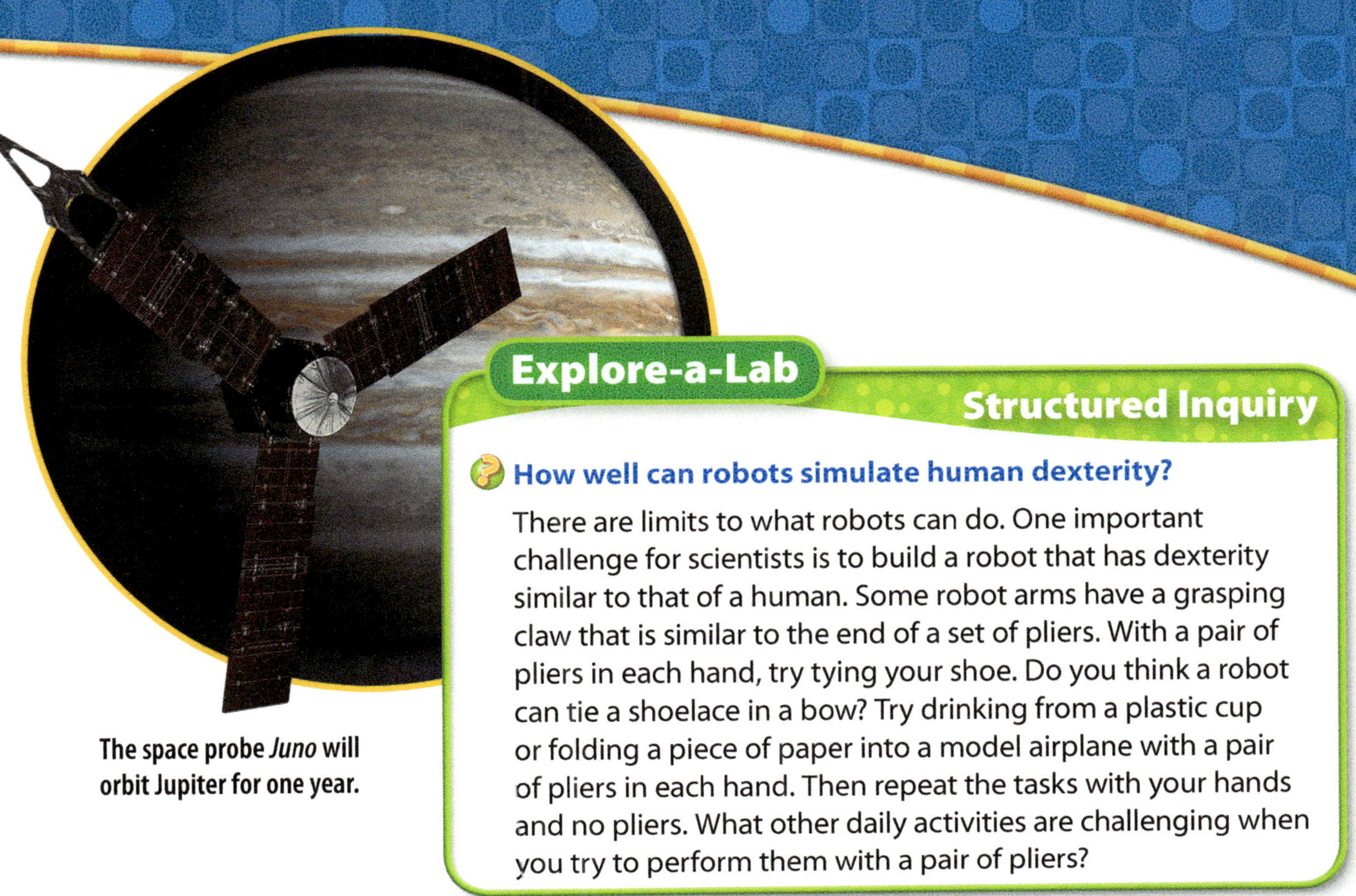

The space probe *Juno* will orbit Jupiter for one year.

Explore-a-Lab

Structured Inquiry

How well can robots simulate human dexterity?

There are limits to what robots can do. One important challenge for scientists is to build a robot that has dexterity similar to that of a human. Some robot arms have a grasping claw that is similar to the end of a set of pliers. With a pair of pliers in each hand, try tying your shoe. Do you think a robot can tie a shoelace in a bow? Try drinking from a plastic cup or folding a piece of paper into a model airplane with a pair of pliers in each hand. Then repeat the tasks with your hands and no pliers. What other daily activities are challenging when you try to perform them with a pair of pliers?

Concept Check

Assess/Reflect

Summary: How do we study space? Humans first used their eyes and reasoning to study space. The early Greek philosophers proposed a model of our Solar System in which Earth was at the center and the other planets and the Sun revolved around it. Known as the geocentric model, this view of our Solar System was replaced by the heliocentric model in which the Sun became the center and served as the focal point around which the planets orbited. Observations made possible with the development of the telescope allowed scientists to collect data that verified the heliocentric model. Scientists use two types of telescopes: optical telescopes and nonoptical telescopes. Optical telescopes that use glass lenses to collect and focus light are known as refracting telescopes. Optical telescopes that use mirrors to collect and focus light are known as reflecting telescopes. Nonoptical telescopes use invisible radiation, such as radio waves, infrared waves, UV waves, and X-rays, to observe objects in space. Space telescopes gather images that are not distorted as a result of dust and gas particles that are found in Earth's atmosphere. Missions with astronauts and robotic missions have also been used to study space.

1. Describe the destination and goals of one future space mission.
2. Why do space telescopes often have clearer images than Earth-based telescopes?
3. Why hasn't a human traveled on a spacecraft beyond the Moon?
4. How does the study of astronomy strengthen your faith in God?

Careers in Science

Extend

Astronautical Engineering

What are some of the ways we are learning more about space? Astronautical engineers are hard at work finding ways for us to explore space in greater depth than we have in the past.

Engineers are people who figure out how to put power and materials to work. The work and ideas of engineers make certain goals attainable. Astronautical engineering brings together the dynamic and cutting-edge fields of advanced science and space technology. These engineers tackle the challenging task of designing and operating many types of spacecraft, missiles, and space vehicles and are in constant demand in research and development groups to help us gain a better understanding of the wonders of our Solar System.

Outer space is increasingly important for our economy and national security as well as exploration. The United States depends on space assets more than any other nation on Earth and leads the world in exploration and utilization of space. Space engineers design and build rockets and space launchers, communications and direct broadcasting satellites, space navigational systems, remote-sensing satellites, manned space vehicles, and planetary probes. They operate complex Earth-orbiting space systems and rovers on Mars from sophisticated ground control centers.

Aerospace engineers typically become specialized in one of two types of engineering, aeronautical engineering or astronautical engineering. Astronautical engineers work with the science and technology of spacecraft and how they perform *outside* Earth's atmosphere. Aeronautical engineers work with aircraft. They design aircraft and propulsion systems while analyzing the aerodynamic performance of aircraft and construction materials. They are concerned with the theory, technology, and practice of flight *within* Earth's atmosphere. Both types of engineers face different environmental and operational issues in designing spacecraft and aircraft. However, the two fields are common in that they both depend on the basic principles of physics.

Today, most of the astronautical engineer's time is spent in an office, more so than in the past. This is due to the fact that current spacecraft design requires the use of sophisticated computer equipment and software design tools, modeling and simulations for tests, evaluation, and training.

Concept Check

1. What are some of the spacecraft an astronautical engineer might design and build?
2. Compare the job of an aeronautical engineer to that of an astronautical engineer.

Science and Technology

Extend

Voyager 1 and *Voyager 2* near the Edge of the Solar System

After 35 years of space travel, the twin *Voyager* planetary probes are nearing the very edge of Earth's Solar System. They will become the first human-made objects to travel between the stars.

Voyager 1 and *Voyager 2* have been traveling through space since 1977. Scientists are not sure exactly when the probes will leave our Solar System. It may take a few months or even a couple of years. Currently the probes are in the outermost boundary of the Solar System, which is in an area called the heliosphere. This final frontier is an immense magnetic bubble surrounding our Solar System, containing solar wind and the entire solar magnetic field. *Voyager 1* has entered a new region of the heliosphere that scientists are calling a "magnetic highway," which allows charged particles from inside the heliosphere to flow outward and particles from the outside galaxy to flow in.

As *Voyager 1*, the outermost of the two spacecraft, gets farther and farther away, it measures more of the higher-energy charged particles thought to originate beyond the Solar System, compared to the lower-energy particles thought to come from the Sun.

Both *Voyagers* will eventually cross the heliopause, which is considered the final boundary between our Solar System and interstellar space. The heliopause is the blurred boundary between the heliosphere and the interstellar gas outside of the Solar System. Scientists did not anticipate this last boundary. We know that the *Voyagers* are still in our Solar System because of the orientation of the magnetic field they detect. The field runs east-west, which is in agreement with the field created by the Sun. Outside the Solar System, it is predicted that the magnetic field will be orientated more north-south.

The *Voyagers* are NASA's longest-running spacecraft. They will keep traveling outward even after they've left the Solar System. It would take them at least 40,000 years to ever come close to another star. However, within a decade, the probes will begin to run out of power to operate their scientific instruments and beam their findings back to Earth. After about seven years, scientists will begin to shut down parts of the probes, and by 2025 the last instrument will be turned off.

Concept Check

1. What is considered to be the outermost edge of our Solar System?
2. How have *Voyager 1* and *Voyager 2* helped us to learn about space?

Chapter 8

Review

Assess/Reflect

Study Guide

Lesson 1

1. The Solar System is made up of the Sun, eight planets, and other smaller objects, such as moons of the planets.
2. Kepler's first law of planetary motion states that the orbit of a planet is an ellipse. Kepler's second law of planetary motion states that the area swept out by the path of a planet in a given time does not change. Kepler's third law of planetary motion describes the relationship between the time it takes a planet to complete one revolution and its semimajor axis.
3. Newton's law of universal gravitation states that the force of gravity between two objects is the product of their masses divided by the square of the distance between them.
4. The terrestrial planets are closer to the Sun, small, and made up of mostly rock, and have surface features like craters and mountains. The gas giants are farther from the Sun, large, and mostly made up of hydrogen, helium, and methane. They have ring systems and have many moons.
5. The Solar System shows evidence of God's Design in that it is highly organized. Further evidence is seen in the many features of Earth that make it suitable to support life, such as its size, distance from the Sun, atmosphere, and the presence of water as a gas, liquid, and solid.

Lesson 2

1. Like planets, dwarf planets orbit the Sun and are nearly round. Unlike planets, they have debris in their path.
2. Comets are made up of a nucleus, a coma, and one or two tails. The nucleus is a rocky core with a layer of ice and frozen gases around it. Comets orbit the Sun at regular intervals that vary in length.
3. The asteroid belt is a region of our Solar System located between the orbits of Mars and Jupiter.
4. Asteroids are small, rocky objects that orbit the Sun. Meteoroids are small bits of rock and dust that cross Earth's orbit. Meteoroids may have come from asteroids or comets.

Lesson 3

1. The geocentric model states that all objects in space move around Earth. The heliocentric model states that the Sun is the center of the Universe and the planets orbit it in circular orbits.
2. Telescopes let us collect data on space objects. Astronaut and robotic missions allow us to gather visual and nonvisual data, perform tests, and bring back samples.
3. Refracting telescopes use glass lenses to collect and focus visual light, while reflecting telescopes use mirrors to collect and focus visual light. Nonoptical telescopes use radio, infrared, and ultraviolet radiation to observe space objects. Space telescopes produce images not affected by Earth's atmosphere.
4. Single-use spacecraft were used for missions to the Moon. The Space Shuttle Program used reusable spacecrafts that allowed a variety of activities to be carried out. A space station allows humans to live and work in space for months at a time. Robotic missions use remote-controlled technology to gather data.

Vocabulary Check

Circle the term in each set of terms that does not belong. Explain your reasoning.

1. reflecting telescope, refracting telescope, nonoptical telescope
2. Kuiper Belt, meteorite, asteroid belt

Multiple Choice

Choose the best answer.

3. According to the geocentric theory, the planets revolve around the __________; according to the heliocentric theory, the planets revolve around the __________.
   A. Moon, Sun
   B. Earth, Moon
   C. Sun, Earth
   D. Earth, Sun
4. What factors influence the strength of gravity between two objects?
   A. size and density
   B. weight and volume
   C. mass and distance
   D. composition and location in the Universe
5. What distinguishes Earth from other planets in our Solar System?
   A. presence of water in all three states of matter
   B. existence of an atmosphere
   C. a rocky surface
   D. a solid core
6. The law that states that the orbits of planets are ellipses with the Sun as one of the foci is __________.
   A. Newton's law of universal gravitation
   B. Kepler's first law of planetary motion
   C. Kepler's second law of planetary motion
   D. Kepler's third law of planetary motion

Check Point

Answer the following questions.

7. Suppose that astronomers find an object between the orbits of Mars and Jupiter. The object is round, has moons, and is surrounded by debris. Should it be **classified** as a planet or as a dwarf planet? Explain.
8. Imagine that a spacecraft named *Juno* is on its way to Jupiter. As the spacecraft approaches Jupiter, it begins to orbit the planet. At what point is *Juno* a space probe?
9. Why does Mercury have such marked, daily temperature extremes?
10. **Compare** a space shuttle with a space station.
11. Evaluate some advantages and disadvantages of using robotic technology to study space.
12. How is Earth's position in the Solar System evidence of God's Design?
13. Using a Venn diagram, **compare** the different kinds of telescopes and their uses. Explain your graphic organizer.

Chapter 9 The Effects of Earth's Movement

Scripture Spotlight

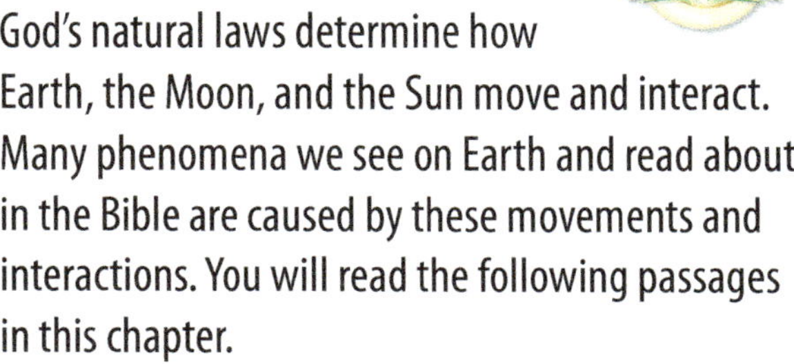

God's natural laws determine how Earth, the Moon, and the Sun move and interact. Many phenomena we see on Earth and read about in the Bible are caused by these movements and interactions. You will read the following passages in this chapter.

Ecclesiastes 3:1–8 (p. 338)
Genesis 1:14–19 (p. 342)
Amos 8:9 (p. 346)

When the Moon, the Sun, and Earth are aligned, the Moon may be between Earth and the Sun and block our view of the Sun. For a few minutes, our source of light and warmth seems to disappear.

The Big Idea

When God created our Solar System, He made Earth and the Moon to move in repeating patterns. The movements of Earth around the Sun and the Moon around Earth cause day and night, seasons, tides, moon phases, and eclipses.

How do Earth and the Moon move in space?

Inquiry Kick-Off Engage

Have you ever witnessed a solar eclipse? Have you seen a lunar eclipse? Because these events don't happen on a regular basis, using directions in your *Science Journal*, you can observe the Sun through a pin-hole camera.

Lesson 1

Objectives

- Compare and contrast rotation with revolution.
- Distinguish between a sidereal day and a solar day.
- Explain how gravity and inertia work together to cause Earth to orbit the Sun.

Vocabulary

solar day

sidereal day

Essential Question

How Does Earth Move in Space?

As you know, Earth orbits the Sun in space and it spins on its axis, but we can't feel this motion. So how do we know that this movement is happening? How do you think life might be different if we could feel this movement? How would life on Earth be different if Earth stayed in one stationary position? What might happen if only Earth moved and the stars and other objects in space stayed still? As you learn more about Earth's movement, think about why it is important for Earth to remain in constant motion.

Earth's Rotation and Revolution Explain

God made order in the Solar System when He created it. Earth and the other planets move in predictable patterns. One predictable motion is Earth's rotation, or spin, on its axis. Earth's axis is an imaginary line connecting the North Pole to the South Pole. Earth rotates once about every 24 hours, or one day. Another predictable pattern is Earth's revolution or orbit around the Sun. Earth completes one orbit about every 365 days.

Earth's rotation causes a phenomenon that has long fascinated people: changing shadows. How does your shadow change throughout the day? What conclusion can you make by considering the changes you observe in your shadow as the day progresses?

What do you think the dials on this clock are showing?

Solar Day versus Sidereal Day

Quantifying the time Earth takes to complete one rotation is a bit more complicated than saying that it takes one day, because Earth is always moving through space in its orbit around the Sun. That is, at a given time of day (such as midnight), Earth will be in a different position in space than it was at the same time the previous day. If a rotating object is moving, determining the time it takes to complete one rotation requires that you acknowledge a reference point from which you are making your determination.

For example, imagine an ice skater spinning as she travels in a circle around her partner. Then imagine that you are watching this performance from a seat that is in the last row of the skating rink. Does the skater complete one rotation every time she faces her partner or every time she faces you? If you timed her rotations with a very precise clock, do you think that you would get the same time from the two different points of reference?

The same question occurs when determining how long Earth takes to complete one rotation. The answer depends on what we use as our point of reference. A **solar day** is a day in relation to the Sun. It is the time Earth takes to make one complete rotation in relation to the Sun. A solar day is about 24 hours. A **sidereal day** is a day in relation to a distant star. It is the time Earth takes to make one complete rotation compared with a distant star. A sidereal day is about 4 minutes shorter than a solar day.

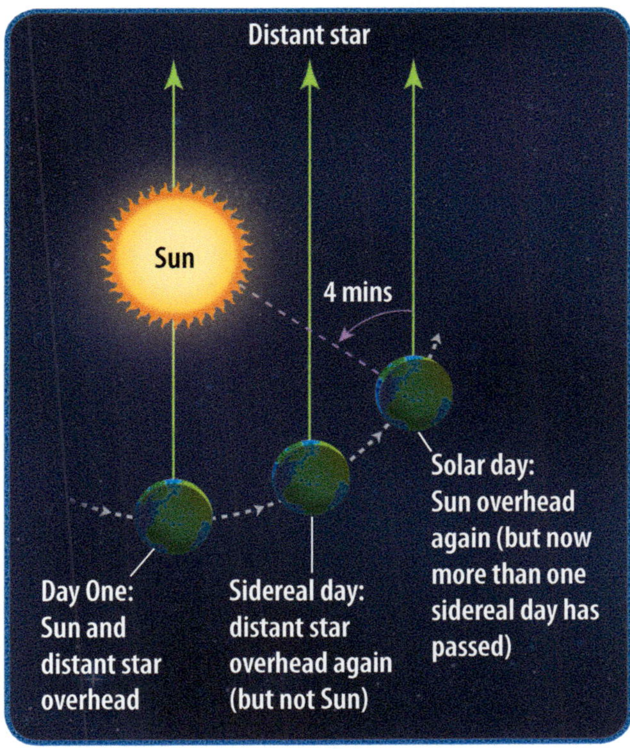

A solar day is the time Earth takes to move from position 1 to position 2. A sidereal day is the time Earth takes to move from one position to another so that a distant star appears in the same place in the sky.

A measurement of the sidereal day is made by noting the time at which a particular star passes directly overhead on two successive nights. The time it takes for this to happen is 23 hours and 56 minutes. This means that a particular star will rise 4 minutes earlier every night. This explains why different constellations are only visible at specific times of the year.

Suppose that you wanted to view a specific star through a telescope. The field of stars appears to move across the sky along the same axis that the Sun appears to move across the sky. To locate where your star will be in the sky at a specific time, would you rely on solar time or sidereal time? If you knew that your star would become visible at 11:15 p.m. on day one, at what time would it become visible on day two?

Record your work for this inquiry. Your teacher may also assign the related Guided Inquiry.

Making a Sun Clock

Is it possible to tell the time by using a shadow created by the Sun?

Materials
- clay
- compass
- pencil
- Sun Clock diagram

Procedure

1. Place the compass on the ground and turn it so that the arrow and the "N" (for "North") line up.
2. Place the Sun Clock diagram on the ground and move it until the arrow that represents your city points the same way as the arrow on your compass.
3. Stand a pencil on today's date and look at the shadow to find out what time it is.
4. Wait 15 min and repeat Step 3.
5. Wait another 15 min and repeat Step 3.

	Actual Time	Sun Clock Time
Initial time		
Time after 15 minutes		
Time after 30 minutes		

Analyze Results

What factors make this clock useful? What problems might result from using a solar clock?

Create Explanations

1. Is it possible to tell the time by using a shadow created by the Sun?
2. According to the Sun Clock, what time was it when you started?
3. Does the Sun appear to move east to west or west to east?
4. Does the pencil cast a longer shadow in winter or in summer? Why?
5. Does a Sun Clock give the same time in San Francisco as it does in New York? Why or why not?

Gravity and Inertia Explain

Both a solar day and a sidereal day are determined by how the Sun and a distant star appear in the sky as Earth travels in its orbit. But what keeps Earth in its orbit so that the length of both a solar day and a sidereal day are constant? You learned that although Kepler discovered much about the orbits of planets, he could not explain why they remained in their elliptical orbits. You also learned that Newton provided an answer with his universal law of gravitation. However, Newton's law is only part of the reason why Earth and the other planets remain in their orbits. If gravity alone were responsible, what should eventually happen to Earth and the other planets as they orbit the Sun?

Faith Connection

Earth moves in regular and predictable patterns. How does this provide evidence of God's Design?

The first part of the answer came from the observations made by Galileo. Galileo had discovered that a moving object, such as Earth revolving around the Sun, will remain in motion until some force acts to stop its motion. Of course, stopping Earth's motion would take an incredible amount of force. But think about a ball rolling along the floor. The ball continues to roll until a force—in this case, gravity and friction—causes it to stop.

Inertia, you may recall, is the tendency of a moving object to remain in motion and an object at rest to remain at rest unless a force acts on them. The law of inertia states that a moving object moves in a straight line unless it is acted upon by a force. All objects have inertia.

But how does inertia act to keep Earth in its orbit? Look closely at this illustration to understand how both gravity and inertia work together to keep Earth and the other planets in their orbits.

The blue arrows indicate how gravity acts to pull Earth toward the Sun. If gravity were the only force operating, Earth would eventually crash into the Sun. But here is where inertia enters the picture. The black arrows indicate how inertia acts to keep Earth moving in a straight line. Recall that inertia keeps an object moving in a straight line unless a force acts on it. In this case, gravity is the force acting on Earth to affect its direction of motion. The combined effects of both gravity and inertia result in Earth remaining in its orbit, as indicated by the ellipse shown in red. While in orbit, Earth undergoes both a solar day and a sidereal day.

Forces and Earth's Orbit

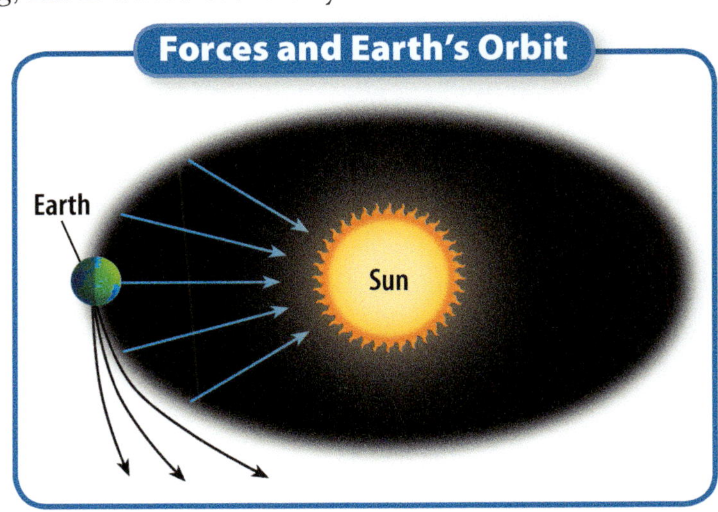

Explore-a-Lab

Structured Inquiry

 **How can you balance the forces of gravity and inertia?**

Using a large kitchen funnel (at least 24 cm opening), keep a marble rolling around inside the funnel at a constant level, demonstrating a balance between gravity (the desire of the marble to roll to the bottom of the funnel) and inertia (the desire of the marble to fly out of the funnel). What variables did you have to adjust to keep the marble at a constant level? Describe any difficulties you encountered in maintaining a balance of the forces.

Concept Check

Assess/Reflect

Summary: How does Earth move in space? Earth's motion in space involves both rotation and revolution. Rotation is the spinning of Earth on its axis. Revolution is the motion of Earth in its elliptical orbit around the Sun. As Earth both rotates and revolves, it completes both a solar day and a sidereal day. A solar day is the time Earth takes to make one complete rotation in relation to the Sun. A solar day has a length of 24 hours. A sidereal day occurs when Earth completes one rotation every time a given point on Earth is at its nearest daily distance to a given star. A sidereal day is the time Earth takes to complete one rotation in relation to a distant star. Both gravity and inertia are responsible for keeping Earth in its orbit. Gravity is a force of attraction between Earth and the Sun, which would result in Earth crashing into the Sun. However, inertia, which is the tendency of an object to remain in its motion, acts to balance the force of gravity so as to keep Earth in its elliptical orbit.

1. Why is a solar day longer than a sidereal day?
2. What would happen if gravity were not a force acting between Earth and the Sun?
3. Imagine that Earth begins to rotate more slowly. How would this affect a solar day? A sidereal day?
4. Imagine that Earth takes longer to complete one revolution. How would this affect a solar day? A sidereal day?

Lesson 2

Essential Question

Why Does Earth Have Seasons?

Why do many flowers bloom only in the spring, while others bloom only in the fall? When do you think these same plants bloom in South America? God gave us four seasons to enjoy, but these seasons are also important to the life cycles of many organisms. How do you think Earth's revolution affects which plants and animals can thrive in certain areas? Why is this revolution such an important part of life on Earth?

Climate and Seasons Explain

When most people in North America think of seasons, they think of four seasons: spring, summer, fall, and winter. What distinguishes one season from another? Why do some areas experience less dramatic seasons than others? You know that *weather* is the condition of the atmosphere at a particular time. What is the weather today? Is it what you would expect for the season? Why or why not? Sometimes the weather doesn't match the season. What do you think causes this?

Objectives

- Describe the tilt of Earth's axis and recognize its relationship to the seasons.
- Compare and contrast solstices with equinoxes.

Vocabulary

orbital plane
solstice
equinox
summer solstice
winter solstice
fall equinox
spring equinox

How does Earth's movement factor into these plants growing only at certain times of the year?

Recall that *climate* refers to the typical weather conditions of a specific season in a specific region over a long period. For example, the winter climate of New England is cold and snowy, even though on any given day the sky may be clear and temperatures could reach 10°C. In contrast, the winter climate of Southern California is cool and rainy, even though on any given day it may be sunny with temperatures above 20°C. While climate and weather may be different concepts, they both are the result of solar energy interacting with Earth.

Solar Energy on Earth

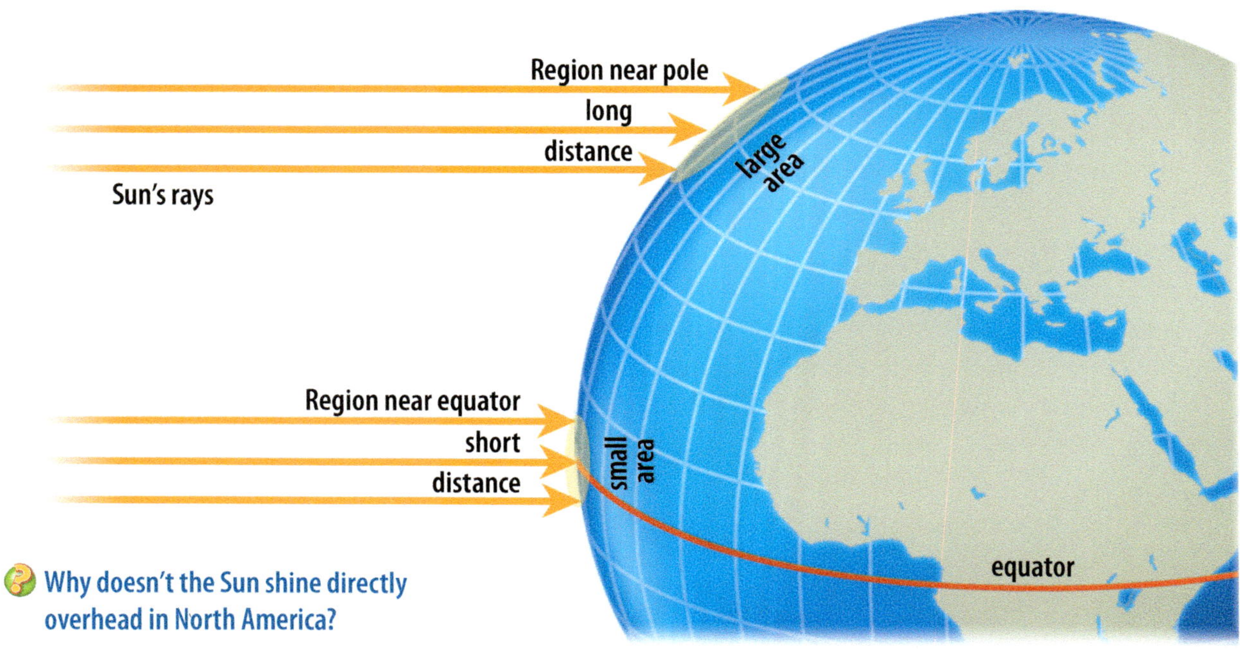

Regions near the North and South Poles have climates that are much colder than regions near the equator. This is because Earth is nearly spherical, and Earth's axis is nearly perpendicular to the direction of solar radiation. This geometry means that the radiation is spread over a larger area in polar regions than it is in areas that are closer to the equator.

Notice in the above diagram that solar energy strikes regions along the equator at almost a 90° angle. The farther you move away from the equator toward either pole, the smaller the angle at which solar energy strikes Earth. At a smaller angle, the same amount of solar energy is spread out over a larger area. The result is less energy absorbed per unit area, which makes for cooler weather and climate conditions.

Axial Tilt of Earth

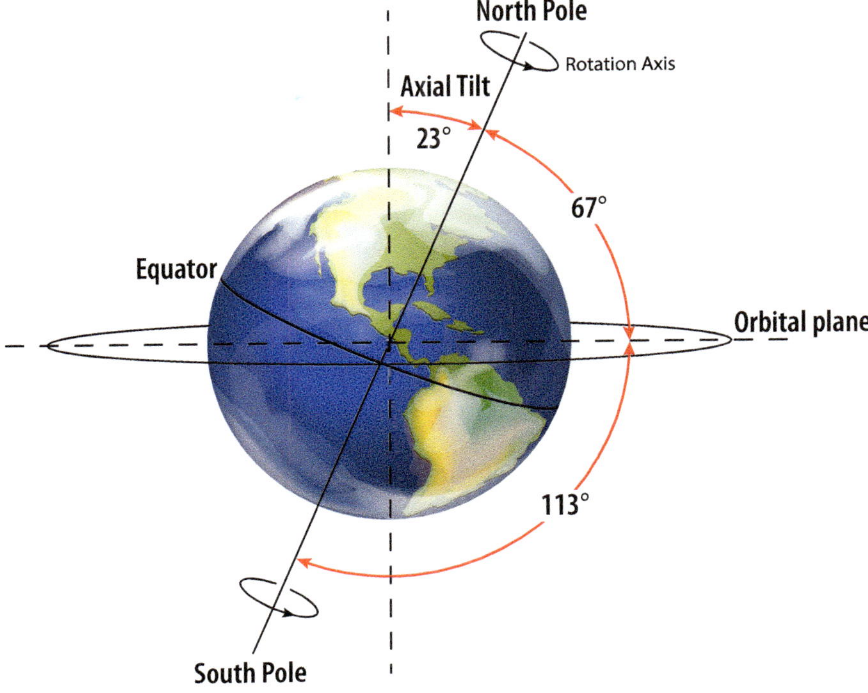

Earth's axis is tilted about 23° relative to the plane of Earth's orbit.

What areas on Earth receive the most direct sunlight?

Earth's Tilted Axis Explain

In the diagram above, Earth's axis is represented as being perpendicular to the direction of solar radiation. That is not an accurate representation. Earth's axis is actually tilted about 23° from the direction of solar radiation and Earth's orbital path. That tilt causes the seasons. Earth's revolution around the Sun and the tilt of Earth's axis explain why seasons occur. Do you think the seasons we experience on Earth today are the same as the seasons God designed Earth to have in the beginning?

Each planet in our Solar System orbits the Sun in what is called its **orbital plane**, which is the plane defined by its orbital path. For Earth, the angle between Earth's axis of rotation and its orbital plane is not 90°, but 113° and 67°.

Because of the tilt of Earth's axis, the intensity of the Sun's energy over any given area differs over the course of a year, especially so the farther an area is from the equator. Why are seasonal differences more extreme the closer an area is to the North or South Pole? During summer, the average angle of incoming sunlight to the surface of a given area is closer to 90° than it is in that area during winter. The difference in the intensity of solar energy that an area receives causes the weather and climate variations that we call the four seasons.

Record your work for this inquiry. Your teacher may also assign the related Guided Inquiry.

Solar Energy

How does the angle at which a location receives sunlight affect how much solar energy it receives?

Procedure

Materials
- penlight or flashlight
- graph paper
- protractor
- metric ruler

1. Place a sheet of graph paper, representing Earth's surface, on a flat surface.
2. Have your partner hold a flashlight 10 cm directly over the paper so that the light beam makes a 90° angle with the surface of the paper.
3. Darken the room and turn on the flashlight.
4. Outline the area covered by the light beam and count the number of squares within that outline. Use your best judgment when determining whether to count a partially covered square. For example, if two squares are only half covered with light, count this as one square; three squares that are one-third covered with light count as one square; and so on. **Record** your results in a table that includes two columns: *Angle of Light* and *Number of Squares Lit.*
5. Have your partner hold the flashlight again 10 cm above the paper, but this time tilt the flashlight so that the light beam forms a 45° angle with the surface of the paper.
6. Repeat Step 4.
7. Repeat the procedure at angles other than 90° and 45°, always keeping the flashlight 10 cm above the paper.

Analyze Results

How is the angle of the light related to the number of illuminated squares on the surface of the paper? Create a **graph** to show how the angle of light affects the surface area getting light.

Create Explanations

1. How does the angle at which a location receives sunlight affect how much solar energy it receives?
2. How would your results change if you tilted the graph paper and not the flashlight?
3. Would your results be different if you held the flashlight 5 cm or 20 cm away from the paper?

The Four Seasons Explain

As you know, most areas in North America experience four distinct seasons. Recall that this is because most of North America is between a latitude of 30° north and 60° north. We call the areas of Earth between a latitude of 30° north and 60° north and between a latitude of 30° south and 60° south *temperate zones*. Areas closer to the equator are known as *tropical zones*. The four seasons are associated with the four positions that Earth has as it orbits the Sun.

Recall that Earth's axis is tilted at a 23° angle. As Earth revolves around the Sun, sometimes the North Pole points toward the Sun, and other times it points directly away from the Sun. The moment when the Pole points directly toward or away from the Sun is called a **solstice.** Between the solstices, Earth moves into a position where neither end of Earth's axis points toward or away from the Sun and the length of day and night are equal. This moment is called an **equinox.**

Earth's Seasons

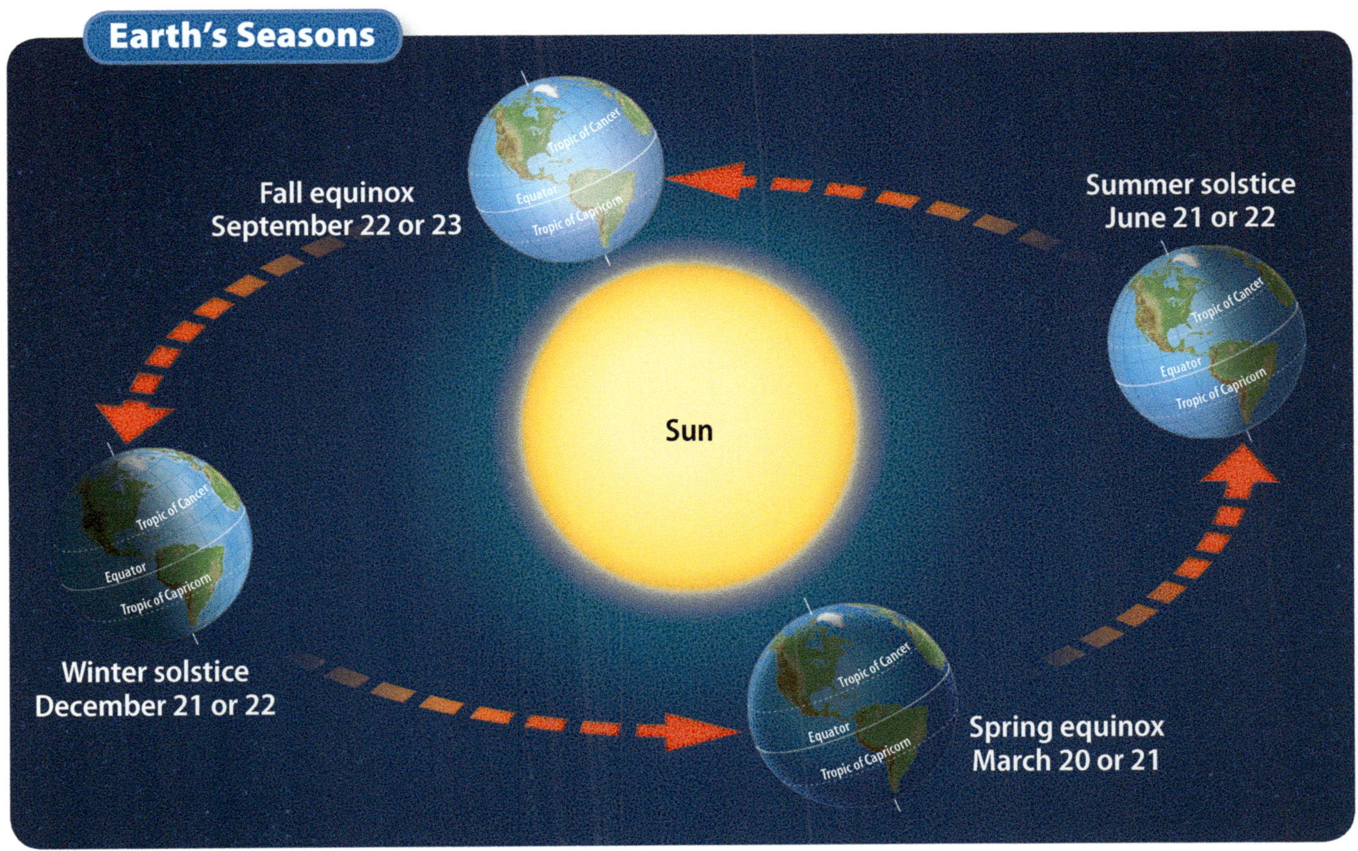

During the solstices, the Sun appears directly overhead farther north or south of the equator than it does on any other day of the year. During the equinoxes, the Sun appears directly overhead at the equator.

Why do solstices and equinoxes each occur twice a year?

Explore-a-Lab

Guided Inquiry

How does the time of sunset change over time?

For a week or two, record the time the sun sets behind a particular object you can see at home. Then use your data to construct a graph showing either a longer day (after 12/22) or a shorter day (before 12/22). How would the sunset times compare if you gathered the data six months from now?

Summer and Winter Solstice

You can pinpoint the moment in time when summer begins—the summer solstice. The **summer solstice** is the day when the Sun appears to move through its highest path in the sky. In the northern hemisphere, the summer solstice occurs on June 21 or June 22. The moment of the northern summer solstice occurs when the Sun appears directly above the Tropic of Cancer, located at a latitude of about 23° north.

The North Pole is tilted most directly toward the Sun at the (northern) summer solstice. From the (northern) spring equinox to the fall equinox, the northern hemisphere receives more direct sunlight and more hours of daylight than does the southern hemisphere. The combination of long daylight hours and more intense sunlight causes warm summer temperatures. With these conditions in the north, what do you think the weather in the southern hemisphere would be like?

Scripture Spotlight

Of the things listed in **Ecclesiastes 3:1–8**, which relate to a season on Earth, and which relate to a different kind of season?

Summer Solstice

Winter Solstice

The South Pole is tilted farthest away from the Sun at the (northern) summer solstice. That moment marks the beginning of winter in the southern hemisphere. The northern and southern hemispheres have opposite seasons.

Notice in the diagram of Earth's Seasons earlier in this lesson the position of the Sun during a summer solstice and during a winter solstice. How would the Sun appear in the sky if an image of an equinox were included?

The cool days of fall are followed by chilly winter. In the northern hemisphere, the winter solstice occurs on December 21 or December 22. The **winter solstice** occurs when the Sun appears to reach its lowest or southernmost point in the sky. The moment of the northern winter solstice occurs when the Sun appears directly above the Tropic of Capricorn, located at a latitude of 23° south. The North Pole is tilted farthest away from the Sun at the winter solstice. The northern hemisphere receives less intense sunlight and fewer hours of daylight than does the southern hemisphere during the winter. South of the equator, such as in Australia or New Zealand, summer begins at the same moment that winter begins north of the equator.

Although the summer solstice and winter solstice occur at different times of the year, Earth receives the same amount of solar energy on both days. Why, then, are areas in the northern hemisphere warm on a summer solstice but cool on a winter solstice, while areas in the southern hemisphere experience the opposite conditions?

Fall and Spring Equinox

Twice during each orbit of the Sun, Earth experiences an equinox. In the northern hemisphere, the **fall equinox** occurs on September 22 or September 23. The fall equinox occurs when the Sun rises exactly in the east because neither pole tilts toward or away from the Sun. All parts of Earth receive nearly equal hours of daylight and darkness. Days are generally getting cooler. Why does the fall equinox in the northern hemisphere coincide with cool, crisp temperatures?

As Earth rotates on its tilted axis in orbit around the Sun, winter turns to spring. The **spring equinox** occurs when the Sun rises exactly in the east because neither pole tilts toward or away from the Sun. In the northern hemisphere, the spring equinox occurs on March 20 or March 21. Just like the fall equinox, hours of daylight and darkness are nearly equal for all parts of Earth. When does the spring equinox occur in the southern hemisphere?

Explore-a-Lab

Guided Inquiry

 **Have you ever heard the legend about balancing an egg on its end during an equinox?**

Try balancing a raw egg on its end. Does the egg balance on one end and not the other? What can you conclude about this legend? Think about the gravitational attraction between the Sun and Earth and between Earth and the egg on an equinox. How might this be used to support the idea that an egg can be balanced only on an equinox? How would you disprove that an egg can be balanced only on an equinox?

Concept Check

Assess/Reflect

Summary: **Why does Earth have seasons?** While climate and weather may be different concepts, they both are the result of solar energy interacting with Earth. The farther you move away from the equator toward either pole, the smaller the angle at which solar energy strikes Earth and the cooler the climate. Earth rotates on its axis at a 23° angle with respect to its orbital plane. This tilt in Earth's axis causes seasonality in the weather and climate. A solstice occurs twice a year when the Sun is at its greatest distance from the equator. The summer solstice is the day when the Sun is at its highest path through the sky. The winter solstice occurs when the Sun appears to reach its lowest or southernmost point in the sky. In the northern hemisphere, the summer solstice occurs on June 21 or 22. An equinox occurs two times a year when the Sun crosses the equator and the hours of daylight and darkness are equal.

1. What are the advantages and disadvantages of Earth having the tilt that it does?
2. How do seasons compare in the northern and southern hemispheres?
3. On a certain day, the time between sunrise and sunset was 12 hours, and the time between sunset and sunrise was also 12 hours. What can you infer about the position of Earth's poles relative to the Sun on this day?
4. Do you think Earth would have seasons if its axis were not tilted? Explain.

Lesson 3

Essential Question

How Do Earth and the Moon Interact?

Throughout history, humans have watched and wondered about the Moon. Sailors have used the Moon as a sign of good weather or bad weather to come. It has even been used to explain odd human behavior. People have asked many questions about this object that orbits Earth. What is the Moon made of? Where does the moonlight come from? What was known about the Moon before astronauts set foot on it? What would happen here on Earth if the Moon disappeared? While many questions have been answered, questions still abound about our nearest celestial neighbor.

Objectives

- Explain evidence for the theories of the Moon's formation.
- Describe features on the Moon's surface.
- Explain the phases of the Moon.
- Compare and contrast the lunar and solar eclipses.
- Explain Earth's tides.

Vocabulary

regolith
highland
maria
crater
eclipse
solar eclipse
umbra
penumbra
lunar eclipse
spring tide
tidal range
neap tide

Earth's Moon Explain

The Moon is easily the brightest object in our night sky. In the Genesis Creation story, the Moon is called the lesser light, and the Sun is called the greater light. The Moon does not give off its own light. We can see the Moon because it reflects light from the Sun.

The Moon is the only place in the Universe other than Earth that humans have visited. In the vastness of outer space, it is our closest and best-known neighbor. How many people have visited the Moon? What other organisms or objects have traveled to the Moon?

This astronaut's footprint will last, undisturbed, for hundreds or even thousands of years.

Why do footprints on Earth disappear much faster than they do on the Moon? Where on Earth do footprints last a long time? Why?

Scripture Spotlight

Read **Genesis 1:14–19**. Why did God create the Moon?

The Moon's orbital path is an average of 384,000 km away from Earth. Altogether, the planets in our Solar System have 166 moons. Earth's Moon is the fifth largest satellite in our Solar System; it is even bigger than the dwarf planet Pluto. In terms of diameter, the Moon is about a fourth the size of Earth. In terms of mass, however, the Moon is only about 1% of the mass of Earth. Why is there such a difference between size and mass? The Moon's gravity is only about one-sixth as strong as Earth's.

The Moon has no atmosphere. Therefore, it has no wind or weather. In sunlight, the surface of the Moon can reach temperatures of 123°C. In shadow, the temperatures can drop to –233°C. How do you think these factors affect astronauts exploring the Moon's surface?

Earth–Moon Data

	Earth	Moon
Diameter	12,756 km	3475 km
Mass	5.9×10^{24} kg	7.3×10^{22} kg
Surface temperature	−88°C to 58°C	−233°C to 123°C
Atmosphere	Nitrogen, oxygen	None

Theories about the Moon's Formation **Explain**

Genesis tells us that God created the Moon on the fourth day. Some scientists have different theories about the Moon's formation. Perhaps the most widely accepted theory is called the *collision ejection theory*. According to this theory, Earth collided with a huge object, possibly the size of Mars, causing the ejection of an enormous amount of material from Earth into space. This material combined and cooled. Earth's gravity captured the material and formed the spherical Moon. Scientists think this theory may also explain why Earth's axis is tilted. Evidence used to support this theory includes:

- The Moon is composed of material similar to Earth's upper mantle.
- The Moon's orbit around Earth is in the same direction as Earth's orbit.
- The Moon is located in the same orbital plane as Earth.
- The Moon has a round shape, so it is not a "captured" object. Captured moons of other planets have irregular shapes.

Even though scientific evidence exists to support their observations, scientists cannot be completely certain about the Moon's formation. As is often the case, evidence used to support one theory can at times be interpreted to support opposing theories. One fact sets Earth's Moon apart: its size relative to Earth is much greater than any other moon relative to the planet it orbits. So, Earth and the Moon are unique. Do you think this is evidence of God's Design?

Surface Features of the Moon

If you could walk on the surface of the Moon, your boots would likely sink into a layer of fine, loose material made of powdered rock. The material that covers the surface of a planet or moon is called **regolith**. On Earth, regolith includes loose rocks, dust, and soil. Why does the Moon's regolith not contain soil? The Moon's regolith varies in depth from 2 m in the maria to 20 m in the highlands. Why do you think the Moon's regolith varies so much in depth?

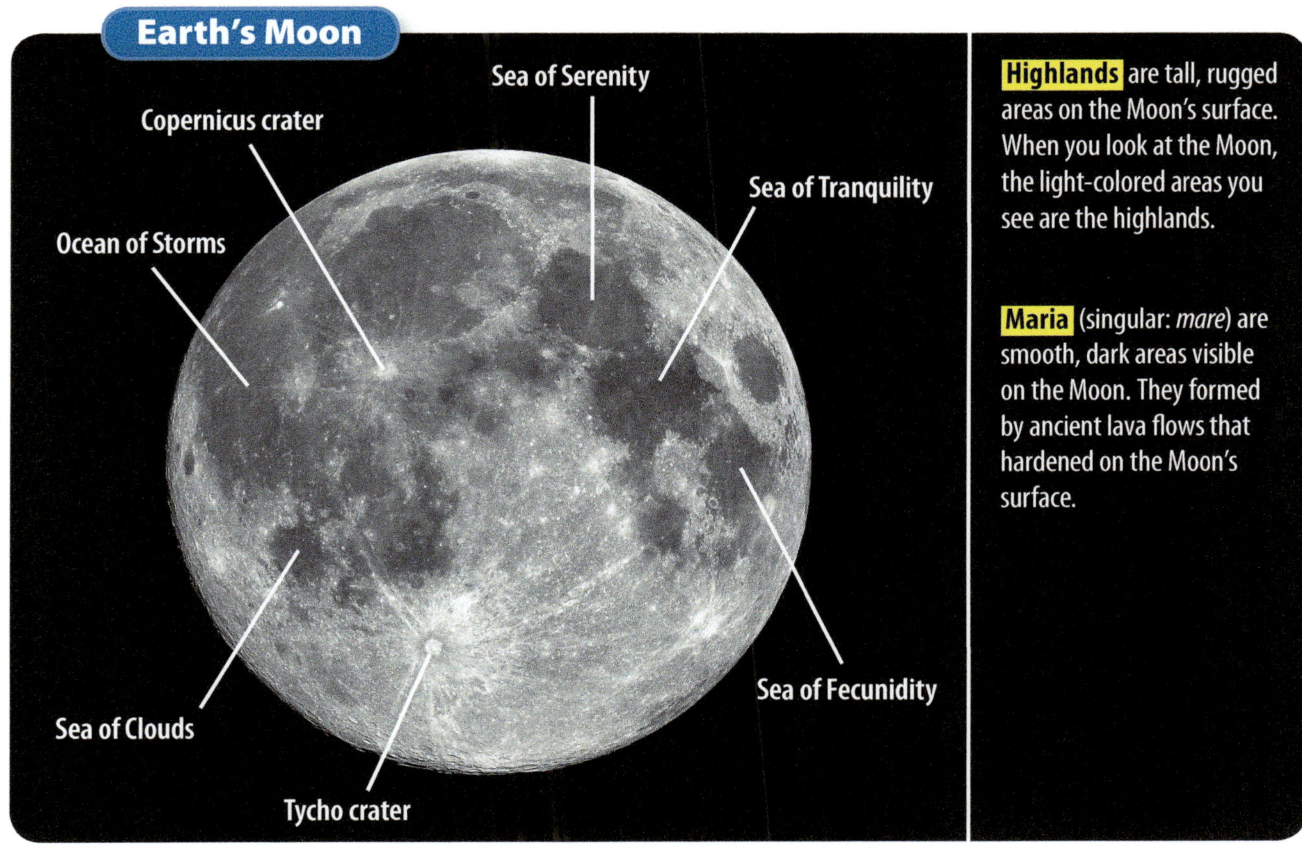

The surface of the Moon also is marked by numerous **craters**, which are depressions caused by impacts with asteroids, comets, and other objects in space. Because the Moon has no atmosphere, those objects do not burn up before they strike the Moon's surface. The absence of an atmosphere also means that the Moon has no wind to erode the craters. Thus, craters that formed long ago are still visible on the Moon's surface.

Craters are often characterized by a distinctive ray pattern. You learned in an earlier science class that when a large object strikes the surface of another object—in this case, the Moon's surface—large amounts of regolith and rock are ejected from the area of impact. This creates a crater ray system. The rays enable geologists to calculate the relative date of the impact by comparing the rays to other surface features.

The Moon's most distinctive crater is named Tycho, after Danish astronomer Tycho Brahe (1546–1601). It has a diameter of 85 km and is located in the Moon's southern hemisphere. What can you conclude about a crater that has twice the diameter of Tycho? Look closely at the picture of the Moon; what other craters can you see?

Motion of the Moon Explain

You just learned about some of the features of the Moon. How are the Moon's features similar to and different from those of Earth? How are their motions alike and different? Scientists can use the Moon's motions to predict the Moon's location in relation to Earth and the Sun. They can use the predictability of the Moon's motion to tell exactly when the Moon can be seen in Earth's sky and where the Moon's shadow will fall on Earth.

Moon Phases

As you know, the shape of Earth's Moon appears to change as you observe it in the sky over a period of about a month. The changes in the Moon's appearance are called *moon phases*. How does the Moon's shape change during its different phases? What happens after the Moon completes one complete cycle of phases? We see different phases as a

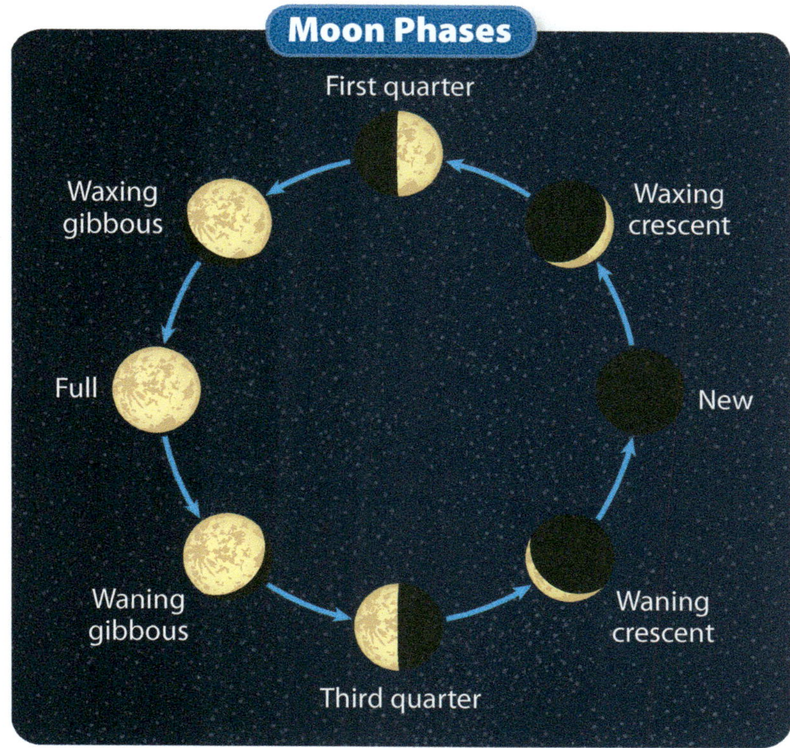

Moon phases as seen from Earth

result of the Moon's revolution around Earth and because it reflects sunlight. Just as on Earth, one-half of the Moon is always lit by the Sun, and we see different parts of this lighted side from Earth. The lighted portion that we see depends on where the Moon is relative to Earth and the Sun.

Recall that it takes the Moon about 27 days to rotate once on its axis. It takes the same amount of time for the Moon to complete one revolution around Earth. These two movements, occurring in the same amount of time, result in an interesting phenomenon—one side of the Moon always faces Earth, and one side always faces away from Earth. Have you ever heard the phrase "the dark side of the Moon"? Why is that phrase inaccurate? What do you think the dark side of the Moon looks like?

Check out your ***Science Journal*** for a Structured Inquiry that explores how to create an eclipse.

Extend

Explore-a-Lab

Structured Inquiry

 **How many times does the Moon rotate during its revolution around Earth?**

Use some coins for a modeling activity. Place a quarter flat on a table or desk. The quarter represents Earth. Place another coin on the table to represent the Moon with its face pointed toward the face on the quarter. Rotate and revolve the second coin around the model of Earth. Keep the head on the second coin continuously facing the same direction toward the quarter. Count how many times the model of the Moon rotates during each revolution. How does this model explain why we always see the same side of the Moon?

Eclipses

You know that the Moon does not give off its own light; it shines because it reflects light from the Sun. What would happen if an object blocked sunlight from reaching parts of the Moon? Could the Moon block sunlight from reaching Earth? When one object partially or totally blocks sunlight from another object, an **eclipse** occurs.

We now know that eclipses occur in space because Earth, the Moon, and the Sun move in regular patterns according to God's natural laws. Why is it possible to predict when eclipses will occur hundreds of years into the future?

This image of the June 22, 2009, eclipse, taken from a satellite above Earth, shows the Moon crossing in front of the Sun.

Why do scientists who study the Sun use eclipses to study the Sun's corona (the Sun's outer ring)?

Solar Eclipse

To see why an eclipse occurs, look at the diagram of a solar eclipse. A **solar eclipse** occurs when the Moon comes between Earth and the Sun. The Moon blocks our view of the Sun (the Sun is eclipsed) because its shadow falls on part of Earth. The darkest part of the shadow is called the **umbra**. Areas within the umbra experience a total solar eclipse; that is, the Sun is almost completely blocked by the Moon. The only part of the Sun that the Moon does not block out during a total solar eclipse is its outer atmosphere, or corona, which you can see in the picture on the chapter opener. The lighter part of the shadow is called the **penumbra**. Areas within the penumbra experience a partial solar eclipse because only part of the Sun is blocked by the Moon in those areas.

Solar eclipses can only occur during new moon phases. But they do not happen during every new moon. The Moon, the Sun, and Earth must be lined up just right for an eclipse to occur. Is it safe to directly view a solar eclipse? Why or why not?

Scripture Spotlight

Read **Amos 8:9.** Do you think the passage could refer to an eclipse that occurred in 763 B.C.? Explain.

Check for Understanding

Why is a lunar eclipse more widely visible on Earth than a solar eclipse?

Lunar Eclipse

Another kind of eclipse is a **lunar eclipse**. A lunar eclipse occurs when Earth is between the Sun and the Moon, casting its shadow on the Moon. Lunar eclipses occur only during certain full-moon phases, when the Sun, Earth, and the Moon are perfectly aligned. During a lunar eclipse, the Moon moves into Earth's shadow. The Moon's orbit is slightly tilted with respect to Earth's orbit. As a result, lunar eclipses do not occur with every full moon.

Recall that during a solar eclipse, the Moon appears as a dark disk over the Sun. But during a lunar eclipse, the Moon is not blacked out—it appears red. Why does the Moon look red during an eclipse? Is it safe to directly view the Moon during a lunar eclipse? Why or why not?

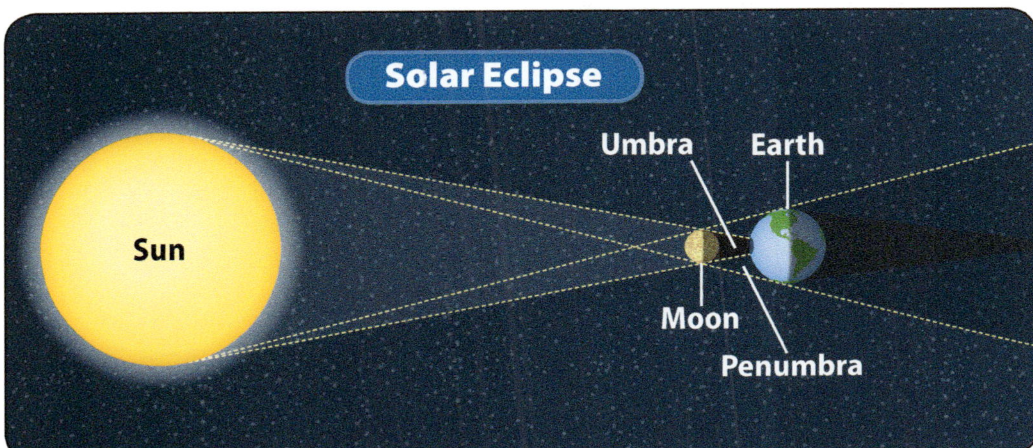

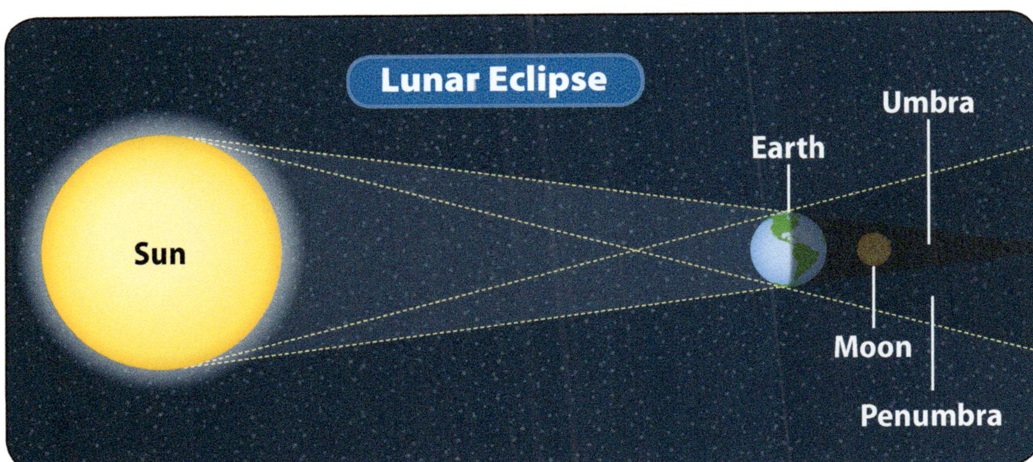

A total solar eclipse and a total lunar eclipse

Lesson Activity

Search the Internet or use other references to find out when and where the next solar eclipse will occur. Then draw a diagram of Earth, the Moon, and the Sun that explains why the eclipse is occurring in this location.

Why are solar eclipses seen only along narrow, curved paths on Earth?

Record your work for this inquiry. Your teacher may also assign the related Guided Inquiry.

The Tides

What is the connection between tides and the Moon's phases?

Materials

- graph paper
- daily tide data
- pencil
- blue and red markers
- tape
- circular self-stick labels

Procedure

1. Obtain tide data for an assigned two-week period from online resources or almanacs. **Record** the data for your weeks in the chart in your ***Science Journal***.
2. For each day in your weeks, record in your ***Science Journal*** the heights for the high and low tides in meters above or below mean sea level. Also include the times.
3. Determine the tidal range for each day by **calculating** the difference between the lowest low tide and the highest high tide for each day. Record the data in the chart.
4. **Graph** your data from Step 2, using red to indicate high-tide values and blue to indicate low-tide values.
5. For your weeks, identify the day on which the Moon was new, in its first quarter, full, or in its third quarter. Place an appropriately trimmed circular label on that day.
6. Combine the data collected by your group with that of the other groups in chronological order to create a continuous tide chart.

Analyze Results

Analyze the continuous tide chart, comparing the high and low tides and the moon phases. What do you notice?

Create Explanations

1. What is the connection between tides and the Moon's phases?
2. If you were to repeat this activity with tide charts from another area of the world, how would your results differ? Explain.
3. How would your continuous tide chart look if you plotted values taken from a shoreline where there are very noticeable changes between low tides and high tides?
4. Did any of your findings surprise you? Explain.

The Moon and Tides **Explain**

If you have ever been to the ocean, you have probably seen an Earth–Moon interaction—the tide. You already know that *tides* are the periodic rise and fall of ocean waters. They are caused by the gravitational attraction between the Moon and Earth and, to a lesser degree, the Sun and Earth. Recall that the force of gravity decreases with distance and increases with mass. The Sun is much more massive than the Moon, but it has a lesser effect on tides because it is so far from Earth. Why do you think some areas have very large tides and other areas have very small tides? How much time would you expect there to be between the tides? Does the time vary? Why or why not? What time of the month would you expect the highest high tides? What about the lowest low tides?

Gravity is a major force responsible for creating tides, but inertia acts to counterbalance gravity. The gravitational force between Earth and the Moon is strongest on the side facing the Moon. Why would this side of Earth have a greater attraction to the Moon than the opposite side? As gravitational forces pull the water closer to the Moon, inertia acts to keep the water in place. But the gravitational force is greater than the inertia. As a result, the water on the side of Earth nearest the Moon is pulled toward the Moon, producing a bulge in the ocean water.

On the opposite side of Earth, the gravitational force of the Moon is less, and inertia is greater than the gravitational force. Recall that you learned in an earlier science class that the water surrounding Earth is pushed outward as a result of Earth's rotation. The water tries to keep going outward in a straight line. What causes the water's tendency to continue in a straight line? This action causes another bulge in the ocean on the side of Earth opposite the Moon.

Tides

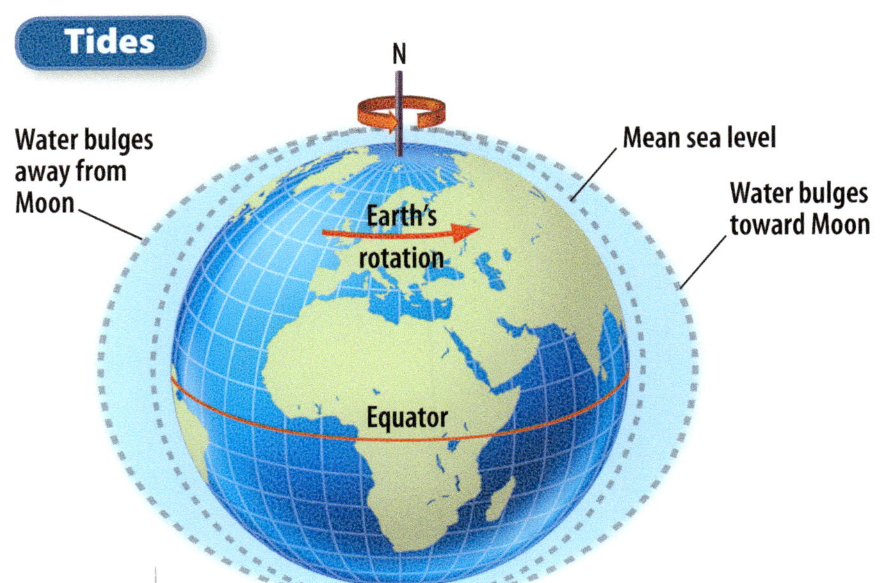

The Sun's Effects on Tides Explain

When the Sun, the Moon, and Earth are aligned in space at the full moon and new moon, the combined gravitational effect of the Sun and the Moon causes spring tides. The name spring tide comes from the fact that the tides "spring" high, not because they occur in spring. A **spring tide** results in the largest daily tidal range. The **tidal range** is the difference between high tides and low tides. Spring tides produce the highest high tides and the lowest low tides for the month.

Tidal ranges are high during spring tides and low during neap tides.

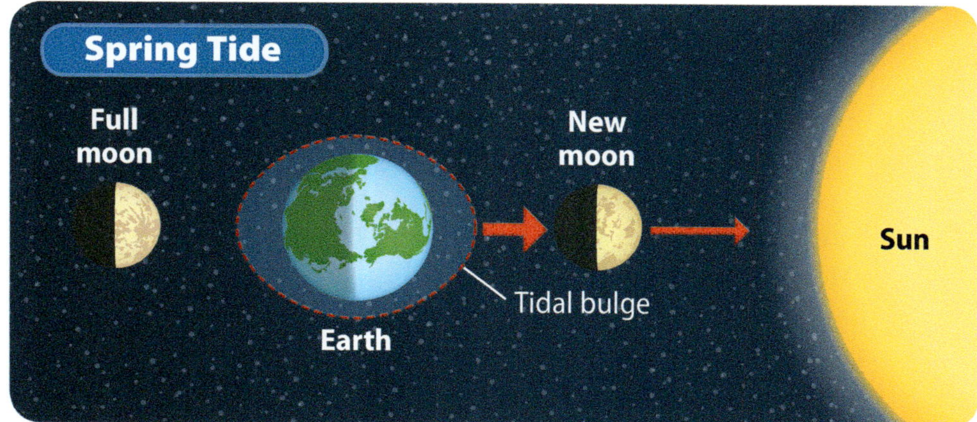

At times, the Sun, the Moon, and Earth form a right angle relative to one another at the first-quarter and third-quarter moons. The gravitational effect of the Sun tends to lessen the gravitational effect of the Moon, causing neap tides. A **neap tide** results in the smallest daily tidal range. Neap tides produce the lowest high tides and the highest low tides.

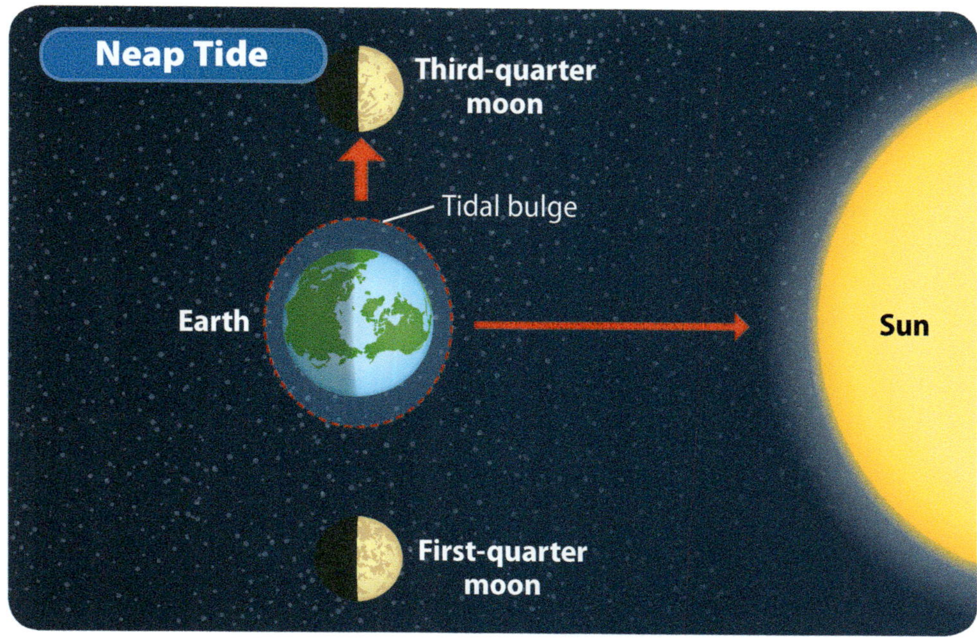

Explore-a-Lab

Structured Inquiry

How many high tides and low tides occur in one day?

Create a posterboard model of Earth and the tidal bulges. Using the diagram on the previous page as a guide, draw and cut out an oval to represent the tidal bulges and a circle to represent Earth. Center the circle over the oval and attach them with a brad. Make an *X* near the edge of the Earth shape to represent a coastal location. Label the *X* "Noon." Line up the *X* with one of the high-tide bulges. Rotate Earth one complete turn. How many high tides did you observe on your model? How many low tides?

Concept Check **Assess/Reflect**

Summary: How do Earth and the Moon interact? The Moon is about 27% of the size of Earth and about 1% of its mass. The Moon has no atmosphere. The most widely accepted theory by non-creationists about the Moon's formation is the collision ejection theory, which states that Earth collided with a huge object, causing the ejection of material from Earth into space. The material that covers the surface of the Moon is called *regolith*, which consists of loose rocks and dust. The Moon's surface also has tall, rugged areas known as *highlands*; smooth, dark areas known as *marias*; and *craters*. The relative positions of Earth, the Sun, and the Moon determine the lunar phases. The relative positions of Earth, the Sun, and the Moon also account for both solar eclipses and lunar eclipses. A solar eclipse occurs when the Moon comes between Earth and the Sun. A lunar eclipse occurs when Earth is between the Sun and the Moon. The gravitational attraction between Earth, the Sun, and the Moon is responsible for the tides. A spring tide has the largest daily tidal range, while a neap tide has the smallest daily tidal range.

1. Why do tides occur?
2. A lunar eclipse occurs when the Moon and the Sun are on opposite sides of Earth. This alignment also occurs during a full moon. Why, then, don't we see a lunar eclipse once a month, when a full moon appears?
3. If the Moon's gravity is the main reason for a high tide on the side of Earth that faces the Moon, why does a high tide also occur at the same time on the opposite side of Earth?
4. Suppose that the Moon did not rotate. How would this affect our view of the Moon from Earth?
5. Suppose that you are viewing the Moon through a telescope. What clues could you use to distinguish highlands from maria?
6. Why do more people see a total lunar eclipse than see a total solar eclipse?

People in Science

Extend

Get to Know
Nicolaus Copernicus

Nikolaus Kopernikus.

At one time, everyone believed that Earth was the center of the Universe. In the early 1500s, a Polish scientist named Nicolaus Copernicus proposed that the Sun was the center of the Universe and that all the planets revolved around it. His idea was new, and, although his model wasn't completely correct, it shaped a strong foundation for future scientists to build on and improve our understanding of the motion of heavenly bodies.

Born in 1473 in Poland, Copernicus traveled to Italy at the age of 18 to attend college at the University of Bologna, where he prepared for a career in the church. As part of his education, he studied astrology—reading the stars to learn about future events—because at the time this knowledge was important for priests and doctors. It was during this period that astronomy, the study of the motion of heavenly bodies, was an important element of preparation for a career in the church.

While attending the university, Copernicus lived and worked with an astronomy professor, doing research and helping him to make observations of the heavens. When he returned to Poland to take up his official clergyman duties, he lived in a room in one of the towers surrounding the town. The tower had an observatory, giving him plenty of time and opportunity to study the night sky.

Though his theory was viewed as revolutionary and met with controversy, Copernicus went on to design and apply a complex mathematical system for proving his theory. In 1513, his dedication inspired him to build his own observatory so that he could view the planets in action at any given time.

In 1514, Copernicus penned a handwritten book that explained his view of the Universe. In it, he proposed that the center of the Universe was the Sun and not Earth. He also suggested that Earth's rotation accounted for the rise and setting of the Sun and the movement of the stars and that the cycle of seasons was caused by Earth's revolutions around the Sun.

At the age of 70, Copernicus published his book *De Revolutionibus Orbium Coelestium* ("On the Revolutions of the Heavenly Spheres"). In the book, Copernicus launched the idea that the planets orbited the Sun rather than Earth, and he presented his model of the Solar System and the path of the planets. Copernicus's ideas, published only two months before he died, took nearly a hundred years to be seriously recognized.

Concept Check

1. Why would people think that priests and doctors needed this information?
2. Why do you think it took so long for Copernicus's theory to be accepted?

Extend

Russia's Robotic Lunar Exploration Program

Destination: the Moon. Russia is developing a renewed robotic Moon-exploration program, building upon the history-making legacy of orbiters, landers, rovers, and sample-return missions the country launched decades ago. The plan was revealed during Microsymposium 54 on "Lunar Farside and Poles—New Destinations for Exploration," held in 2014 in The Woodlands, Texas.

Russia launched its last Moon mission in August 1976, when it was still the Soviet Union. That mission, called *Luna 24*, was a successful mission. It was the last in the Luna series and featured a spacecraft that landed on the Moon and returned samples of the Moon's surface.

Five missions are in the planning stages at the time of this writing. Depending on the success of the first three missions, two additional missions will be executed.

The five possible Moon missions will launch as follows:

2015—Luna 25 (Luna Glob Lander): A small lander on the Moon's South Pole that will analyze lunar soil and the outermost region of the atmosphere. This spacecraft will experiment with lunar landing system technology, communication systems, and longtime operations.

2016—Luna 26 (Luna Glob Orbiter): An orbiter for the Moon in a 100-km polar circular orbit that will globally map the lunar surface, measure the outermost region of the atmosphere and plasma around the Moon, and carry out investigations of landing sites for future lunar exploration.

2017—Luna 27 (Luna Resource-1): A large lander sent to the Moon's South Pole to study lunar soil and the outermost region of the atmosphere. It will also test for volatiles (substances that will evaporate at relatively low temperatures) in the lunar subsurface. This lander will also test a drilling system for cryogenic (the freezing of material) sampling of the Moon's soils.

2019—Luna 28 (Luna-Resource-2): A "to be determined (TBD)" mission involving cryogenic delivery of lunar samples back to Earth. This mission will help develop return flight system technology for transiting between the Moon and Earth.

2020—Luna 29 (Luna-Resource-3): Another TBD mission. This spacecraft will carry a Lunokhod—a large, long-distance Moon rover. Once on the move, the wheeled device will study the lunar surface at a distance of about 30 km and conduct cryogenic preservation of the lunar subsurface.

Concept Check

1. Why is the first mission in the Russian Moon-exploration program titled *Luna 25*?
2. Describe some of the goals for the first three missions.

Chapter 9 Review Assess/Reflect

Study Guide

Lesson 1

1. Earth's rotation is its spinning on its axis. Earth's revolution is its movement around the Sun in an orbit.
2. A sidereal day is a day in relation to a distant star. A solar day is a day in relation to the Sun.
3. Inertia acts to keep Earth moving in a straight line, while gravity acts to pull Earth toward the Sun.

Lesson 2

1. Seasons are caused by the tilt of Earth's axis relative to the plane of its orbit around the Sun. The tilt of Earth's axis results in an uneven distribution of solar energy in the northern and southern hemispheres, except during equinoxes.
2. Equinoxes mark the beginning of spring and fall, when neither the North nor the South Pole is tilted toward or away from the Sun, and all locations on Earth have nearly equal hours of daylight and darkness. Solstices mark the beginning of winter and summer, when the North and South Poles are tilted maximally toward or away from the Sun.

Lesson 3

1. Genesis states that God created the Moon on the fourth day. Some scientists have different theories about the Moon's formation. The most widely accepted theory is the collision ejection theory. Evidence to support this theory is listed on page 342.
2. The Moon has various features on its surface, including highlands, maria, and craters. See the illustration on page 343.
3. As the Moon revolves around Earth, it appears to change shape over the course of a month. These apparent changes are called moon phases. See the illustration on page 344.
4. A solar eclipse occurs when the Moon casts a shadow on Earth. A lunar eclipse occurs when Earth casts a shadow on the Moon.
5. Tides are the result of the gravitational interaction and relational positions of the Moon, the Sun, and Earth's oceans. When the Sun, the Moon, and Earth are aligned in space, the largest tidal range occurs, called a spring tide. When the Sun, the Moon, and Earth form a right angle, the tidal range is the smallest, called a neap tide.

Vocabulary Check

Fill in the blank with the correct vocabulary word.

1. During a(n) _________, the Sun appears to reach its farthest point north or south of the equator.
2. During a(n) _________, the Sun appears directly over the equator.
3. Collisions with asteroids and comets have created _________ on the Moon's surface.
4. A _________ is a day in relation to the Sun, while a _________ is a day in relation to a distant star.
5. A(n) _________ occurs when the Moon comes between Earth and the Sun, blocking our view of the Sun.
6. The lightest part of the shadow cast by the Moon on Earth is the _________.

Multiple Choice

Choose the best answer.

7. When are the Sun, the Moon, and Earth aligned in a row?
   A. during the first quarter
   B. during a full moon and new moon
   C. during a solar eclipse but not a lunar eclipse
   D. during a lunar eclipse but not a solar eclipse
8. On June 21 or June 22 each year, the Sun appears directly over which of these?
   A. the equator
   B. the North Pole
   C. North America
   D. the South Pole
9. How do the Moon's highlands differ from its maria?
   A. Highlands are smooth areas.
   B. Highlands are rugged and tall.
   C. Highlands are covered by regolith.
   D. Highlands appear dark when viewed from Earth.
10. During which moon phase is it most difficult to see the Moon?
    A. new moon **C.** third quarter
    B. full moon **D.** waning crescent

Check Point

Answer the following questions.

11. Explain the effects of the Moon and the Sun on Earth's tides.
12. One day, at a certain time, a student measured the shadow of a fence and found that it was 2 m long. The next day, she measured the shadow of the same fence at a different time and found that it was 3 m long. What can you infer about the time of day at which the student measured the shadows? Explain.
13. Compare an umbra with a penumbra.
14. Describe how a lunar eclipse occurs.
15. Identify the forces that affect Earth's orbit and explain how they work together.
16. Explain the collision ejection theory of the Moon's formation.

Stars, Galaxies, and the Universe

Scripture Spotlight

God created the Universe. We do not know how He did it, but the Bible tells us many things about the events that happened during Creation. According to **Psalms 33:6**, "By the word of the Lord were the heavens made; and all the host of them by the breath of his mouth." You will read the following passages in this chapter.

Job 9:9 (p. 361)
1 Corinthians 15:40–41 (p. 367)
Genesis 1:1–5 (p. 384)

The Big Idea

God created the Universe by simply speaking. God designed the billions of galaxies that make up the Universe and the billions of stars that are found in each galaxy.

How does the design of the Universe provide evidence of a Creator?

Inquiry Kick-Off

Engage

How would you describe what you see in the sky on a clear night when you are away from city lights? Why do you think the ancient Greeks thought "Milky Way" was an appropriate name for our galaxy? What would you name the galaxy? In your *Science Journal*, you will investigate the density of our galaxy.

Science Journal

Lynds 1688 is the main cloud in the Rho Ophiuchi cloud complex.

Lesson 1

Objectives

- Describe common constellations.
- Distinguish between absolute magnitude and apparent magnitude.
- Explain how parallax enables scientists to determine star distances.
- Compare the Sun's properties with those of other stars.
- Identify factors that affect a star's life cycle.

Vocabulary

circumpolar constellation
asterism
apparent magnitude
absolute magnitude
parallax
fusion
main sequence
red giant
white dwarf
black dwarf
supergiant
supernova
neutron star
pulsar
black hole
quasar

Essential Question

How Can We Describe Stars?

How many times does the Bible mention the stars? In **Job 38:31**, God asks Job, "Canst thou bind the sweet influences of Pleiades, or loose the bands of Orion?" The Pleiades is a group of seven stars. Astronomers have discovered that, in addition to the seven stars that are easily visible, nearly 500 stars make up this cluster. The stars of the Pleiades are in our home galaxy, the Milky Way. They are an average distance of about 900 trillion km, or about 95 light-years (ly), from Earth. The cluster has a diameter of about 151 trillion km, or about 16 ly. How do the stars in the Pleiades compare to our Sun? How are stars different from one another? How can astronomers know how far away stars are and what they are made of?

Constellations Explain

You might recall that a *star* is a body of hot gases that generates light, thermal energy (heat), and other forms of energy through the process of nuclear reactions. The Pleiades is a cluster of stars. The group of stars known as Orion is a constellation, which is different from a star cluster. Recall that a *constellation* is an area of the sky where a recognizable pattern of stars can be seen.

Pleiades Constellation

While the star cluster Pleiades contains an estimated 500 stars, some star clusters may contain thousands or even millions of stars.

This star map shows the stars that make up some constellations and star groups.

Which of these can you find in the night sky?

Today scientists recognize 88 constellations. Most were named after familiar objects or characters in ancient myths, such as Orion, who was a hunter in Greek mythology. While the ancient Greeks saw this constellation as a hunter, other civilizations connected the stars into a different picture. For example, the Japanese say this constellation is a long sleeve of a kimono. People in ancient India saw it as a deer. Why did people see different pictures in the pattern of these stars?

Orion is one of the constellations that is easy to identify because it contains two of the brightest stars in the sky—Betelgeuse and Rigel. Orion can also be spotted from anywhere on Earth. Orion will appear upside down, however, to those looking at this constellation in the southern hemisphere compared with those viewing it from the northern hemisphere. No matter where people see Orion, they may think that the stars in this constellation are near each other or are arranged in the same plane in the sky. But they are not. Why do the stars in Orion or any other constellation appear to be in the same plane in the sky?

Although Orion can be seen from both hemispheres, it is most visible only from January to March in the northern hemisphere. Other constellations that can be seen from both hemispheres, such as Hercules, are most visible only in July. Why do these constellations appear during only certain months?

Record your work for this inquiry. Your teacher may also assign the related Guided Inquiry.

Star Brightness

How can you model the brightness of stars?

Materials

- meterstick
- masking tape
- penlight
- flashlight
- colored pencils

Procedure

1. Form a team of three. Use a meterstick to measure a distance of 1 m from the wall. Mark this distance with a strip of masking tape. Use the meterstick again and measure 3-m, 6-m, and 9-m distances from the wall. Mark each distance with a strip of tape.
2. Darken the room.
3. Have one person in your group be the observer. Two other people, the light holders, will hold either the penlight or flashlight at different distances. The observer will **observe** and **compare** the brightness of each light by ranking the brightness of each light on a scale of 1–5, with 1 being the brightest and 5 being the dimmest. **Record** the data in your ***Science Journal***.
4. Have the light holders stand side by side on the 1-m mark. When the observer says "light on," the light holders should turn on their lights at the same time, shining them toward the observer.
5. The observer should look at the lights just long enough to compare their brightness and rank the brightness of each light.
6. Repeat Steps 3–5, having the lights held at 3 m, 6 m, and 9 m from the wall.
7. Repeat Steps 3–6, but using different students as observer/recorder and as light holders.

Analyze Results

Plot your results in a bar **graph**. Plot the comparative brightness of each light against its distance from the observer. Use a different colored pencil for each type of light.

Create Explanations

1. How can you model the brightness of stars?
2. Which light appeared brighter in each trial? Which light was actually brighter in each trial?
3. Which trial(s) modeled absolute brightness? Which trial(s) modeled apparent brightness? Explain.
4. Predict what results you would obtain if the lights were held 12 m from the wall.

Other constellations are visible year-round. Due to Earth's rotation, these constellations appear to circle the North and South Poles. They are called **circumpolar constellations**. In the northern hemisphere, circumpolar constellations include Ursa Major (the Big Bear), Ursa Minor (the Small Bear), and Draco (the Dragon). The Big Dipper, which is an **asterism**, or pattern of easily recognizable stars, is part of Ursa Major. In the southern hemisphere, circumpolar constellations include the Southern Cross, Centaurus, and Carnia.

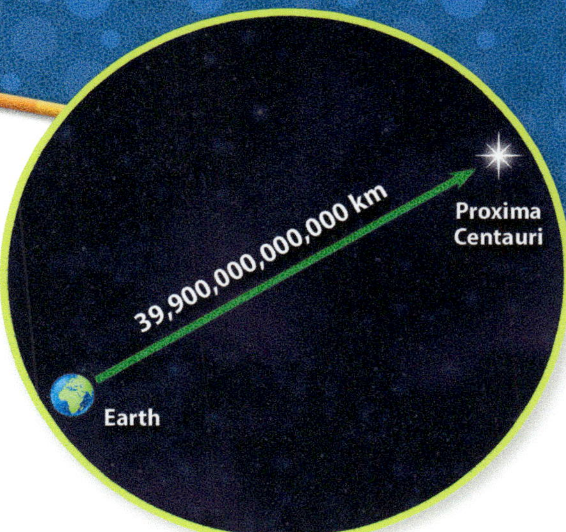

Proxima Centauri is about 4.2 ly from Earth.

 How many kilometers are there in one light-year?

How far away are the stars? Earlier, you learned about a unit of distance called an astronomical unit (AU). One AU is equal to about 150 million km. An AU is convenient for describing distances within the Solar System. But the vast distances of interstellar space are usually given in light-years. A light-year (ly) is the distance light travels in one year. It is equivalent to 63,240 AU or about 9,500,000,000,000 km. The closest star to Earth (other than our Sun) is Proxima Centauri; it is about 4.24 ly, or 268,000 AU, away.

Explore-a-Lab

Structured Inquiry

How can the same set of stars appear to change positions?

Work with a partner. Cut 10 pieces of string to different lengths, ranging between 5 cm and 30 cm. Tape a small, plastic foam ball to the end of each piece of string. Tape the free ends of the strings to a meterstick. Be sure to space the strings at various points along the meterstick. Have your partner hold the meterstick so that the foam balls hang freely. Imagine that each ball is a star. Close one eye and look at the hanging balls. Sketch a pattern that the balls make. Repeat this process but switch places with your partner. Stand in the same place and hold the meterstick in the same way as your partner did, while your partner looks at the hanging balls from a different angle and makes a sketch. Compare your sketches. What can you conclude?

Math in Science

The bright star in Orion is called Betelgeuse, which is about 650 ly from Earth. How many kilometers is Betelgeuse from Earth? How many AU away from Earth is it?

Scripture Spotlight

Read Job 9:9. What constellations are mentioned?

Star Magnitude Explain

If you look at a star map showing the constellation Orion, you will notice that the outline of the constellation is made of only a few stars. Actually, a constellation covers a large region of the sky and may contain millions of stars.

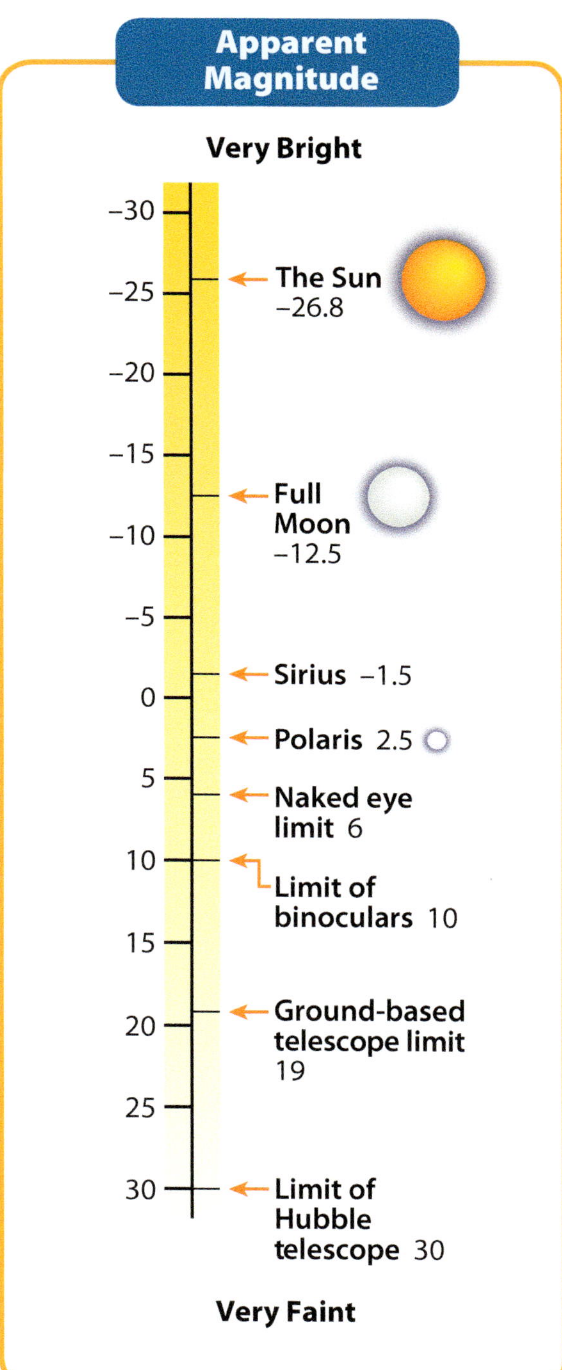

How many stars can you see when you look up at a clear night sky? Hundreds? Thousands? Millions? The number you see depends on the conditions. You see far more if you are away from city lights than you do if you are near cities. Scientists have estimated the number of stars in the Universe to be about 100 sextillion stars. Some scientists think that number should actually be tripled to account for cooler, dimmer stars that are not easily seen. Will we ever know how many stars there are?

On a clear night, what do you notice about the stars? Do they all look the same? The brightness of a star as it appears to an observer on Earth is called its **apparent magnitude**. In determining apparent magnitude, scientists needed a reference point. They chose the star Vega and assigned it a magnitude value of 0. Stars that appear brighter than Vega have apparent magnitudes of less than zero.

Apparent Magnitude Scale

How does the magnitude of the North Star, or Polaris, compare with the magnitude of the full Moon on this scale?

For example, Sirius, the brightest star other than our Sun, has an apparent magnitude of −1.5. This value indicates that Sirius has a brightness that is one and a half times the brightness of Vega. Stars that are very dim have apparent magnitudes of up to 28. Such stars are 28 times dimmer than Vega.

Apparent magnitude is useful in describing stars, but only as the stars appear from Earth. If a relatively dim star is closer to Earth than a relatively bright star that is farther away, it may appear to be the brighter star. To account for the decrease in brightness that accompanies an increase in distance from Earth, scientists describe the brightness of stars also in terms of absolute magnitude. **Absolute magnitude** describes how bright a star would be if all the stars were the same distance from Earth. The Sun's absolute magnitude is +4.8, while its apparent magnitude is −26.8. How is this possible?

Explore-a-Lab

Structured Inquiry

How can you estimate the number of stars seen with an unaided eye?

Use a bathroom tissue tube or paper towel tube to make a "telescope" to observe stars. Measure the diameter of the tube. Cut its length so that it is three times its diameter. Face each of the four compass directions. Hold the tube three-quarters of the way up from the horizon in each direction and count the number of stars you see through the tube. Hold the tube halfway up from the horizon and repeat the count. Repeat the procedure again with the tube pointed one-third of the way up. You can use a protractor to determine the angle or make your best estimate. Repeat your observations for the other directions. Add up the number of stars for all 12 sightings. You would need 144 tubes to cover the whole sky. Therefore, you have observed one-twelfth of the sky. Multiply the number of stars you counted by 12 to estimate the total number of stars in the sky. Add up and compare the three measurements in each direction. Why do you see more stars in certain directions?

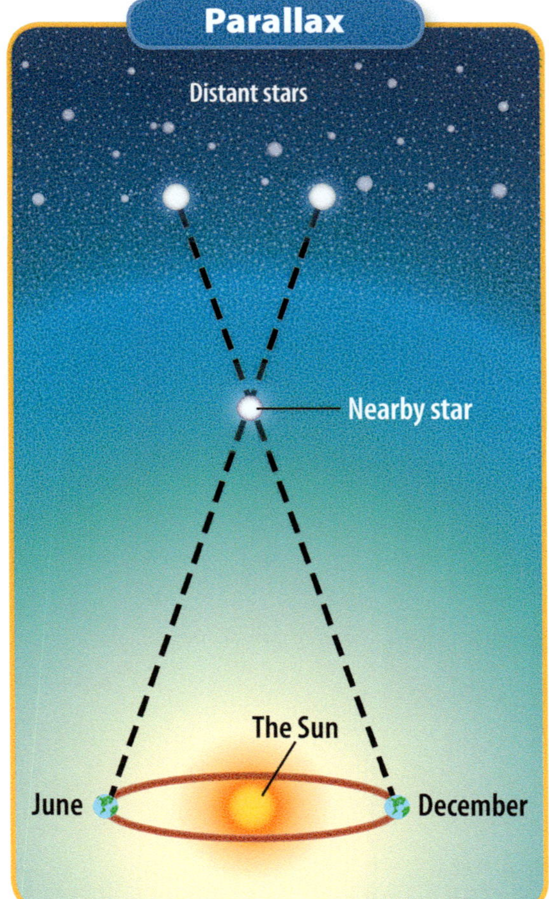

Parallax Explain

How can anyone figure out how far a star is from Earth? Clearly, we cannot use measuring tools such as a tape measure, so scientists use techniques such as parallax. **Parallax** is the apparent change in position of an object when viewed from different locations, or points of view. To understand how parallax works, study the illustration on this page. As Earth revolves around the Sun, an observer's point of view changes. The nearer stars will appear to have moved more compared with the more distant stars. Stars that are closer to Earth tend to have larger shifts, and stars that are farther away tend to have smaller shifts. Very distant stars cannot be measured using parallax because their shift in position is too small to observe. How does using parallax give us an idea of the distance to stars?

Why does the small shift of very distant stars make parallax an ineffective tool to measure their distance? How do you think scientists measure the distances to very distant stars?

As Earth revolves around the Sun, stars that are closer to Earth appear to move relative to more distant stars.

Is the measured star farther from Earth during the winter or the summer? Explain.

Explore-a-Lab

Structured Inquiry

How does the parallax of stars at different distances compare?

Look at the object that your teacher displays. Closing your left eye, hold one arm out in front of you. Use your thumb to block your view of the object. Keeping your arm and thumb still, open your left eye and close your right eye. Observe the distance that your thumb appears to move when you switched eyes. Draw a sketch of your observations. Now hold your thumb close to the end of your nose. Close your left eye and use your thumb to block the view of the same object. Open your left eye and close your right eye. Observe the distance that your thumb appears to move when you switched eyes. Draw a sketch of your observations. Did your thumb have the greatest shift when it was closer to or farther from your nose? Why?

The Sun—An Average Star

Core	• Central part of the Sun • Area in which nuclear fusion occurs • Temperates reach 15,000,000 °C
Photosphere	• Surface of the Sun • Produces the visible light we see
Chromosphere	• Layer of hot gas beyond the photosphere • Bright red halo extending beyond the photosphere • Red color due to presence of hydrogen gas
Prominences	• Flames of gas that shoot out from the chromosphere • Extend hundreds of thousands of kilometers from the surface
Corona	• Thin transparent zone beyond the chromosphere • Made up of hot gas and plasma

Although the Sun may appear to have a solid surface, it is made completely of gases. These gases are arranged in layers. You know from a different chapter that the outermost layer of the Sun is called the *corona*. This name comes from its crown-like appearance when a solar eclipse occurs. Particles from the corona travel far out into space, reaching even Earth and beyond. How far into space do you think these solar particles travel? The table describes the other layers that make up the Sun, beginning with the core, which is the source of all the Sun's energy.

Astronomers study the Sun to learn about what makes it and other stars work. It is the only star we can study at close range. The Sun is mostly composed of hydrogen, about 75% by mass, and around 25% helium. Like all stars, the Sun generates its energy from nuclear reactions in its interior. This process, called **fusion**, occurs when nuclei of lightweight elements, such as hydrogen, combine to form heavier elements, such as helium. Fusion releases a tremendous amount of energy in the form of heat and light. It can occur only at the extreme temperatures and pressures found in the cores of stars. The process is shown below. How does fusion differ from the *fission* reaction in a nuclear reactor?

Depending upon the age and mass of a star, the energy output of stars may result from the fusion of hydrogen nuclei (or protons), from helium fusion, or from the carbon cycle. This latter process is more common in stars that are much more massive than our Sun. For brief periods in their life cycles, some stars fuse elements heavier than carbon, but the fusion of elements heavier than iron consumes energy rather than liberating it. What do many scientists believe that these processes eventually do to a star?

During fusion, stars fuse hydrogen and helium to create heavier elements.

Faith Connection

For a planet to support life, many factors must exist within very precise parameters. Our Sun is just the right kind of parent star. If our Sun had 20% more mass, Earth would be hotter than Venus. If it had 20% less mass, Earth would be colder than Mars. Either way, life on Earth would be impossible. This is evidence of a Creator with a precise design.

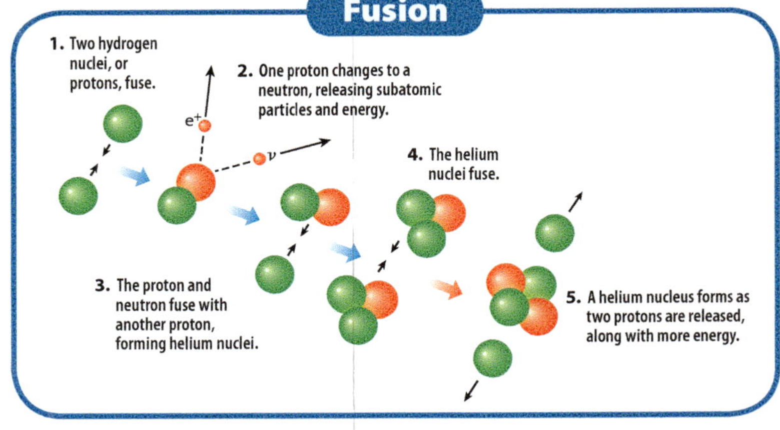

H-R Diagram Explain

In 1911, a Danish astronomer named Ejnar Hertzsprung created a graph that compared the brightness and temperatures of stars. In 1913, an American astronomer named Henry Norris Russell made some similar graphs. Their graphs showed that the hottest stars are often the brightest, and the coolest stars are often the dimmest. The combination of their ideas is now called the *Hertzsprung-Russell (H-R) diagram*. The H-R diagram is an essential tool for astronomers. It shows the relationship between a star's temperature and its absolute magnitude.

Most stars, including the Sun, are represented in a region of the diagram that extends from the upper left to the lower right. This region is called the **main sequence**. The stars in the upper left tend to be hot, bright, and blue. The stars in the lower right tend to be cool, dim, and red. The Sun is a yellow star with medium temperature and brightness. Where do you find it on the H-R diagram?

About 10% of stars fall outside the main sequence. They might be hot, dim, white dwarfs or bright, cool, red giants. The largest red giants, the supergiants, are about 70 times more massive than the Sun. The radius of a large supergiant can be about 7 AU. That is seven times the distance from Earth to the Sun, or about halfway between the orbits of Jupiter and Saturn!

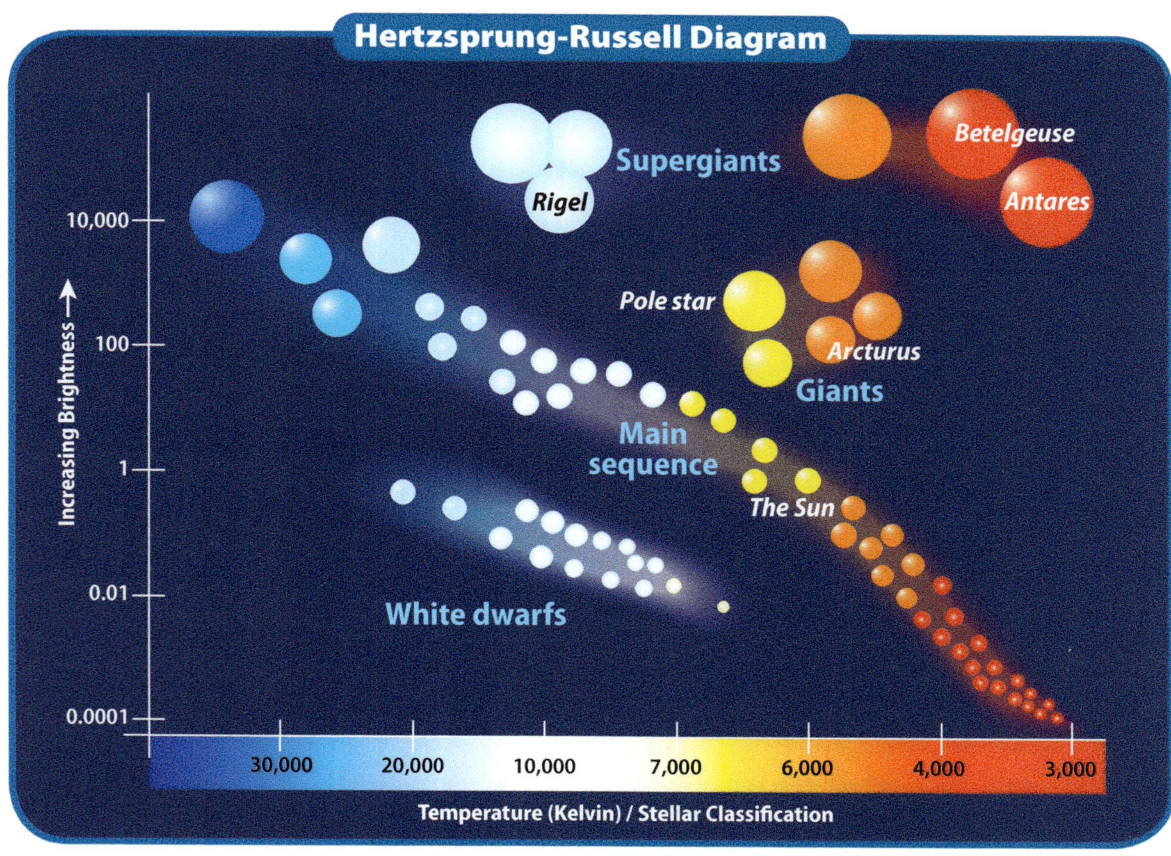

Lesson Activity

Use the Internet or other resources to learn the name, temperature, and absolute magnitude of a star. Write the information down. Cut a representation of the star from a sheet of construction paper. On the cutout, write the name, temperature, and brightness of the star. Use tape to attach the star cutout at the proper location on the large H-R diagram posted in the classroom. Be ready to explain to the class why you placed your star in that particular location. Once all the stars are in place, sketch the H-R diagram on a sheet of drawing paper or graph paper, including the axes and the stars' names.

 **How are the stars distributed on the H-R diagram?**

Life Cycles of Stars

The H-R diagram shows a star's brightness and temperature, but it is useful in another way as well. It also describes how stars change over time. Many scientists believe that stars follow a sequence of events throughout their existence. These events explain the changes in a star's density, color, and temperature throughout its life cycle.

We know that God created the Sun and the other stars. How did He do it? Scientists believe that stars begin their life cycle as a cloud of dust and gas called a *nebula*. Gravity causes the cloud to contract. Eventually, hot balls of gases might form from the nebula. Each ball of gas will become a star when temperatures in its core become hot enough for fusion to begin. Once fusion begins, the remainder of a star's life cycle depends largely on its mass.

Many Creation scientists argue against this idea of the birth of a star. They say that gases tend to expand, not contract, based on the ideal gas laws. It would be more logical to assume that an intelligent agent triggered the gravitational collapse leading to star formation.

Scripture Spotlight

1 Corinthians 15:40–41 discusses how God's Creation glorifies Him. How does this passage relate to what you are learning about stars?

The image of these pillars of gas in the Eagle Nebula was taken by the Hubble telescope in 1995.

Small and Medium Stars

Based on current theories, many scientists think that stars with masses similar to the Sun eventually use up all the hydrogen in their cores. The core contracts and its temperature increases. In response, the outer layers expand and cool, and the star becomes a **red giant**. In the red giant phase, stars fuse helium into carbon. When the helium is used up, the star becomes a **white dwarf**, in which its core becomes even denser and hotter. It no longer produces energy, and its outer layers drift into space. Scientists think that, eventually, the star will cool and no longer give off light, becoming a **black dwarf**.

Large and Massive Stars

How is the life cycle of a star that is more massive than the Sun different? What is it about these stars that causes their life cycles to be different? According to accepted theories, these stars burn more energy and use up hydrogen at a faster rate. They fuse heavier elements together in their cores. This type of star becomes a **supergiant**. When fusion stops, the outer layers explode as a brilliant **supernova**. Depending on the mass of the remaining core, the star becomes one of two things. It might become a very dense, very small **neutron star**. If a neutron star is spinning, it is called a pulsar. A **pulsar** is an object in space that spins very rapidly. As a result of this extremely rapid spinning, a pulsar sends out a beam of radiation. Radio telescopes on Earth can detect the pulses, or rapid clicks, generated by this pulsating beam of radiation.

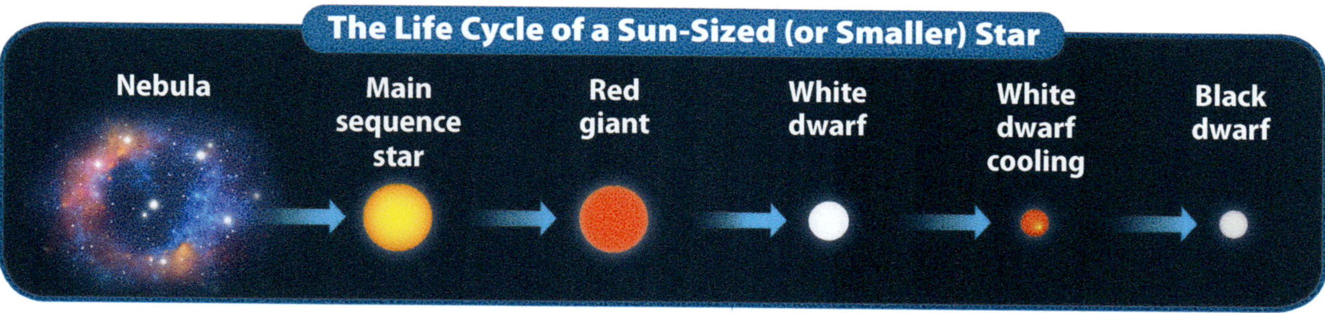

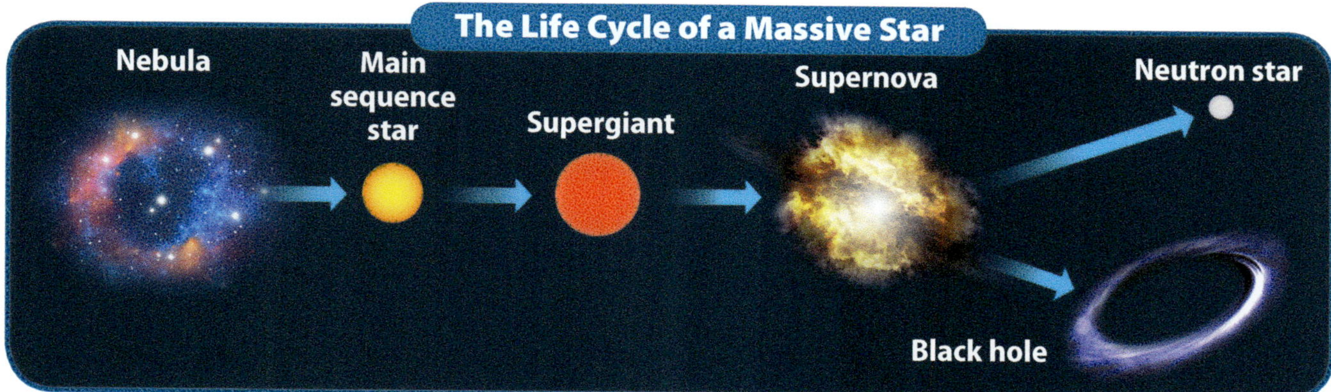

If the remains of a supernova are extraordinarily massive, however, they do not form a pulsar. Rather, they collapse to form a black hole.

A **black hole** is theorized to be an area of space with an incredibly strong gravitational pull. Not even light can escape the gravity of a black hole. Why can't scientists observe a black hole directly? Scientists believe there are different sizes of black holes. Some can be as small in diameter as one atom. Yet these tiny black holes can have the same mass as a mountain! Other black holes, called supermassive black holes, are estimated to have a mass of more than one million Suns put together! Scientists believe there is a supermassive black hole in every large galaxy. In fact, scientists have even named the supermassive black hole in the center of our galaxy Sagittarius A. How do scientists discover black holes?

This image is an artist's rendition of a black hole and the objects around it.

Black holes have been connected to another kind of object found in space—quasars. A **quasar** is a source of light that is extremely far away in space. In fact, quasars are the most distant objects discovered by astronomers. Quasars are billions of light-years away.

Most quasars are so massive that they are larger than our entire Solar System. A quasar is also one of the most powerful energy sources in the Universe. Most scientists think that supermassive black holes are the "engines" that power quasars.

Concept Check Assess/Reflect

Summary: How can we describe stars? A constellation is an area of the sky where a recognizable pattern of stars can be seen. Today scientists recognize 88 constellations. Stars are incredibly massive balls of hot gases that emit huge amounts of energy through fusion. Stars come in a vast array of sizes, colors, temperatures, and levels of brightness. Scientists study stars' magnitudes and temperatures. They calculate distances to stars using a technique called parallax. A relationship between the magnitude and temperature of a star is shown in an H-R diagram. Scientists believe that stars follow a life cycle, forming as a nebula and following paths determined by their masses.

1. How are circumpolar constellations different from other constellations?
2. Would constellations appear the same from any position in a galaxy? Explain.
3. Why do astronomers use absolute magnitude rather than apparent magnitude to plot stars?
4. How does the Sun compare with other stars in terms of brightness and size?
5. Create an illustration to show the life cycle of a medium-sized star and a massive star.

Essential Question

What Are Galaxies?

Objectives

- Compare and contrast the different types of galaxies.
- Identify four space telescopes that scientists use to study galaxies.
- Identify the location of the Solar System in the Milky Way Galaxy.
- Describe the vastness of the Universe.

Vocabulary

galaxy
Milky Way Galaxy
spiral galaxy
barred spiral galaxy
elliptical galaxy
irregular galaxy
Local Group

Have you ever wondered what is the most distant object in space that you could see with just your unaided eyes? Would it be a planet? Would it be a comet? Might it be a star? Actually, what you would see would be none of these. Rather, the most distant object that you could see with just your eyes is a giant structure that contains billions of stars.

Discovering Galaxies Explain

The most distant object in space that you can see with your unaided eyes is the Andromeda Galaxy. A **galaxy** is a vast group of stars, dust, and gas held together by gravity. Scientists estimate that there are about 100 billion galaxies in the Universe. Smaller galaxies may contain tens of millions of stars. Larger galaxies most likely contain trillions of stars. The galaxy that contains the Sun and our Solar System is the **Milky Way Galaxy**. The Milky Way most likely has about 100 billion stars. Why can't scientists be more accurate than this?

At one time, people thought that space was limited to the Milky Way. They thought that the Milky Way made up the entire Universe. The discovery of galaxies outside the Milky Way is relatively new to science. In the 1920s, astronomer Edwin Hubble used a large telescope to view other galaxies. What did this scientist observe, and what did he help prove about the Milky Way?

You can see the Milky Way Galaxy from Earth. People once thought that there was nothing in space beyond the Milky Way.

Types of Galaxies Explain

You've learned that stars have various properties. Some are large and relatively cool. Others are relatively small and bright. Galaxies have various properties, too. They are often classified according to shape. What are the main galaxy shapes?

Spiral Galaxy

A **spiral galaxy** has arms that spiral outward from a central bulge, similar to the arms of a pinwheel. Stars, dust, and gas make up the arms. Spiral galaxies tend to consist of a mixture of young and old stars. In the past, the Milky Way Galaxy was considered a spiral galaxy; however, it is now considered to be a barred spiral galaxy.

Barred Spiral Galaxy

A **barred spiral galaxy** has arms that spiral outward from a large central bar. Barred spiral galaxies have a mixture of both young and old stars.

Elliptical Galaxy

An **elliptical galaxy** is shaped like a large ellipse. Some elliptical galaxies resemble footballs, while others are rounder, resembling basketballs. They exhibit a vast range of sizes. Elliptical galaxies have little gas and dust between the stars. Without this gas and dust, many scientists believe that new stars cannot form.

Irregular Galaxy

An **irregular galaxy** has an irregular shape that does not fit easily into other classification schemes. Usually, it is made mostly of younger stars. This kind of galaxy may have an odd shape because it recently collided with another galaxy or because it has not yet developed a structured shape. Do you think an irregular galaxy will keep its shape or change? Why? The Magellanic Clouds are an example of an irregular galaxy. This is also one of the closest galaxies to our own.

Spiral galaxy: Andromeda Galaxy

Barred spiral galaxy: Milky Way Galaxy

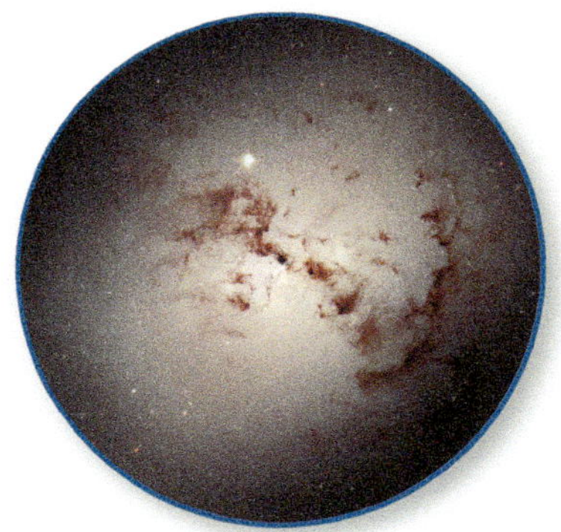

Elliptical galaxy: Galaxy NGC 1316

Irregular galaxy: Small Magellanic Cloud

Record your work for this inquiry. Your teacher may also assign the related Guided Inquiry.

Identifying Galaxies

How can the different types of galaxies be distinguished?

Materials

- colored pencils
- deep field survey image
- galaxies key
- hand lens

Procedure

1. Look over the galaxies key provided to learn how to recognize the different kinds of galaxies. Make a sketch of each galaxy in your ***Science Journal***.
2. Look carefully at the deep field survey image and count how many galaxies you can see. **Record** this number in your ***Science Journal***.
3. Find Galaxy A on the deep field survey image. Using the hand lens, carefully **observe** this galaxy. Record what kind of galaxy it is and draw a sketch of it that shows its shape and orientation. Use colored pencils to provide the correct color to your sketch.
4. Repeat Step 3 with the remaining nine galaxies labeled on the deep field survey image.

Create Explanations

1. How can the different types of galaxies be distinguished?
2. In looking at the overall deep field survey image, what seems to be the most common type of galaxy?
3. Which type of galaxy was the easiest to identify? Why?
4. Which type of galaxy was the most difficult to identify? Why?

Explore-a-Lab

Guided Inquiry

 **How can you make a spiral galaxy?**

Use a hole punch to punch out pieces of construction paper to represent stars. Fill a large bowl halfway with water and pour the stars into the water. How can you make the stars form a spiral galaxy? How did your actions model a spiral galaxy?

Studying Galaxies Explain

Why do scientists study galaxies? What tools do they use? What do they hope to learn? The large telescope that Edwin Hubble used to observe space was an Earth-based telescope, located on top of Mt. Palomar in California. The high elevation of the mountain allowed the telescope to avoid light pollution from nearby cities, rain clouds, and some distortion from the atmosphere. Using this telescope, Hubble and other scientists learned about the galaxies they could see. However, the Earth-based telescopes were unable to show the detail of galaxies that scientists needed. A better tool was needed.

Objects in space emit light that is a mixture of visible light and light at other wavelengths on the electromagnetic spectrum. The wavelengths of light that objects emit depend on the temperature and composition of the object. Scientists use telescopes that detect different wavelengths of energy to gather different kinds of information about stars and galaxies, such as what they are made of and their apparent ages.

Space telescopes enable scientists to observe parts of the Universe that cannot be seen by the unaided eye. By using different types of telescopes, scientists can deduce the temperatures of the stars that make up a galaxy. Scientists have also discovered that the Universe appears to be expanding, an idea based on the observation that galaxies appear to be moving away from each other. (This is discussed more in the next lesson.) Because space is a vacuum and there is no air in space, space telescopes can detect light and other energy from the most distant objects without distortion. How does this help scientists?

Scientists are using several space telescopes to improve their ability to make better observations of galaxies to learn more about them. The table on the next page describes four of these telescopes.

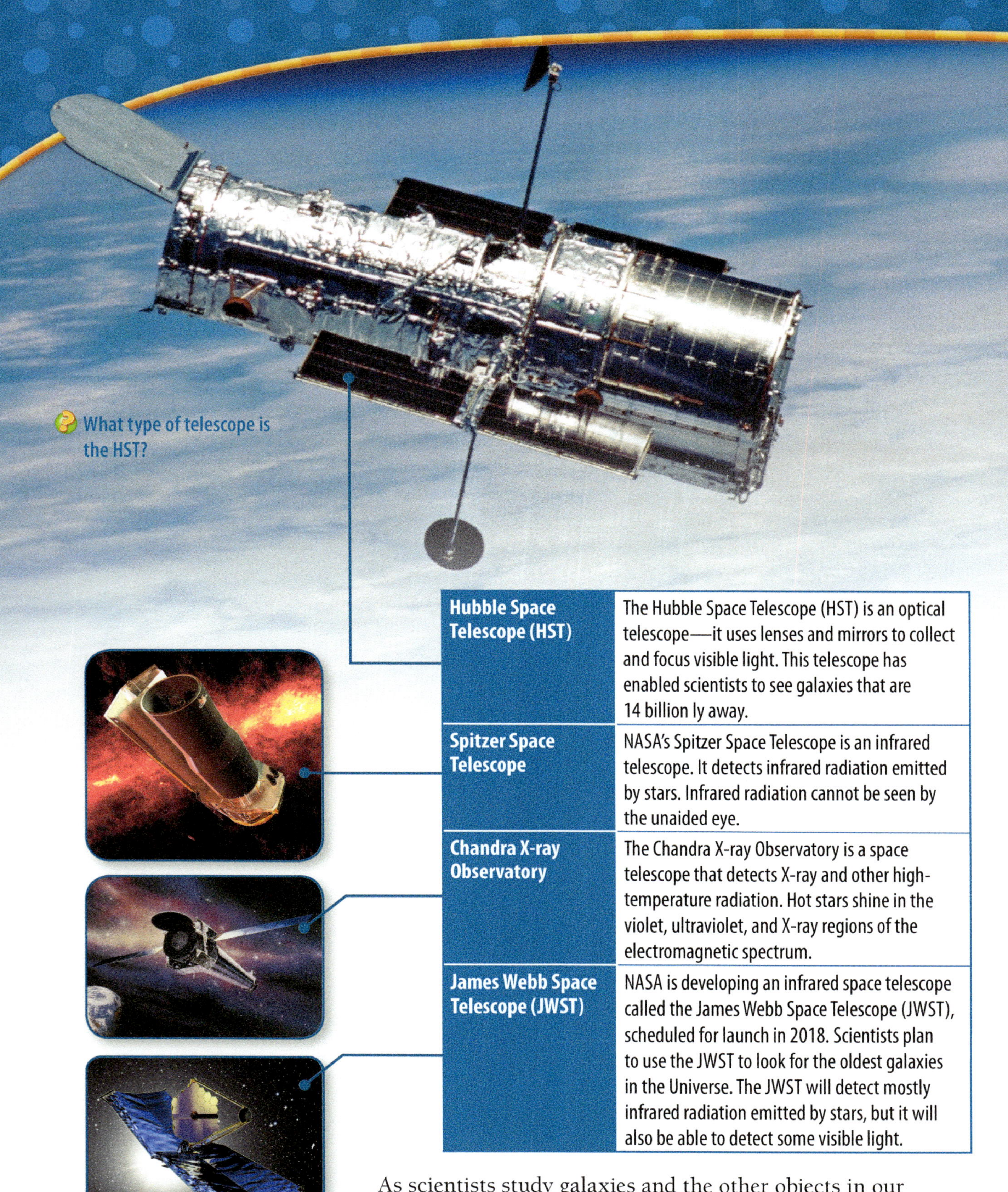

What type of telescope is the HST?

Hubble Space Telescope (HST)	The Hubble Space Telescope (HST) is an optical telescope—it uses lenses and mirrors to collect and focus visible light. This telescope has enabled scientists to see galaxies that are 14 billion ly away.
Spitzer Space Telescope	NASA's Spitzer Space Telescope is an infrared telescope. It detects infrared radiation emitted by stars. Infrared radiation cannot be seen by the unaided eye.
Chandra X-ray Observatory	The Chandra X-ray Observatory is a space telescope that detects X-ray and other high-temperature radiation. Hot stars shine in the violet, ultraviolet, and X-ray regions of the electromagnetic spectrum.
James Webb Space Telescope (JWST)	NASA is developing an infrared space telescope called the James Webb Space Telescope (JWST), scheduled for launch in 2018. Scientists plan to use the JWST to look for the oldest galaxies in the Universe. The JWST will detect mostly infrared radiation emitted by stars, but it will also be able to detect some visible light.

As scientists study galaxies and the other objects in our Universe, they are able to better understand God's vast Creation. What do you think scientists will learn about God's Universe in the years to come? What questions should they ask? You could be one of the scientists asking the questions and searching for the anwers.

Your Place in Space

The images on the next two pages show the Milky Way Galaxy. Our galaxy is about 100,000 ly across. It has a pinwheel shape and a central bulge. Scientists theorize that a black hole exists in the center of the bulge. Our Solar System is located about 27,000 ly from the central bulge. Just as Earth revolves around the Sun, the Sun revolves around the center of the Milky Way. The Sun and the Solar System travel around the center of the Milky Way at about 792,000 km/hr. It takes our Solar System about 225 million years to complete one revolution around the center of the Milky Way.

Galaxies are usually millions of light-years apart. However, they do tend to group together. The Milky Way belongs to a cluster of galaxies called the **Local Group**. Why do you think this cluster of galaxies is so named? The Local Group includes more than 50 galaxies, most of which are small and dim. The closest galaxies to our own are the Magellanic Clouds. They are about 160,000 to 200,000 ly away and are visible only from Earth's southern hemisphere. Why can't viewers in the northern hemisphere see the Magellanic Clouds?

Faith Connection

Earth seems to be located in a special place in our galaxy. It is not too near the center of the galaxy where many dangers exist—deadly radiation, possible collisions, and exploding supernovas. Not only is our location on one of the galaxy's spiral arms, but it is also a perfect vantage point for observing and studying the rest of the cosmos! This is evidence of our Creator's desire for us to be safe and explore our Universe.

Math in Science

Imagine taking a trip from Earth to the star Betelgeuse. When you return to Earth, you realize you left your suitcase there! If Betelgeuse is 650 ly away, how many light-years will you have traveled once you get back home with your suitcase? How many kilometers will you have traveled through space?

The Vast Universe Explain

Have you ever considered how small we are compared with the Universe? The Hubble Space Telescope (HST) has helped astronomers view the Universe far beyond Earth. We now have a glimpse of how vast a Creation God has made.

From our Solar System's location in the Milky Way spiral, the HST can look out into deep space. Astronomers can see that the Milky Way Galaxy is just one of many galaxies found within our Local Group.

Beyond the Magellanic Clouds, our closest galaxy neighbors are Andromeda (M31) and M33. Andromeda (M31) is about 2.5 million ly away, while M33 is about 3 million ly away. Even farther beyond these galaxies are still more galaxies. The HST has helped astronomers to gain a view of countless numbers of galaxies, like the sands of the shore, in deep space. How were all these galaxies created?

Our Solar System is found on the Orion arm of the Milky Way Galaxy.

What is the reason for the name "Milky Way Galaxy"?

Concept Check Assess/Reflect

Summary: What are galaxies? A galaxy is a vast group of stars, dust, and gas held together by gravity. Our Solar System is part of the Milky Way Galaxy, which contains about 100 billion stars. The Milky Way Galaxy belongs to a cluster of galaxies known as the Local Group. Galaxies are classified according to their shape and include spiral, barred spiral, elliptical, and irregular galaxies. Those classified as irregular galaxies do not have a shape that can easily fit into one of the other classifications. Scientists use space telescopes, such as the HST, Spitzer Space Telescope, and the Chandra X-ray Observatory, to study galaxies. Such observations of galaxies have shown that the Universe is expanding.

1. Describe Earth's location in the Milky Way Galaxy.
2. Explain why observers can see the spiral arms of the Andromeda Galaxy more clearly than the spiral arms of the Milky Way Galaxy.
3. What can scientists learn by studying galaxies?
4. Compare and contrast the different types of galaxies.

Lesson 3

Essential Question

How Did the Universe Begin?

Objectives

- Distinguish between historical and experimental science.
- Describe the big bang theory.
- Identify the evidence that supports the big bang theory.
- Identify the evidence that contradicts the big bang theory.
- Arrange the stages in the formation of the Solar System.

Vocabulary

historical science
experimental science
big bang theory
redshift
blueshift
dark matter
protoplanet

The Universe is all the matter, energy, space, and time that God created and sustains for humans to inhabit, discover, and observe. Humans and other living things are part of the Universe. Why did God create the Universe? God acts in the Universe, but He is not limited by it. Why is this? The Bible teaches that He can act outside of time and space and that He surpasses the Universe. What do you think this means? What effect does this have on us?

Historical and Experimental Science Explain

The essential question of this lesson is one that is asked by many people. Questions such as this are part of historical science. Before we can answer this question, we need to understand the difference between experimental and historical science.

Historical science studies past events by examining their results. We may be limited in our study of historical events by how long ago an event happened, whether anyone was there to see it happen, and whether observers left any kind of record of the event. We may also be limited in our study of historical science because the event is too large for us to control. For example, no one can experiment with a worldwide flood or the impact of an asteroid with Earth. These are questions of history, and, even with eyewitnesses' accounts of past events, we cannot know every detail with certainty.

Experimental science, the science with which you are most familiar, deals with events in which the beginning conditions are known. Investigations of this type can be repeated to allow us to observe the event many times. What do historical and experimental science have in common? How are they different? How do they both help to explain the Universe to us?

This image from the Hubble telescope is the deepest look at the visible Universe and shows approximately 10,000 galaxies!

This is an edge-view representation of our Milky Way Galaxy, just one galaxy of thousands that make up the Universe.

Origin of the Universe

The origin of the Universe is a topic of historical science. We can't repeat it or perform an experiment to find out what happened. In the Bible, God has given us a record of what happened, though. It tells us that God made the Universe by speaking. He didn't give us all the details about Creation that we might want to know, but we can continue to study the things He made to look for clues. Can humans learn enough from studying the Universe to be certain of how He created it? Explain.

In the past, as people began to figure out scientific explanations for things, some reasoned that if a phenomenon had a "natural" cause, that must mean that God wasn't involved in the process. Sometimes human knowledge has been used to deny God's power. This is not logical. Speculation about how something might have happened does not mean that it happened without God. Remember, ideas about how an event might have happened do not mean that the event actually happened that way. Why is it important to keep these ideas in mind as we study scientific theories of how the Universe might have been created?

The Pulsating-Universe Theory

Several theories try to explain the origin of the Universe. One of these, the pulsating-universe theory, states that the Universe is eternal, so it doesn't have an origin. Over the ages, it slowly goes through a cycle of expansions and contractions similar to an accordion expanding and contracting.

Record your work for this inquiry. Your teacher may also assign the related Guided Inquiry.

The Scale of the Universe

How do scientists measure the vastness of the visible Universe?

Materials

- meterstick
- masking tape
- calculator
- grains of salt
- bead, small
- button, mediumsized
- ping-pong ball
- softball
- grapefruit
- basketball
- pizza tray, large, round

Procedure

1. The measurements in the data table in your ***Science Journal*** show the diameters of different objects in the Universe. You will be doing calculations with this data to determine the relative sizes of these objects.
2. Use the data table in your ***Science Journal*** to **record** your calculations. **Use the numbers** in the table to calculate how many times bigger the diameter of each object is compared with the previous object by dividing the diameter of the larger object by the diameter of the smaller object. The first calculation is done for you.
3. **Measure** and clear out a circle that is 10 m in diameter.
4. Imagine this area represents the Sun. Which object would you use to **model** Earth? Record your answer in the data table.
5. Now imagine that same area represents the Solar System. Which object would you use to model the Sun? Record your answer.
6. Repeat the activity for the remaining objects listed in the data table. Use the same 10-m-diameter circle to represent each new object.

Analyze Results

Which two objects had the greatest difference in diameters? The least difference in diameters? **Compare** your results with those of other students. What may have caused any differences?

Create Explanations

1. How do scientists measure the vastness of the visible Universe?
2. When we understand the vastness of the Universe, how does this help us have evidence regarding the origins of life?
3. Now suppose that the 10-m-diameter area represents the visible Universe. Is it possible to select an object to represent Earth within the Universe? Explain.
4. Were any of your results unexpected or surprising in any way? Explain.

The Big Bang Theory

Another theory is the big bang theory. Although it is the most well-known theory, it is not necessarily correct. According to the **big bang theory**, the Universe formed about 14 billion years ago when, in mere fractions of a second, it expanded from a tiny, hot, dense point to thousands of times the size of the Solar System.

In the first seconds following this expansion, many scientists think that the Universe was incredibly hot. By the time it was three seconds old, basic particles that make up atoms had formed. After about one to three minutes, protons and neutrons came together to form hydrogen.

Next came helium. According to the theory, as the Universe continued to expand and cool, matter began to clump together. The first stars and galaxies formed about 100 million to 300 million years after the big bang. About 8.7 billion years after the big bang, it is theorized that the Sun began fusing hydrogen into helium and became a star.

Several lines of evidence have been offered to support this theory. These include background radiation in space, the abundance of light elements (hydrogen and helium), and a phenomenon known as the redshift.

According to the big bang theory, the Universe was denser and hotter in the past. It has cooled as it has expanded.

Background Radiation

In 1964, scientists accidentally discovered radiation coming from all directions in space. Scientists proposed that this was cosmic background radiation left over from the big bang. According to this theory, radiation from the original explosion was distributed in every direction as the Universe expanded. As a result of the big bang, cosmic background radiation now fills all of space.

Hydrogen and Helium

The observation that hydrogen and helium make up about 98% of the matter in the Universe is also used to support the big bang theory. According to this theory, the Universe was originally incredibly hot and dense, so much so that none of the molecules or atoms that we know of could have existed. Very quickly, however, this matter expanded and cooled to the point where electrons, neutrons, and protons were able to condense. According to the big bang theory, these subatomic particles fused to form hydrogen and helium. Today these two elements make up most of our Universe, an observation that is consistent with the big bang theory.

Redshift

One important idea of the big bang theory is that the Universe has been expanding ever since the big bang. To understand how scientists have observed that the Universe is expanding, you must understand that, like sound waves, light waves can be compressed or stretched out. The shorter a light wave is, the bluer it appears; stretching it out moves it toward the red end of the spectrum. In the early twentieth century, scientist Edwin Hubble noticed that the farther away a galaxy, star, or other light source is in space, the more red it appears to be. Hubble called this color shift toward the red end of the light spectrum the

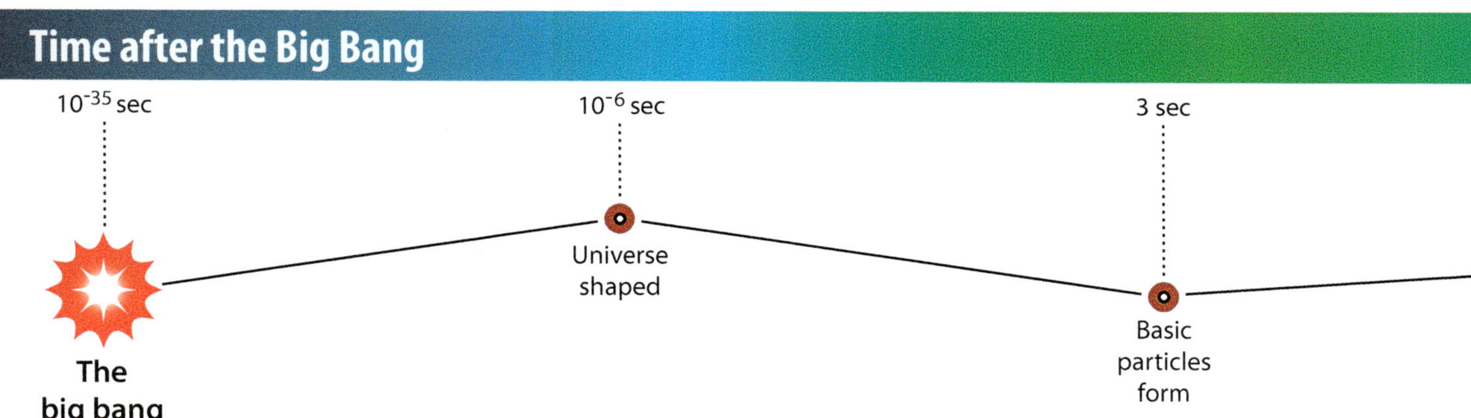

redshift. This characteristic of the galaxies suggested to Hubble that the galaxies were moving away from each other. If an object in space is moving toward the observer at a very high speed, the light from the object will appear bluer than it would otherwise. This effect is called a **blueshift**.

How could the redshift be explained? The solution Hubble and others came up with is that the Universe is expanding. Even though light travels very fast, it still takes a long time for it to travel across the vast distances of space. During the time light is traveling, the space through which it travels expands, stretching the light waves out, which shifts them toward the red end of the spectrum. The farther away a light source is, the longer it takes for the light to reach Earth. The longer the time the light has been traveling, the greater the expansion of space it has experienced and the more its wavelengths are stretched out toward red.

Although many scientists believe that this theory explains most of the observations made concerning the Universe, there is evidence that causes many scientists to question this current understanding.

Check out your *Science Journal* for a Structured Inquiry that explores the concept of a redshift. **Extend**

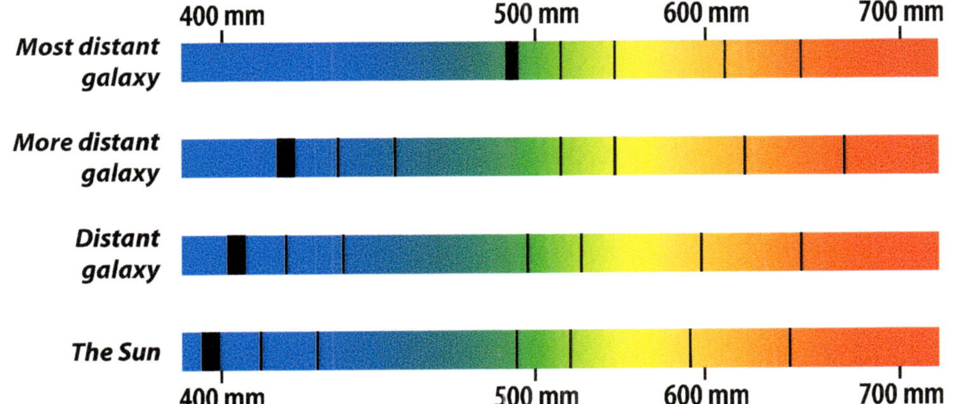

The bottom spectrum is from the Sun. The spectra above come from galaxies that are more and more distant from Earth. Notice the dark lines in the spectra. They shift farther to the red end of the spectra with increasing distance from Earth.

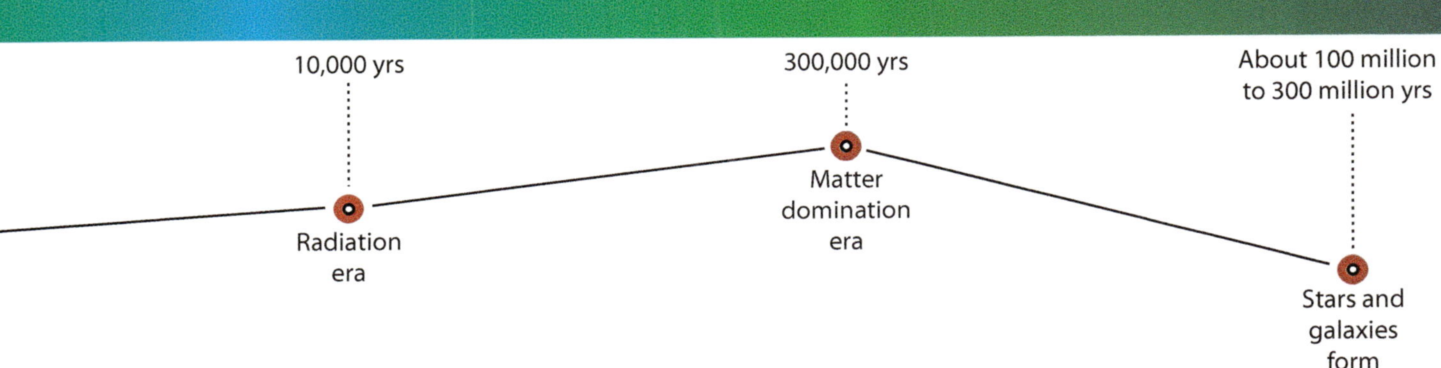

Arguments against the Theory

We should not forget that the theory of the expanding Universe is based on a specific interpretation of the redshift. This interpretation may not be correct. No scientist can be 100% certain. Ideas regarding the expanding Universe are likely to change as more evidence is discovered, so it is good to keep an open mind. Why is it a good idea to be open-minded with respect to these ideas about the Universe? One major agreement between the big bang theory and the Bible story of Creation is that both agree that the Universe is not eternal—it had a beginning. The Bible says the Universe was created by the eternal God who is not limited in space and time. Knowing that God exists allows Christians to believe in a sufficient cause to explain the Universe. Those who do not consider God's role in Creation struggle to suggest a sufficient cause for the big bang. They appeal to things such as the temporary appearance of energetic particles that are created in the vacuum of space, but no one knows whether these have the power to make a Universe as amazing as ours.

Scripture Spotlight

Read **Genesis 1:1–5**. How is Earth described at the beginning of the Creation story? Do you think there is any way that God could have described the Creation of the Universe in the Bible?

We have faith that God created the Universe. The Bible tells us some things about God's Creation of the Universe, but it does not answer all our questions.

While some people believe that God created the Universe using a process similar to the big bang, some scientific problems exist with this theory. For example, if the big bang theory is true, the Universe should contain more matter than scientists have been able to find. The "missing" matter has been called **dark matter**, because no one can see it. It is possible that someone will discover this missing matter, but so far no one has. Another contradiction comes from the current theory that the expansion of the Universe is speeding up. For the big bang theory to be true, this should not happen. In fact, until recently, scientists speculated that the expansion may slow down and eventually collapse. What can you learn from studying these theories?

It is important to understand the big bang theory so that we can understand alternative theories of the origin of the Universe and why people believe them. While scientific theories always leave room for uncertainty, one thing is certain: the Universe shows us how great God must be to have planned and created it.

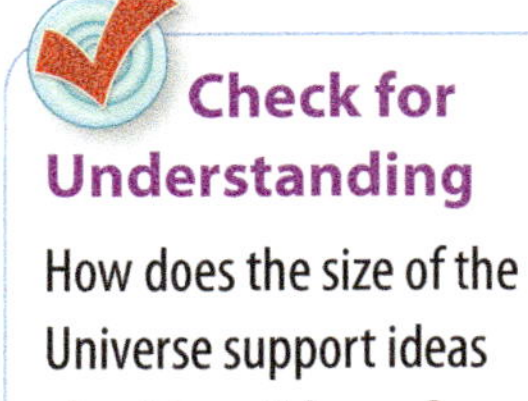

Check for Understanding

How does the size of the Universe support ideas about how it began?

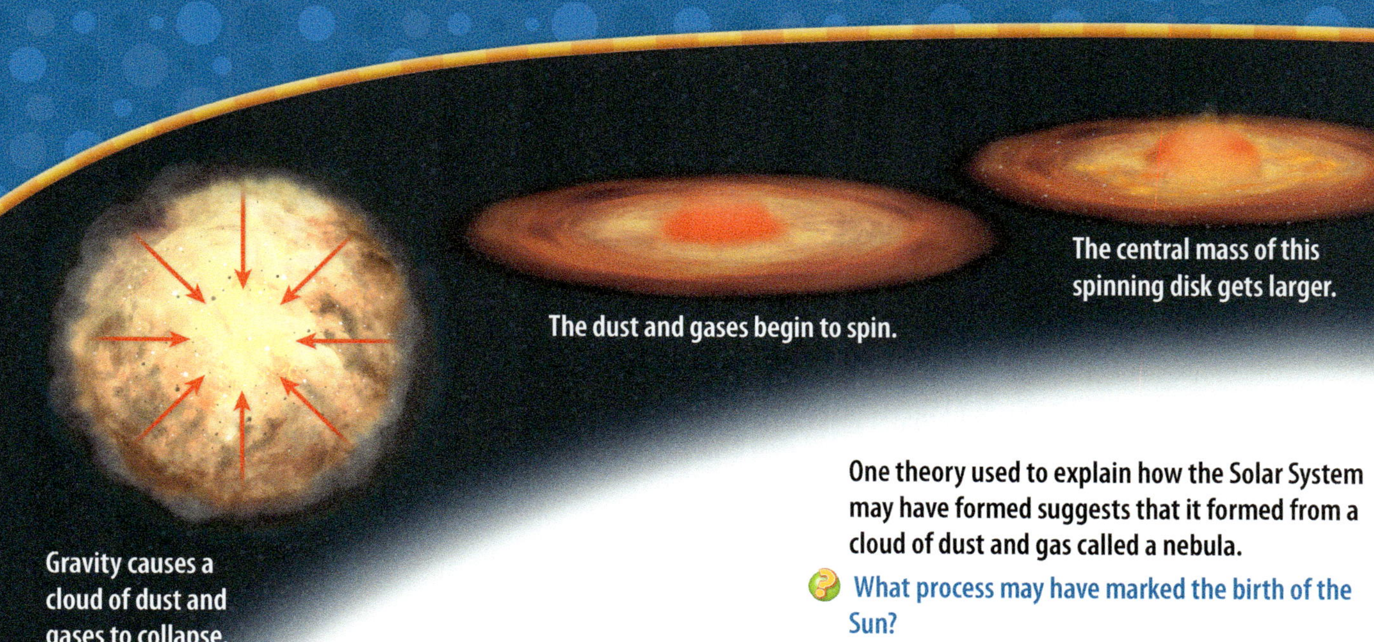

Gravity causes a cloud of dust and gases to collapse.

One theory used to explain how the Solar System may have formed suggests that it formed from a cloud of dust and gas called a nebula.

What process may have marked the birth of the Sun?

Formation of the Solar System Explain

Just as there are different ideas about the beginning of the Universe, there are also different ideas about where our Solar System came from. The Bible tells us in **Genesis 1:14–19** that on the fourth day God created ". . . lights in the firmament of the heaven to divide the day from the night; and let them be for signs, and for seasons, and for days, and years . . . and that He created the stars also." We don't know exactly how God created the Solar System, but we know that He did.

Some scientists, however, theorize that the Solar System formed about 4.6 billion years ago from a cloud of dust and gas called a nebula. These scientists believe that the nebula contained mainly light elements, such as hydrogen and helium, along with some dust and other elements. According to this theory, scientists believe that gravity caused the materials in the slowly rotating cloud to come together. It collapsed upon itself and flattened into a disk. As density increased, particularly at its center, temperatures in the cloud increased, as did the rate of rotation. It is believed that the center of the disk was the hottest, densest area. Here, matter clumped together to form a hot ball of gases. When temperatures inside this ball of hot gases became high enough for fusion to occur, a star was born—our Sun.

According to this theory, the vast majority of the mass in the Solar System came together to form the Sun. The remaining mass began to cool. It is suggested that gravity caused the matter to clump together into protoplanets. **Protoplanets** are relatively small, round objects in space that have some structured interior layers. Eventually, the protoplanets merged together to form the planets in our Solar System. According to this theory, what did the matter that did not gather together into planets form in our Solar System?

This theory goes on to suggest that temperatures close to the Sun were extremely hot, so light gases could not condense into planets.

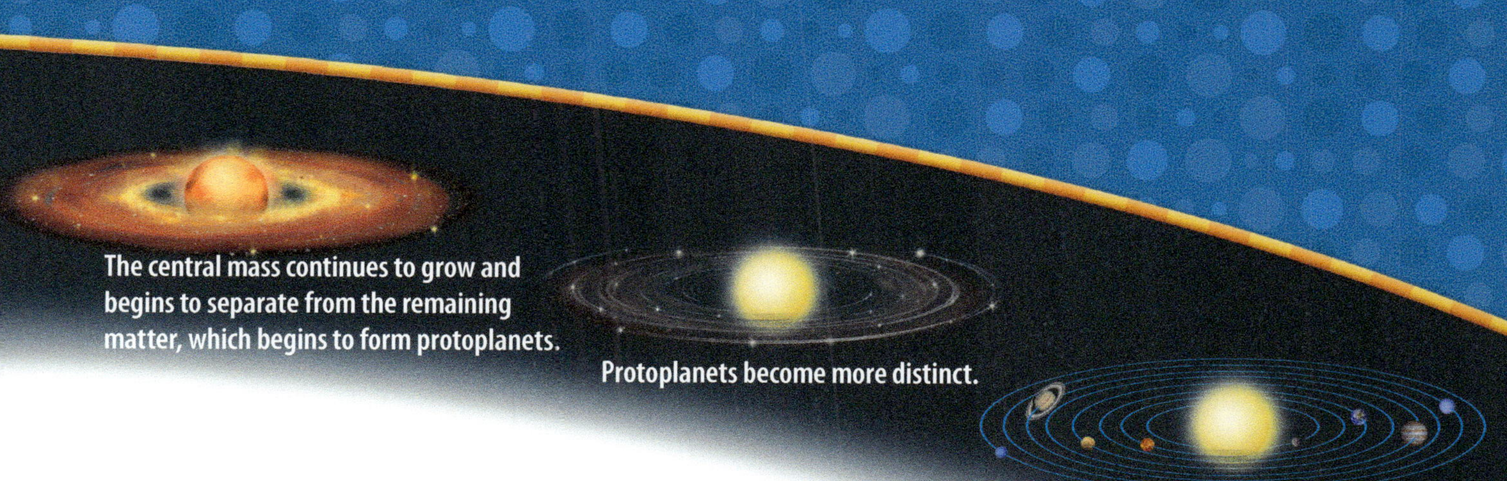

Only particles of dust made of heavy elements, such as silicon and iron, could condense close to the Sun, forming the rocky planets. Because the heavier elements make up only a small fraction of matter in the Universe, rocky planets—Mars, Earth, Venus, and Mercury—are smaller than those that formed from dust and gases that could condense into planets in cooler regions farther from the Sun.

Remember that studying the origin of the Solar System is historical science, so scientists can only offer theories. The Bible teaches us that God created the Solar System. An Intelligent Creator would have been able to plan the vast Solar System in a way that supports life. What reasons do you think God had for creating stars, planets, and other galaxies?

Concept Check Assess/Reflect

Summary: How did the Universe begin? Our Universe is vast and seemingly infinite. Many theories exist about how our Universe began. Historical science studies past events by examining their results. Experimental science deals with events in which the beginning conditions are known. Several theories try to explain the origin of the Universe, including the pulsating-universe theory and the big bang theory.

The big bang theory is one of the most widely accepted theories about how our Universe began its expansion into what it is now and about how it continues to expand. Some scientists believe that the redshift is one piece of evidence that supports the big bang theory. Other scientists believe that the interpretation of the scientific data does not explain some scientific questions, like dark matter. The Bible teaches that God is the Creator and Sustainer of our Universe.

1. What is the big bang theory?
2. What evidence seems to support the big bang theory?
3. What evidence contradicts the big bang theory?
4. According to the big bang theory, how might the composition of Earth differ if it formed beyond the orbit of Saturn?

People in Science

Extend

Called to Serve

Edwin Hubble's work in the field of astronomy enabled humankind to understand the world beyond the planet Earth. His studies identified galaxies outside of our own, expanding our knowledge and enabling us to explore all of God's Creation.

Get to Know Edwin Hubble

Edwin Hubble is known as the father of observational cosmology. He pioneered the study of distant stars and gave rise to the idea of the big bang theory, paving the way for modern astronomy.

Hubble's fascination with science and undiscovered worlds began as a child when he read *20,000 Leagues Under the Sea* and *From the Earth to the Moon* by Jules Verne. Growing up in Illinois, Hubble excelled at academics and sports. He studied mathematics and astronomy before accepting a Rhodes scholarship to Oxford, where he studied law. Realizing his first love was astronomy, Hubble returned to the University of Chicago in 1917 and completed a doctorate in astronomy.

While working at the Mount Wilson Observatory in California, Hubble took numerous photographs of Cepheid variables, determining they were outside our galaxy and identifying the existence of several other galaxies similar to our Milky Way. He created a classification system for the galaxies he observed. This classification system sorted the galaxies by content, distance, shape, and brightness. During these observations, Hubble noticed redshifts in the emission of light from the galaxies. He saw they were moving away from each other at a rate constant to the distance between them. These observations led to the formulation of Hubble's Law in 1929, which helped astronomers determine the age of the Universe and prove that the Universe was expanding.

Hubble's Law was a breakthrough discovery for the astronomy of that time. It changed the belief that the Universe was stationary and showed that the Universe itself was expanding. Einstein had originally predicted an expanding Universe about a decade prior to Hubble but gave in to the belief of the day and adjusted his calculation. Hubble's work proved that Einstein was right in the first place.

Concept Check

1. How did Hubble's work change people's beliefs about the Universe?
2. Explain Hubble's Law.

Science and **Technology**

Extend

Stellar Spectroscopy

You can use a prism to spilt sunlight into different colors, called the spectrum. You have learned that the shorter the wavelength, the greater the angle of refraction.

In much the same way, starlight can pass through a prism and form a spectrum. The analysis of the light observed from an object is called *spectroscopy*. Spectroscopy is the measure of the amount of light received at each wavelength. Stellar spectroscopy is the study of the spectra specific to starlight.

The importance of stellar spectroscopy comes from its use as a powerful tool in the study of astronomy. Once starlight passes through a prism, astronomers observe the wide range of wavelengths and examine the exact intensity of the light at each point on the spectrum. From this information, they can determine the physical features of a star, such as its temperature, velocity, and composition. Astronomers can infer the mass of a star, its distance from Earth, and many other pieces of information.

Color- enhanced spectrum of a star

Three basic types of spectra were identified by German physicist Gustav Kirchoff in the mid-1800s: continuous, emission line, and absorption line. His investigations on the properties of spectra determined that the three kinds of spectra are produced under different physical conditions. Kirchoff's observations led him to develop three laws for radiation, beginning the field of chemical analysis of stars.

Stellar spectroscopy enables astronomers to use the information to learn more about the stars they observe and classify them according to distinct features in their spectra. This opens up vast amounts of knowledge about the Universe far beyond our planet.

Concept Check

1. How do stellar spectroscopy and spectroscopy differ?
2. What types of information can be learned from stellar spectroscopy?

Chapter 10 Review Assess/Reflect

Study Guide

Lesson 1

1. The constellation Orion, the hunter, contains two of the brightest stars in the sky: Betelgeuse and Rigel. Orion can be seen from both hemispheres and is most visible from January to March in the northern hemisphere. Ursa Major, Ursa Minor, and Draco can be seen year-round as they appear to circle the North Pole. The Southern Cross, Centaurus, and Carnia can be seen year-round circling the South Pole.
2. Apparent magnitude is the brightness of a star as it appears on Earth. The absolute magnitude describes how bright a star would be if all the stars were the same distance from Earth.
3. Scientists use parallax to measure the distance between Earth and nearby stars by using the apparent change in position of a star when viewed from different locations.
4. The Sun, our nearest star, is made completely of gases arranged in several layers. It is a medium-sized, medium-aged star in comparison two other stars in the Universe.
5. Factors that affect a star's life cycle are its size and mass.

Lesson 2

1. A spiral galaxy has arms that spiral outward from a central bulge and is composed of a mixture of young and old stars. A barred spiral galaxy has arms that spiral outward from a large, central bar and has a mixture of young and old stars. An elliptical galaxy is shaped like an ellipse and contains little gas and dust, which prevents the formation of new stars. An irregular galaxy does not have a shape that is easily classified and is usually made up of younger stars.
2. The Hubble Space Telescope is an optical telescope that has enabled scientists to see galaxies that are 14 billion ly away. The Spitzer Space Telescope is an infrared space telescope that detects infrared radiation. The Chandra X-ray Observatory is a space telescope that detects X-ray and other high-temperature radiation. The James Webb Space Telescope is an infrared space telescope that is in development by NASA. Scientists will use it to look for the oldest galaxies in the Universe.
3. Our Solar System is in one arm of the Milky Way Galaxy. It is about 27,000 ly away from the central bulge.
4. More than 100 billion galaxies make up the Universe. Astronomers have identified galaxies about 14 billion ly away.

Lesson 3

1. Using experimental science, we can investigate and observe an event. These events are repeatable. We can study the results of past events using historical science, but we cannot repeat the events.
2. The big bang theory states that the Universe formed 14 billion years ago when it expanded from a tiny, hot, dense mass to become thousands of times the size of the Solar System. Very quickly, the base particles of atoms formed, followed by the formation of hydrogen, and then helium. Over time, as the Universe expanded and cooled, matter clumped together, eventually forming stars and galaxies.
3. Evidence that supports the big bang theory includes the existence of background

radiation in space, the abundance of the light elements hydrogen and helium, and the evidence of redshift, suggesting that the Universe is expanding.

4. Evidence that contradicts the big bang theory includes indications that the expansion of the Universe is speeding up and that more dark matter should exist than scientists have been able to find.
5. One theory of how the Solar System formed proposes that gravity caused material in a nebula to collapse. The Sun and the planets formed out of the collapsed nebula. Other material that did not form the Sun or planets formed other objects in the Solar System.

Vocabulary Check

Fill in the blank with the correct vocabulary word.

1. According to theory, after a(n) __________ uses up its helium, a white dwarf forms.
2. __________ is the effect produced by an object that is moving very fast away from Earth.
3. A(n) __________ galaxy can be either round or oval in shape.
4. Scientists call the missing matter in the Universe __________.
5. Most stars tend to be found in a region on the H-R diagram called the __________.
6. The Milky Way is found in a cluster of more than 50 galaxies called the __________.
7. The process of __________ occurs when nuclei combine in the core of a star.

Multiple Choice

Choose the best answer.

8. What type of galaxy are the Magellanic Clouds?
   A. barred **C.** irregular
   B. elliptical **D.** spiral
9. Suppose fusion in a supergiant stopped. According to theory, what would form next in the star's life cycle?
   A. supernova **C.** nebula
   B. black hole **D.** red giant
10. Which star is closest to Earth after the Sun?
   A. Betelgeuse **C.** Polaris
   B. Proxima Centauri **D.** Sirius

Check Point

Answer the following questions.

11. How did Edwin Hubble expand our knowledge of astronomy?
12. Miriam lives in New Zealand. She watches the stars nightly. Every evening, all year long, she can see the constellation Crux (the Southern Cross). What term can Miriam use to **classify** Crux?
13. **Compare** the Sun with a red supergiant.
14. Describe the evidence that scientists use to explain an expanding Universe.
15. Suppose you discover a star that is 80 times more massive than the Sun. **Predict** the remaining steps in that star's life cycle, according to theory.

When you are at a carnival, you are surrounded by color, movement, and sounds. Although they seem very different, all of them are produced by energy.

Unit Overview

The world is filled with sights, sounds and constant activity. But what are light and sound, really? How do objects move and interact with each other? Physical science studies how matter and energy follow God's natural laws. God can alter the laws as we understand them when He performs miracles, but fortunately matter and energy follow His natural laws every day. That makes it easy to study changes in matter.

People use physical science to try to understand why things work the way they do. For example, why is it easier to drive a nail through wood with a hammer than with your fingers? What happens inside an electrical outlet that makes a lamp work?

In this unit, you will learn the answers to all of these questions and more. You'll see how physical science connects seemingly unrelated events: the wind blowing your hair, a truck rumbling by, a pizza fresh out of the oven, and sunlight reflecting off a window. God's natural laws explain why all these events occur.

Your teacher may assign an Open Inquiry lab and a Lifestyle Challenge activity. Use your ***Science Journal*** to record your work.

Chapter 11

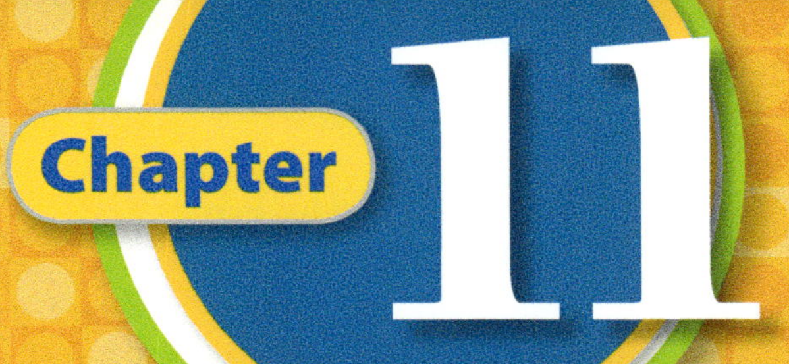

Energy Waves

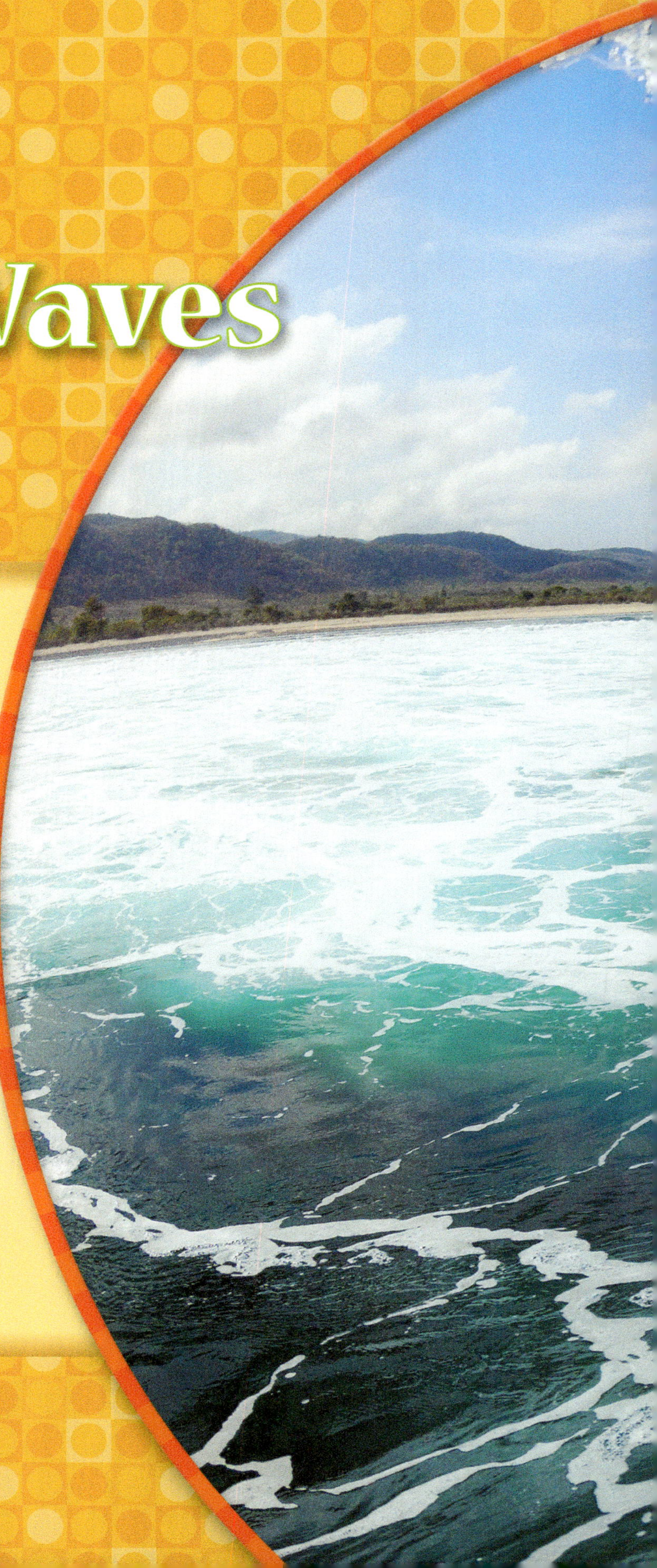

Scripture Spotlight

God sustains the Universe through His natural laws governing matter and energy. By studying these natural laws, you can better understand how God cares for us every day. You will read the following passage in this chapter.

Psalms 150:3–5 (p. 406)

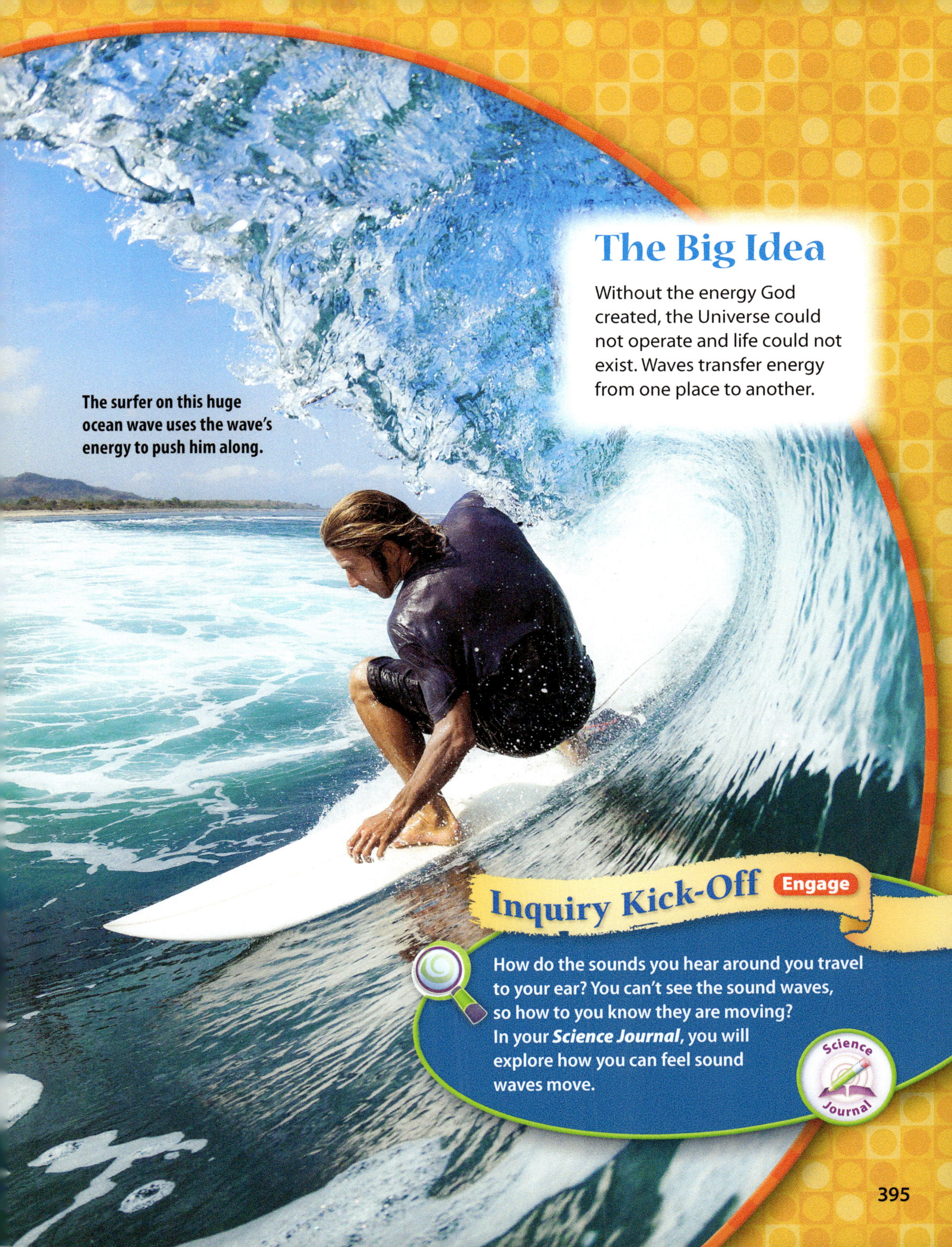

The Big Idea

Without the energy God created, the Universe could not operate and life could not exist. Waves transfer energy from one place to another.

The surfer on this huge ocean wave uses the wave's energy to push him along.

Inquiry Kick-Off

Engage

How do the sounds you hear around you travel to your ear? You can't see the sound waves, so how to you know they are moving? In your ***Science Journal***, you will explore how you can feel sound waves move.

Lesson 1

Essential Question

What Are Transverse and Longitudinal Waves?

Objectives

- Explain how energy is transferred by waves.
- Compare and contrast transverse and longitudinal waves.
- Compare mechanical and electromagnetic waves.
- Explain the properties of waves.

Vocabulary

medium

transverse wave

longitudinal wave

mechanical wave

The wave that propels a surfer toward the shore may have begun far out to sea, or its origin may have been much closer to shore. Wind energy transferred some of its energy to the water's surface, creating a wave. How did a small wave created by wind energy become so large? Where did all that energy come from? Where did the wind's energy come from in the first place? What happens to the energy when the wave crashes on the beach? If you were on the beach when the wave crashed, what other forms of energy could you detect? Do any of these waves make a difference in our lives? Explain.

Waves Explain

Ocean waves, sound waves, light waves, radio waves, heat waves and many other types of waves surround us. A *wave,* you will remember, is a repeating back-and-forth or up-and-down motion that transfers energy. Scientists describe waves as any disturbance that transfers energy through matter or empty space. Energy has many forms, including the sound energy you hear at the beach as the waves crash on shore and the light energy you see as the sunlight reflects off the water. Both sound energy and light energy are carried away from their sources by waves. But how are these waves different? How do these waves travel?

Waves carry energy.

Where does the energy of these water waves come from?

Waves and Energy

Did you ever notice that a ball or duck floating on the surface of the ocean does not move forward with a wave? Instead it bobs ups and down. That is because even though waves carry energy, the material through which waves travel does not move. Recall that all matter is made up of particles. The matter through which waves move is called a **medium**. Some waves transfer energy by the vibration of particles that make up the medium. Can waves move through solids, liquids, and gases? How do you know? Through what kinds of media do sound waves travel?

When a particle in a medium vibrates, it can bump into a particle next to it. When it does so, the particle passes energy along to the second particle, and the second particle will vibrate in a manner similar to the way in which the first particle vibrates. In this way, energy is passed through a medium.

Transverse Waves

Waves can be classified according to the direction in which particles of the medium vibrate compared with the direction in which the waves travel. Waves in which the motion of the particles is perpendicular to the motion of the waves are called **transverse waves**. The waves that the surfer in the Chapter Opener is riding carry energy as transverse waves. The diagram below shows the movement of particles in a transverse wave.

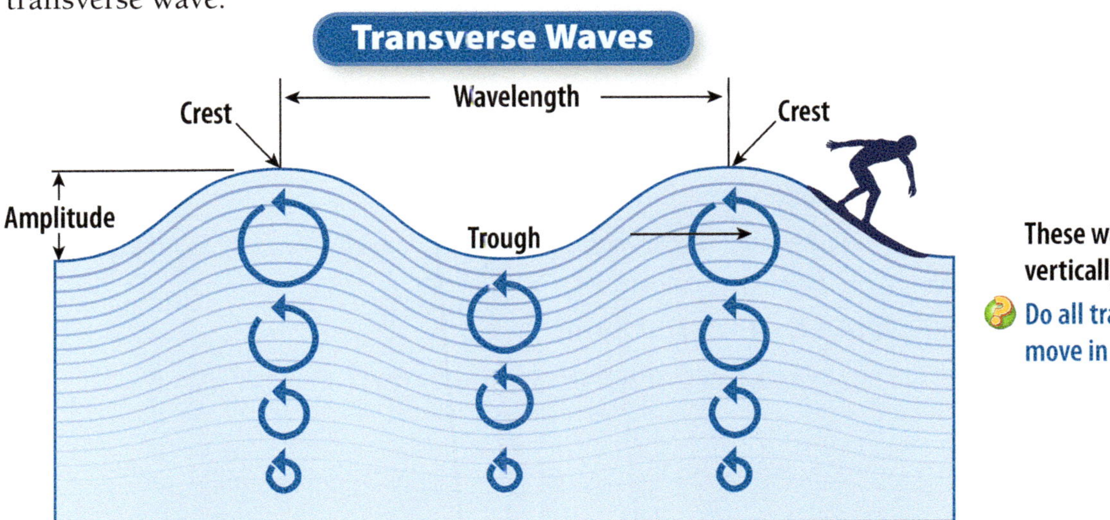

These water waves move vertically.

Do all transverse waves move in a vertical plane?

Compare the movement of the wave with the movement of the surfer. The surfer on the ocean's surface moves up as the top of the wave passes and down as the low point on the wave passes on its way to the shore. If waves are not transferring matter, such as the surfer, what do they transfer? Why does the surfer go up and down sometimes and then move forward other times? Develop a theory to explain how riding a wave works.

Record your work for this inquiry. Your teacher may also assign the related Guided Inquiry.

Make Waves

What wave features can you see in water waves?

Procedure

Materials
- large shallow pan
- pencil with eraser
- water
- bits of notebook paper
- flexible drinking straw

1. Place about 1–2 cm of water in the pan. Allow the water to stop moving.
2. Dip the eraser end of the pencil into the water at the center of the pan in one quick down-and-up motion. **Observe** and **record** what happens to the water.
3. After the water comes to a rest, dip the eraser into the water several times in a row. Record your observations.
4. Drop two or three small bits of notebook paper onto the surface of the water between the center and the edge of the pan. Repeat Step 3. Observe and record the motion of the paper.
5. Allow the water to stop moving. Use a drinking straw to blow on the water at one end of the pan. Observe and record the motion of the water.

Analyze Results

How does the water move the energy you used to produce the wave from one place to another?

Create Explanations

1. What wave features can you see in water waves?
2. What happens to the water in the pan as the wave carries energy from the center to the edge? How do you know?
3. What has to happen for a wave to form?
4. What does and does not travel outward with the wave? How do you know?

Some transverse waves—light waves, radio waves, and other electromagnetic waves—do not need a medium through which to travel. They can also travel through a vacuum. How would you be affected if these waves were unable to travel through a vacuum?

Longitudinal Waves

Waves that vibrate parallel to the motion of the wave's movement are known as **longitudinal waves**. Making waves with a rope will help you understand something about how waves travel. Complete the Explore-a-Lab below.

Explore-a-Lab

Structured Inquiry

 **How can you model waves with a rope?**

Have someone hold one end of a 2-m length of rope while you hold the other end. The other person should not move that end. What different movements can you make with the rope? How can you change the spacing of the "hills and valleys" in the rope? What is being transferred down the "hills and valleys" in the rope to the other person? How is it transferred? Sketch what you observe happening. What conclusion can you draw? What has to happen for a wave to form? What different kinds of waves can you make?

Sound waves travel as longitudinal waves. Longitudinal waves can only travel as they pass from one atom to the next atom in matter. If there is no matter, there is no sound. Many movies and TV programs about space adventures depict the sounds of spaceships streaking by, laser cannons being fired, and objects exploding. Are these accurate depictions of scientific principles?

These sound waves pass energy from one atom to the next in the matter they travel through.

Why do you think a sound speaker vibrates as it produces sound?

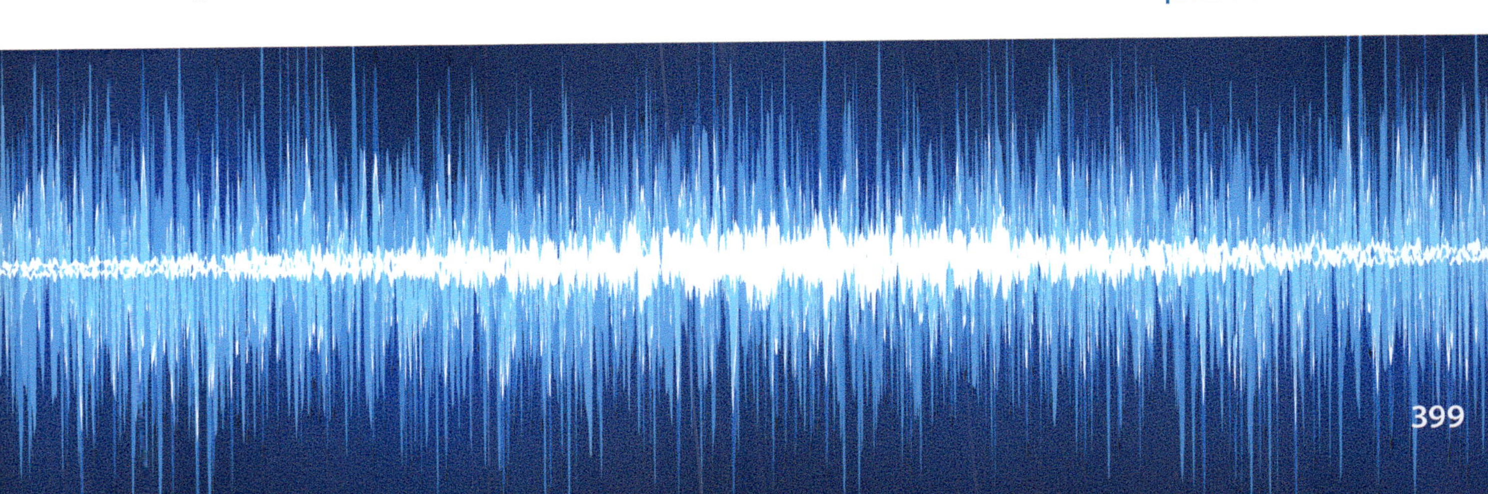

Properties of Waves Explain

Suppose you are the surfer in the Chapter Opener picture. Before you can ride a wave, you must paddle out into the ocean, going up and over each incoming wave. Why do you think some waves are long, gentle swells while others are short and choppy? When surfers describe waves, what kind of waves are they talking about?

Remember when you learned about waves earlier in your study of science that transverse waves have high points called *crests* and low points called *troughs*. Longitudinal waves do not have crests and troughs. Instead they have compressions and rarefactions. *Compression* describes the area where the coils are packed together; *rarefaction* is the area where the coils are spread apart. What part of longitudinal waves corresponds to the crest of transverse waves? What part corresponds to the troughs of transverse waves?

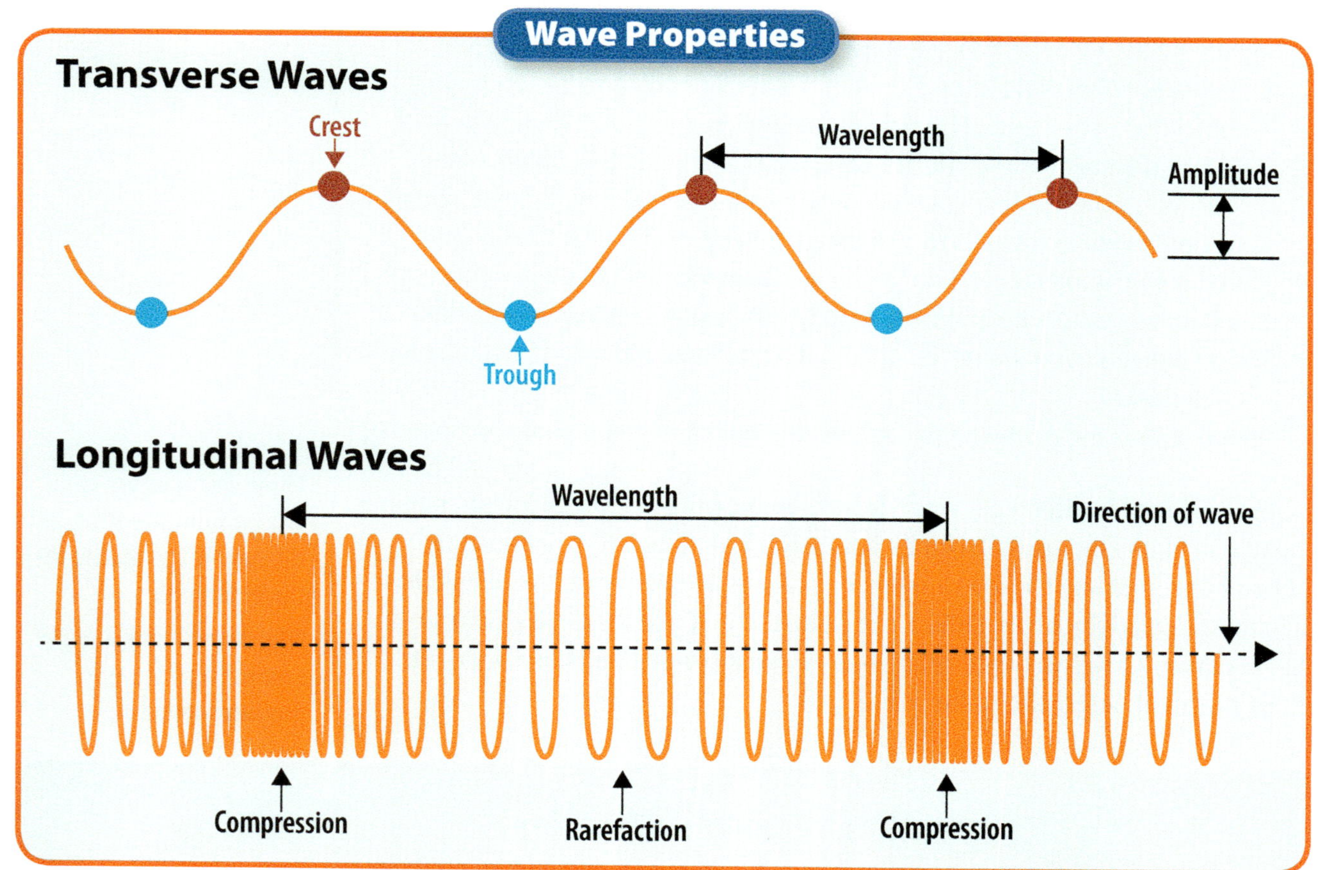

Transverse waves and longitudinal waves share several properties.

Why do you think longitudinal waves are also called compression waves?

Wavelength

You know that the *wavelength* of a transverse wave is the distance between two consecutive crests or two consecutive troughs. In a compressional wave, wavelength is the distance between the beginning of a compression and the beginning of the next compression. If one wave has a greater wavelength than another, what do you know about these waves?

Amplitude

Half the vertical distance from the trough to the crest of a wave is its *amplitude*. Construct a model that demonstrates the amplitude of a large wave and the amplitude of a small wave. Which do you think has more energy? Use your model to explain your answer.

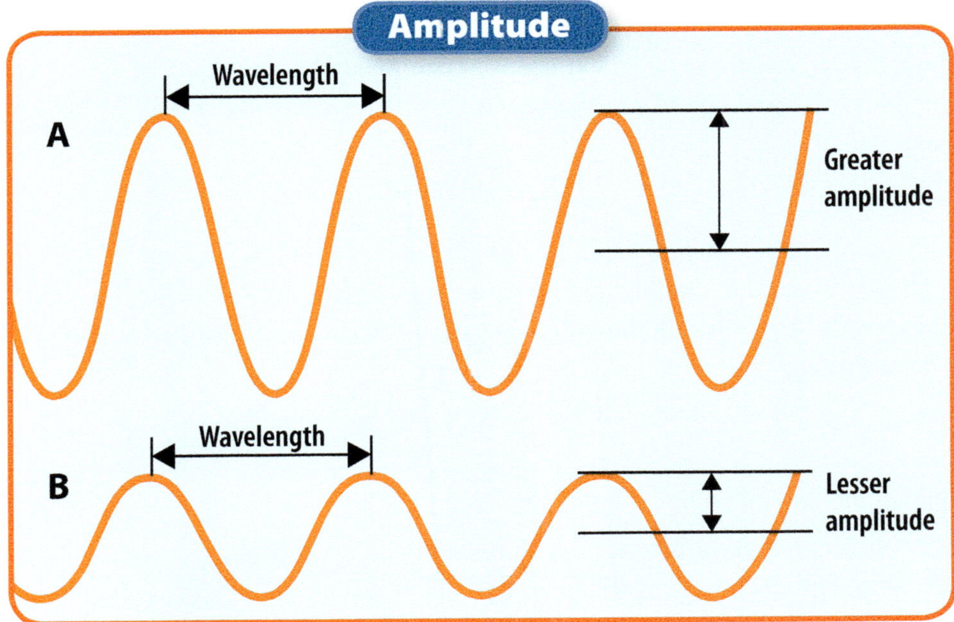

While the wavelengths of A and B are the same, their amplitudes are different.

Frequency

Recall that the number of waves that pass a specific point in a given time is called *frequency*. Imagine you are floating on the ocean on your surfboard. Each time a wave crest passes, your surfboard rises. When a wave trough passes, your surfboard moves down. If you moved up and down three times in one minute, what would the frequency be?

Scientists measure frequency with a unit called a hertz (Hz). One wavelength passing a certain point in one second is one hertz (1 Hz = 1/s). If two wavelengths pass a certain point in one second, what would the frequency be?

Wave Frequencies

Type	Hertz
Earthquake waves	0.5
Microwaves	100,000,000,000
Radio waves	100,000,000
Sound (middle C)	512
X-rays	10,000,000,000,000,000,000

Wave Speed

Another property of a wave is speed. Recall that *speed* is distance traveled over time, so the speed of a wave can be determined by measuring the distance a crest or trough travels in a given amount of time. You can calculate the speed of a wave using this equation:

$$\text{speed } (v) = \text{wavelength } (\lambda) \times \text{frequency } (f), \text{ or}$$
$$v = \lambda \times f$$

Notice that in the equation, the symbol λ is used for wavelength. Why are the units for speed m/s? For example, if a wave has a wavelength of 5 m and a frequency of 4 Hz, then its speed is calculated this way:

$$v = 5 \text{ m} \times 4 \text{ Hz} = 20 \text{ m/s}$$

At a constant speed, the frequency of a wave is inversely proportional to its wavelength. Recall that in an inversely proportional relationship, as one quantity increases the other decreases. The equation for wave speed can be rearranged to determine this relationship:

$$\lambda = v/f, \text{ or}$$
$$f = v/\lambda$$

What does this tell you about the measures of frequency and wavelength for waves? What must happen for the frequency of a wave to increase?

Math in Science

You conduct an experiment with a spring toy and find that the wavelength is 0.6 m and the frequency is 3.5 Hz. What is the speed of the wave? In a second trial, you calculate the speed of the wave to be 2.16 m/s and the frequency to be 1.2 Hz. What is the wavelength? Now consider two waves on identical springs, one with a frequency that is twice the other. How do their wavelengths compare?

Explore-a-Lab

Structured Inquiry

 **How does removing air affect sound waves?**

Gather a battery-powered radio, one plastic garbage bag, a vacuum cleaner with hose, several pillows, wire screen about 30 cm × 30 cm, and duct tape. Turn on the radio and tune in to a station with continuous music. Put the radio in the plastic bag. Tape the wire screen into a cylinder shape and attach it to the end of the nozzle on the vacuum cleaner hose. Put the nozzle and screen into the plastic bag and seal the bag so air cannot get into or out of the bag. If the vacuum is too loud to hear the radio, pile pillows around the vacuum to muffle the sound. Why did the volume of the music decrease as air was pulled from the bag? What does this tell you about how sound travels?

Mechanical and Electromagnetic Waves Explain

You have learned that waves can be classified as transverse or longitudinal. What is the basis for this classification? Waves can also be classified another way—mechanical or electromagnetic.

What would happen to the movement of a floating object if it were placed on one of these waves? Why?

Mechanical Waves

Mechanical waves cannot transmit energy in a vacuum. They require matter to transport their energy from one location to another. A sound wave is an example of a mechanical wave. Other examples include water waves and jump-rope waves. What do these waves have in common? If you were on the Moon, why couldn't you hear explosions taking place right next to you?

What kinds of electromagnetic waves does the surveillance camera use?

Electromagnetic Waves

Electromagnetic waves are a group of transverse energy waves that can transmit energy in a vacuum. There are many types of electromagnetic waves including:

- Heat and light waves can move through the vacuum of outer space.
- Electromagnetic waves, such as infrared lights, provide heat.
- Without visible light waves, we would not see anything.
- Radio waves and microwaves enable us to communicate over great distances.
- Gamma waves are used in treating cancer and other illnesses.
- X-ray waves allow doctors to see what is happening inside the body.

Concept Check **Assess/Reflect**

Summary: What are transverse and longitudinal waves? Waves transfer energy, not matter. Waves can be classified as transverse waves that vibrate perpendicular to the motion of the wave or as compressional waves that vibrate parallel to the motion of the wave. Properties of waves include their wavelength, amplitude, frequency, and speed. Waves also can be classified as mechanical or electromagnetic. Mechanical waves, such as ocean waves and sound waves, need a medium through which to travel, but electromagnetic waves, such as radio waves and light waves, can travel in a vacuum.

1. What happens to a boat sitting on a lake when a wave passes under it?
2. Compare crests and troughs with compressions and rarefactions.
3. What happens to the wavelength if you increase the frequency of a wave?
4. How are mechanical waves produced?

Lesson 2

Essential Question

What Are the Properties of Sound?

Just before the picture below was taken, the glass was in one piece. As the glass was bombarded with sound waves, its molecules began to vibrate. Why did this happen? What do you think the waves that hit the glass looked like? The camera caught the action just as the glass shattered. How is it possible for waves that you can't see to cause this kind of damage? Glass objects can be shattered by the high note of an opera singer or the sonic boom of a jet. Seeing the damage that is possible, why do you think it is important to use ear protection at loud events?

Sound waves carry energy.

Into what kind of energy is sound energy changed to break this glass?

Objectives

- Explain how sound travels through different types of matter.
- Describe the properties of a sound wave.
- Describe the behavior of sound waves.
- Explain the Doppler effect.

Vocabulary

ultrasound

infrasound

intensity

Doppler effect

Properties of Sound Explain

When you drop a book, why do you hear it hit the floor? As you have learned, sound waves are longitudinal waves. When a radio plays, the music causes the speaker to vibrate and make longitudinal waves that travel though the air to your ear. When the waves push on your eardrum, your eardrum vibrates. The vibrations cause nerve impulses to be sent to the brain, and you hear sound. The air that the sound waves travel through does not move with the sound waves. Only the particles in the air vibrate back and forth. What could happen to your eardrum if the air did move along with the sound waves?

Sound waves travel at different speeds through different materials. For example, sound waves travel through air at 33 m/s. They travel through steel 15 times faster than they do through air. Why is that? Do you think sound will travel faster through pure water or seawater? Explain.

Scripture Spotlight

Read **Psalm 150:3–5**. What are some ways in which the sounds described differ? Why are sounds such as these important?

Pitch

Recall that the highness or lowness of a sound is called *pitch*. The pitch of a sound wave is caused by the frequency of the sound wave and, like frequency, is measured in hertz (Hz), or the number of waves that pass a certain point per second. Low pitches have a frequency of less than 200 Hz, or vibrations per second, while high-pitched sounds have a frequency greater than 400 Hz. How could you describe the relationship between the frequency of sound and its pitch?

Study the two sound waves below. Which wave has a longer wavelength? Which wave has a higher frequency? Which wave represents a sound with a higher pitch?

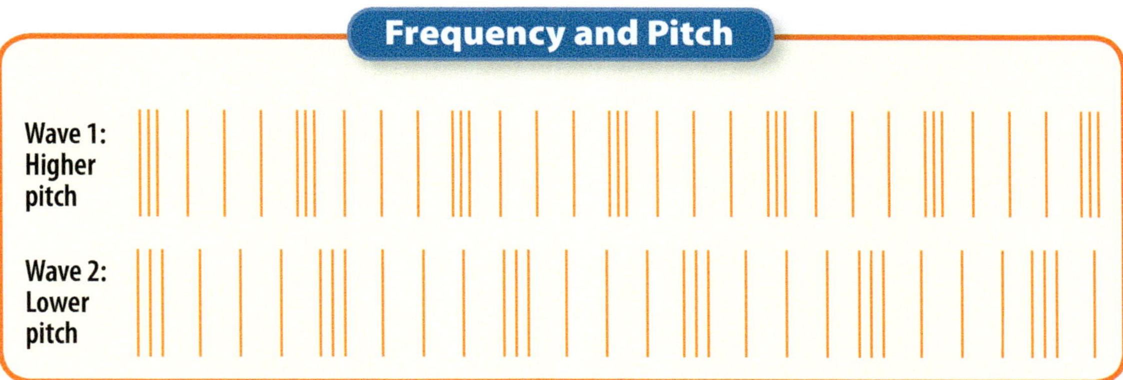

Explore-a-Lab

Structured Inquiry

How does the length of a straw affect its pitch?

Flatten one end of a straw with scissors, and cut off the corners of the flattened end to make a point. Wet the pointed end of the straw with your mouth, and with the straw still in your mouth, blow and make a buzzing sound. Once you can make it buzz, cut about 2 cm from the end you do not blow into and make it buzz again. What happens to the sound produced? Cut the straw at least two more times, and record what happens each time. Make a different straw with a pointed end. Lengthen it by placing rolled up paper over the end and blowing. How do you describe the sound that is produced? What is the relationship between the length of the straw and the pitch of the sound produced? Why does this relationship exist?

Record your work for this inquiry. Your teacher may also assign the related Guided Inquiry.

Gluba

What causes the pitch of sound to change?

Procedure

1. Place about 2.5 cm of the pipe or tube into the glove opening and attach it with a rubber band. The glove should hang freely from the pipe.
2. Cut a slit in the end of one of the glove fingers.
3. Insert a straw at least 2.5 cm into the slit and tape securely with masking tape.
4. Stretch the glove over the pipe opening and up along the side of the pipe, forming a tight membrane like the head of a drum.
5. Blow into the straw and inflate the glove. **Observe** what happens as you increase the amount of air in the glove.

Materials

- rubber band
- scissors
- straw, plastic
- masking tape
- latex glove
- 1″ PVC pipe (15-cm length)

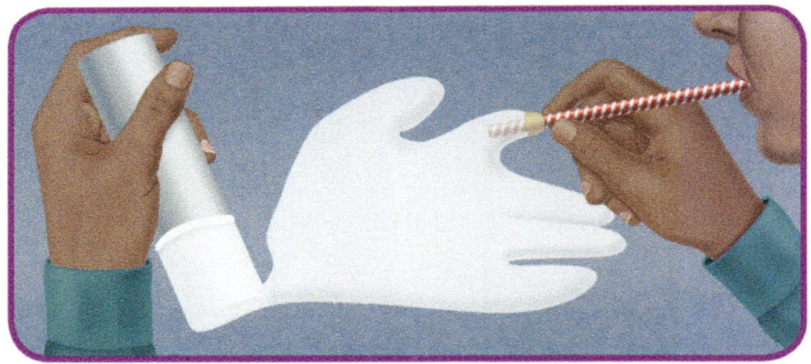

Analyze Results

Explain what causes the sound as you blow air through the glove.

Create Explanations

1. What causes the pitch of sound to change?
2. What happens to the pitch if you stretch the glove tighter? Why?
3. What happens to the pitch if you loosen the glove? Why?
4. What happens if you put your finger over the top of the tube as you play?

The size of the tuning fork affects the pitch it produces.

Pitch can be demonstrated with tuning forks. Some forks are small, while others are large. Each tuning fork produces only one pitch of sound. Which tuning fork do you think produces high-pitched sounds, and which produce low sounds with low pitches?

The human ear is sensitive to sound that ranges from approximately 20 to 20,000 Hz. Sound that has a frequency greater than 20,000 Hz is called **ultrasound**. Humans cannot hear ultrasonic sounds, but some animals can hear these sounds quite well. Which animals can hear ultrasounds? If we could hear ultrasonic sounds, would they have a high or low pitch? How do you know? Whales communicate with each other through ultrasound and with frequencies that humans can hear.

Sound that has a frequency of less than 20 Hz is called **infrasound**. Biologists have discovered that elephants use these waves to communicate with each other; they can hear infrasound over distances greater than 16 km. If humans cannot hear these sounds, how do biologists know elephants use them?

Intensity

The loudness or quietness of a sound is determined by the amplitude, or size, of the wave. The larger the amplitude is, the louder the sound. If two sound waves vibrate at the same rate, which will have the louder sound? When you hit a key softly on the piano, its string vibrates and makes a quiet sound. How could you make the sound louder? Will the pitch change?

The **intensity** of sound is the amount of energy that is transported per unit time past a given point of the medium through which the sound travels. Intensity is measured in units called decibels (dB). The table below shows the intensity level of several common sounds. For every increase of 10 decibels, the sound is actually 10 times greater. How much louder is the dishwasher than the vacuum cleaner? What about the busy city street—how does it compare with the dishwasher and the vacuum cleaner? Sounds that are at 120 dB or higher can be painful.

Intensity of Common Sounds

Sound	Decibels	Sound	Decibels
Threshold of hearing	0	Vacuum cleaner	70
Rustle of leaves	15	Dishwasher	80
Whisper	20	Busy city street	90
Library	40	Power lawn mower	100
Light traffic	50	Jackhammer	110
Normal conversation	55	Rock band	115
Background music	60	Thunder overhead	120

Math in Science

The table shows the recommended time of exposure to sounds at different decibel levels before risking hearing damage. How long could you safely remain at a rock concert without risking hearing loss?

Recommended Noise Exposure Limits

Sound level (dB)	Time (hr)
90	8
95	4
100	2
105	1

Bats emit high-pitched sounds that produce echoes, which help them hunt as they fly.

Behavior of Sound Explain

When sound waves hit a solid surface, they *reflect*, or bounce back. A reflected sound wave is called an *echo*. What happens when someone shouts across a canyon or toward a mountain?

Bats use echoes to find food and to avoid running into things when they fly. They send out sounds at about 80,000 Hz. How can they use these sounds to locate an object? Recall that the process is called *echolocation*. Why do bats depend on echolocation?

Sonar is another example of echolocation. Sonar equipment sends out sound waves through water and measures how long it takes for the sound to reflect. How can scientists use sonar to measure how far below the surface an object is?

Explore-a-Lab

Structured Inquiry

 **How do sound waves reflect?**

Stand outside a building that has a large, flat wall. (Hint: the bigger, the better). Move so that you are 3 m away from the building. Face the wall and clap loudly. Record what you heard. Repeat the process when you are 10 m, 15 m, and 20 m from the building. What did you hear when you were farther from the building that you did not hear when you were closer? Then repeat the procedure, but this time record the amount of time between your clap and the echo you hear. Use the echo timings to estimate distance.

Explore-a-Lab

Structured Inquiry

How far will sound carry?

With a partner, go outside. An ideal location is a football field. Play a high note on an instrument such as a xylophone. Have your partner raise a hand to indicate that the sound was heard. Ask your partner to back up 10 m and play the sound again. Repeat until the note cannot be heard. Note the distance between you and your partner. Then repeat the activity while playing a low note. What was the difference in the results?

Sound waves also *refract,* or bend. They travel faster in warm air than in cold air. Imagine that you are camping on one side of a small lake. During the day, you see campers on the opposite side of the lake, but you do not hear them. At night, these campers start a camp fire. After a short while, you can hear the sounds they make. Why can you hear the sounds at night but not during the day?

As sound waves travel through any medium, some energy is lost due to *absorption*. Recall that *absorption* occurs when wave energy is transferred to matter as a wave passes through it. When sound is absorbed, it seems to disappear. What happens when the girl in the photo below shouts into a thick sponge? Absorption reduces the size of the waves. Why? What then happens to the sound? The walls and ceilings of sound studios are covered with material that absorbs and reduces echoes that would reflect from bare walls and ceilings.

How much sound is absorbed is related to the frequency and pitch of the sound. For example, your voice at its loudest will not carry for 2 km before the sound waves are absorbed by the air particles. However, the low-pitched infrasounds of an elephant can travel 8 km.

Soundproofing materials absorb sound, preventing it from traveling.

Doppler Effect Explain

Think about a police car that is parked with its siren going, as shown in the top diagram below. Look at the sound wave created by the siren on that parked car. Assume there are people on either end of the street. How does the frequency of the sound being emitted compare with the frequency that each person hears?

Now look at the bottom image where the police car is moving to the right with the siren going. What happens to the distance the wave has to travel before it is observed by the person at the right end of the street? Will it take more or less time to reach the person at the left end of the street? How will the person at the right end of the street perceive the frequency of the siren compared with its actual sound frequency? How do you know? What can you say about the frequency of the sound wave that the person at the left end of the street will observe?

The **Doppler effect** is an observed change in the frequency of a wave when the source or the observer is moving. We are most familiar with the Doppler effect as it relates to sound waves. Do you think it applies to other types of waves, too?

How does the frequency of the sound waves change as the car moves?

Doppler Effect

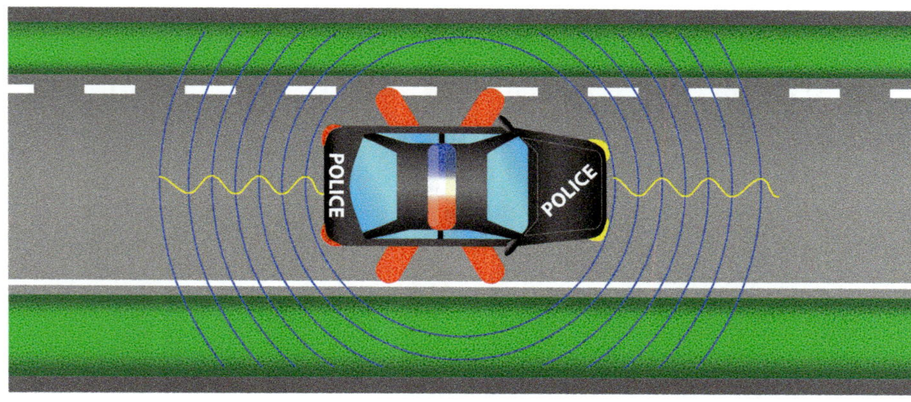

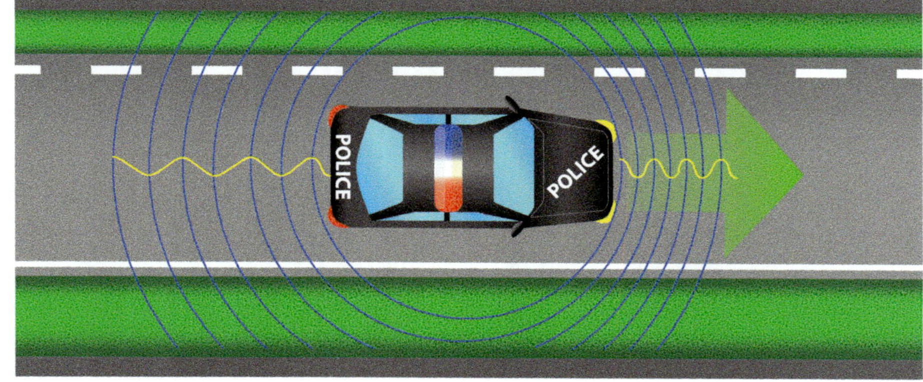

Explore-a-Lab

Structured Inquiry

Why does the sound of a car going by change from high pitched to low pitched?

Gather construction paper, a ruler, scissors, tape, and a toy car. Use construction paper to make five strips that are 2.5 cm wide each with a length of 28 cm, 25.5 cm, 23 cm, 20.5 cm, and 18 cm. Tape the ends of each strip together to make a loop. Arrange the loops so that they are all centered around the car, with none of the loops touching each other or the car. Draw a picture of what you observe. What do these loops represent? Push the car forward slowly. Draw a picture of what you observe. What happens to the sound waves? What does this represent? How does this change affect the pitch of the sound?

Concept Check

Assess/Reflect

Summary: What are the properties of sound? Sound waves are compressional waves that need a medium through which to travel. The speed at which they travel depends on the medium. Pitch refers to the high or low quality of a sound and is determined by the frequency of the sound wave. The amplitude of the wave determines how loud the sound is. Sound intensity is measured in decibels. Sound waves can reflect, creating an echo, or refract. They can also be absorbed. The Doppler effect explains why the same sound can be perceived at different frequencies when the source of the sound or the observer is moving.

1. Why must astronauts use microphones and headphones in outer space?
2. Why do you think sound travels faster in denser materials?
3. How is an echo different from echo location?
4. Why can we not hear a dog training whistle, and yet the dog responds to it?
5. A train is moving from east to west. Explain how its whistle will be heard by the engineer, a person to the east of the train, and a person to the west of the train. Draw an illustration that explains your answer.

Lesson 3

Essential Question

What Is the Electromagnetic Spectrum?

Objectives

- Describe the electromagnetic spectrum.
- Explain how a photon is related to the different wavelengths of light.
- Identify how various types of electromagnetic waves are used.

Vocabulary

radiant energy

photon

Look carefully at the narrow shaft of white light that travels through the prism in the picture below. What do you see happening to the light as it enters the prism? Why does this happen? Why does the white light separate into all the colors of the *visible spectrum*? Why is the red light on the top and the blue light on the bottom? Do you think you could position another prism in the visible light spectrum and cause it to bend so that the colors would go back to the shaft of white light? Why or why not? Each color of light is created by energy waves that have a specific wavelength and frequency. What other energy is present here that you can't see?

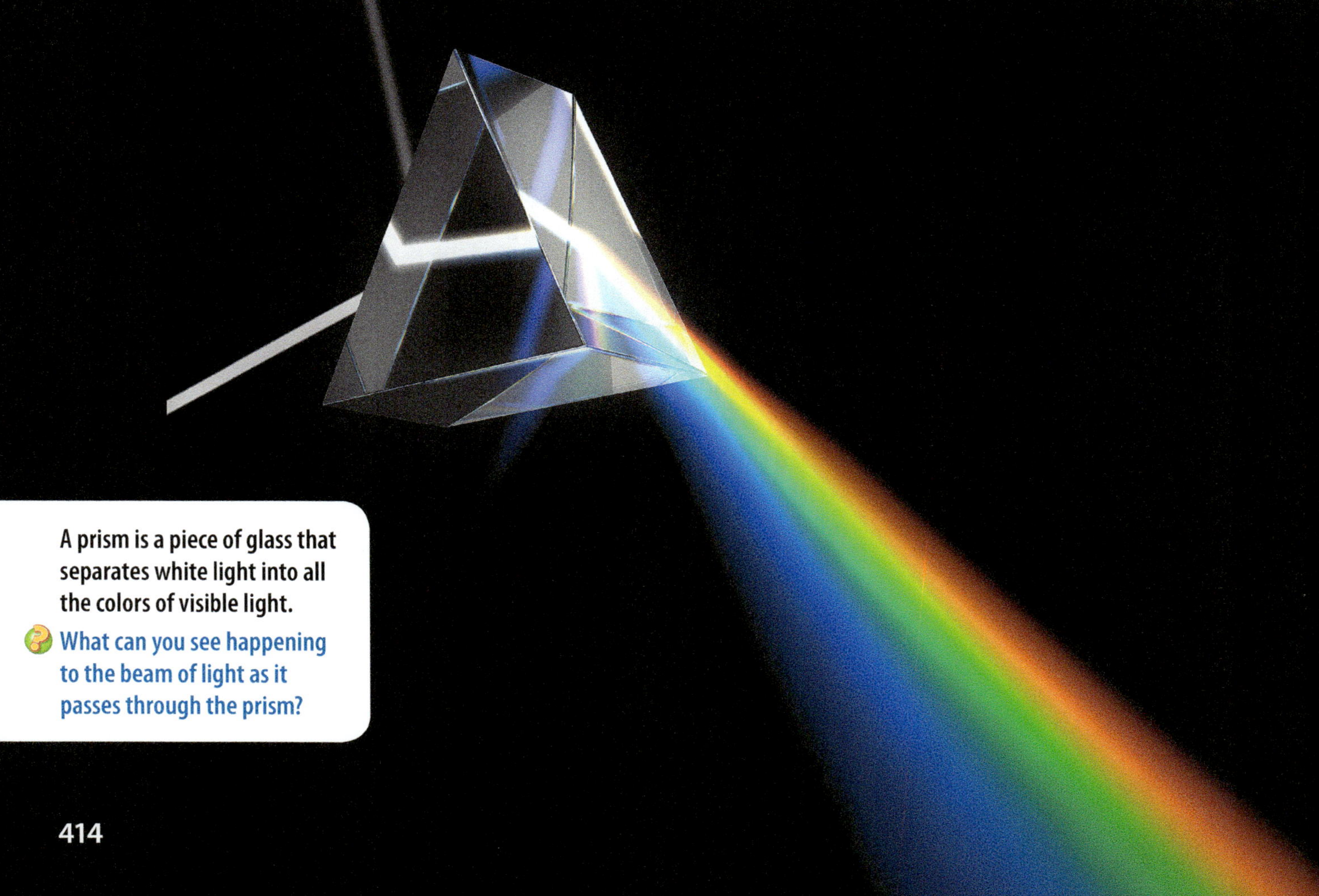

A prism is a piece of glass that separates white light into all the colors of visible light.

What can you see happening to the beam of light as it passes through the prism?

Electromagnetic Spectrum Explain

As you know, the electromagnetic spectrum is the range of all types of electromagnetic waves. Electromagnetic waves are also known as *electromagnetic energy*, *electromagnetic radiation*, or *radiant energy*. **Radiant energy** is energy transmitted by electromagnetic waves. Study the diagram on the bottom of this page.

What pattern do you see in the organization of the electromagnetic spectrum? Why is it that humans can see visible light energy but can't see infrared or ultraviolet energy? Bees and other insects are able to see ultraviolet light, and many snakes are able to "see" infrared light. Why do you think they are able to do this? What type of radiant energy can you detect on a warm sunny day?

Although electromagnetic radiation can be described in terms of waves, it can also be described in terms of photons. **Photons** are particles, or bundles of electromagnetic waves, that travel in wavelike patterns at the speed of light (300,000 km/s). How can all types of radiation travel at the same speed if the waves have differing wavelengths and frequencies?

Each photon has a certain amount of energy, and the different types of radiation are defined by the amount of energy in their photons. For example, radio waves have low-energy photons, and gamma rays have the most energy of all types of radiation.

Like all waves, electromagnetic radiation can be described in terms of energy, wavelength, or frequency. You will recall that *frequency* is measured in cycles per second, or hertz, and wavelength is measured in meters. Energy is measured in electron volts (eV). These quantities for measuring radiant energy are related to each other in a precise way.

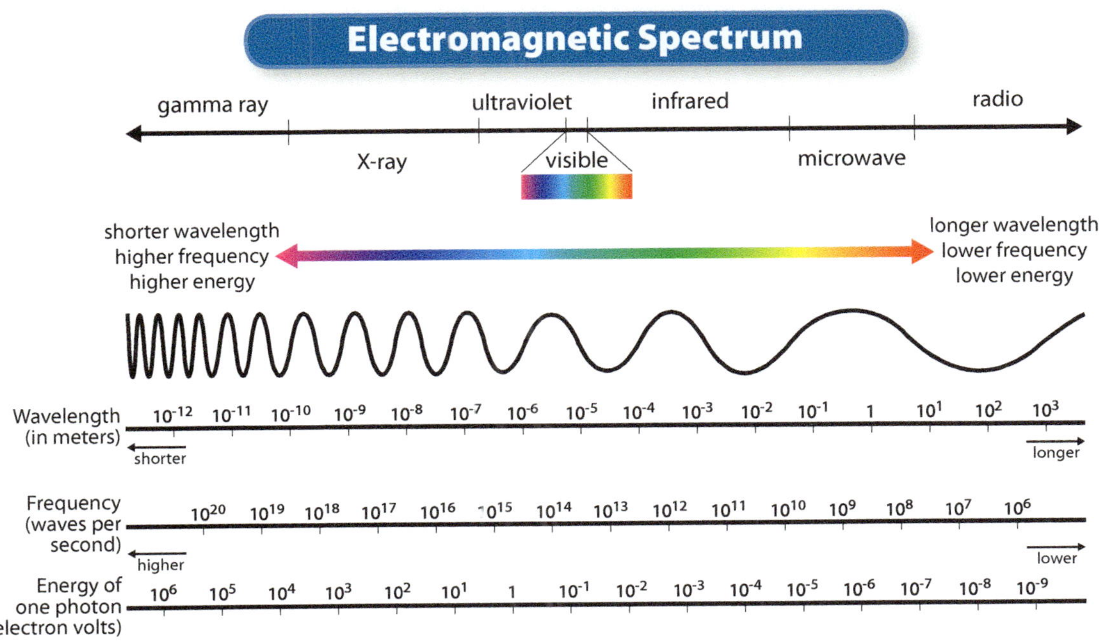

Explore-a-Lab

Structured Inquiry

 **What color of waves make up the light you see?**

Hold a prism with one finger on top and one finger on the bottom, positioned 10 cm in front of the eye. Look through one side in the direction of the light source: fluorescent light, halogen light, or incandescent light. Record your observations, listing the colors that are visible from left to right. Repeat with the other light sources. Were the colors always in the same order? Always in bands? Why or why not? Did the bands always form the same shapes? Explain. Compare these waves with sound waves. How are they alike and different? Explain your answer with a theory.

Uses of Electromagnetic Radiation Explain

Electromagnetic radiation is the energy of the Universe. This energy powers the processes that form galaxies and stars. Besides the starlight we see in the night sky, objects in space emit large amounts of energy representing the whole electromagnetic spectrum. How can we detect this energy?

Electromagnetic Radiation

Type	Uses
Gamma	Detecting and treating cancer; sterilizing food and medical equipment
X-ray	Medical imaging of teeth and other parts of the body; observing objects at airport security
Ultraviolet	Marking security information; building fluorescent lamps; detecting forged bank notes; disinfecting water
Visible	Creating photographs; lighting homes, buildings, and streets
Infrared	Cooking; creating images in night-vision goggles; TV remote controls; mapping the dust between stars in space; security systems; infrared photography
Microwave	Cooking food; transmitting sound via satellites to phones; learning about the structure of nearby galaxies
Radio	Broadcasting TV and radio FM stations; bringing sound and music to television

Record your work for this inquiry. Your teacher may also assign the related Guided Inquiry.

Spectroscope

What colors of light make up different kinds of light?

Materials

- colored pencils
- diffraction grating—2 cm square
- fluorescent light
- incandescent light
- masking tape
- paper tube (toilet paper tube)
- pencil
- poster board (5 cm × 10 cm)
- scissors

Procedure

1. Use the pencil to trace around the end of the paper tube on the poster board. Make two circles and cut them out. The circles should be slightly larger than the diameter of the tube's opening.
2. **Measure** and cut a 2-cm square hole in the center of one of the poster board circles you cut out in Step 1.
3. Tape the diffraction grating square over the hole.
4. Tape the circle with grating inward to one end of the tube.
5. Measure and cut a slot that is about 2 mm × 2 cm in the center of the other circle of poster board. Place the circle with the slot against the other end of the tube and tape it in place (see the diagram).
6. Place the circle with the slot against the other end of the tube. While holding it in place, **observe** fluorescent light. Be sure to look through the grating end of the spectroscope. The spectrum will appear off to the side from the slot. Rotate the circle with the slot until the spectrum is as wide as possible. Tape the circle to the end of the tube in this position.
7. **Record** the spectrum you see by drawing a color picture in your ***Science Journal***.
8. Repeat the process with an incandescent light and two other types of light. Nighttime street lighting works well.

Analyze Results

Compare and discuss the different spectra that you saw.

Create Explanations

1. What colors of light make up different kinds of light?
2. How does the spectrum of fluorescent light compare with that of incandescent light?
3. How does the diffraction grating produce the spectrum?
4. Do you think that there are other electromagnetic waves that make up the spectra you observed? Why or why not?

Lesson Activity

Keep a journal for one day to learn how you are affected by electromagnetic radiation. Record each time you experience, either through observation or contact, electromagnetic energy. Record what the incident is and what type of electromagnetic energy was involved. At the end of the activity, create a class tally to determine the most common activity that involved exposure to electromagnetic energy.

When are you affected by electromagnetic radiation?

What can we learn about our Universe from studying these energy waves? What types of celestial objects release radio waves? What is the source of ultraviolet energy? The Universe is the biggest producer of gamma rays. How are they produced?

Without electromagnetic waves, life as we know it could not exist. What benefits do we receive from electromagnetic waves? What harm can be caused by these energy waves? During a typical day, what invisible energy wave do you use the most? How can you protect yourself from harmful electromagnetic energy?

Summary: What is the electromagnetic spectrum? Electromagnetic waves are transverse energy waves that can travel through matter or a vacuum. They are organized by wavelength into the electromagnetic spectrum and include (from longest wavelength to shortest wavelength) radio waves, microwaves, infrared light waves, visible light waves, ultraviolet light waves, X-ray waves, and gamma waves. The properties of these waves make them very useful in our lives. In addition, these energy waves are critical to life on Earth and the processes in the Universe.

1. Why can we also say that the electromagnetic spectrum is organized by wave frequency?
2. Give an example of how electromagnetic energy is harmful and how it is beneficial.
3. What does the name "short-wave radio" imply about the frequencies at which it operates? Why might this not be the best name for this type of radio?

Lesson 4

Essential Question

What Are the Properties of Light?

Visible light, often called white light, allows you to read this sentence, to pick out a matched pair of socks, and to enjoy the many sights that make up your day. Visible light, like all electromagnetic energy, travels in a straight line as transverse waves. And like all electromagnetic energy, light can travel through matter as well as empty space (a vacuum). Light travels at 300,000 km/s. At that speed, you could go around Earth's equator seven times in one second. Light only takes eight minutes to travel the 150,000,000 km from the Sun to Earth. Imagine that you turn on a flashlight. How far should the light travel after just one minute? Why doesn't the light actually travel that far?

Reflection Explain

When light meets a mirror, it is *reflected* off the mirror. What would happen if you darkened the room and shone a flashlight on a mirror at an angle? You would be verifying the **law of reflection**, which states that a beam of light will reflect off a surface at the same angle at which it struck the surface.

When a beam of light approaches a surface, such as a mirror, it is called an *incident ray*. The light that is reflected is called the *reflected ray*. The line that is perpendicular to the surface of the mirror is called the *normal line*. The angle between the incident ray and the normal line is called the *angle of incidence*, and the angle between the reflected ray and the normal line is called the *angle of reflection*. What does the law of reflection tell us about the angle of incidence and the angle of reflection?

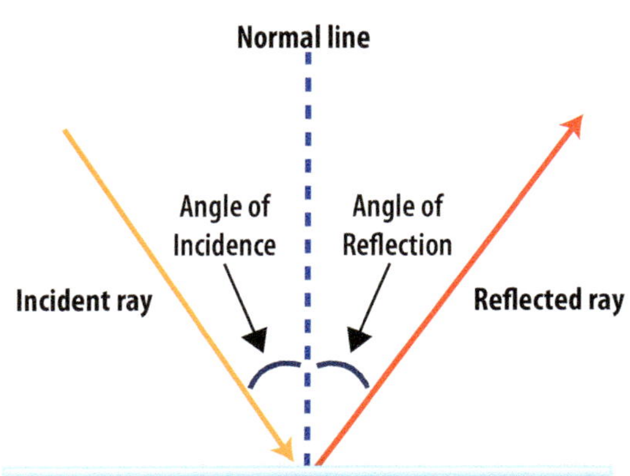

Objectives

- Compare and contrast reflection, refraction, and diffraction of light.
- Compare and contrast the behavior of light with convex and concave mirrors and lenses.
- Explain how polarized light is created.
- Describe how light waves interact.
- Describe how light interacts to produce color.

Vocabulary

law of reflection
index of refraction
diffraction
interference
opaque object
translucent material
transparent material

Explore-a-Lab

Structured Inquiry

What are the properties of a reflected beam of light?

Stand a mirror at one end of a table so that it is perpendicular to the tabletop. Lay a laser pointer at the other end so that the light shines along the table on the way to the mirror. Observe the reflected light. Move the laser pointer to several positions, paying close attention to what the beam of light does. What happens to the beam of light when you move the laser pointer? What can always be said about how a beam of light reflects from the mirror?

Refraction

When you look at an object through different mediums, it may appear to bend. This optical illusion can be explained by refraction. Notice in the photo that the spoon does not appear to be straight because the light from the bottom part of the spoon is bent, or refracted. Why is this considered an optical illusion?

Refraction occurs because the speed of light depends on the medium through which the light waves are traveling. In space, light waves travel at 300,000 km/s. But light travels more slowly through materials such as air and water. Why do you think this happens? Would ice water refract light more than hot water? Explain. How could you verify your explanation?

The way in which a material refracts light is known as its **index of refraction**. The index of refraction can be calculated by dividing the speed of light in a vacuum by the speed of light in the medium. When would the index of refraction be large? When would it be small? What property of the matter will increase the index of refraction?

Through which material in the glass—air or water—does light travel more quickly?

Index of Refraction

Material	Index
Vacuum	1.00
Air	1.00
Ice	1.31
Water (20°C)	1.33
Sugar solution (30%)	1.38
Sugar solution (80%)	1.49
Sodium chloride	1.54
Diamond	2.41

Explore-a-Lab

Structured Inquiry

 **Why does energy bend as it moves between mediums?**

Roll a can of fruit juice down a ramp into water. At the same time, roll an identical can down the same ramp into sand. What happens when the can leaves the ramp? Draw your observations. Repeat and roll the can down the ramp onto concrete or tile, carpet, and grass. How did the results differ in the different mediums? Why do you think that happened? How does this explain why light waves bend as they move through different mediums?

Diffraction

Suppose a friend calls you from around the corner of a building. You can hear the friend's voice, but you cannot see her. Both sound and light are forms of energy that travel in waves. However, the wavelengths of these two types of waves differ. Sound waves have long wavelengths, and light waves have short wavelengths. As a result, sound waves can change direction more easily as they pass around a barrier in their path. The bending of waves around a barrier or through an opening is known as **diffraction**. The amount of diffraction for a wave depends on its wavelength and the size of the barrier or opening. Why are the edges of a shadow blurry? How is diffraction like refraction? How is it different?

The light in the picture has passed through several diffraction gratings, which split and diffract light into several beams that travel in different directions.

Reflection and Mirrors Explain

Print some letters from the left to the right on an index card. Hold the card in front of a mirror and read the letters from the image in the mirror. What is unusual about the letters? Where have you seen something similar happen? How does the path of light cause this unusual image?

Reflection and Mirrors

You learned earlier in a science class that reflection of light is the reason why you can see objects in a mirror. Mirrors reverse images. Mirrors that curve in at the center are called *concave mirrors*. When light hits a concave mirror, it converges or comes together toward the center, making the object appear magnified. Mirrors that curve outward at the center are called *convex mirrors*. The image reflected by a convex mirror appears smaller and farther away. The image, however, covers a wide area.

Think about a silver teaspoon. Which surface is the concave mirror and which is the convex mirror? Which type of mirror do you think stores often use to watch for shoplifters? Many truck and car side-view mirrors are convex to allow drivers to see a wide area. Why would this be important to the driver?

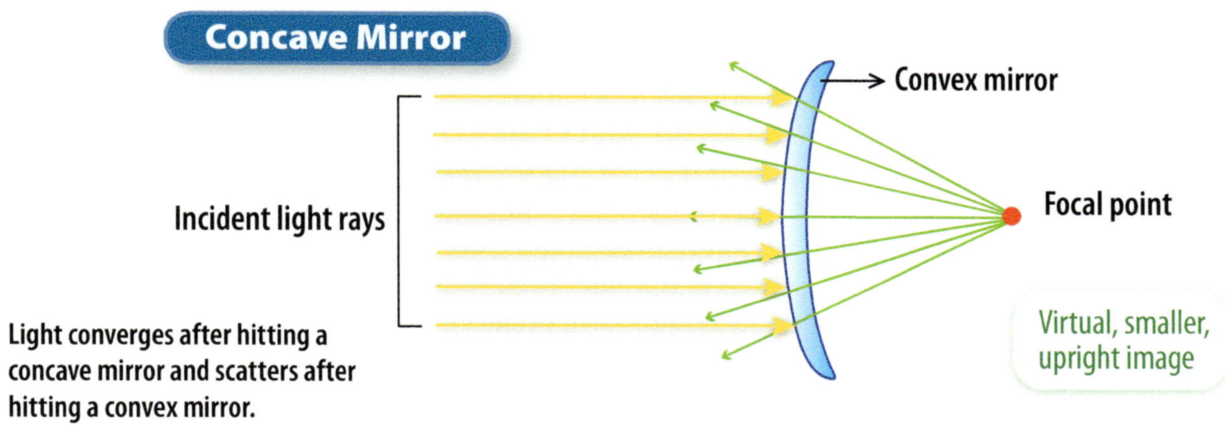

Light converges after hitting a concave mirror and scatters after hitting a convex mirror.

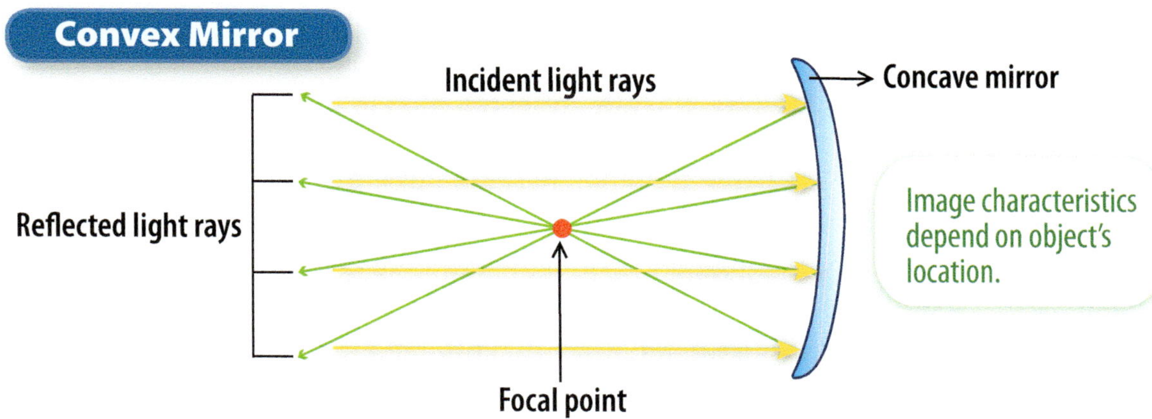

Refraction and Lenses

Unlike mirrors that reflect light, lenses refract light. Just as there are convex and concave mirrors, there are also convex and concave lenses.

A convex lens is thicker in the middle than at the edge. Notice how the light rays refract as they pass through the lens. In what two places does the light bend? As you can see, the convex lens bends light rays together at one location called the *focal point*.

Convex lenses usually magnify objects. When you look through the magnifying lens, objects look larger than they are. A three-power (3X) lens makes objects appear three times larger; a two-power (2X) lens doubles the size of objects. Why can a drop of water work as a magnifying lens? Two convex lenses may be used together to form a telescope. What do you think is the purpose of the second lens?

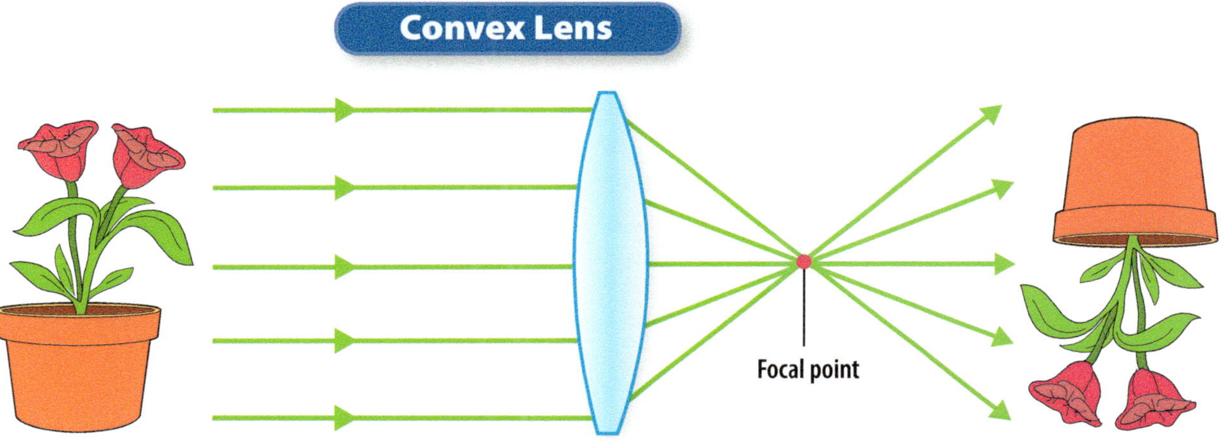

Convex lenses magnify objects.

Concave lenses are thinner in the middle than at the edges, causing light to spread out, or scatter. How do the images appear? They are used in many types of optical equipment, including cameras, projectors, microscopes, and telescopes.

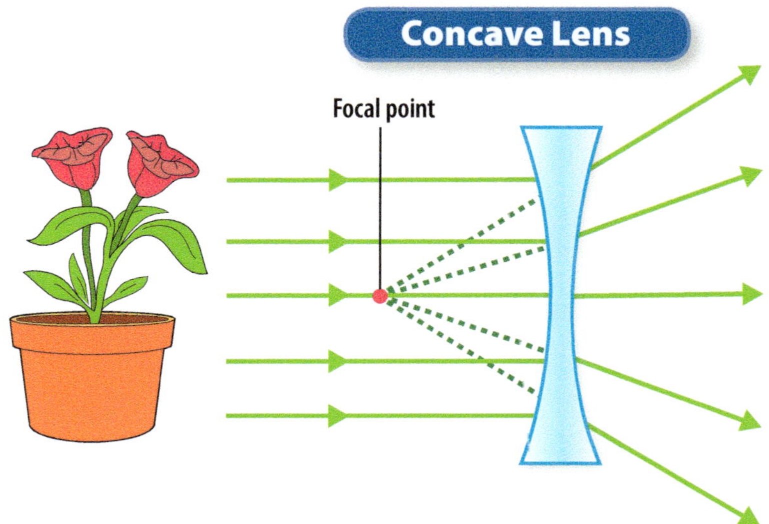

A concave lens causes light rays that pass through it to scatter.

Prisms

Refraction is also involved when light passes through a prism to produce a spectrum of colors. This spectrum occurs because the colors that make up white light have different wavelengths, so they are refracted at different angles. Shorter wavelengths are always refracted more than longer wavelengths. Which color would be refracted the least? The most?

Rainbows occur when sunlight shines through droplets of water. Each drop of water works as a tiny prism that refracts the sunlight into the visible spectrum. The raindrops, the Sun, and you must form the correct angle for you to see a rainbow.

Explore-a-Lab

Structured Inquiry

 **How does the position of the prism change what it refracts?**

Hold the prism in sunlight, and make a rainbow (spectrum) of colors on the wall or on a piece of paper. Draw a picture of what you see. Turn the prism upside down in the sunlight. Did anything change? If so, draw a picture of any differences.

Polarized Light

If you have shopped for sunglasses lately, you may have noticed that some are marked as being polarized. Polarized sunglasses have lenses that reduce the glare from sunlight.

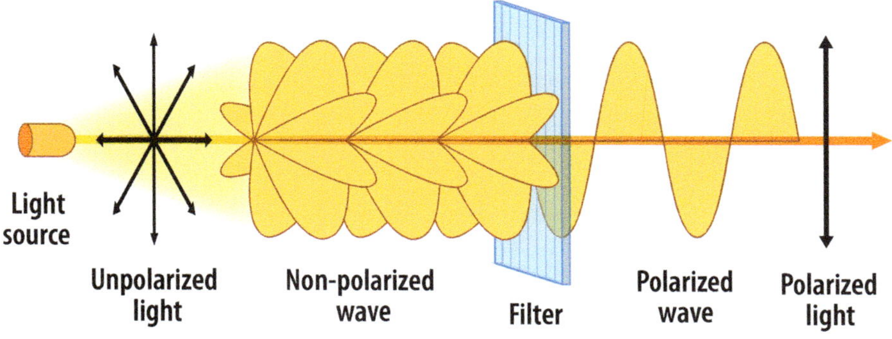

After passing through a filter, polarized light waves all vibrate in the same plane.

Explore-a-Lab

Structured Inquiry

 **How is polarized light affected when it passes through transparent plastic materials?**

You will need 2 sheets of polarizing filters and a protractor. Place the filters on top of each other, and look through both of them toward a window. Rotate one of the filters until no light passes through. Then place the protractor between the filters and look at the window. What happens? What pattern of light is created? Why does the plastic change the results?

Electromagnetic waves vibrate in numerous planes, somewhat similar to what is shown in the diagram. Light waves that are vibrating in more than one direction are called *unpolarized light*. Some examples of objects that give off unpolarized light include the Sun and the lights in a classroom.

The polarized lenses of sunglasses can transform unpolarized light into polarized light. The process of transforming unpolarized light into polarized light is called polarization. As you can see in the diagram, the lenses in polarized sunglasses reflect light so that only vertically vibrating light waves pass through them.

In general, polarized light is easier on your eyes. When might this be helpful or important? Polarized light even provides more depth perception when making observations using a microscope. If you have ever seen a 3-D movie, then you may have seen the results of light polarization. Both the glasses you wear and the projectors have filters. They work together to make the viewer see the movie in 3-D.

Wave Interactions Explain

Chances are that a light wave will encounter another light wave as it travels through a medium or space. When two or more light waves meet or overlap, the result is known as **interference**. Interference is the wave interaction that occurs when two or more waves overlap. There are two types of interference—constructive interference and destructive interference.

Constructive Interference

When the crests of the two light waves overlap, the amplitude of the combined wave is greater than the amplitudes of either of the original waves. This is called *constructive interference*. Look at the graph below. How can you find the amplitude of the new wave?

With constructive interference, the amplitude of the combined wave is greater.

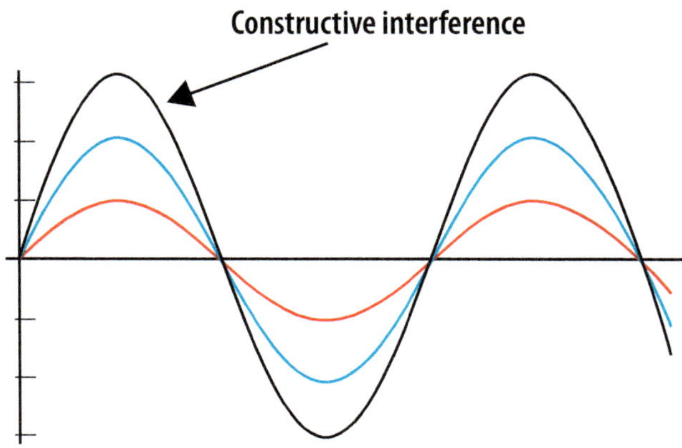

Destructive Interference

When the crest of one wave overlaps with the trough of another, the resulting amplitude of the combined waves is less than the amplitude of either of the original waves. This is called *destructive interference*. The waves could even cancel each other out completely. How could this happen? To find the amplitude of the combined wave, subtract the amplitudes of the original waves.

The two waves involved in destructive interference do not need to have equal amplitudes.

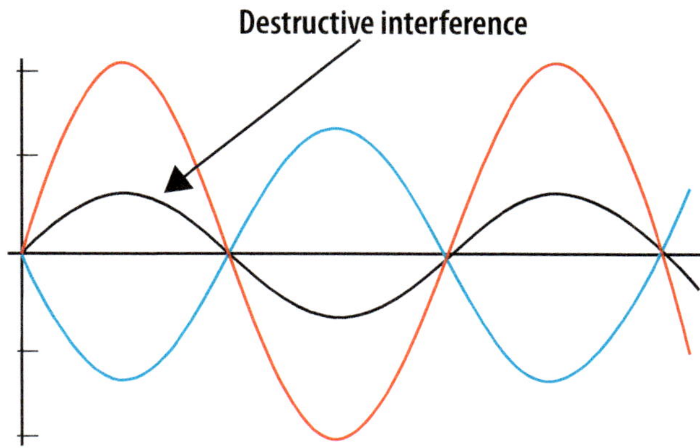

Standing Waves

A standing wave is formed when a wave interferes with its reflection. A standing wave is called that because it causes the medium to vibrate as if it were standing still. It appears to be a single stationary wave, but it is really two waves traveling in opposite directions.

Record your work for this inquiry. Your teacher may also assign the related Guided Inquiry.

Creating Interference

What are two types of interference?

Procedure

1. Peel the paper from the acrylic plastic and smooth off all edges with sandpaper if necessary. Be careful not to scratch the surfaces. Clean the top and bottom surfaces with alcohol and a soft cloth. Press the two plates tightly together and tape around the edges to hold them in place. Tape a sheet of dark construction paper to one plate to make the interference patterns more visible.

2. Hold the plates, with the dark-paper side on the bottom, in any strong source of white light. Observe the rainbow-colored interference patterns. The patterns will change as you bend, twist, or press on the plates. Notice that the patterns strongly resemble the contour lines on a topographic map.

3. Place the red plastic between the light source and the plates. What patterns did you see when twisted compared with when they were bent?

Materials

- 2 sheets acrylic plastic, ¼ or ⅛ inch (.64 or .33 cm) thick and approximately 1 foot (30 cm) square. (Size is not critical.)
- 1 piece of dark construction paper
- one 3 × 5 inch (8 × 13 cm) piece of transparent red plastic
- electrical or duct tape
- light source, such as a desk lamp

Analyze Results

Compare the patterns with and without the red plastic.

Create Explanations

When you open a package of new, clean microscope slides, you can often see colored interference patterns created by the thin air space between the glass slides. Why does this happen?

1. What are two types of interference?

2. What patterns of waves do you see when you twist or bend the plates?

3. Why do the light patterns change as you press the two pieces on acrylic plastic together?

Explore-a-Lab

Structured Inquiry

What does a standing wave look like?

Tie one end of a 2-m of rope to a wall. Move your end to create a wave. What happens when the wave reaches the wall? Continue sending waves from your end, increasing the interference until the rope appears to stand still. What is actually happening?

Visible Light and Color Explain

You already know that white light is made up of all the colors of light. What are two objects in your classroom that are lit by the same light? Are they the same color? Why or why not? Why do some objects block light while other objects allow light to pass through? Why is a strawberry red and a banana yellow? Why do some objects appear to be different colors when viewed under different lights?

Primary Colors of Light

Visible light contains six basic colors: red, orange, yellow, green, blue, and violet. Together these colors make the visible spectrum. This visible spectrum is responsible for all of the colors we see. Why do we see various colors?

Colors result from specific wavelengths of the light. If the wavelength changes, the color changes. What happens if there is no light? The wavelength of red light is about twice as long as the wavelength of violet light. Although this difference is very small, it is detectable by the eye. Scientists estimate that the human eye can detect more than two million colors. How do we know that all the other rainbow colors have wavelengths between red and violet? We can also see colors that are not one of the six basic colors. How does this happen?

The image shows how red, green, and blue light mix to form other colors. These three colors are called the three *additive primary colors of light*. They can be combined to make all of the colors we see.

Primary Colors of Light

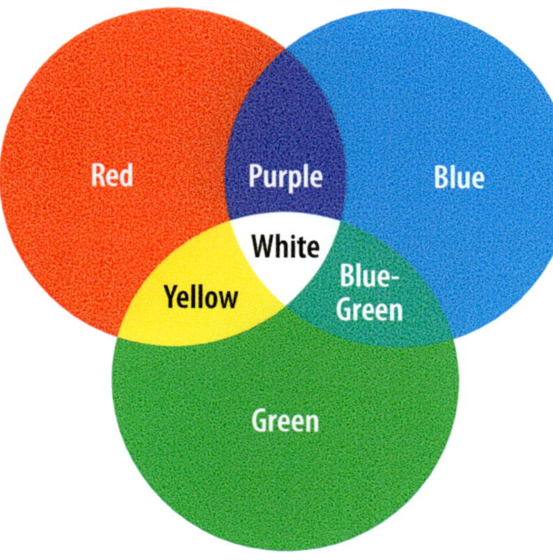

Red, green, and blue are the primary colors of light used to make thousands of different colors of light.

Colors of Opaque Objects

You know that white light is made of all the colors of light. And you have learned that light reflects off objects, such as the wall of the classroom. What colors of light are being reflected by the wall? How does the reflected light from the wall differ from the light reflected from your desk? How do these differences in color relate to the wavelength of light?

A leaf like the one in the picture is an example of an **opaque object**, which is an object that does not allow light to pass through it. What are some other examples of opaque objects? When visible light hits an opaque object, some wavelengths of visible light are absorbed by the object while other wavelengths are reflected. Only the reflected wavelengths reach your eyes. What colors are absorbed and reflected from the leaf to give it its green color? If an object absorbs all the light that hits it and does not reflect any light, what color is the object? If an object reflects all the colors of visible light, what color will the object appear to be?

Check out your ***Science Journal*** to create color wheels that display blended colors when they spin.

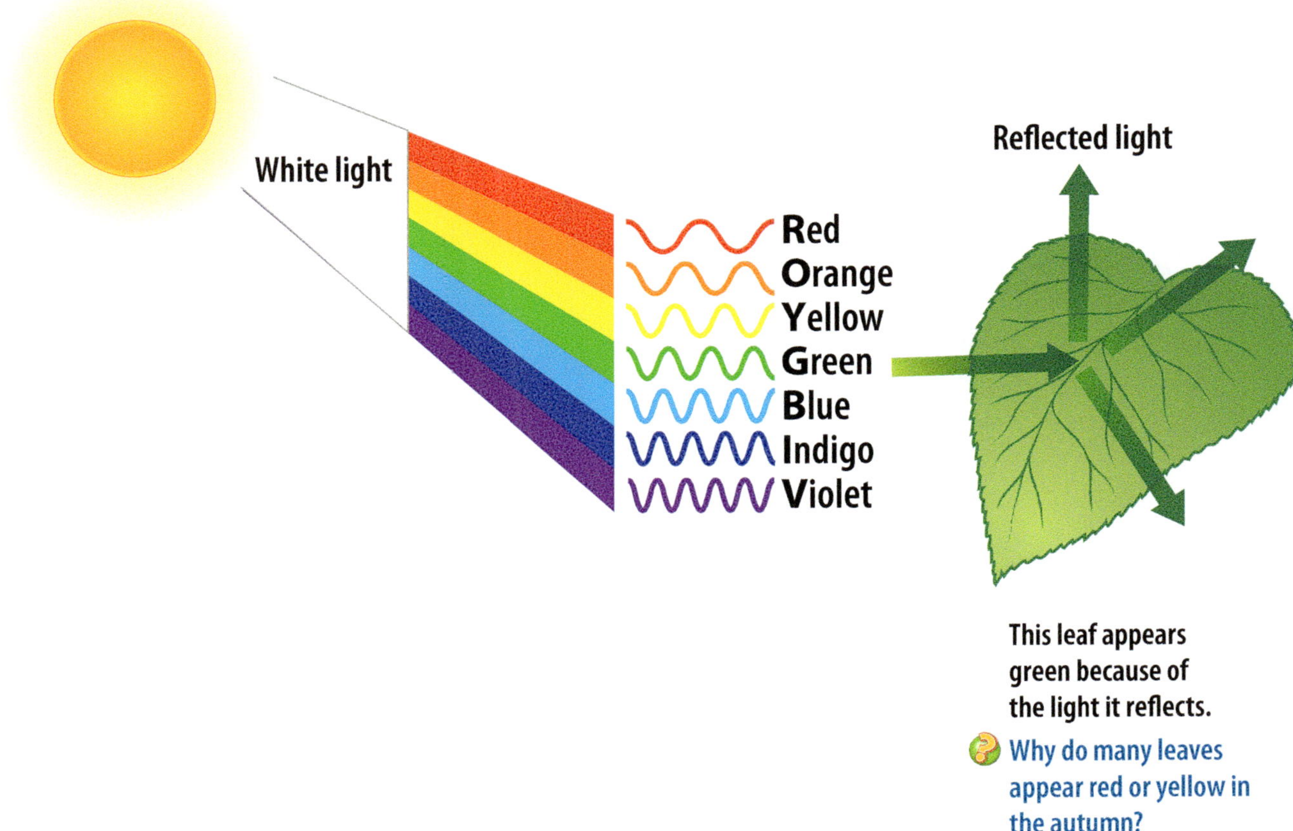

This leaf appears green because of the light it reflects.

Why do many leaves appear red or yellow in the autumn?

The glass in stained-glass windows allows only certain colors of light to pass through.

What colors of light does the green glass allow to pass through?

Colors of Translucent and Transparent Objects

How does the transmission of light differ between a piece of white paper and a piece of green cellophane? The green cellophane is a **translucent material**, which means that it allows some wavelengths of light to pass through it. A **transparent material** allows all wavelengths of light to pass through it. What are some examples of translucent materials? Of transparent materials? Which type of material do you think would heat up faster? Why?

Primary Colors of Pigments

You may recall from your art class that some paint colors are made from combinations of colors, too. These colors in paint, called *pigments*, are made differently from those in light. The three primary colors of paint are magenta, yellow, and cyan. Notice how the mixing of magenta and cyan forms purple, while cyan and yellow mix to form green. In paint, what happens if you mix all three colors together? The colors are called the *subtractive primary colors*. They work just the opposite from colors of light.

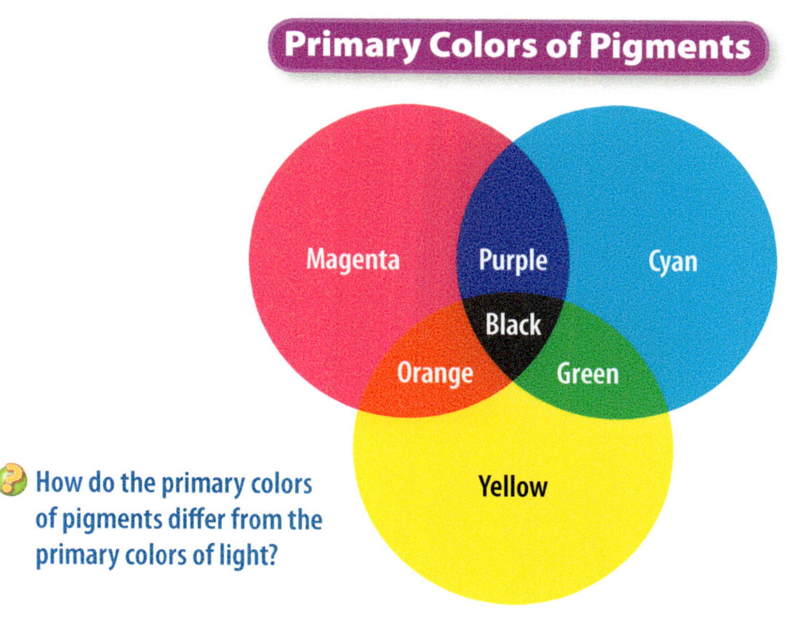

How do the primary colors of pigments differ from the primary colors of light?

Explore-a-Lab

Guided Inquiry

 **How can you combine colors to make other colors?**

A newspaper uses four colors of ink: cyan, magenta, yellow, and black to make all of the colors it needs for production. Look at the colors of the "ink" in the vials your teacher has prepared. Use these four colors to make "inks" that match your teacher's colors. Are these colors additive or subtractive?

Concept Check

Assess/Reflect

Summary: What are the properties of light? Light travels in a straight line until something interferes with its path. Its intensity can be altered by the material it goes through. Its path can be changed by reflection, refraction, or diffraction, depending on what obstacles are in its path. Concave and convex mirrors and lenses can change the image of an object. Light can be polarized by a lens so that all of its vibrations are in one plane. Light waves can experience constructive interference that increases amplitude, destructive interference that decreases amplitude, or can even become a standing wave. All color that we see comes from different wavelengths of visible light waves. The way light waves combine to make colors is different than the way pigments combine.

1. How is the visible spectrum different from the electromagnetic spectrum?
2. Some people like to look at their funny images in the mirrors at amusement parks. Why might you look different in mirrors at an amusement park from the way you look in mirrors at home?
3. Compare refraction with diffraction.
4. The intensity of the light coming through your window is bothering you while you are studying. What could you do to the window to solve this problem? Use your understanding of light to explain why that would work.
5. White light shines on an object. It absorbs all green light while reflecting blue and red wavelengths. What color will the object appear to be?

Careers in Science

Extend

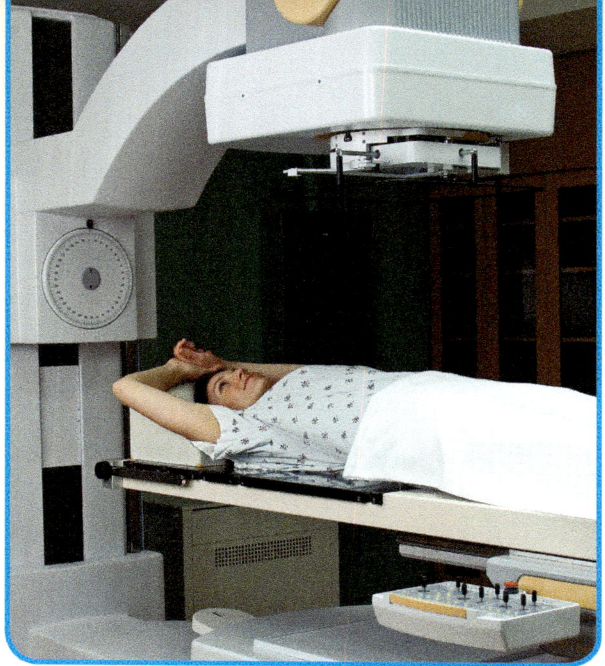

Radiation Oncologist

Radiation oncology consists of the research, management, and sharing of knowledge concerning the causes, prevention, and treatment of cancer. Radiation oncologists use various treatment methods, such as chemotherapy and radiation therapy. They treat children and adults, woman and men. The most commonly treated cancers are lung, breast, head and neck, prostate, cervix, uterine, and colorectal.

Radiation oncology involves three different medical specialties that focus on the treatment of cancer patients:

- Radiation Oncologist—a medical doctor who completes training to specialize in the management of cancer patients, specifically using radiation therapy. Radiation oncologists use cutting-edge technology and work in teams with other doctors to create and deliver radiation therapy to patients.
- Radiation Therapist—a health professional who designs, plans, and provides the radiation dose to patients.
- Radiation Oncology Medical Physicist—a scientist who creates, implements, and monitors the delivery of radiation therapy, taking into account the protection and safety of patients and others involved in the treatment process.

Radiation is the emission (sending out) of energy from any source. X-rays are an example of radiation, but so is the light that comes from the Sun. Radio waves, phones, televisions, infrared lamps, and satellites, as well as microwave ovens, all emit radiation. This type of radiation belongs to the category known as *nonionizing radiation*. It is considered nonionizing because the individual waves have too little energy to cause ionization (the removal of electrons from atoms, breaking molecular chemical bonds, which give matter structure). Nonionizing radiation does not have enough energy to directly damage DNA. Aside from the Sun's ultraviolet rays, nonionizing radiation is not known to increase cancer risk.

Ionizing radiation, on the other hand, is capable of removing electrons from atoms and breaking chemical bonds, creating highly reactive ions. It can affect DNA. Ionizing radiation is used by the radiation oncologist in the diagnosis and treatment of certain kinds of cancer. During radiation therapy, high doses of ionizing radiation (much higher than those used for X-rays) are directed at the cancer, resulting in the death of the cancer cells. The radiation oncologist focuses the radiation on the cancer cells as much as possible, which means fewer normal cells are exposed to damaging radiation that may increase cancer risk.

Concept Check

1. What are some of the treatment methods a radiation oncologist uses when treating patients for cancer?
2. How is nonionizing radiation different from ionizing radiation?

Use of Light in Photography and Film

Science and Technology

Extend

The use of light is what creates the images in a photo. Photographers manipulate and record light rays that enter the camera. There are many controls and methods for controlling light available to today's photographer.

Photography is derived from the Greek words *photos* ("light") and *graphein* ("to draw"). The word was first used by the scientist Sir John F.W. Herschel in 1839. It is a method of recording images by the action of light, or related radiation, on a sensitive material. All camera technology is based on the law of optics, which was first discovered by Aristotle. He was the first to observe that when the distance between the aperture and the light reflected from an object's surface increased, the image was magnified.

A camera is an optical device. It is lightproof, with a lens that captures incoming light and directs the light toward film (optical camera) or the imaging device (digital camera). Camera controls are all about modifying light, either how much light strikes the film or how sensitive the film is to light. The two primary ways to control the light striking the film are shutter speed and aperture. The aperture is an opening in an optical device that determines the diameter of the bundle of light rays entering the camera. Shutter speed controls how *long* light is allowed to strike the film, while aperture controls how *much* light strikes the film. These two controls work together with the ISO setting (that controls how sensitive the image sensor or film is to light) to fine tune the light recorded by the camera.

In filmmaking, the use of light can affect the meaning of a scene. A popular technique in film lighting is three-point lighting. Three lights in different positions combine to create a dramatic image. The soft frontal light is known as the fill light; the strong light at the back is known, unsurprisingly, as the backlight. The key light is usually the strongest light in the scene and has the most influence on how the subject will appear. It is placed behind the camera and, usually, off to either the left or right to create a shadow.

Concept Check

1. What are two primary ways photographers control light striking the film (or image sensor) in a camera?
2. Describe how filmmakers use lighting to create the mood of a scene.

Chapter 11 Review Assess/Reflect

Study Guide

Lesson 1

1. Energy, not matter, is transferred by waves. Sound waves transfer energy by the vibration of particles that make up the medium through which they travel.
2. The vibration of a transverse wave is perpendicular to its motion, but the vibration of a compressional wave is parallel to its motion.
3. Mechanical waves require a medium through which to travel, but electromagnetic waves can travel in a vacuum.
4. Wavelength is a measure of the distance between two corresponding parts of consecutive waves. Amplitude is a measure of half the vertical distance of the wave. Frequency refers to how many waves pass a point in a given amount of time. Wave speed is equal to wavelength times its frequency.

Lesson 2

1. Sound cannot travel through a vacuum, and it travels at different speeds through different materials.
2. The properties of a sound wave include its pitch, loudness, and intensity.
3. When sound waves hit a solid surface, they reflect. This is called an echo. Sound waves also refract. They travel faster in warm air than in cold air. As sound is absorbed by a medium, some energy is lost and the size of the waves decreases.
4. The Doppler effect explains that the perceived frequency of a sound decreases as the source of the sound moves away from the observer.

Lesson 3

1. The electromagnetic spectrum organizes electromagnetic waves according to their wavelength or frequency.
2. Photons are particles that travel in wavelike patterns at the speed of light. The different kinds of electromagnetic radiation are defined by the amount of energy in their photons.
3. Electromagnetic energy is used for heat, light, communication, and medical technologies. These uses are summarized on page 416.

Lesson 4

1. The law of reflection states that the angle at which a light ray hits the surface from the normal line is the same angle the reflected ray forms with the normal line. Light waves change direction and refract when they travel from one material to another because they change speed based on the material. Diffraction is the bending of waves around a barrier or through an opening.
2. Concave mirrors reflect light so that it converges at the center and magnifies objects. Convex mirrors make objects appear smaller and farther away. Convex lenses refract light to magnify objects, and concave lenses make objects appear smaller.
3. Lenses can polarize light by only allowing through the light that vibrates in a single plane.
4. When waves interact, constructive or destructive interference occurs; if the interference is from the wave's reflection, then a standing wave can form.

5. Color is produced by light waves of different wavelengths. All colors of light are created by combinations of red, blue, and green light waves. The primary colors of pigments are magenta, cyan, and yellow.

Vocabulary Check

Match each statement with the correct vocabulary word.

Matching Words

A. Doppler effect	**E.** decibel
B. ultraviolet	**F.** diffraction
C. translucent	**G.** radiant energy
D. compressional	

1. travels in waves and can travel through a vacuum
2. explains why sound can be perceived at different frequencies when the sound or the observer is moving
3. when light bends or spreads out as it turns or passes through a small opening
4. a unit used to measure sound intensity
5. the type of wave that can be used to disinfect water
6. the type of object that only allows some light to pass through

Multiple Choice

Choose the best answer.

7. Which of these electromagnetic waves has the highest energy?
   A. visible light waves
   B. microwaves
   C. X-ray waves
   D. infrared light waves

8. Five waves pass a certain point in 10 seconds. What is the frequency?
   A. 0.2 Hz
   B. 0.5 Hz
   C. 2 Hz
   D. 50 Hz

9. What is the spread-out portion of a compressional wave called?
   A. rarefaction
   B. reflection
   C. refraction
   D. rebound

10. Which describes a convex lens?
   A. It is thinner in the middle.
   B. It magnifies the image.
   C. It alters the color of the image.
   D. It scatters light.

11. What medium would sound travel fastest in?
   A. water
   B. steel
   C. wood
   D. feathers

Check Point

Answer the following questions.

12. Explain why compressional waves are unable to move through a vacuum.
13. **Compare** how an increase in wavelength affects sound and light waves.
14. Sometimes it is very hard to hear in a gymnasium because of echoes. How can that be changed?
15. Scientists say we actually see the light of the Sun before it rises above the horizon. What can you **infer** is happening? Explain.

Chapter 12 Motion and Forces

Scripture Spotlight

Think about all the ways you can move your body. How do all the objects around you move? What forces act on those motions? How do you think God planned for motion and forces during Creation? By studying forces and motion, you can better understand God. In this chapter, you will read the following passages.

Ezekiel 10 (p. 439)
Isaiah 13:13 (p. 462)
Psalm 26:1 (p. 468)

The thrill of participating in sporting activities is enhanced by learning to control the forces that influence motion.

The Big Idea

God's natural laws control forces in the Universe and how forces affect objects and other forces.

What forces are acting on this snowboarder?

Inquiry Kick-Off Engage

Just as in snowboarding, learning to control forces can help you achieve many tasks. What other forces can your body produce? How can you control those forces? In your ***Science Journal***, you will have to control one of your body's forces to win a friendly competition.

Lesson 1

Essential Question

How Can Motion Be Described?

Objectives

- Describe how a reference point is used to observe motion.
- Calculate the motion of an object in terms of its speed.
- Distinguish between average speed and instantaneous speed.
- Analyze a graph to determine the speed of a moving object.

Vocabulary

reference point
qualitative
quantitative
average speed
instantaneous speed
frame of reference

How can you tell whether you are moving? Which of your senses gives you clues? Think about riding in a car. Which senses give you clues when you look out the window? Which senses give you clues if you shut your eyes? Have you ever thought you were moving when you weren't? How can you describe your motion to someone else so they can understand what you mean? How might a scientist describe this same motion?

Motion Explain

It's easy to tell that you are moving in a car, especially when your eyes are open. All you have to do is look out the window and see that you are passing by the road, trees, or other cars passing by. What the road, trees, and other cars represent are reference points. A **reference point** is another object that appears to stay in place. Usually a reference point is stationary. However, a reference point may also be another object that is moving. Another car that is traveling along the same road can be a reference point. When could you not use this car as a reference point to determine that your car is moving?

You may recall that motion is an object's change in position relative to a reference point. Motion may be either a change in the distance or a change in the direction from a reference point. Usually you can observe the motion of an object, but not always. Most glaciers only move a few centimeters a day. The motion of plates in Earth's crust is rarely seen. The plates only move an average of about 2 cm per year. Without a reference point, this motion would be undetectable. During what kind of event would movement of Earth's crust be visible?

You can judge your speed by how fast the landscape moves past your window.

How could you calculate your speed without using the speedometer?

Distance and Time

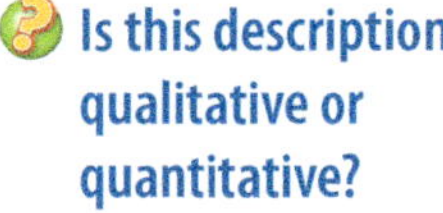

To say something is moving fast or slow is a **qualitative** description because it has no numbers. A more accurate description could use measured quantities. It would be a **quantitative** description if it uses numbers. Motion involves both distance and time. By using both distance and time, you can be more specific in describing the motion of an object. For example, you would be more specific by saying that your car is traveling 50 km/hr rather than by saying that your car is going slower than the other cars on the road. Saying that your car is going 50 km/hr involves numbers and is a quantitative description. Saying that your car is going slower than the other cars does not involve numbers and is a qualitative description. Scientists usually describe motion in quantitative terms. Why is it better for speed limits to use quantitative measurements rather than qualitative measurements?

To describe motion in quantitative terms, both the distance traveled and the time it took to travel that distance can be measured. Suppose it takes 30 min to walk from your house to a park that is 2100 m from your house. You could tell someone that you walked the 2100 m in 30 min. These two values can be used to determine speed. Recall that *speed* is the distance traveled divided by the time it took to travel that distance, as shown in this equation:

$$\text{speed} = \frac{\text{distance}}{\text{time}} \text{ or } s = \frac{d}{t}$$

To calculate your speed during the walk, divide the total distance of 2100 m by the total time of 30 min:

$$\text{speed} = \frac{\text{distance}}{\text{time}} = \frac{2100 \text{ m}}{30 \text{ min}} = 70 \text{ m/min}$$

Our Milky Way Galaxy is moving towards the Andromeda Galaxy, a neighboring galaxy shown above, at about 112 km/s.

Is this description qualitative or quantitative?

Scripture Spotlight

Read **Ezekiel 10**. How do you think the wheels were able to go in all directions without turning? Try to demonstrate this movement with your body.

Record your work for this inquiry. Your teacher may also assign the related Guided Inquiry.

Measuring Motion

How do you determine speed over a crooked path?

Procedure

Materials
- masking tape
- stopwatches or timers
- meterstick
- compass
- graph paper
- toy car
- string

1. Working in a group, use masking tape to mark a starting point for the toy car. Then mark three other points where the path will change direction and the finish point of the trip. Make a tape line that connects the three points. **Measure** and **record** the length of each section.
2. Make a labeled **diagram** of the path showing the measurements between each point.
3. Use a compass to determine the overall direction of a line between the start and finish points of the car's path. Record the direction on the diagram.
4. Tie the string to the front of the toy car.
5. Pull the toy car along the taped path. The car should follow the taped line as close as possible. Time how long it takes for the car to make each leg of the trip. Record the time for each leg of the trip on the diagram.
6. Repeat Step 5 two more times.

Analyze Results

Calculate: **(a)** the total distance; **(b)** the speed for each section of the trip; **(c)** the total distance divided by the total time for the entire trip; **(d)** the direction from the starting point to the finish point. **Construct a graph** of distance versus time for the car.

Create Explanations

1. How do you determine speed over a crooked path?
2. Did the object travel the same speed for each section of the trip? How does the graph show this?
3. How does your calculation of total distance divided by total time compare with the speeds in each section of the trip?
4. Do you think the object traveled at a constant speed in any one section? What methods or measurements could you have used that would enable you to determine whether the speed at any instant was equal to the average speed for that section?
5. Was the direction important in calculating the speeds?

Average Speed versus Instantaneous Speed

Chances are that the object in the Structured Inquiry did not move at a constant speed all of the time. Your calculations gave you an average. Even in the supposed walk from your house to the park discussed earlier, can you imagine that you would have walked faster at certain times and slower at other times. As you walked to the park, your speed was not always 70 m/min. Therefore, 70 m/min represents your average speed. **Average speed** is the total distance traveled divided by the total time it took to travel that distance, as shown in this equation:

$$\text{average speed} = \frac{\text{total distance}}{\text{total time}}$$

This corresponds to a walking speed of about 2 km/hr, or about 1.2 m/s. What would be your average speed if you walked the 2100 m in 40 min?

Explore-a-Lab

Structured Inquiry

 **How do you find average speed?**

Find the distance and time to walk the length of your classroom. Then find the time to run back the same length. Calculate the speed for the walking section and for the running section. Then calculate the average speed for the entire distance. Did you expect the average speed for the entire trip to be halfway between the walking speed and the running speed? Why or why not? Why wasn't it?

During a walk, you probably move faster at some times and slower at others. The speed at any given moment is called the **instantaneous speed**. Runners like to know their instantaneous speed and may use a GPS device or phone application that shows their approximate instantaneous speed. Suppose you had used such a device on your walk to a park, and it recorded that your position changed by 20 m in 0.25 min. If you were walking at constant speed, your instantaneous speed at any moment during that time would have been:

$$\begin{aligned} s &= d/t \\ &= 20 \text{ m}/0.25 \text{ min} \\ &= 80 \text{ m/min} \end{aligned}$$

Why must your instantaneous speed be less than this value at other moments along your walk to the park?

Math in Science

During a bicycle ride, the cyclist takes 6 s to cover the first 20 m. She then travels 30 m in 5 s at constant speed.

1. What is the average speed of the bicyclist during any part of the first part of the ride? The second part of the ride?

2. What is the average speed for the entire ride?

Frame of Reference

Imagine you are on a field trip, sitting beside a friend on a school bus. If you determine motion from the reference point of your friend, you are sitting still. However, if the reference point is the road beneath you, you could be moving forward at 50 km/hr or more. Determining the motion of an object depends on the **frame of reference**, which is the position of an object in relation to other objects in the area. All motion is relative to a frame of reference. If you could look down at the road through a hole in the floor of the bus, you would seem to be sitting still, but what direction would the road seem to be moving?

Sometimes drivers will speed up in a whiteout snowstorm or dense fog without realizing it. This is because they have lost their frame of reference. They don't see passing objects such as trees, poles, guardrails, and so on to give a sense of speed. They can be tricked by this lack of input into thinking they are at rest, and that will tend to make them speed up. What happens when you are riding your bike faster than you thought you were and you have to turn a sharp corner?

Suppose you are sitting on a subway car or rapid-transit car waiting for it to leave the station. Suddenly you feel like you are moving forward. Are you? How do you know? Why might you feel as if you are moving when you are not? Will a frame of reference always help you determine whether you are moving? Why or why not?

Motion is determined by a frame of reference.

What could this train passenger use as a frame of reference to detect motion?

Explore-a-Lab

Structured Inquiry

How can you use a frame of reference to estimate distance?

Have a blindfolded student sit on a chair and hold the hands of other students standing beside the student at shoulder height. Have four students lift the four corners of the chair so that the student is slowly raised several centimeters off the floor, gently swaying the chair. At the same time, have those students holding the hands of the seated student simultaneously slowly lower their hands to knee height. Ask the seated student to estimate the distance he or she is from the floor based on how it feels to the student. Repeat with other students if desired. How can the senses be tricked in this manner? What is the blindfolded student being guided to use as a frame of reference?

Displacement Explain

Motion is not always in a straight line, and sometimes the important thing is where a journey ends, not all the detours along the way. People refer to measuring this distance using the expression "as the crow flies." You may recall that the length and direction of a line joining the starting position to the finish "as the crow flies" is called the *displacement*. In what ways is displacement different from distance?

Consider again your walk from your house to the park that was previously discussed. Suppose that the path traveled was three blocks east and four blocks north. The total distance over the seven blocks was 2100 m. The purple arrow in the figure below represents the displacement of this trip. It is described in words as "fifteen hundred meters thirty-seven degrees northeast." Would the displacement be different if the trip to the park started out by going four blocks north and ended going three blocks east?

Displacement

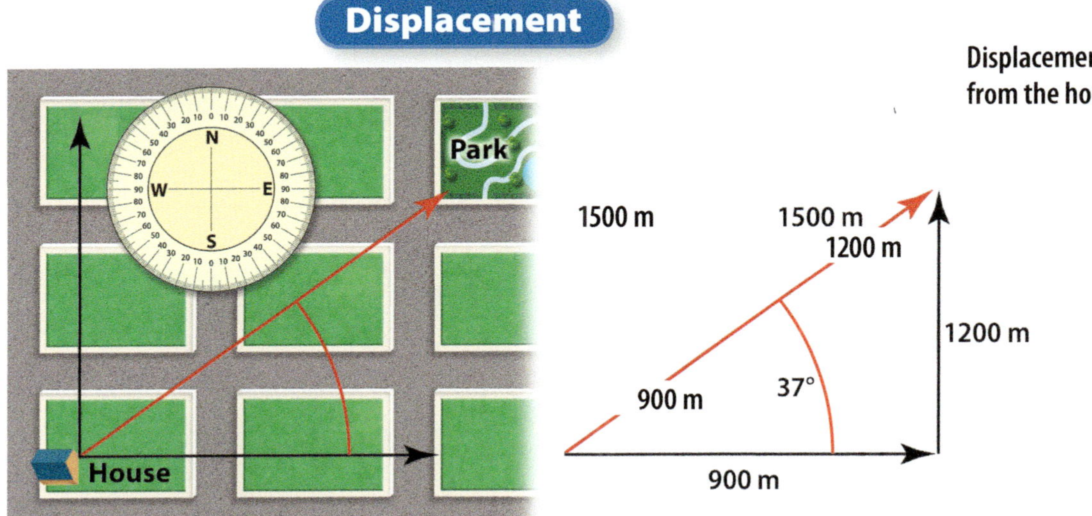

Displacement of the walk from the house to the park

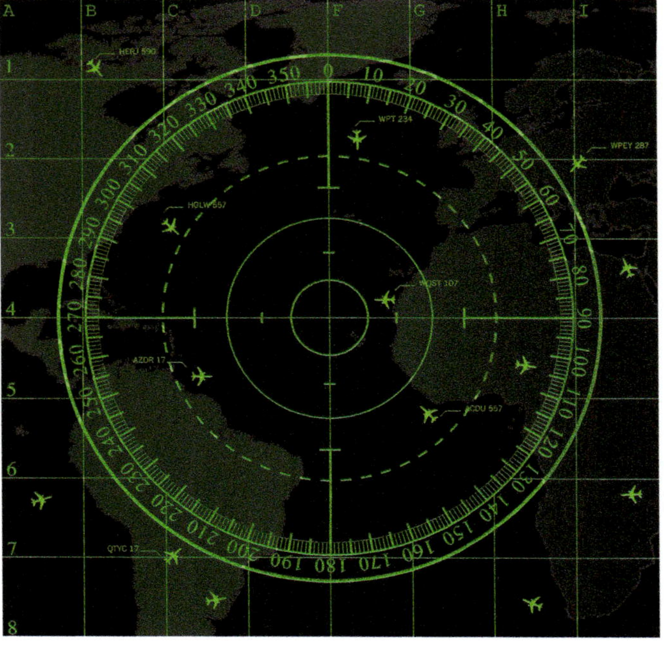

This radar screen is centered on the Atlantic Ocean to show the direction and position of planes.

Velocity

An aircraft controller is responsible for the safety of all aircraft in the vicinity. To do this, it is not enough to know the positions and altitudes of the planes. The controller must also know precisely the speed, direction, and velocity of all planes around the airport. Recall that *velocity* is the measurement of the speed and direction of an object. It is found using this equation:

$$\text{velocity} = \frac{\text{displacement}}{\text{time}}$$

To calculate your average velocity during the walk to the park, divide the displacement of 1500 m at 37° NE by the total time of 20 min:

$$\text{velocity} = \frac{\text{displacement}}{\text{time}} = \frac{1500\text{ m}}{20\text{ min}} = \frac{75\text{ m}}{\text{min}} \text{ at } 37^\circ \text{ NE}$$

What would be your *average velocity* if you took 50 min to walk from the house to the park?

Lesson Activity

Use your measurements for the Structured Inquiry already completed in this lesson to calculate:

a. the displacement of the toy car

b. the average velocity of the toy car

Graphing Speed

A distance versus time graph can illustrate the speed of an object. Suppose you decided to jog to the park rather than walk. The data table below represents this trip with 5-min walks at the beginning and at the end. Draw a line graph of your trip. Put time on the horizontal (*x*) axis, using intervals of 5 min up to 25 min. Put distance on the vertical (*y*) axis, using intervals of 300 m up to 2400 m. Add a title to the graph and label each axis, including the units used.

Time (min)	0	5	10	15	20	25	30
Distance (m)	0	350	1050	1750	2100	500	0

You can see from your graph that the line changes at several points. These changes represent changes in speed during the course of your trip. The first part of the graph and the last part of the graph should be similar in the amount they are slanted. The middle part of the graph should have a steeper slant. The steeper slant of the line is due to a larger distance being covered during the time elapsed and therefore represents a faster speed. How would the middle portion of the graph change if you ran faster?

The graph below shows the average speed of two objects. Again you can see that the steeper the slant of the line, the faster the speed of the object. The fastest commercial jet was the Concorde. It could maintain a cruising speed of 600 m/s. What would the slant of its speed look like on this graph? What would the slant of your speed of 4 m/s look like from your walk to the park described earlier?

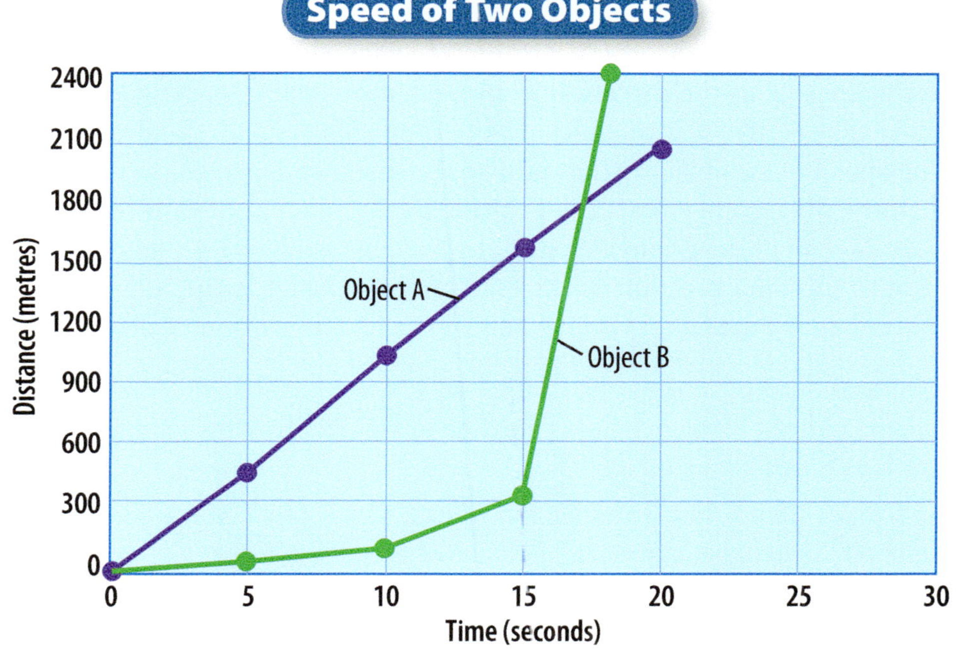

The graph shows the distance traveled from a reference point by two objects.

The lines on the graph show a different type of motion for object B than for object A. How would you describe that motion? You can tell that the speed is constant because the line is straight.

$$\text{speed} = \frac{\text{distance}}{\text{time}} = \frac{(75\text{ m} - 50\text{ m})}{15\text{ s} - 10\text{ s}} = \frac{25\text{ m}}{5\text{ s}} = 5\text{ m/s}$$

The instantaneous speed of object B is changing during that time. However, it is possible to calculate the *average* speed of object B from 5 s to 15 s using a similar process. How does the average speed of object B during that time compare with object A's speed? When was object B going fastest?

Summary: How can motion be described? Motion involves a change in position, either a change in distance or in the direction from a reference point. A statement about motion is more descriptive if it includes numbers so that there are quantitative measurements. The speed of an object is the ratio of the distance traveled divided by the time taken. A frame of reference fixes the position of the reference point relative to other objects. Displacement describes the straight-line distance and direction from the starting position to the final position of travel. Average velocity is calculated by dividing the displacement by the time. Graphs can illustrate the motion of objects.

1. Using the sidewalk as a point of reference, suppose you were inline skating north at 3 m/s. What is a point of reference that could have you not moving? Could have you moving south at 8 m/s?
2. A puppy runs 15 m west and then runs back 3 m east in 6 s total. Make a simple sketch of the motion. What is the average velocity of the puppy?
3. Explain how speed is indicated on a distance versus time graph and how the numerical value can be calculated from the graph.

Lesson 2

Essential Question

How Are Acceleration and Momentum Determined?

Objectives

- Demonstrate how to calculate acceleration.
- Explain the law of conservation of momentum.
- Distinguish between elastic and inelastic collisions.

Vocabulary

slope
momentum
elastic collision
inelastic collision

In a circular waterslide, the initial rush as you first accelerate down the tube is surpassed by the feeling of acceleration around turns and in further drops. What keeps you moving in flatter, straight sections? How is your velocity changing in your downward rush to the pool? Why do small children sometimes get stuck on the way down a waterslide?

Change in Velocity Explain

Most things don't move at the same speed forever. They usually do not even move in the same direction for very long. Whenever you have a change in either the speed or direction of an object, you have acceleration. *Acceleration*, you may recall, is the rate of change in speed or direction or both. In other words, acceleration is the rate of change in velocity. Why do you suppose the gas pedal in a car is called the accelerator? Why could the brake pedal also be called an accelerator?

Acceleration, with its changing speed and changing direction, is part of the exhilaration of the waterslide experience.

Swimmers at the start of a race

Can an object be accelerating even though its speed stays the same? Think about an object moving in a circle at constant speed. When you ride a Ferris wheel at its top speed, can you feel the acceleration? Any change in direction of motion is acceleration. If you ride a bicycle around a corner, you are accelerating even if your speed does not change.

Acceleration is calculated by dividing the change in velocity by the time over which the change occurred.

$$\text{acceleration} = \frac{\text{final velocity} - \text{initial velocity}}{\text{time}}$$

When competitive swimmers on the starting blocks hear the sound of the horn, they leap forward with all their strength. This gives them the best start for their race. Suppose a swimmer accelerates for a half-second (0.5 s) to reach a velocity of 3 m/s forward. What is the swimmer's average acceleration during that time?

The initial velocity was zero, and the final velocity was 3 m/s forward, so:

$$\begin{aligned} \text{acceleration} &= \frac{\text{final velocity} - \text{initial velocity}}{\text{time}} \\ \text{acceleration forward} &= \frac{(3\text{ m/s} - 0\text{ m/s})}{0.5\text{ s}} \\ &= \frac{3\text{ m/s}}{0.5\text{ s}} \\ &= \frac{3\text{m}}{0.5\text{ s} \times \text{s}} \\ &= 6\text{ m/s}^2 \end{aligned}$$

The swimmer had an acceleration of 6 m/s^2 forward at the start of the race. Velocity is expressed in meters per second, while time is expressed in seconds. Therefore, acceleration is expressed in meters per second per second, or meters per second squared (m/s^2). This means that in the case of the swimmer, he is changing his velocity by 6 m/s. The direction is also given for acceleration, just as it is for velocity and for displacement.

Record your work for this inquiry. Your teacher may also assign the related Guided Inquiry.

Accelerated Motion

How can you measure, describe, and predict accelerated motion?

Materials

- marble track
- marble
- meterstick
- stopwatches
- wood blocks
- modeling clay
- graph paper

Procedure

1. Working in a group, mark off the provided track so that it is at least 2 m long. Mark off 1.0-m distances down the track and record the end distance.
2. Use blocks to set the height of one end of the track so the marble will roll consistently, but slowly enough that it can be timed accurately. Use clay to stabilize the ends of the track.
3. One person in the group holds the marble at the top of the track. One member of the group should be positioned at the 1.0-m mark with a stopwatch, and another member of the group should be positioned at the 2.0-m mark with a stopwatch.
4. When the marble is released, the timers should start their stopwatches. When the marble passes their position they are to stop their stopwatches.
5. **Record** the time from each distance in your ***Student Journal***.
6. Repeat Steps 3 and 4 two more times.
7. Raise the starting end of the ramp 5 cm and repeat Steps 3–5.

Analyze Results

Graph your results and find the average time for the marble to run the track. Why do you think you had a variety of times? What variables could you control? What variable could you not control?

Create Explanations

1. How can you **measure, describe**, and **predict** accelerated motion?
2. What similarities and differences do you see when you compare your graph with those of your classmates? Why would that be?
3. What does the shape of the graph tell you about how the velocity of the marble changed as it rolled down the track?

Tables and Graphs of Motion Explain

As you have experienced in your work on motion, a table of values is another way of describing motion. Suppose someone dropped a rock down a deep mineshaft. The table below shows the distance downward of the falling rock for the first 5 s. Can you see from the numbers that the rock is falling a greater distance every second? This indicates that the rock is accelerating downward as it falls. What do you think the distance will be at 6 s? What is the average speed of the rock during the first 5 s of its fall down the mineshaft?

Distance a Rock Falls over Time

Time (s)	Distance (m)
0	0.0
1	4.9
2	19.6
3	44.1
4	78.4
5	122.5

What would a graph of its motion look like? As you can see below, the line graph shows the distance versus the time. The curved line on a distance-versus-time graph tells you the object is accelerating. A distance-versus-time graph of a drag car starting a race would also have a curved line. How does a graph describe motion differently than a table does?

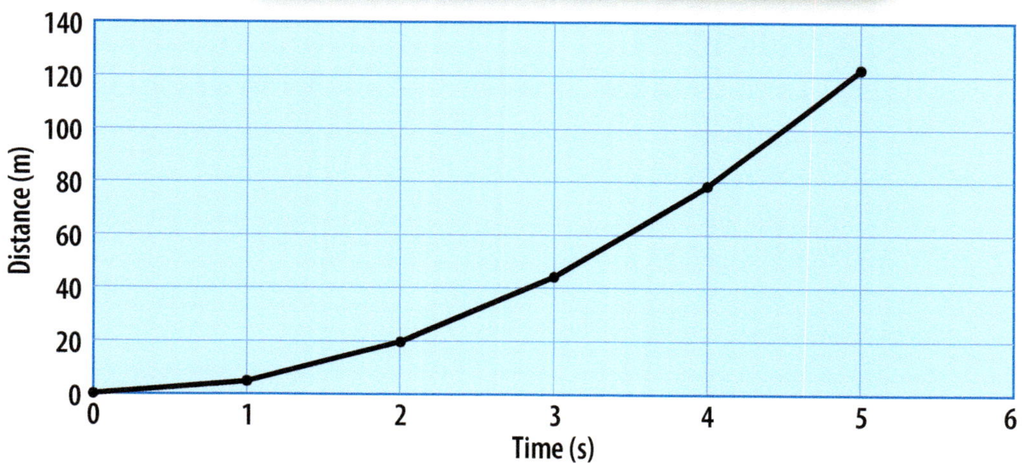

Lesson Activity

Construct a displacement-versus-time graph using the data from your toy car in the earlier Structured Inquiry. What does the shape of the graph line tell you about the motion? Can you tell from your displacement-versus-time graph whether the object had constant acceleration, or do you need to look at your velocity-versus-time graph?

 **What can you learn from a graph?**

The top graph on this page displays the downward velocity of the rock as it falls. What can you say about the increase in velocity each second? Can you predict the velocity at 6 s?

The falling rock had constant acceleration, so its velocity-versus-time graph produces a straight line. How does that line compare to your velocity-versus-time graph of the marble rolling down a track?

Not all acceleration is constant. A drag car in a race would accelerate at a constant rate. Near the end of the race, it would not be accelerating as rapidly as at the beginning, partly because of greater air resistance at high speed. The velocity-versus-time graph of its motion would be a curve because its acceleration was changing. Can you see that the slope of the line decreases over time in the bottom graph on this page?

Negative Acceleration

After passing the finish line, the drag car would begin to slow down rapidly. The velocity-versus-time graph would curve downward instead of upward. When an object is slowing down it has a negative acceleration. The word *deceleration* is a term often used for negative acceleration. It is the opposite of acceleration. What device is used to change the acceleration of the drag car as it passes the finish line?

Velocity vs. Time for a Falling Rock

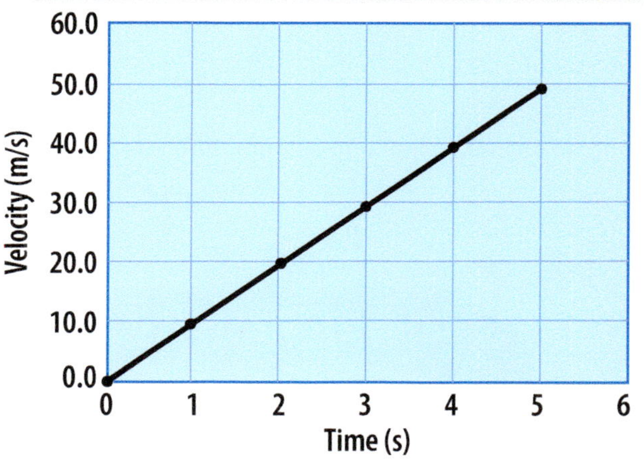

Velocity vs. Time for a Drag Car

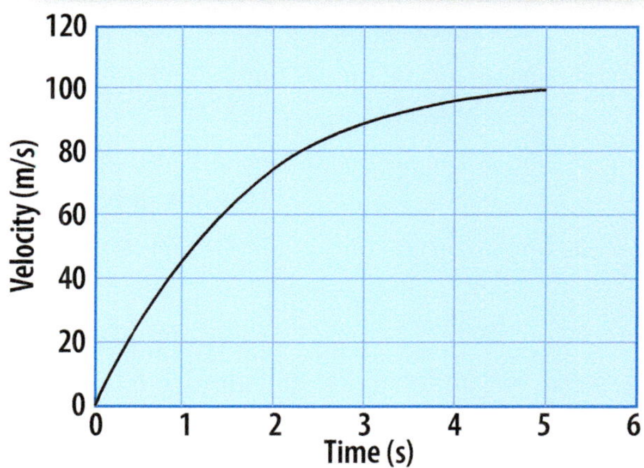

The graph of velocity vs. time for a drag car shows that the acceleration gradually decreased.

Slope

The **slope** of a line graph is the ratio between the change in the *y*-value (rise) and the *x*-value (run). The slope of a velocity-versus-time graph can tell you the acceleration of the object, which you may have discovered in your Structured Inquiry of the object rolling down a track. The steeper the slope, the greater is the acceleration. The image below shows a calculation of the slope of the velocity-versus-time graph for a falling rock. Would you get the same slope if you chose different points?

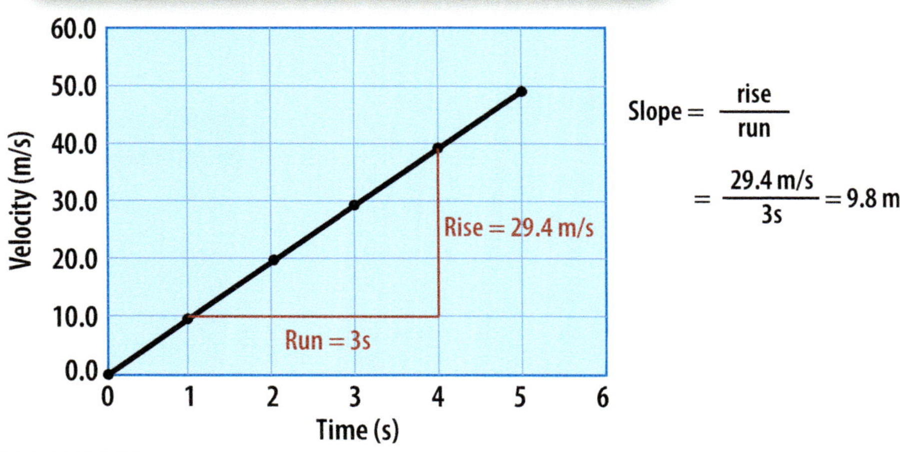

Lesson Activity

Choose two different points other than the ones used in the graph above and use them to calculate the slope. Also calculate the acceleration from the data using the formula for acceleration given earlier in the lesson.

 **How do these answers compare with the calculation on this page? Why is that?**

Momentum Explain

An object in motion has momentum. **Momentum** is the force that an object has due to its motion. It is calculated by finding the mathematical product of the mass and the velocity of an object.

$$\text{Momentum} = \text{mass} \times \text{velocity}$$

In this equation, the mass of an object is expressed in kilograms and its velocity is expressed in meters per second. Remember velocity includes both speed and direction. Therefore, like velocity and acceleration, momentum must indicate direction. The direction is always the same as the object's velocity. For example, if a mass of 40 kg were moving south at 2 m/s, its momentum would be 80 kg·m/s south. If later it was moving west at 0.5 m/s, it would have a momentum of 20 kg·m/s west.

What would be your momentum if you were running north at 2.5 m/s? In terms of the strength of its force, do you think the mass or the velocity plays a more significant role in an object's momentum? Why?

Suppose a 1250 kg automobile moving at a velocity of only 0.01 m/s north bumped you. You would definitely feel the momentum of the car as it gave a noticeable push. How could a car with less mass give you the same push? Suppose a 0.15-g bumblebee hit a motorcyclist traveling at 30 m/s north in the face. The bee has a tiny mass. But at that velocity, the effect of its momentum can be painful. It doesn't matter whether the bee or the rider is moving fast; what matters is their velocity relative to each other.

Conservation of Momentum

The bee and rider both had momentum before impact. What happens to that momentum after they collide? All of the bee's momentum is transferred to the rider, causing the pain. Similarly, some of the rider's momentum is transferred to the bee at the point of impact. The bee would have no chance of surviving all that momentum. This example illustrates the law of conservation of momentum. This law states that *any time objects collide, the total amount of momentum stays the same.* In other words, momentum is always conserved. What are some other examples of momentum being conserved?

Check out your ***Science Journal*** that investigates the conservation of momentum.

Extend

Math in Science

Solve the following problems about momentum.

1. Calculate the momentum in kg m/s.
- **a.** a 70-kg cheetah traveling at 25 m/s south
- **b.** a 45-kg student bicycling at 8 m/s east
- **c.** a 95-kg passenger on an airplane that is stopped

2. At what velocity must a 3000-kg wrecking ball travel forward to achieve a momentum of 15,000 kg m/s?

3. Which has a higher momentum—a 0.058-kg tennis ball traveling at 40 m/s or a 0.62-kg basketball traveling at 4.4 m/s?

Sometimes conservation of momentum is not obvious. Is momentum conserved if you are standing on the ground and throw a heavy box forward? What moves backward? You and Earth do, but Earth is so massive that the extremely tiny movement is undetectable. What happens if the box hits the ground and stops? Earth moves forward again the same tiny amount it moved backward. Conservation of momentum always holds. What would happen if you wore inline skates and threw a heavy box forward?

Explore-a-Lab

Structured Inquiry

 **Can you explain the motion of the tennis ball using conservation of momentum?**

Drop a basketball and a tennis ball from waist height and note how high each bounces. Then place the tennis ball on top of the basketball and drop them simultaneously from waist height. Note how high the tennis ball bounces. Repeat with the two balls together again, but this time notice how high the basketball bounces.

Collisions Explain

When objects collide, they may bounce off one another. In what kinds of situations have you seen this? These types of collisions are examples of elastic collisions. An **elastic collision** occurs when objects collide so that kinetic energy is conserved along with momentum. Recall that *kinetic energy* is the energy of an object due to its motion. When two objects collide, the total combined kinetic energies of the objects remain the same before and after the collision, but the individual kinetic energies change. Some of the kinetic energy from one object is transferred to the other object. For example, in an elastic collision, one molecule has a kinetic energy of four units, and another molecule has a kinetic energy of five units before the collision. If the first molecule has a kinetic energy of six units after the collision, the second molecule will have a kinetic energy of three units. The sum

Collisions between hard spheres, such as in the Newton's cradle, are close enough to perfect for scientists to use them in studying elastic collisions.

What is the path of the flow of kinetic energy in this Newton's cradle?

will be nine units of kinetic energy both before and after the collision. Few examples of perfect elastic collisions exist because some kinetic energy is lost as sound or heat energy in the collisions. Collisions of gas molecules are an example of something that is close to a perfect elastic collision.

Not all objects that bounce off each other have the same total kinetic energy as they did before they collided. An example would be a bowling ball hitting bowling pins. Some of the kinetic energy of the rolling bowling ball is transferred to the pins, while the rest is converted to other forms. What other forms of energy result from a bowling ball hitting the bowling pins? The objects in this collision do not have the same total kinetic energy as they did before they collided. This is known as an inelastic collision. An **inelastic collision** is one in which some of the kinetic energy converts to another form of energy. What evidence of the conversion of kinetic energy to other forms of energy are seen in a car crash?

Momentum is conserved in an inelastic collision, but kinetic energy is not.

You can summarize the relationship among collisions and the laws of the conservation of energy and the conservation of momentum in this way:

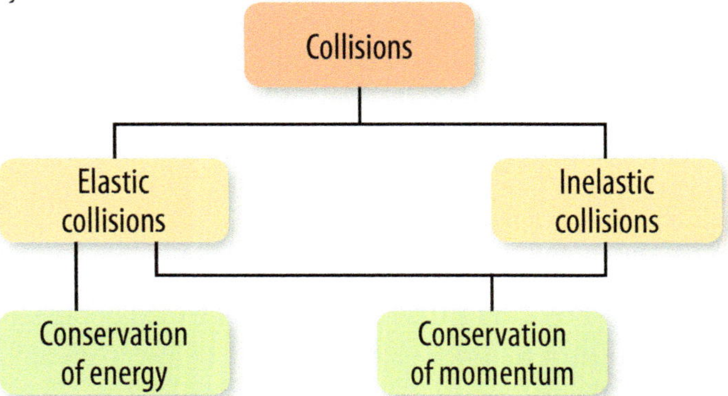

Summary: How are acceleration and momentum determined? Acceleration is a measure of the rate of change in velocity. The displacement-versus-time graph of accelerated motion would be a curved line. To calculate acceleration, find the difference between the final velocity and initial velocity. Then divide by the time.

$$\text{acceleration} = \frac{\text{final velocity} - \text{initial velocity}}{\text{time}}$$

The slope of the velocity-versus-time graph gives the acceleration. An object moving at constant speed can be accelerating if its direction is changing. The momentum of an object is the product of its mass and its velocity.

$$\text{Momentum} = \text{mass} \times \text{velocity}$$

Momentum is conserved in all interactions. This means that the total momentum of a system before an interaction will equal the total momentum of the system after the interaction. In an elastic collision kinetic energy is also conserved; however, in inelastic collisions kinetic energy is not conserved.

1. How would you describe acceleration in your own words to a younger student?
2. A 250-kg snowmobile is traveling north at 12 m/s. Suppose that 15 s later it is going south at 18 m/s.
   a. What is the acceleration of the snowmobile?
   b. What is its momentum at the end of the 15 s?
3. The line on a velocity-versus-time graph slopes down and to the right. What does this tell you about the motion being depicted?
4. In an elastic collision, one molecule has a kinetic energy of three units and another molecule has a kinetic energy of five units before the collision. If the first molecule has a kinetic energy of six units after the collision, what will be the kinetic energy of the second molecule?

Lesson 3

Essential Question

How Do Forces Affect Motion?

In previous lessons, you learned how to describe motion. But why do objects move, and why do they stop moving? Why do they change directions? Suppose you wanted to cause a sled to move easier or faster. What would you do? If you were on the sled heading for a tree, what would you do to change the direction of the sled? How can you push or pull something without any contact with it? What keeps Earth spinning on its axis and orbiting the Sun? What keeps it from slowing down? In this lesson, you will develop your ability to explain and control motion, not just describe it.

Objectives

- Distinguish between net force, balanced forces, and unbalanced forces.
- Distinguish between contact and noncontact forces.
- Describe Newton's laws of motion.

Vocabulary

net force
balanced forces
unbalanced forces
contact force
noncontact force
action-reaction force pair

Force Explain

You have already learned that scientists define *force* very simply. A *force* is just a push or a pull. For example, you use force when you pull a sled through the snow. You also use force if you push the sled. A force can change the acceleration of an object by changing either its speed or direction. Force is measured in the SI unit of newtons, abbreviated N. A newton equals the amount of force it takes to accelerate a 1-kg mass at a rate of 1 m/s^2. What force is needed to give a 1-kg mass an acceleration of 4 m/s^2? Any time you see an object change its speed or direction, you can conclude that a force was acting on it. When a batter swings a baseball bat for a hit, is the force a push or a pull, or both?

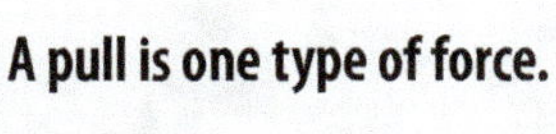
A pull is one type of force.

Faith Connection

In the same way that physical forces cause movement, the Holy Spirit enables spiritual "movement" or growth. How has the Holy Spirit moved you?

Balanced and Unbalanced Forces

Often several forces act on an object at the same time. For example, your pull on the sled rope is not the only force acting on a sled when you pull it. What are some other forces acting on the sled? The combination of all the forces acting on an object is known as the **net force**. The forces that result in the net force can be of the same magnitude or of different magnitudes. For example, the two forces can both be 4 N, or one force can be 4 N, while the other force is 6 N. These forces can also be acting from the same direction or from different directions.

Explore-a-Lab — Structured Inquiry

How does net force affect motion?

With a small group of students, hold a tug-of-war contest using a thick rope. When opposing groups pull with equal forces, what is the net force? How can you tell? If one group pulls with greater force so the forces are unbalanced, what happens to the rope? How does the net force affect the motion of the rope?

For example, both forces can be pushing an object upward, or one force can be pushing it upward, while the other force is pushing it downward. Illustrations can be very helpful in analyzing how the forces acting on an object produce the net force. An arrow is used to indicate the direction of the force, and the amount of force applied is given in N. The length of the arrow is an indication of the strength of the force. For example, consider the following illustration.

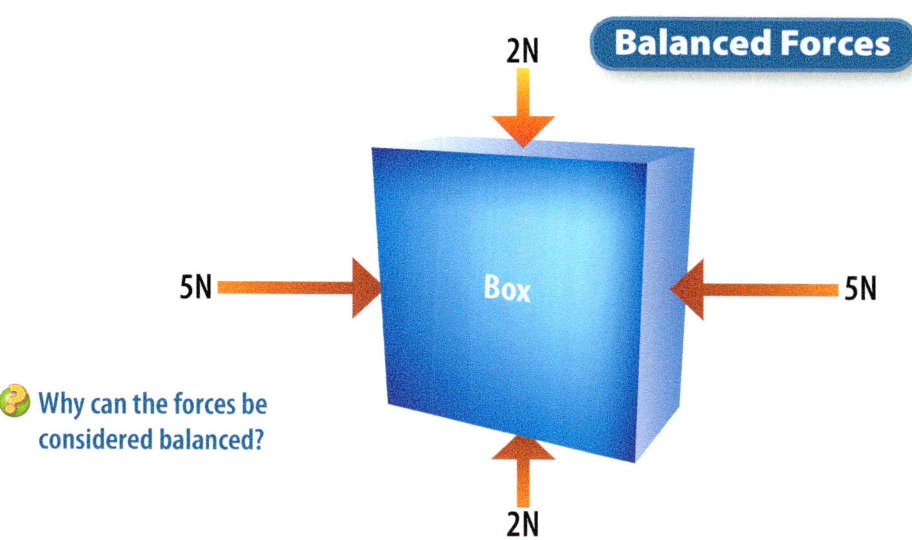

Why can the forces be considered balanced?

Notice that 5 N are pushing the object to the right and 5 N are pushing the object to the left. In addition, 2 N are pushing upward on the object, while 2 N are pushing downward on the object. In this case, the forces are said to be balanced. If equal forces act in opposite directions, they are called **balanced forces**. What will happen to this object if a balanced force is being applied? What is the net force in this case?

In the top figure below, 2 N are still pushing upward on the object, but now 5 N are pushing downward on the object. There are also 2 N pushing the object to the right and 2 N pushing the object to the left. The horizontal forces are balanced. However, the vertical forces are not. When unequal forces act in opposite directions, there are **unbalanced forces**. There also is an unbalanced force if only one force acts on an object. What is the net force on the box? In what direction will the box accelerate?

Both examples illustrate unbalanced forces that will cause a change in an object's motion. The object may speed up, slow down, or change direction.

Notice in the bottom example that neither the vertical forces nor the horizontal forces are balanced. There is a net upward force of 2 N and a net force to the right of 2 N. How would you describe the direction in which the box will accelerate due to these forces?

Unbalanced Forces

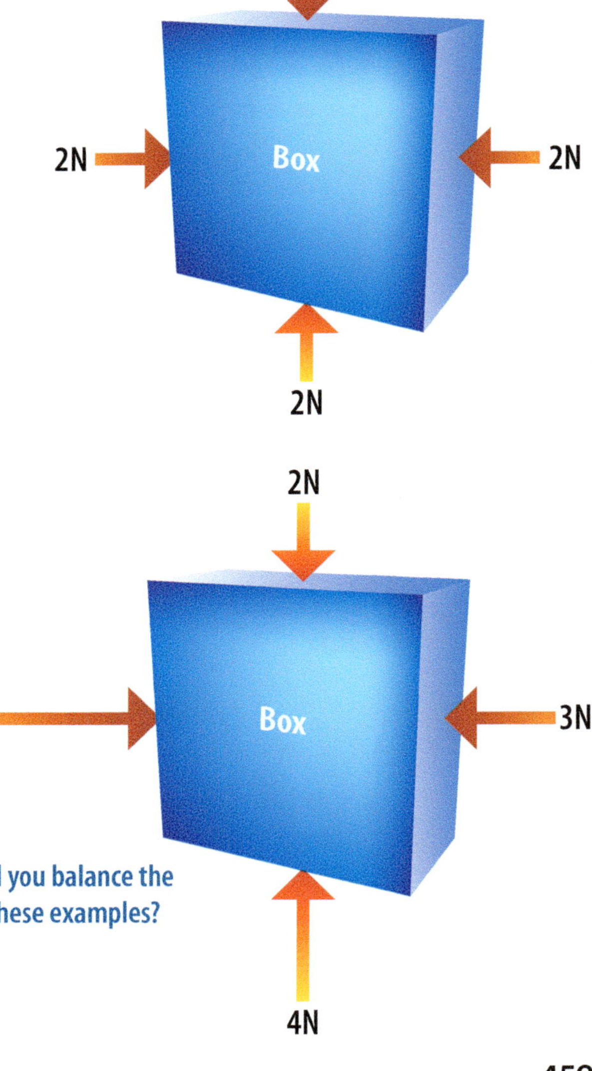

How could you balance the forces in these examples?

Record your work for this inquiry. Your teacher may also assign the related Guided Inquiry.

Unbalanced Forces

How do unbalanced forces affect motion?

Materials

- wooden block
- stopwatch
- meterstick
- string
- pulley
- ring stand
- pulley clamp
- masking tape
- weights
- graph paper

Procedure

1. Set up a stand and pulley on the end of a lab bench or table so that a weight hanging from a string draped over the pulley can pull the block horizontally.
2. Attach a string to the block and run it along the table and over the pulley. **Measure** the distance from the tabletop to the floor. **Record** this distance in the data table in your ***Science Journal***. Measure this distance on the tabletop. Put a piece of tape at this start position. While one person holds the block, attach a small weight to the other end of the string. Mark the start position with masking tape and then allow the block to slowly move down the bench until the weight touches the floor. Mark the end position the same way. Record the displacement between these two positions.
3. Return the block to the start position and time how long it takes for the block to go from the start to the end position when the force of the weight is pulling on the string. Record that time. **Calculate** the speed of the block and record the speed in the data table in your ***Science Journal***.
4. Repeat Step 3 for two more trials.
5. Increase the weight and repeat Steps 3 and 4.
6. Repeat Step 5 three more times.
7. Complete the remaining calculations outlined in your ***Science Journal***.

Analyze Results

Construct a graph of the average speed of each set of trials.

Create Explanations

1. How do unbalanced forces affect motion?
2. What can you conclude from your data and graph about the effect of an unbalanced force on the acceleration of the block?
3. Would it have been better to average the values for the three trials at each weight before graphing them, instead of graphing all the trials? Why or why not?
4. What can you **conclude** from your data and graph about the effect of mass on the acceleration of the block?

Contact and Noncontact Forces

If you push or pull on an object, you usually have to touch it. You can also exert a force with something that touches the object, such as using a cord or a push-pole. Forces exerted by objects touching are called **contact forces**. Pulling a sled with a rope is an example of using a contact force. You exert a contact force on the rope, and the rope exerts a contact force on the sled.

Give an example of a force that can act without contact. When forces affect objects from a distance without a physical connection, they are called **noncontact forces**. Magnetic forces are noncontact forces. Earth's magnetism moves compass needles without touching them. How does a magnet affect a small nail as the magnet moves toward the nail?

Check out your *Science Journal* to investigate how to measure momentum being conserved.

Explore-a-Lab

Structured Inquiry

 **Why would the distances be different?**

Put a small magnet on a table and see whether you can get it to move by moving another magnet right under the table. How could you get it to move by pushing it and also by pulling it with the lower magnet? Measure and record how far away the two magnets must be from each other so that there is no apparent effect. Repeat the measurement and recording when one magnet is suspended from a string and the other is brought near it.

Static electricity is another noncontact force. Have you ever had flyaway hair? Static electricity causes the ends of your hair to push one another away without them touching. The hairs fly away from one another because of the negative charges that build up on each hair. Recall that similar charges repel one another. Static electricity is also the noncontact force that causes an inflated balloon that you rub vigorously on a carpet to stick to a wall. How does static electricity act as a noncontact force in this case?

Static electricity generated on the slide pushes strands of hair away from each other.

Scripture Spotlight

Read **Isaiah 13:13**. What will God do with His force? Describe the force you think He will use.

Explore-a-Lab

Structured Inquiry

 **What happens when charged particles move closer together?**

Blow up a balloon and tie it off with a string so there is about 1 m of string hanging free. Give one side of the balloon an electrical charge by rubbing it vigorously against your hair or a wool cloth for half a minute. Mark carefully with a felt pen the place on the balloon where you rubbed. What happens to your hair when you hold the rubbed piece near your head? Why? When you hold your balloon and a classmate's balloon close using the strings, how do the marked portions of the balloons behave? Why? What happens when you put your hand near the marked portion? How do you explain that? Measure and record how far apart the balloons need to be so that the movement of one does not affect the other. How does that compare with the distances in the magnets activity? Why? Compare your distances with those of other groups of students. If your distances are different, why could that be?

Gravity, a third noncontact force, is a force of attraction between objects due to their masses. The Sun's gravity pulls on Earth from a great distance and keeps it in its orbit. The Moon's gravity also pulls on Earth and is the main cause of ocean tides. How does gravity act as a noncontact force differently from magnetism and static electricity?

Laws of Motion Explain

Early Greek philosophers thought that once an object was set into motion, the reason the object eventually stopped was due to a absence of a continuous force. Galileo pondered the idea, taking friction into consideration. He deduced that a force is not needed to keep an object in motion, but rather a force is needed to stop an already moving object. Isaac Newton built on Galileo's work and eventually developed three laws of motion.

Explore-a-Lab

Structured Inquiry

How does an accelerometer show acceleration?

Get a canning jar with lid, a candle, string, a fishing weight, a lighter, and water. Cut a piece of string so that it is about 1 cm shorter than the height of the jar. Tie the fishing weight to one end of the string. Light the candle and use melted candle wax to attach the other end of the string to the center of the bottom of the lid. Allow the wax to harden. Check to make sure that the string is securely attached to the lid. If it is not, add more wax and allow it to cool. Fill the jar with water. Place the fishing weight into the water and screw the lid on tightly. Hold the jar steady in front of you and start walking. Observe the position of the weight as you accelerate and as you move at a constant speed. What direction does the weight move as you accelerate from a resting position? What is the position of the weight when you are walking at a constant speed in a straight line? Does the weight move toward or away from the direction of acceleration? Why?

Newton's First Law of Motion—Law of Inertia

You may have to push or pull a sled to get it started moving. Objects, such as a sled, have a tendency to resist any change in motion. Therefore, a sled that is not moving will tend to stay at rest. Similarly, an object that is in motion will tend to stay in motion. This property of matter is called inertia. Recall that *inertia* is the tendency of an object to resist any change in motion, either by changing its speed, or its direction if it is moving. This concept of inertia is known as Newton's first law of motion, or the law of inertia.

Newton's Second Law of Motion—Law of Acceleration

Imagine that a friend gets on a sled and asks you to give a push. No problem—you apply a little force and you get the sled and your friend moving. From what you learned in the Structured Inquiry, what will happen if you push twice as hard? How would the motion be different if there were two friends on the sled instead of just one?

The example of the sled and your friends illustrates that acceleration depends on both mass and the force applied to that mass. This relationship is known as Newton's second law of motion, which states that the acceleration of an object is inversely proportional to the mass of the object and directly proportional to the unbalanced force that is applied. This means that the greater the mass, the less the acceleration, but the greater the unbalanced force, the greater the acceleration. This law can be stated mathematically with the following equation, where F is the net force, m is mass in kg, and a is acceleration in m/s^2.

$$a = \frac{F}{m}, \text{ or } F = m \times a$$

In what units is force measured?

Math in Science

Suppose a loaded grocery cart had a mass of 50 kg.

a. What force is needed to give the cart an acceleration of 0.20 m/s^2 south?

b. What would be the acceleration of the cart if you pushed it east with a force of 25 N?

Newton's Third Law of Motion—Law of Action-Reaction

So far, you have been looking at what happens to an object when either a balanced force or an unbalanced force is applied to it. For example, suppose that you are applying an unbalanced force to a sled to get the sled and your friend on it moving. But you are not the only one applying a force. At the same time, your friend and the sled are applying a force on you. You feel the rope pulling back every time you pull forward. Your forward pull and the rope's backward pull form an **action-reaction force pair**. The two forces are equal but act in opposite directions. They also act on different objects. Whenever one object applies a force to a second object, the second object applies an equal force in the opposite direction to the first object. This is known as Newton's third law of motion. You have probably heard the simplified version of Newton's third law: "For every reaction there is an equal and opposite reaction." How does this apply to the propulsion of a jet ski?

Action-reaction forces can also be used to explain rocket propulsion. The rocket forces gases backward at high speed. The reaction force is that gases push the rocket forward. This is consistent with the law of conservation of momentum, which you learned about in the previous lesson. How do action-reaction forces explain how an inflated balloon behaves when the untied top is released?

Explore-a-Lab

Structured Inquiry

 **How does a balloon rocket demonstrate Newton's third law of motion?**

Gather several differently sized balloons, a drinking straw, 3 m of nylon fishing line, transparent tape, and a chair or table. Working with a partner, attach one end of the string to the table or chair. Partially blow up your balloon and use several pieces of tape to attach the straw to the balloon. Blow up the balloon to full capacity and twist the end so no air leaks out. Work with your partner to thread the string through the straw. The open end of your balloon should be facing the loose end of the string. Have you partner hold the string taut. Release the balloon and measure how far it travels. Vary the amount of air you put in the balloon and record your results. How will using a differently sized balloon change your results? Try it and compare your observations. What other variables could you change? How does this experiment model Newton's third law of motion?

Concept Check

Assess/Reflect

Summary: How do forces affect motion? The net force is the sum of all the pushes and pulls on an object. A balanced force means the net force is zero. Unbalanced forces are required to produce motion. Forces exerted on objects by touching them are contact forces. Some forces, such as gravity and magnetism, are noncontact forces and act on objects without physical contact. An object tends to remain at constant velocity unless an unbalanced force acts upon it, which is Newton's first law of motion. Newton's second law of motion states that the acceleration of an object is directly proportional to the net force acting on it and inversely proportional to its mass. Newton's third law of motion says that for every force acting on an object, there is an equal force acting in the opposite direction on another object.

1. A 40-kg sack of rice experiences simultaneous forces of 10 N north, 20 N south, 392 N downward, and 392 N upward. What is the net force on the bag of rice? What is the acceleration of the sack of rice?
2. Compare the effects of the inertia of an object in motion with the inertia of the same object at rest.
3. What are the two action-reaction force pairs when a ball is hung by a string? Describe the direction of each of the four forces and the objects that exert those forces. Include whether they are contact or noncontact forces.
4. Use Newton's second and third laws to justify the law of conservation of momentum.

Lesson 4

Essential Question

What Are Some Forces around Us?

Objectives

- Describe how friction affects motion.
- Distinguish between kinetic friction and static friction.
- Explain the advantages and disadvantages of friction.
- Explain the law of universal gravitation.
- Explain why objects that are thrown follow a curved path.
- Describe common forces that act on objects.

Vocabulary

kinetic friction
static friction
elastic force
buoyant force
centripetal force
mechanical force
magnetic force
electrical force
nuclear force

Who doesn't like to be in control of things? If you controlled all forces, you could control all motion. Would that give you superhero powers? The better you understand the forces involved, the more you will be able to control your physical world. What forces are acting on you at this very moment? What forces would you like to control if you could pick your top three?

Friction Explain

Think about the last time you rode a bike. How did you slow it down? What made the wheel grab the road and push you forward? Why didn't the wheel just spin in place when you pushed on the pedal? Now think about the game of air hockey. What allows the disk to glide so easily over the tabletop? What if the disk were twice as heavy? Would it still glide so easily? Why or why not? *Friction* is a force that opposes the motion of an object. It occurs when two surfaces touch each other. Would friction be considered a contact or a noncontact force? Rub your hands back and forth together. Does friction change when you reverse directions?

Winter in some places brings ice and snow. That also means that objects slide better because of reduced friction. Ice skating, sledding, skiing, and snowboarding are now possible. In warmer climates, people may water ski, surf, and wakeboard using the low friction between water and solid objects. What other sports use low friction?

Air hockey uses low friction in an entertaining contest.

How is the amount of friction changed in air hockey?

Record your work for this inquiry. Your teacher may also assign the related Guided Inquiry.

Friction Factors

What are some of the factors that affect the force of friction between two objects?

Materials

- wood block with hook
- string
- spring scale
- round pencils
- weights
- graph paper

Procedure

1. **Measure** and **record** the weight (N) of a wood block.
2. Measure and record the force needed to start the block moving from rest as the force of static friction.
3. Measure and record the force needed to keep the block moving at a constant velocity once it has started moving. Call it the force of sliding friction.
4. Turn the block onto its narrow side. Then measure and record the force needed to keep the block moving at a constant velocity once it has started moving. Call it the force of low-surface-area sliding friction.
5. Measure and record the force needed to keep the block moving at a constant velocity when it is sitting on a row of spaced parallel pencils oriented perpendicular to the line of travel. Call it the force of rolling friction.
6. Add the data from Steps 1 and 3 as trial 1 in the data table for part B of this activity in your ***Science Journal***.
7. Add weight to the block approximately equal to the original weight of the block and record the total weight in the data table.
8. Measure and record the force needed to keep the block moving at a constant velocity once it has started moving.
9. Repeat Step 8 at least two more times.

Analyze Results

What is the ratio of the force of sliding friction to the force of rolling friction in your experiment? **Construct three line graphs**, one for each set of data you collected.

Create Explanations

1. What are some of the factors that affect the force of friction between two objects?
2. How does the force of rolling friction compare with the force of sliding friction?
3. Based on your data, how does the area in contact with the surface affect the force of friction?

Types of Friction

The force opposing motion when two solid surfaces in contact are moving past one another is called **kinetic friction**. *Kinetic* is a word that refers to motion. There are three types of kinetic friction.

- *Sliding friction* occurs when two surfaces rub against each other.
- *Rolling friction* occurs when an object rolls on a surface.
- *Fluid friction* occurs with motion of an object in a fluid.

What are everyday examples of each type of friction?

It is harder to get an object moving than it is to keep it moving. That is partly due to the object's inertia. Another factor is **static friction**, which is the friction between two surfaces when they are not moving relative to each other. The force of static friction is greater than the force of kinetic friction. Once an object on a steep slope starts sliding, it will keep sliding until the surface or the angle of the slope changes. Why is it a good idea to take small steps and go slowly when walking on an icy sidewalk? Why is sand often put on icy roads? What happens to static friction once an object starts to move?

Scripture Spotlight

Read **Psalm 26:1**. What kind of spiritual friction could help David not slip? What kind of friction has kept you on a strong spiritual path?

Explore-a-Lab

Guided Inquiry

 **How can gravity overcome static friction?**

Put a wood block on a board approximately 2 cm × 19 cm × 40 cm. Measure how high one end of the board has to be raised for the force of gravity to overcome the force of static friction so that the block will begin to slide. Record this height. Modify one side of the board so it can be raised at least 1.5 times the original height before the block begins to slide. Record this height. Now modify the other side of the board so it can be raised only 0.5 times the original height before the block begins to slide. Record this height.

Advantages and Disadvantages of Friction

Friction is useful when you put on the brakes to stop. It helps you hold onto things. It keeps your socks up and your shirttail tucked in. It enables you to walk and to change direction. It holds buildings together by keeping nails and screws in wood. How do football receivers increase the force of friction between their hands and the ball?

Friction is not as useful when you want to move a heavy object across a concrete floor. It takes extra work to move almost anything because friction opposes the motion. The work done against friction may result in excess heat and cause damage. For these reasons, the technology of decreasing friction in some places and increasing it in others becomes very important.

Factors Affecting Friction

Skimboarding is a relatively new activity that turns low fluid friction values into a great ride.

Smooth surfaces have less friction than rough ones, so polishing a surface will decrease friction. People sprinkle sand on icy pathways to increase friction. The rougher the surfaces involved, the stronger the frictional forces will be. Have you seen a baseball pitcher rub and rough up a new ball so it will curve more due to increased fluid friction when it is thrown?

The force of friction depends on the force with which the two surfaces are pushed together. Doubling the force doubles the force of friction. If you were pulling two friends instead of one on a sled, you would need to pull about twice as hard.

Different materials of the same smoothness have different forces of friction. Running shoes and rubber tires use special rubber compounds. These are chosen to give high friction. Winter tires have rubber that has more friction on ice than summer tires. This difference is indicated by a number called the *coefficient of friction*. The higher the number, the greater the frictional forces. If you found the slope of the graph in the Structured Inquiry on sliding a block, you found the coefficient of sliding friction for those two surfaces. This coefficient is the ratio of the force of friction over the force pushing the surfaces together. The coefficient of static friction for a rubber tire on dry asphalt is twice the coefficient for wet asphalt. Which will be higher for the same material, the coefficient for static friction or the coefficient for sliding friction? Why?

Less effort is needed to move objects on wheels, because the force of rolling friction is less than the force for sliding friction. Bicycle wheels and inline skates have bearings with smooth, steel balls in them. These balls roll inside the bearing with little friction as the wheel turns. Where else are bearings used around you?

The balls in this ball bearing are held between the inner and outer rings by a cage.

In what types of items are ball bearings used to reduce friction?

Fluid friction balances the force of gravity when skydivers reach their maximum velocity.

What do they do to increase their fluid friction?

Friction is also decreased by lubricating the surfaces. Oil and grease make things slide over one another with much less friction. Oil is a fluid, because it flows. Both air and water flow, so they are also fluids. Why is fluid friction an example of kinetic friction?

Airplanes experience fluid friction as they fly through the air. This force of fluid friction is called *air resistance*, or *drag*, and opposes the thrust provided by the engine in an action-reaction force pair. When you ride your bike, you encounter fluid friction as air resistance. What are some methods planes and cyclists use to reduce air resistance?

Other Types of Forces Explain

In addition to friction, many other forces also act on objects. Some of these forces include elastic force, buoyant force, centripetal force, mechanical force, magnetic force, electrical force, and nuclear force. As you will remember, some forces are classified as contact forces, while others are considered noncontact forces. Think about the examples of forces that follow and decide which are contact forces and which are noncontact forces.

Elastic Force

Elastic force is the force a solid exerts to return to its original shape after it has been deformed. If you stretch a rubber strap by pulling on it, it will pull back until it returns to the original length. If you compress or stretch a spring, it will return to its original size when you release it. What happens if you push or pull or twist a spring too much? Why does this happen?

Buoyant Force

Buoyant force is the upward force of a fluid or a gas on an object. An object floats in a fluid if the upward force of the fluid on the object is greater than its weight. Helium-filled balloons float because of a buoyant force of air that pushes them upward. Boats float because of the buoyant force of water that pushes them upward. What will happen to the buoyant force on a boat if more people get into it and the boat sinks deeper into the water?

Explore-a-Lab

Structured Inquiry

How does a boat's design affect its buoyancy?

Use an 8 cm × 12 cm piece of aluminum foil to build a boat and test it to see how many pennies it will support without sinking. Use a second identical piece of aluminum foil and your experience to build another boat that will hold even more pennies. Test your new design.

Centripetal Force

If you spin an object tied to a string around your head, the string stays taut. The force with which you pull to keep it going in a circle is called the **centripetal force**. If you let go of the string, which way will the object go? How do Newton's laws of motion explain this? What are some everyday uses of centripetal force?

What would happen to the seats on this ride if the ride slowed?

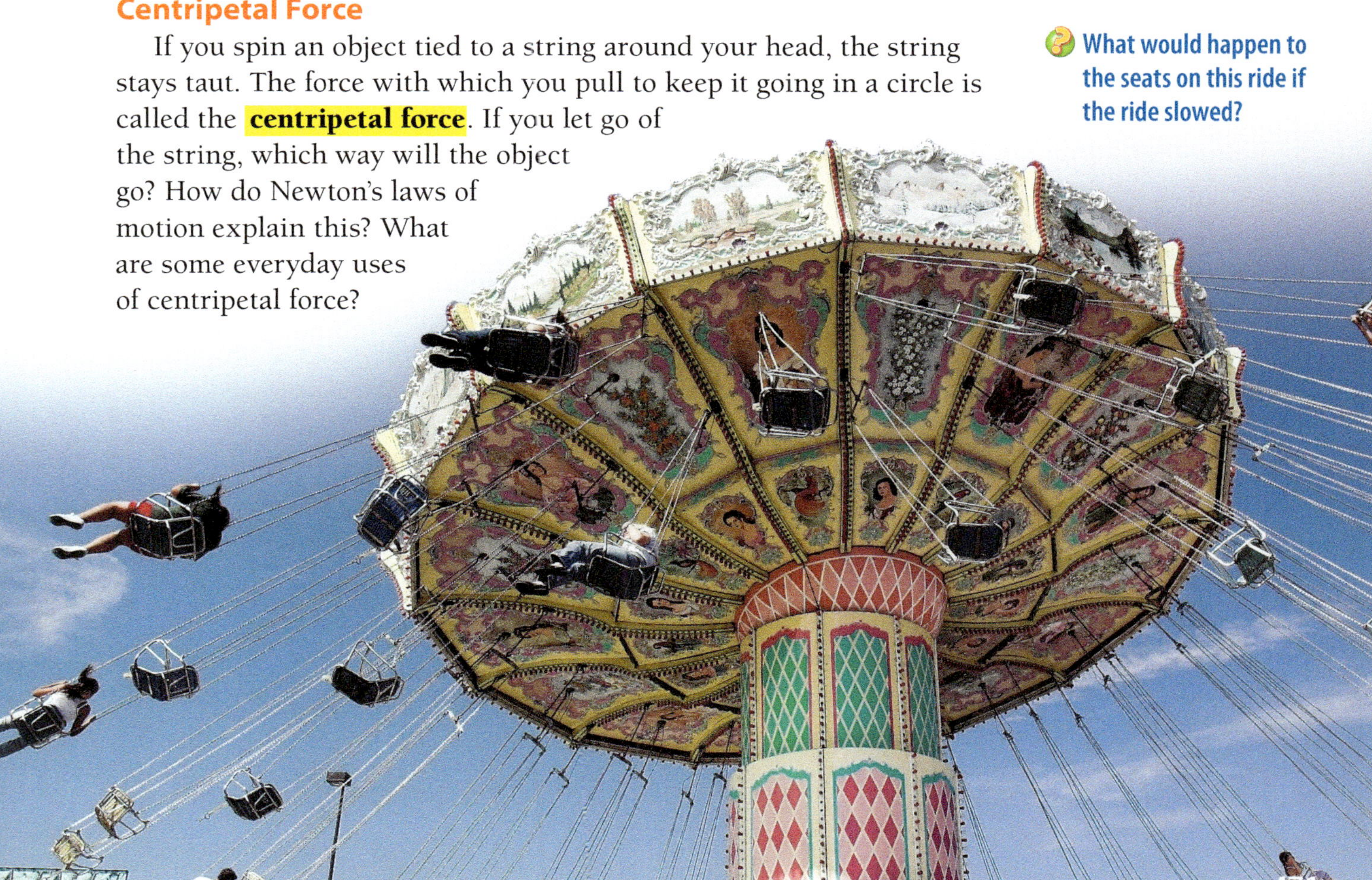

Explore-a-Lab

Structured Inquiry

How does centripetal force affect a moving ball?

Do this activity in a gym or outdoors. Working in a small group, tie a string around a tennis ball so you have about 1 m of string hanging loose. Securely tape the tied portion onto the ball with masking tape. Using the diagram here as a guide, discuss with your group which way the ball will go when the ball is swung and released. It is OK to have differing opinions in the group. You are testing to see the result. Slowly swing the ball around your head until it is moving in a circle at constant speed. Being careful not to hit anyone or anything fragile, let go of the string at a predetermined point of the swing. Record the direction in which the ball goes from the release point. How did the direction of the ball at the moment of release affect its path? Why?

Top-down view of tennis ball being whirled on a string

Mechanical Force

With **mechanical force**, the object exerting the force is in physical contact with the object being pushed or pulled. Mechanical forces are also called contact forces. The force of a string pulling an object around in a circle and buoyant force lifting a balloon are both mechanical forces.

Magnetic Force

Magnetic force is an attraction or repulsion exerted on moving electrical charges and on certain metals such as iron and nickel. Earth's magnetism shields us from deadly cosmic rays and traps electrical particles streaming from the Sun to make the auroras. What are other common uses of this force? Look around you; what can you see that uses magnetic force?

Electrical Force

Electrical force is the force that acts between electrically charged objects. You have learned that like charges repel, and unlike charges attract. The chemical bonds that hold atoms and molecules very close together without quite touching are due to electrostatic attractions and repulsions. Clothes coming out of a dryer may show electrical attractions and repulsions. A sock may fly off one garment and onto another. If the sock has a negative charge, what is the charge on the garment that attracts it? How can the electrical charge of an object change?

Nuclear Force

Nuclear force affects subatomic particles such as protons and neutrons. These forces are very strong but only act for extremely short distances, holding the protons and neutrons in place within the core of an atom. Atomic energy uses these powerful forces to make electricity. Nuclear forces also produce the explosive power in atomic bombs. What could be some of the challenges in harnessing nuclear forces to produce electricity?

Gravity Explain

A force that is always acting on every object in the Universe is gravity. Recall that *gravity* is a force of attraction between objects due to their masses. The force of gravity is responsible for an object's weight. You have previously learned that *mass* refers to the amount of matter in an object. Mass is measured in kilograms. *Weight* is a measure of the gravitational force on an object and is measured in newtons (N). An astronaut has the same mass on Earth as on the Moon. However, on the Moon's surface, astronauts have only about one-sixth of their weight on Earth. What explains this difference in weight?

You would find slight differences in your weight at different places on Earth. The density of Earth varies slightly, so the gravitational pull is greater where the minerals in the crust have more mass per cubic centimeter. Geologists can use these differences to search for certain minerals. They fly over areas with sensitive gravity meters hooked to computers.

A greater variation in gravity occurs with altitude. The farther you are from Earth's center, the less the force of gravity. What would you find if you weighed yourself on top of Mount Everest? Why?

The Law of Universal Gravitation

The *law of universal gravitation* states how all masses attract each other. The gravitational force is directly proportional to the product of their masses and inversely proportional to the square of the distance between their centers. The diagram illustrates how the law of universal gravitation works.

You can see that doubling one of the masses doubles the force of gravity. Doubling both of the masses would quadruple the gravitational force. What would be the effect of tripling one of the masses? What if both masses were tripled?

Effect of Mass on Gravitational Force

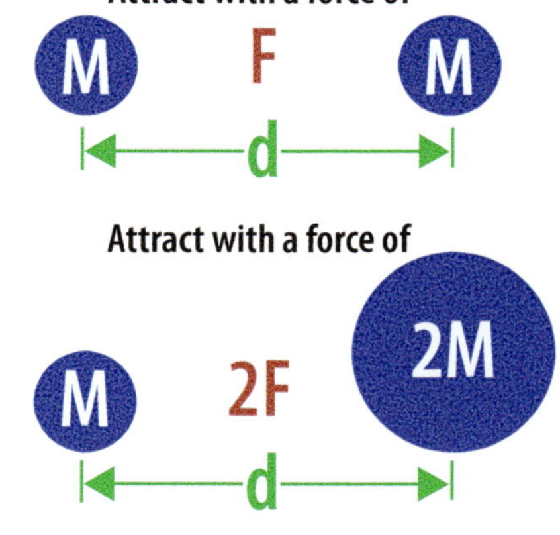

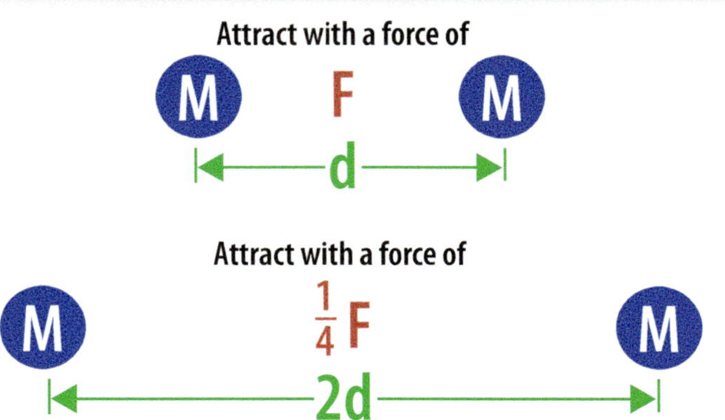

The diagram shows that doubling the distance decreases the force by a factor of four. That is the effect of gravitational force being inversely proportional to the square of the distance. If the masses were three times as far apart, the force of attraction would only be one-ninth as much. What would be the force of attraction if the masses were four times as far apart?

Venus and Earth have approximately the same mass. However, Earth is farther from the Sun than Venus, so the gravitational force between the Sun and Earth is much less than the force between the Sun and Venus. Venus travels faster in its path around the Sun than Earth does, so it has a greater tendency keep going in a straight line and fly off into space. It takes more force to get it to change its velocity into the tighter circle than it does to keep Earth in orbit. Fortunately, with Venus being closer to the Sun, the stronger gravitational pull exactly balances the tendency of Venus to keep going away, and that keeps Venus in orbit. Why do you suppose the Moon affects tides more than the Sun?

If you throw a ball into the air, the gravitational attraction of Earth pulls it back down. At the same time, the gravitational pull of the ball on Earth pulls Earth upward with the same force in an action-reaction pair. However, because the mass of Earth is so much larger, the effect on Earth cannot be detected. How does Earth's gravity affect the Sun and the Moon?

In what direction is the force greater in order to create a flat parabola? A high parabola?

Paths of Projected Objects

Once you throw a ball to a friend and it leaves your hand, its inertia tends to keep it going in a straight line at constant speed. However, other forces are at work, such as gravity and friction. Gravity pulls the ball downward. Air resistance gradually slows its forward velocity.

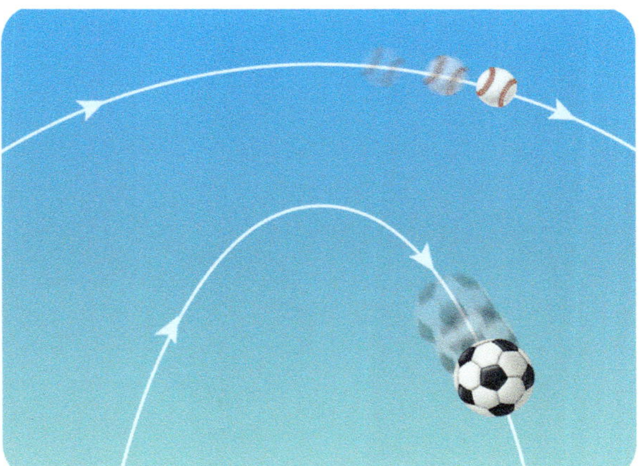

A thrown object takes a downward-curving path. The shape of the curving path is called a *parabola*. In a parabola, the sides of the curve are straighter than a half circle would be. All objects thrown, shot, launched, or projected will approximately follow a parabolic path. The two paths shown in the diagram look different. The shape of a ball's path depends on the angle and the speed at which it was launched.

Explore-a-Lab

Structured Inquiry

 **Why doesn't the marble follow a straight path like Newton's first law predicts?**

Roll a marble off a table and make a sketch of its path from the table to the floor. Repeat with the marble rolling off the table faster. Repeat with the marble rolling even faster. What do you notice about the three paths? What is the name for the shape of each path?

 Concept Check Assess/Reflect

Summary: What are some forces around us? Frictional forces oppose the motion of a body. Static frictional forces are greater than the forces of kinetic friction. Friction has advantages and disadvantages, and there are ways to increase or decrease the force of friction. Due to the force of gravity, weight varies in different places, but mass stays the same. The strength of gravitational force increases with the masses of the objects involved and decreases according to the square of the distance between them. Objects projected on Earth follow a parabolic path. Elastic forces tend to return an object to its original shape. Buoyant force is an upward force exerted by a fluid on an object immersed in it. Centripetal force is the force pulling an object into a circular path. With mechanical force, the object exerting the force is in physical contact with the object being pushed or pulled. Magnetic force is an attraction or repulsion exerted on moving electrical charges and on certain metals. Electrical force is the force that acts between electrically charged objects. Nuclear forces are noncontact forces that act over very short distances within atoms.

1. You are assigned the task of moving a heavy wooden pallet across a concrete floor. It is too heavy for you to slide as it sits. Briefly describe at least four ways to reduce friction enough so that you can move the pallet.
2. A classmate thinks that Newton's laws of motion only work with mechanical forces and that his law of universal gravitation is only for noncontact forces. How could you help your classmates see that Newton's laws of motion apply to all forces?
3. How would the gravitational pull of the Moon on Earth be different if the Moon were three times as massive but twice as far away?
4. Newton's first law may seem to predict that an arrow shot horizontally east would travel straight in that direction forever. Explain why that is not the case.

People in Science

Extend

Ray Hefferlin

Dr. Ray Hefferlin was born in Paris, France, and moved with his family to the United States in 1936. He studied physics at Pacific Union College in California and received his Ph.D. from the California Institute of Technology (Caltech), one of the most prestigious schools for physics in the United States. In 1955, he joined the physics faculty at Southern Adventist University. He retired in 1997.

Dr. Hefferlin balanced his time between research and teaching. He taught courses on physics, astronomy, and mathematics. His classes had the reputation of being very challenging. He was awarded several teaching honors, such as the Pegram Award, the Professor of the Year Gold Medal Award of the Council for the Advancement and Support of Education, and others.

The focus of his research was originally spectroscopy. Spectroscopy is a process that uses the dispersion of light waves to determine the physical properties of an object, such as its temperature, mass, and composition. It is used in astronomy to determine the size, mass, and make-up of distant astronomical objects, such as planets, stars, and asteroids.

Dr. Hefferlin's thesis focused on spectroscopy, but he expanded his research to the construction and testing of particular molecules, called diatomic and tritomic molecules. He has published hundreds of articles, papers, and books. Although he is now retired, he still conducts research at Southern's facilities.

Called to Serve

Dr. Hefferlin and his wife lived in Russia, donating their time to help coordinate Russian church events.

Concept Check

1. How can spectroscopy tell us what things are made of?
2. How do you think Dr. Hefferlin's work relates to his faith?

Careers in Science

Extend

Crash Test Investigator

Most people are familiar with crash test dummies, but many people are not aware of what can be tested using these scientific devices and what can be learned.

Crash test investigators, also called crash forensic experts or accident reconstructionists, use the information gathered from crash test dummies and accident reconstructions to determine how accidents occurred. Crash test dummies can be used to identify the force and direction of an accident and to determine the potential trauma of an accident victim. These tests can be done under safe, controlled conditions in a laboratory.

At accident scenes, investigators can measure the skid marks that appear. They can use their knowledge of physics to determine how fast the car was driving just by the length of the skid mark. They may also take photographs of the scene, interview witnesses and drivers, review police reports, and collect forensic evidence, such as paint chip samples. All this data will be reviewed and assessed to determine the exact cause of an accident.

You need strong math and physics skills to follow this career path. In addition to these skills, you may need to train in law enforcement or have a specialized engineering degree, such as forensic engineering. Potential employers include insurance agencies, police departments, and law firms.

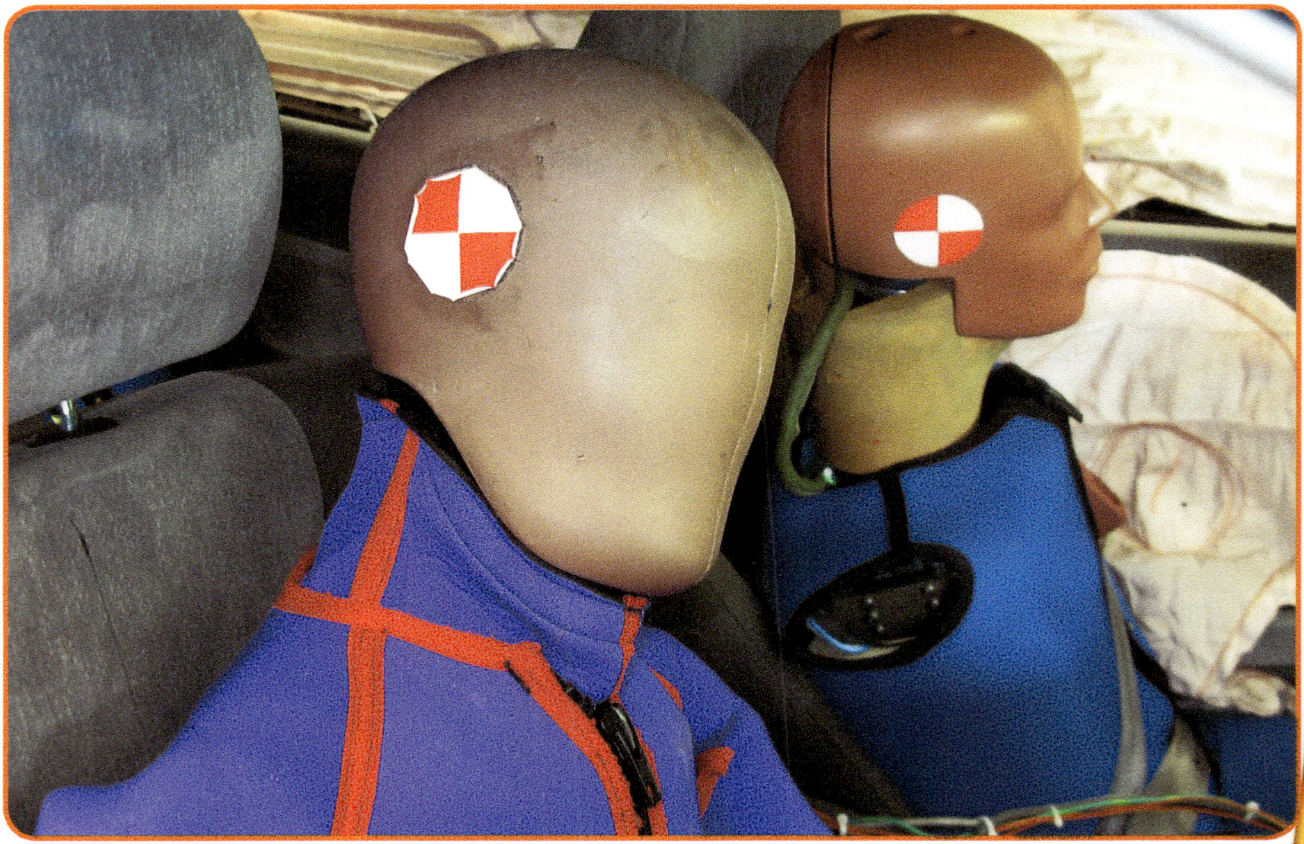

Concept Check

1. Why are math and physics skills important in this line of work?
2. How does the work of a crash test investigator help people who have been in an accident?

Chapter 12 Review Assess/Reflect

Study Guide

Lesson 1

1. Motion is the change in position of an object relative to a reference point.
2. Speed is found by dividing the distance traveled by the time taken.
3. Average speed is the total distance divided by the total time. Instantaneous speed is the speed at a given moment.
4. The line of a graph changes at several points. A steeper slant of the line is due to a larger distance being covered during the time elapsed and therefore represents a faster speed.

Lesson 2

1. Acceleration is the rate of change of speed or direction, or both. Acceleration may be calculated using the formula $\text{acceleration} = \frac{\text{final velocity} - \text{initial velocity}}{\text{time}}$.
2. The momentum of all objects in a system before an interaction between them is equal to the total momentum of all objects in the system after the interaction.
3. Elastic collisions conserve both momentum and kinetic energy. Inelastic collisions only conserve momentum because some energy is converted to another form.

Lesson 3

1. The net force is the combination of all forces acting on an object. Balanced forces are equal forces that act in opposite directions. Unbalanced forces are unequal forces that act in opposite directions.
2. Contact forces involve contact between the object exerting the force and the object being changed by the force. Noncontact forces do not involve contact between the force and the object that is being changed by the force.
3. Newton's first law states that all objects tend to stay at rest if they are at rest and tend to keep moving if they are already moving until acted on by a force. Newton's second law states if a net force acts on an object, the object will accelerate in the direction of the force. The acceleration will be directly related to the size of the force and the mass of the object the force is acting on. Newton's third law states that for every force, an equal reaction force acts in the opposite direction on the object exerting the force.

Lesson 4

1. Friction is a force opposing motion between two surfaces in contact.
2. Kinetic friction is the force that tends to stop the motion of one object against another. Static friction is the force that tends to prevent stationary objects from moving.
3. Friction is useful in holding objects in place and slowing motion when needed. It is a disadvantage when it causes us to use more force to do things.
4. The law of universal gravitation states that all objects attract each other with a force that is proportional to the product of their masses and inversely proportional to the square of the distance between them.
5. Projected objects on Earth follow curved paths called parabolas because of Earth's gravitational attraction and inertia. Gravity pulls a moving object downward and inertia keeps it moving in a straight line.
6. Common forces that act on objects include elastic force, buoyant force, centripetal force, mechanical force, magnetic force, electrical force, and nuclear force.

Vocabulary Check

Match each term with the correct statement.

A. quantitative	**E.** deceleration
B. slope	**F.** newton
C. elastic collision	**G.** static friction
D. average speed	**H.** buoyant force

1. The total distance covered divided by the total time taken
2. Upward force when an object is immersed in a fluid
3. A description that uses numbers
4. The rate at which the velocity of an object is decreasing
5. Gives a 1-kg mass an acceleration of 1 m/s^2
6. The rise divided by the run on a line graph
7. A force opposing motion of an object at rest
8. Both momentum and kinetic energy are conserved

Multiple Choice

9. What is your average speed if you go 10 m and 14 m in 100 seconds?
   A. 140 m/s
   B. 4.2 m/s
   C. 1.4 m/s
   D. 0.24 m/s

10. Suppose you are riding in a car. It seems that the person sitting next to you is sitting still but the trees and power poles at the side of the road are moving to the rear. Which would not be your frame of reference?
   A. the trees
   B. the road
   C. the car seat
   D. the power poles

11. What quantity will be the same before and after an inelastic collision?
   A. momentum
   B. kinetic energy
   C. net force
   D. speed

12. An object tends to remain at constant velocity because of its
   A. gravity.
   B. inertia.
   C. weight.
   D. net force.

13. What is the measure of the force of gravity on an object called?
   A. mass
   B. weight
   C. pressure
   D. buoyancy

Check Point

14. A football player does a shuttle-run in which he sprints 20 m south, then 15 m north, then 10 m south, and then 5 m north. If his total time is 20 s, what are his displacement and average velocity?
15. What net force would give an 8.0-kg bowling ball an acceleration of 1.5 m/s^2 southward?
16. When you jump on a trampoline, your weight extends the springs around the outside rim. Then their elastic force propels you upward. What is the reaction force to the upward force of the trampoline mat on your body as you jump?
17. Mars has two moons. If Earth had a second moon that was three times the mass of our Moon and the same distance away, how would the second moon's gravitational force compare with that of our Moon?

Chapter 13 Work, Power, and Machines

Scripture Spotlight

Do you think Adam and Eve used any machines in the Garden of Eden? What machines would have made life better in the Garden? Some of the machines you will learn about in this chapter were used in the Bible. In this chapter, you will read the following passages.

1 Peter 1:3–5 (p. 484)
Exodus 35:30–35 (p. 492)
2 Chronicles 26:15 (p. 506)

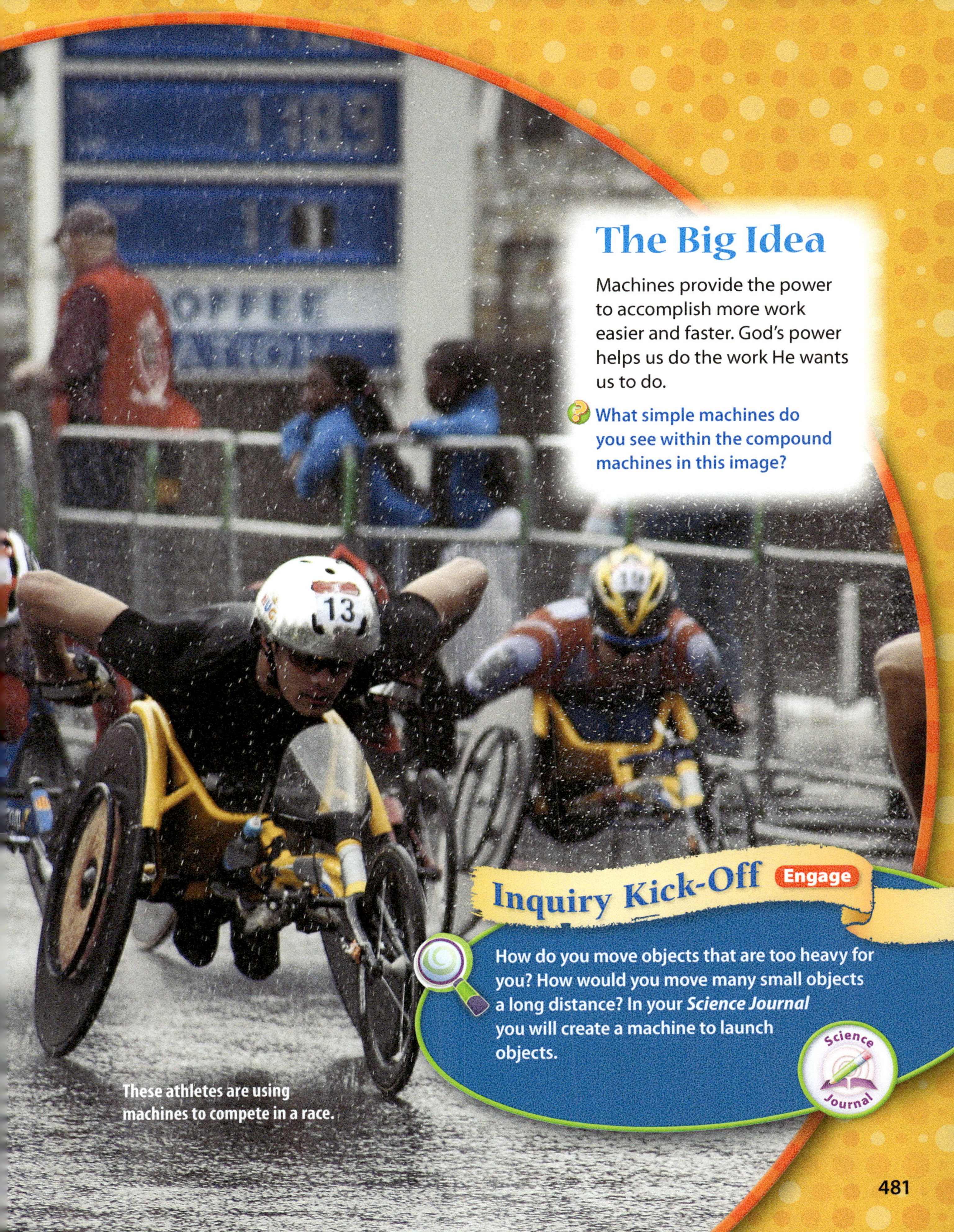

These athletes are using machines to compete in a race.

The Big Idea

Machines provide the power to accomplish more work easier and faster. God's power helps us do the work He wants us to do.

What simple machines do you see within the compound machines in this image?

Inquiry Kick-Off Engage

How do you move objects that are too heavy for you? How would you move many small objects a long distance? In your ***Science Journal*** you will create a machine to launch objects.

Science Journal

Lesson 1

Essential Question

How Are Work and Power Related?

Objectives

- Explain what is required for work to be done.
- Describe how work and power are related.
- Explain the relationship between machines, work, and power.
- Describe how to calculate mechanical advantage and efficiency.

Vocabulary

power

mechanical advantage

input force

output force

efficiency

How many meanings can you think of for the word *work*? Construction companies use heavy equipment to accomplish much of the *work* they do. Homework assignments and tests are often referred to as being *work*, even hard *work*. You might ask someone, "Does that pen *work*?" The word *work* can be a noun or verb and has many different meanings. Think about the activities you have done over the last couple of days. Which activities would you classify as work? Do you think a scientist would also classify those activities as work? Why or why not?

Doing Work Explain

Recall that to a scientist, *work* occurs when force is exerted on an object, and the object moves in the direction of the force. The second part of that statement is very important. No matter how hard you think you might be working, for work to be done, an object must move.

Heavy equipment is used to make work easier.

Visualize what you have learned about work. Imagine a boy pushing with all his strength against a brick wall. The brick wall does not move. Now imagine that the same boy lifts his book bag and holds it above his head until his arms get tired and he puts it down. At which point did the boy do work? Use your previous knowledge about work to explain your reasoning to a partner.

Carrying a box down a hall and putting it on a shelf represents nearly the same amount of work on the box as lifting the box onto a shelf of equal height.

Calculating Work

Scientists can quantify the amount of work a person does. For example, a girl lifts a box that weighs 10 N onto a shelf that is 1 m high. To calculate the amount of work she does, you can use the following equation:

$$\text{Work (W)} = \text{Force (F)} \times \text{distance (d)}$$
$$W = 10 \text{ N} \times 1 \text{ m} = 10 \text{ N/m}$$

Note that force is measured in newtons (N) and distance is measured in meters (m). Another way of expressing this is to use a unit called a joule. A *joule* is a measurement of the amount of work done. One joule (J) is equal to 1 N/m. So, the girl lifting the box did 10 N/m, or 10 J, of work.

Now suppose that the girl picks up the same box weighing 10 N. Remember that weight is measured in newtons because weight is a measure of force—it measures the force of gravity pulling an object to Earth's center. The girl holds the box 1 m high and walks it down a long hallway and places it on a shelf that is 1 m high. How much work did the girl do? To do work, the motion must be in the same direction as the force. The girl used an upward force to lift the box, and then she moved horizontally. The force was not in the direction of her movement. Her arms might be tired from carrying the box down the long hall, and she had to overcome inertia and air resistance as she carried the box. But she did nearly the same amount of work when she carried the box a long distance as she did when she simply lifted the box and placed it on a shelf.

Scripture Spotlight

In **1 Peter 1:3–5**, who are kept through the power of God and what is He keeping them from?

Explore-a-Lab

Guided Inquiry

How much work does it take to lift an object?

Lift several objects from the floor to the top of a desk or table. Objects might include a bag of rice, a container of water, and a textbook. Use the work equation $W = F \times d$ to determine the amount of work you did to lift each object. Make a graph that shows your results. Keeping the objects the same and the distance they moved the same, explain how you could decrease the amount of work needed to accomplish each task.

Work and Power **Explain**

A concept related to work is power. **Power** is the rate at which work is done. It is measured in watts. A *watt* (W) is defined as 1 J of work per second. Can you think of some objects that generate watts and have power? A large truck engine and a car engine both produce power. The truck engine is more powerful than the car engine because it can do more work each second it is working.

How does the relationship between work and power apply to everyday situations? Imagine you had to dig a hole for a swimming pool. You could use your hands, which are a type of machine, but that would take a very long time. A shovel would give you more power, but it would still take a long time to dig the large hole. You could use a backhoe and do the same amount of work in a short time. Why does the backhoe complete the job more quickly?

A truck does more work and has more power than a car.

Calculating Power

Scientists can quantify power, just as they can quantify work. For example, a student weighs 300 N. He wants to walk up a flight of stairs to the roof of his apartment building, which is 15 m high. It will take him 50 seconds to climb the stairs. To calculate power, you must first calculate the work he does:

$$\text{Work (W)} = \text{Force (F)} \times \text{distance (d)}$$

Calculate the work done by the student. Then use the following equation to find the power required to reach the roof:

$$\text{Power (P)} = \text{Work (W)} \div \text{time (t)}$$

How much work did the student do to reach the top of his apartment building? How much power was used?

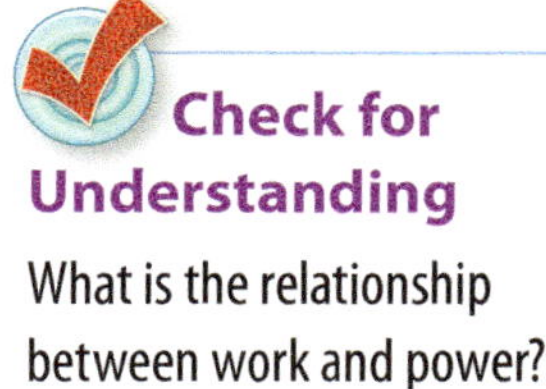

Check for Understanding

What is the relationship between work and power?

Math in Science

A girl lives in the same apartment building as the student who weighs 300 N. She uses an elevator rather than the stairs to reach the roof of the building. She weighs the same amount as the boy. It takes the girl 10 sec to reach the roof. In which situation was more power used to reach the roof? Show your calculations.

Machines, Work, and Power Explain

Machines are devices that multiply force or increase speed and make work easier. Examples of machines include pencil sharpeners, screwdrivers, hammers, ramps, and even scissors. You'll learn more about machines in the following lessons.

For now, let's focus on how machines affect work and power. Picture a group of students working in shop class. They are trying to remove an engine from a car. This isn't an easy job. All wires, hoses, and tubes must be identified and disconnected. What other things must be done before the engine is lifted out of the car and suspended above the floor? Think about the parts of an engine as well as safety measures. Once everything is disconnected, a pulley can safely lift the engine out of the car.

A pulley is a type of machine that can be used to lift heavy objects.

How do machines affect force?

Record your work for this inquiry. Your teacher may also assign the related Guided Inquiry.

Measuring Human Power

How can you measure human power?

Procedure

1. Use masking tape to mark a line 2 m long on the floor or a tabletop. Put a short piece of tape at one end to mark the starting point and a piece of tape at the other end to mark the finish line.
2. Attach a spring scale to the block of wood. Use the spring scale to pull the wood block over the 2-m distance. Your partner should read the **measure** on the spring scale to determine the force you exert and use the stopwatch to time how long it takes you to pull the block over the distance. **Record** your results in your ***Science Journal***.
3. Repeat Step 2 two more times and average your results.
4. Next, use the bathroom scale to find your weight. Remember that you need your mass in newtons. You can convert kilograms to newtons using this ratio: 1 kg = 9.8 N (approximately).
5. Measure and record the distance from the top of the stool to the floor.
6. While your partner times you for 10 seconds, count how many times you step on and off the stool. Record your results.
7. **Calculate** the average work you did and the power you exerted. **Compare** your results with those of other students.

Materials
- stopwatch
- meterstick
- metric bathroom scale
- small stool
- wooden block
- spring scale
- masking tape

Analyze Results

What relationship do you see between work and power? How did work and power differ between the two activities?

Create Explanations

1. How can you measure human power?
2. What factors determine the amount of work done? What factors determine power?
3. Why did you do multiple trials to determine the amount of work you completed?
4. What would happen to your results if you pulled the wooden block a distance of 3 m?

Machines can make work easier and increase force in one other way: they can increase the distance over which you apply a force. For example, you can lift a heavy box and place it on a shelf. Or, you can use a machine called a ramp to lift the box and push it a certain distance up to the shelf. You do the same amount of work in both cases, but you do not need as much force when you use the ramp.

Mechanical Advantage Explain

The ability of pulleys or other machines to multiply force is called mechanical advantage. **Mechanical advantage** (MA) is identified as the number of times a machine multiplies the effort put into it. It identifies how effective the machine will be at performing the work desired.

How do you calculate mechanical advantage? First, you determine the input force (IF) and the output force (OF). The **input force** is the amount of force applied to a machine. The **output force** is the force applied by the machine to an object. Picture a simple bottle opener. You secure the bottle opener to the rim of a bottle cap and pull up. That is the input force. The bottle opener applies force to the cap and the cap comes off. How would you classify this force?

Let's say that the IF applied to the bottle opener is 50 N and that the OF is 500 N. You can use the following equation to calculate the mechanical advantage of the bottle opener:

$$\text{mechanical advantage} = \text{output force} \div \text{input force}$$
$$MA = OF \div IF$$

Do the calculations to determine the mechanical advantage of the bottle opener. What would happen to the mechanical advantage of the bottle opener if you applied more force?

In general, machines make work easier when the output force is greater than the input force. Keep in mind, though, that work involves both force and distance. You may be getting more force out of a machine than you put into the machine, but you are applying your force over a greater distance.

Many bottle caps, like the one pictured, have been replaced with twist caps.

Do you think more force is required to open a bottle with an opener like the one shown or with a twist cap? Explain.

A plane is a type of machine that has low efficiency due to many factors, including weight and wind resistance.

Efficiency Explain

The mechanical advantage calculated for each machine, called the *theoretical mechanical advantage*, rarely equals the *actual mechanical advantage* obtained from a machine. This is because other forces such as friction, gravity, and air resistance interfere with the working of the machine. Because there is a difference between theoretical advantage and mechanical advantage, another value is used to measure the work of machines. This value is called efficiency. **Efficiency** is the measure of the actual work of a machine compared with the work put into it. Efficiency (E) is usually represented as a percent (%).

Efficiency can be calculated by dividing the amount of work gotten out of a machine, called work output (WO), by the amount of work put into the machine, called work input (WI). Let's use the example

Math in Science

The power-steering system in a riding mower has a mechanical advantage of approximately 70. A person uses 44 N of force to turn the steering wheel of the riding mower. Calculate the output force that makes the front wheels turn.

of a ramp to calculate efficiency. A ramp is a kind of machine that people use to make loading objects easier. Suppose it takes 25 J of work to push a box up a ramp; that is the work input. Suppose you only get 15 J of work from the effort; that is the work output. You would use the following equation to calculate the efficiency of the ramp:

$$\text{efficiency} = \text{work output} \div \text{work input} \times 100\%$$
$$E = WO \div WI \times 100\%$$

Use the equation to calculate the efficiency of the ramp. Share your results with a classmate. Do you have the same results? What might account for any differences in your answers? Would making the ramp more slippery help increase efficiency? Would that cause problems?

An efficiency of 60% means that 60% of the work put in helps you, and 40% is lost. What is this work lost to? Where does this work go? No machine is 100% efficient, but the machines rated closest to 100% are the most helpful. Many household appliances are tested to determine their efficiency. Why do you think engineers and manufacturing companies are constantly researching ways of making these items more efficient?

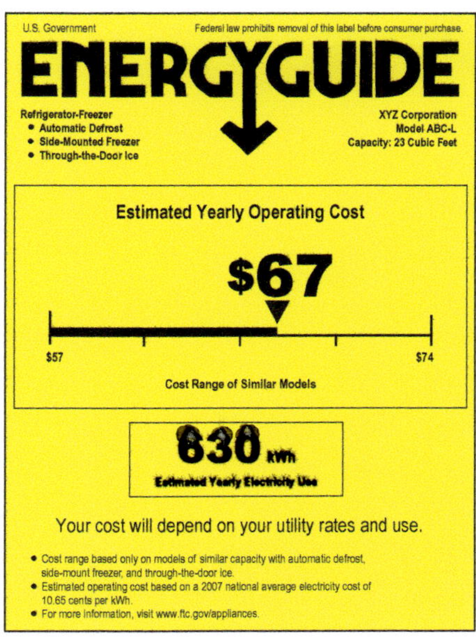

New appliances sold today contain information about their operating costs and energy use.

Lesson Activity

Work in small groups to conduct an analysis of automobile power and fuel efficiency. Research both foreign and domestic vehicles. Include a variety of models, such as small cars and large SUVs. Research at least six different models. As a class, compile your results in a master chart. What relationship, if any, do you observe between the power of a vehicle and its fuel efficiency? Which vehicle would you choose for your own personal use? Explain your answer. What are the most efficient transportation vehicles?

How efficient are powerful vehicles?

Trains are probably the most efficient means of transportation there is. Why might this be? Airplanes can approach the efficiency of cars in seat miles per gallon, especially if only one or two people are in the car.

When might an airplane have greater efficiency than a car? Why?

Concept Check Assess/Reflect

Summary: How are work and power related? Work is the force acting on an object multiplied by the distance the object is moved. When work is done on an object, the object always moves. Power is the rate at which work is done. The more power available to do a job, the quicker the work will be done. Machines are devices that make work easier. Mechanical advantage describes how effective a machine is at performing the work desired. Forces such as friction, gravity, and air resistance interfere with the efficiency of a machine.

1. You lift a box weighing 7 N onto a shelf. What else do you need to know before you can calculate the work you did?
2. Design a machine made of simple parts that would help you move a box up a short flight of stairs. Describe the materials you would use. Make a model or sketch of your machine.
3. Think about the factors that affect the efficiency of a machine. Then brainstorm a way to reduce the impact of one of these factors to improve efficiency.
4. There are some very efficient machines that have a high mechanical advantage that can raise the productivity of companies. Why is it that many companies do not always use these machines?

Lesson 2

Essential Question

What Are Inclined Planes, Wedges, and Screws?

Objectives

- Describe common types of simple machines.
- Determine the effort force, resistance force, mechanical advantage, and efficiency of an inclined plane.

Vocabulary

effort force

resistance force

You go to a water park to cool off on a hot summer day. The last thing on your mind is science. But the second you place your feet on the steps to the waterslide, you are using simple machines. Look at the photo on this page. Look around your classroom. What examples of simple machines can you see?

Simple Machines Explain

Simple machines are machines with one or two parts that change the strength or direction of a force or the distance over which a force is applied. The force that must be applied to cause these changes is the input force, also called the **effort force**. The force that resists these changes is the output force, also called the **resistance force**. Resistance force is sometimes called load force. Regardless, many machines are useful because the effort force is smaller than the resistance force. When this occurs, work is done that may have been impossible to do without the machine. What kinds of simple machines have you used today?

The slide at this water park is made of simple machines.

What types of simple machines can you identify in the photo?

Scripture Spotlight

In **Exodus 35:30–35**, God calls on artisans to create the tabernacle. What tools do you think they would need to create the art God seeks?

Inclined Plane Explain

You will remember that an *inclined plane* is a slanted surface that decreases the amount of force needed to do work. A ramp is an example of a common inclined plane. The ability of an inclined plane to make work easier is directly related to its length compared with height. An inclined plane with a gentle slope makes work easier than an inclined plane with a steep slope. You have probably experienced this yourself if you have ever pushed a box up a ramp. Does it take more power to climb a steep ladder or a ladder with a gentle slope to reach the top of a slide at a water park? Why?

Explore-a-Lab

Guided Inquiry

Can you build a perpetual-motion machine?

Inventors have long sought to make perpetual-motion machines that can run forever without any additional inputs of energy. Use the library or online resources to find out more about perpetual-motion machines. Make a series of drawings that illustrate several of these machines. With each drawing, include an explanation of why the machine did not work. Choose one of the machines you drew to build. Try to modify the design to make it maintain motion longer. Do you think a perpetual-motion machine could ever work? Explain your answer.

Types of Inclined Planes

A ramp and a playground slide are inclined planes that stay in one place. That is, you can push a wheelchair up a ramp, but the ramp itself doesn't move. Some inclined planes move while they do work. Examples include the screw and the wedge.

Screw

A screw is an inclined plane wrapped around a post. The incline forms a long spiral that goes from the tip of the post to nearly its top. Each time the incline wraps around the post it forms a thread. Pitch is the distance between each adjacent thread on a screw. The greater the pitch, the easier it is to turn a screw. What happens when pitch is smaller?

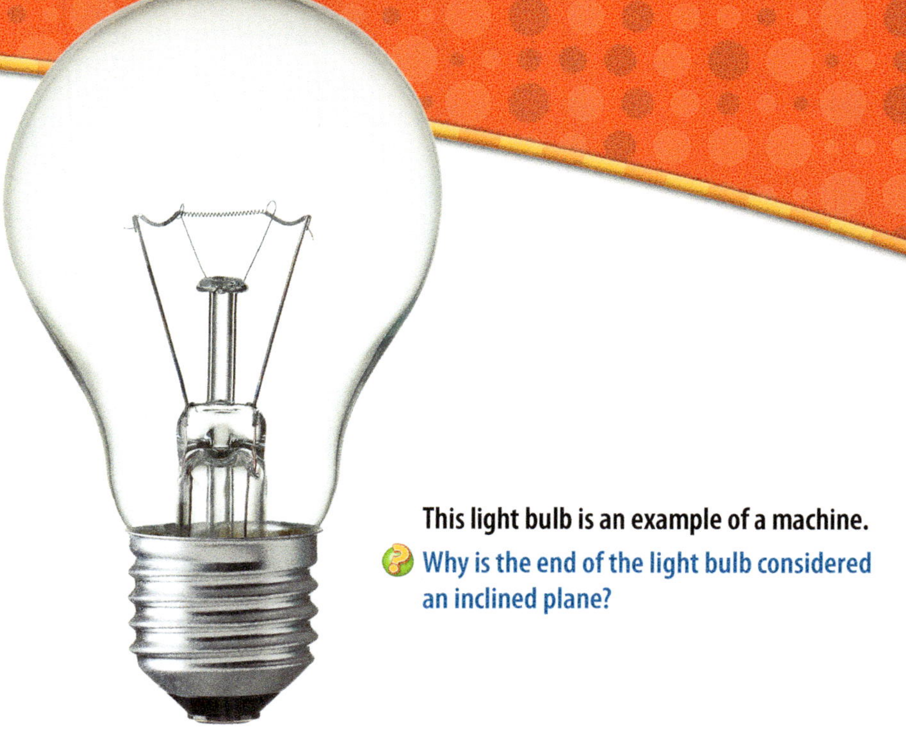

This light bulb is an example of a machine.

Why is the end of the light bulb considered an inclined plane?

When you picture a screw, you might think of a screw used to fasten a shelf to a wall. Take a moment to think about other screws you may have seen. What about the lid of a jar? Lids often contain screws that make it easier to open jars. If there were no friction, would a screw or jar lid work the same?

Explore-a-Lab

Structured Inquiry

 **What is the relationship between pitch and force?**

Gather together the following materials: a block of wood, a screwdriver, a ruler, and two screws with different pitches. Working with a partner, use the ruler to determine the pitch of each screw. Use the pitch to predict which screw will be easier to screw into the block of wood. Test your prediction. Use a stopwatch or clock to measure how long it takes to screw each screw into the wood. Record your results. Include a statement that summarizes what you learned about the relationship between pitch and force.

Wedge

Another type of inclined plane is a wedge. A wedge is two inclined planes attached back to back. Like a screw, a wedge moves through something when force is applied. A common example is a knife. A knife can neatly slice through an apple when force is applied.

Check for Understanding

Compare and contrast a screw with a wedge. How are they alike? How are they different?

Explore-a-Lab

Structured Inquiry

What features make an efficient wedge?

Obtain a wooden wedge, a plastic wedge, and a foam wedge from your teacher. The wedges should all have the same slant. Use each wedge to cut a stick of cold butter in half. Observe which wedge was easiest to use. Predict what would happen if the wedges were bigger. Test your prediction. Then write a brief statement explaining why engineers must consider the properties of the tools they use in their work.

Mechanical Advantage: Inclined Plane

Recall that mechanical advantage is a measure of the output force, or resistance force, divided by the input force, or effort force. To calculate the mechanical advantage of an inclined plane, you divide the length (L) of the slanted plane by the height (H) of the raised end, as shown in this formula:

$$MA = L \div H$$

Look at the ramp in the picture. What is its mechanical advantage? What do you think would happen if you increased the length of the inclined plane? Make a prediction, and then do the calculations to see if you were right.

This ramp has a length of 3 m and a height of 1 m.

Why is a ramp like this one built with a shape that isn't straight? What effect does it have on riders?

Record your work for this inquiry. Your teacher may also assign the related Guided Inquiry.

Finding Mechanical Advantage

What is the mechanical advantage of an inclined plane?

Materials

- board, 1 m long by 2.5 cm thick by 20 cm wide
- 4 books
- brick
- meterstick
- spring scale
- string
- talcum powder

Procedure

1. Tie the string to the brick. Tie the other end of the string to the spring scale and **measure** the weight of the brick. **Record** this and further data in the table in your ***Science Journal***.
2. Place one end of the board on a stack of three books.
3. Record the length and height of the inclined plane. **Calculate** the theoretical mechanical advantage of the inclined plane.
4. Holding the spring scale, pull the brick up the inclined plane at a constant speed. Read the spring scale while the brick is moving. Record your results.
5. Increase the height of the inclined plane by adding another book to the stack. Repeat Steps 3 and 4.
6. Decrease the height of the inclined plane by removing all but one of the books. Repeat Steps 3 and 4.
7. Sprinkle talcum powder on the board and repeat Step 4 for an inclined plane with three books, four books, and one book.
8. Create a **graph** or multiple graphs that show your results.

Analyze Results

Use your graph(s) to analyze your results. What patterns do you see?

Create Explanations

1. What is the mechanical advantage of an inclined plane?
2. How did the theoretical mechanical advantage of the inclined plane compare with its actual mechanical advantage?
3. What happened to the mechanical advantage of the inclined plane as its height increased?
4. What effect did sprinkling talcum powder on the inclined plane have on its mechanical advantage? Why?

Efficiency: Inclined Plane Explain

Recall that efficiency is the measure of the actual work of a machine compared with the work put into it. It is found by dividing work output by work input. Consider the case of the ramp that has a length of 3 m and a height of 1 m. A student uses 400 N of force to push a box weighing 900 N up the ramp. What is the efficiency of the ramp?

First, you must find the work put into the ramp, or work input. Work input (WI) is the effort force multiplied by the distance over which the force is applied. In other words, how much work did the student do as he pushed the box up the length of the ramp? Then, you must find the work output of the ramp. Work output (WO) is the resistance force multiplied by the distance over which resistance is raised. How much force must be overcome to raise the box to a certain height?

1. Use the following equation to determine work input:

$$WI = \text{Effort } (E) \times \text{distance } (d)$$

2. Determine work output using this equation:

$$WO = \text{Resistance } (F) \times \text{distance } (d)$$

3. Use your calculations of work input and work output to determine efficiency:

$$E = WO \div WI \times 100\%$$

What is the efficiency of the ramp? Now go back to the data you recorded in the Structured Inquiry and calculate the efficiency of that inclined plane.

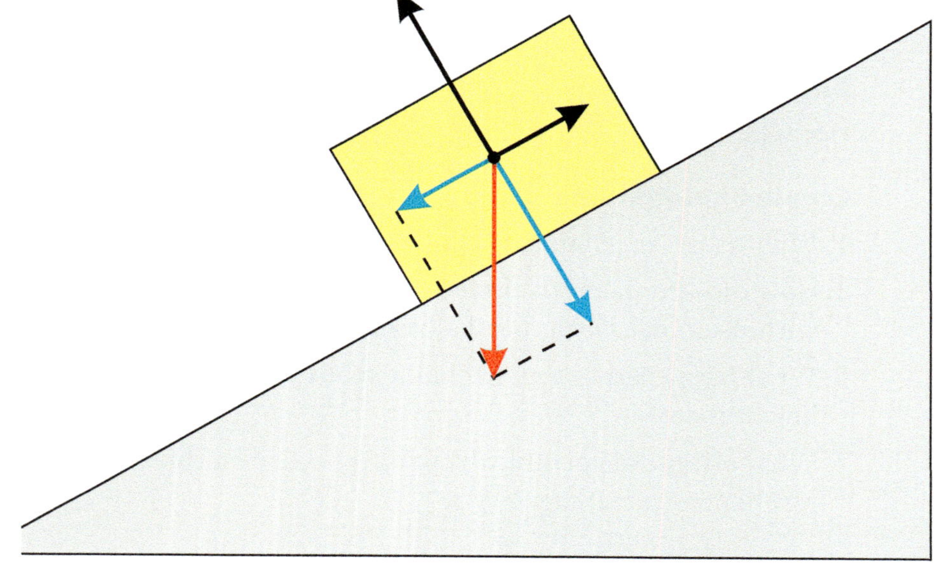

To find the efficiency of this ramp, you first have to find work input and work output.

Explore-a-Lab

Structured Inquiry

How can you increase the efficiency of an inclined plane?

Use a 1 m piece of board and a wood block to make an inclined plan with a height of 5 cm–10 cm. Tie a string to a second wood block. Attach the other end of the string to a spring scale. Pull the block up the ramp. Calculate the amount of work done, the mechanical advantage, and the efficiency of the inclined plane. With your partner think about several ways you can increase the efficiency of the inclined plane without changing its height or length or changing the block being pulled up the inclined plane in any way. Pick the three best ways of increasing the efficiency of the inclined plane and try them. For each solution calculate the amount of work done, the mechanical advantage and the efficiency of the inclined plane each time. Which solution was best? Would this solution work in most applications of inclined planes? Explain.

Concept Check

Assess/Reflect

Summary: What are inclined planes, wedges, and screws? A simple machine has one or two parts that change the strength or direction of a force or the distance over which a force is applied. An inclined plane is a simple machine that has a slanted surface that decreases the amount of force needed to do work. Screws and wedges are types of inclined planes. To find the mechanical advantage of an inclined plane, divide the length of the slanted plane by the height of the raised end. To find the efficiency of an inclined plane, divide work output by work input.

1. An effort force of 200 N is applied to a machine. The machine applies a resistance force of 180 N. Is the machine useful? Explain.
2. Look around the classroom or your home and classify at least five common objects as a wedge or a screw.
3. An inclined plane has a mechanical advantage of 1.5 and a height of 2 m. What is its length?
4. Construction workers are using a ramp to slide scraps of wood and torn shingles from the roof of a garage to the ground. Which ramp is likely to be more efficient: a ramp that has been sanded smooth or one that is unsanded? Why?

Lesson 3

Objectives

- Describe the classes of levers.
- Explain how to determine the effort arm, resistance arm, and mechanical advantage of a lever.
- Compare a wheel and axle with a lever.
- Compare and contrast fixed and moveable pulleys.
- Identify the simple machines that make up a compound machine.

Vocabulary

fulcrum

effort arm

resistance arm

Essential Question

What Are Levers, Wheels and Axles, and Pulleys?

You've learned about inclined planes, screws, and wedges. Three more types of simple machines are: levers, wheels and axles, and pulleys. The wheel and axle and the pulley are types of levers, just as screws and wedges are types of inclined planes. What examples of levers, wheels and axles, and pulleys can be found in your classroom? What are examples of these machines that you use at home?

Lever Explain

Have you ever played baseball or rowed a boat? If so, you've used a simple machine called a lever. A lever is a rigid bar or board that rotates on a fixed point. The point on which the lever rotates is called the **fulcrum**.

A lever has an **effort arm**, which is the distance between the fulcrum and the point where effort force is applied. It also has a **resistance arm**, which is the distance between the fulcrum and the point where resistance force is exerted. You supply the effort force. What creates the resistance force when you swing a baseball bat?

Types of Levers

The fulcrum is found in different places on different classes of levers. The locations of the effort force and resistance force differ, too. Scientists classify levers into three groups based on these differences. The groups are first-class levers, second-class levers, and third-class levers. As you read about the different levers, search your classroom or school for examples of each type. Sketch the levers or take pictures. Share the pictures with a classmate, and ask him or her to identify the types of levers.

First-Class Lever

In a first-class lever such as a crowbar, the fulcrum is positioned between the effort force and the resistance force. You can better remember this relationship with the help of this diagram:

Effort Force — Fulcrum — Resistance Force

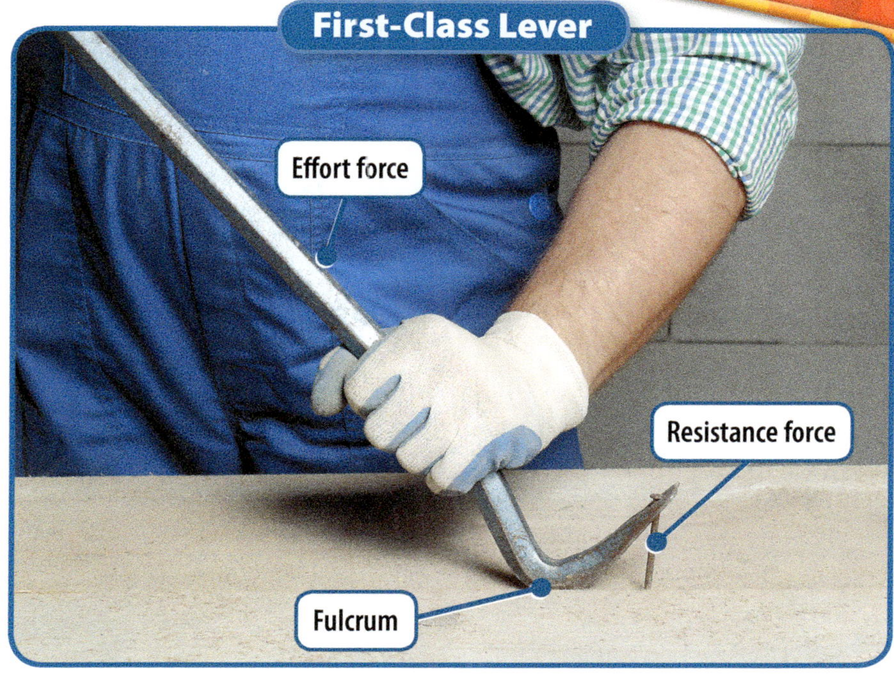

A crowbar is a first-class lever.

Could this machine do work if the effort force and resistance force were in the same direction? Explain.

In a first-class lever, the effort force is in the opposite direction from the resistance force. That is, you push down on the crowbar and it pushes up on a nail. What does the board represent?

Second-Class Lever

In a second-class lever such as a wheelbarrow, the resistance force is found between the fulcrum and the effort force. Here's a simple diagram to help you remember this relationship:

Fulcrum — Resistance Force — Effort Force

In a wheelbarrow, the fulcrum is the wheel. You supply the effort force when you lift the handles of the wheelbarrow. The resistance force, or the load in the wheelbarrow, is positioned between the handles and the wheel. How does the direction of the effort force compare to the direction of the resistance force?

A wheelbarrow is a second-class lever.

What would happen to effort force if the load in the wheelbarrow increased?

Third-Class Lever

In a third-class lever such as a set of tweezers, the effort force is found between the fulcrum and the resistance force. Here's a diagram to help you remember this relationship:

Fulcrum — Effort Force — Resistance Force

The fulcrum is the joined edge of the tweezers. The effort force is usually applied along the length of the tweezers. The resistance force is exerted at the open end of the tweezers. In this type of lever, the resistance force is not greater than the effort force. Does this lever provide mechanical advantage? Why or why not?

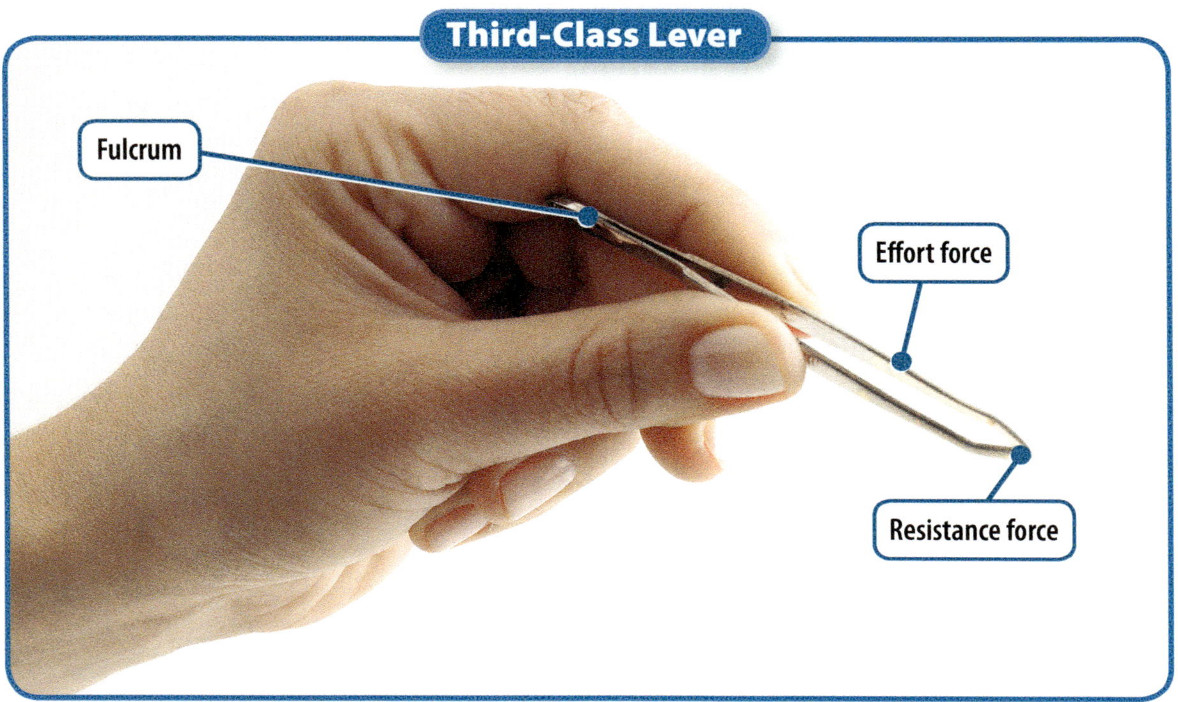

A set of tweezers is a third-class lever.

Why do you think third-class levers are particularly helpful for picking up things that are small or easily breakable?

Record your work for this inquiry. Your teacher may also assign the related Guided Inquiry.

Balancing Act

How can a lever be used to measure mass?

Procedure

Materials
- balance
- dime
- metric ruler
- nickel
- paper (20 cm x 28 cm)
- quarter
- marker

1. Make a lever by folding a sheet of paper into a strip 3-cm wide × 28-cm long.
2. Make a line 2 cm from one end of the strip of paper. Label the line *Resistance*.
3. Slide the other end over the edge of the table until the strip begins to teeter on the edge. Mark a line across the paper strip at the table edge. Label this line *Effort*.
4. **Measure** the mass of the paper to the nearest 0.1 g. Write this mass on the *Effort* line.
5. Center the dime on the *Resistance* line. Locate the fulcrum by sliding the paper strip until it begins to teeter on the edge. Mark the balance line. Label it *Fulcrum 1*.
6. Measure the lengths of the resistance arm and effort arm to the nearest 0.1 cm. **Calculate** the mechanical advantage of the lever.
7. Multiply the mechanical advantage by the mass of the lever to find the approximate mass of the coin.
8. Repeat Steps 5–7 with the nickel, and then the quarter. Mark the fulcrum for the nickel as *Fulcrum 2* and the fulcrum for the quarter as *Fulcrum 3*.

Analyze Results

Draw simple diagrams of your levers and identify the effort arm, resistance arm, and fulcrum.

Create Explanations

1. How can a lever be used to measure mass?
2. In this activity, is the effective total length of the lever a constant or a variable? Explain.
3. What are the dependent variables? What are the independent variables?
4. What provides the effort force? What provides the resistance force?

Mechanical Advantage: Lever

To find the mechanical advantage of a lever, you divide the length of the effort arm (EA) by the length of the resistance arm (RA). Suppose you had a lever with an effort arm of 100 cm and a resistance arm of 60 cm. You would use the equation below to calculate its mechanical advantage.

$$MA = EA \div RA$$

What is the mechanical advantage of the lever?

Think back to what you learned about a third-class lever and mechanical advantage. How would you use this equation to determine if a lever is third class? What type of advantage does it give?

Math in Science

Find an example of a lever in your classroom. Classify it as a first-class, second-class, or third-class lever. Then use appropriate tools to measure the effort arm and the resistance arm. Calculate the mechanical advantage of the lever.

Wheel and Axle Explain

A wheel and axle is a modified lever made of an axle, or rod, that goes through the center of a wheel. The axle and the wheel turn together. The point of rotation is the fulcrum. Effort force is usually applied to the wheel. The axle exerts resistance force.

The wheel and axle of a car is a common example of this type of simple machine. What other parts of a car use a wheel and axle to do work? Picture the steering wheel—the column of the wheel is the axle, and the steering wheel is the wheel. Older models of cars had handles to roll down windows or dials to turn the radio knobs. These are examples of wheels and axles. What other everyday examples of this machine can you think of?

Mechanical Advantage: Wheel and Axle

In a wheel and axle, the radius of the wheel is the effort force and the radius of the axle is the resistance force. The mechanical advantage of a wheel and axle is calculated by dividing the radius of the wheel (WR) by the radius of the axle (AR).

$$MA = WR \div AR$$

Suppose you had a wheel with a radius of 90 cm and an axle with a radius of 20 cm. What is the mechanical advantage of the wheel and axle?

Note that the wheel is larger than the axle and always moves a greater distance than the axle. What would happen if the axle were larger than the wheel?

Check out your ***Science Journal*** to design and build a wind turbine.

A bike pedal is a wheel and axle.

Identify the part of the bike on which the pedal exerts force.

Pulley Explain

A pulley is a grooved or flat wheel with a rope, belt, or chain that passes over it. It is often used to lift heavy weights, such as an elevator. The more ropes supporting the weight, the easier the pulley lifts the weight. The fewer ropes supporting the weight, the harder the pulley lifts. There are three kinds of pulleys: fixed pulleys, movable pulleys, and compound pulleys.

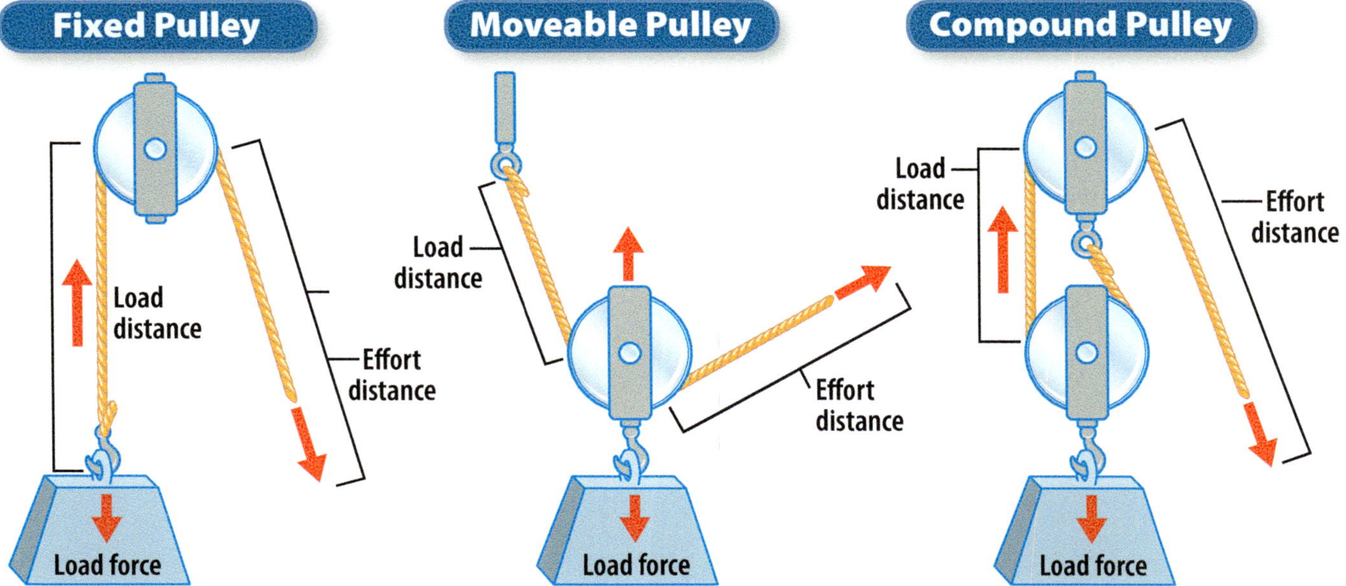

Multiple pulleys working together increase mechanical advantage.

What would be the mechanical advantage of a pulley system with five supporting ropes?

Fixed Pulley

In a fixed pulley, the wheel and axle remain in place. They are attached to an unmovable object, such as a ceiling or pole. This pulley is a modified first-class lever with the wheel turning on an axle. The axle is the fulcrum and the ropes on either side represent the effort force and the resistance force. You pull down on one end of the rope,

Explore-a-Lab

Structured Inquiry

 **How much help can a pulley provide?**

Fill a 1-gal jug with water. Tighten the lid. Tie a rope about 2 m long to a fixed horizontal bar. Loop the free end of the rope through the handle of the container and over the bar. Pull on the rope and note how much force it takes to lift the container. Then loop the rope through the handle of the container and over the bar again. Compare how much force is needed to lift the container. Predict what will happen if you add another loop through the handle and over the bar. Test your prediction.

supplying the effort force. The other end of the rope is attached to a weight, and it pulls up on the weight, exerting a resistance force.

A fixed pulley cannot increase effort force. Instead, it changes the direction of the effort force. This can be useful when you want to raise an object, such as a flag, high above your head.

Movable Pulley

In a movable pulley, one rope end is attached to an immovable object, but the wheel is not attached in place. It can move freely along the rope. A weight is attached to the wheel. The immovable object—a beam, for example—supports half of the weight. You pull on the other end of the rope, adding effort force equal to the remaining half of the weight. Because half of the weight is already supported by the beam, you do not need to use as much effort force as you would with a fixed pulley. In this way, a movable pulley reduces the amount of force needed to lift an object.

Check for Understanding

How are compound machines different from simple machines?

Mechanical Advantage: Pulley

Recall that a fixed pulley does not change the amount of force needed to do work. So effort force is always equal to resistance force. The mechanical advantage of a fixed pulley is therefore 1.

For a movable pulley, the mechanical advantage is 2. Why? The immovable object supports half the weight of the object being lifted. So if an object weighing 6 N is attached to a pulley that in turn is attached to a beam, the beam supports 3 N and you supply an effort force of 3 N to lift the object. Write the equation that shows the mechanical advantage of a moveable pulley.

Pulleys can work together in a complex system. To find the mechanical advantage of a pulley system, you can count the number of ropes that support the weight.

Explore-a-Lab

Structured Inquiry

How is a movable pulley better than a fixed pulley?

Use an empty thread spool, a pencil, and 1 m of string to create a fixed pulley. Place the pencil inside the spool and make sure the spool can freely turn. Then run the string over the spool. Attach a bottle of water to one end of the string and a spring scale to the other end. Ask a partner to hold the pencil horizontally by both ends as you pull on the spring. Measure and record the amount of work needed to lift the bottle. Then revise your design to create a moveable pulley. You may need to use extra string or tape. Record and compare the amount of work needed to lift the bottle in each trial. What is the mechanical advantage of each type of pulley?

Scripture Spotlight

What kind of devices did skillful men make in **2 Chronicles 26:15**? Do you think these were simple or compound machines? Why?

Compound Machines Explain

A family drives up to the departure entrance at an airport. The mother and daughter step out of the car and walk into the airport through the automated doors. They use a machine to print out their tickets and then ride up an escalator to their gate. In a matter of minutes, the mother and daughter have used several compound machines.

A compound machine is a machine made of two or more simple machines. They are combined so that tasks can be accomplished that would be more difficult or even impossible without them. For example, the compound machines that the mother and daughter used include the car, the automated doors, the ticket machine, and the escalator. What simple machines can you find in a car that are examples of compound machines?

The compound machine you probably use most is your body. It is a specially designed lever system that helps you move. Each movement that you make with your arms and legs involves third-class levers. Your muscles supply the input force, and your joints are the fulcrums. The output force is the load being moved, such as your body or a book. How can you improve the mechanical advantage of your arms? What could you do to improve the mechanical advantage of your legs?

Your body is a compound machine that helps you do work.

What simple machines that make up the body are visible here?

Lesson Activity

Carefully examine objects in the classroom. Find at least three compound machines. Sketch the machines and then circle and label the simple machines in each one. Share your sketches with another pair of students and compare your results.

 **What compound machines do you see every day?**

Scissors are an example of a compound machine because they are a combination of first-class levers that share the same fulcrum. The edges of the blades are inclined planes. Both a shovel and an axe are also compound machines. The blade of each tool is a wedge. The handle of each tool acts as a lever. In both cases, the wedge cuts through the material, and the handle allows the user to move the wedge more easily.

A pencil sharpener is also a compound machine. Inside are two sets of spiral cutting blades that are turned by a wheel and axle—the handle. Even a fishing pole is a compound machine. It is a third-class lever with a wheel and axle—the fishing reel—attached to hold the line.

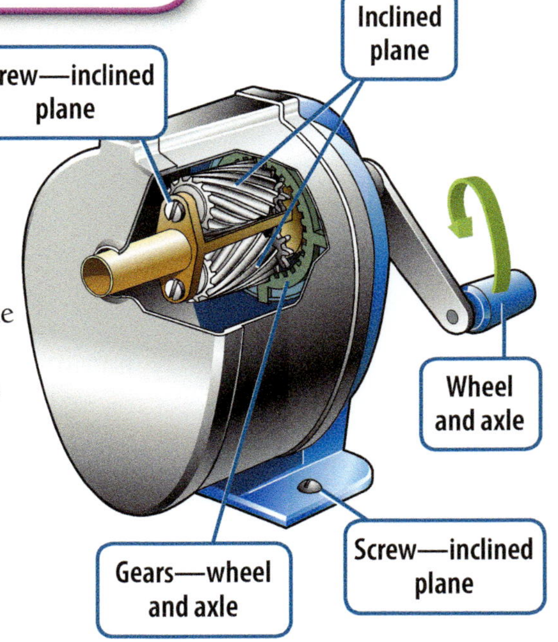

 Concept Check Assess/Reflect

Summary: What are levers, wheels and axles, and pulleys? A lever is a rigid bar or board that rotates on a fixed point called the fulcrum. Scientists classify levers as first-class, second-class, and third-class levers based on the locations of the fulcrum, effort force, and resistance force. A wheel and axle is a modified lever made of an axle, or rod, that goes through the center of a wheel. A pulley is a grooved or flat wheel with a rope, belt, or chain that passes over it. It is often used to lift heavy weights. A compound machine is a machine made of two or more simple machines. Everyday examples include scissors and doorknobs.

1. A book is sitting near the edge of a table. Design a way to raise the book off the table using a simple machine.
2. You are playing soccer and pull back your leg to kick the ball. Where is the fulcrum in your kicking leg?
3. What could you do to increase the mechanical advantage of a wheel and axle?
4. A pulley system is used to raise a load that weighs 1200 N. A force of 400 N is used to raise the load. How many supporting ropes does the pulley system have?

People in Science

Extend

Get to Know
George Stokes

George Stokes was born in 1819 in Ireland to a minister and the daughter of a minister. He had, understandably, a very religious upbringing and early education.

Stokes began his formal education in Dublin, and later he continued his studies in Bristol, Ireland. In 1837, he moved forward with his education at Pembroke College, Cambridge. During his tenure at Cambridge, he contributed to the understanding of mathematical physics, publishing original works as early as 1840. Over the years, his work numbered more than a hundred.

Stokes is known for his extraordinary mathematical skill in combination with superior experimental abilities. He was especially interested in fluid mechanics, the forces in fluids and how forces act upon fluids. Based on his contributions, the science of hydrodynamics would later expand significantly. His theories helped explain many natural phenomena, such as clouds, ripples and waves in water, and river water flow.

In 1845, Stokes expanded his area of research into the study of light. He published papers on the dynamic theory of light diffraction and light spectrum bands. He continued his research and analysis of light into 1852. His papers were read by other well-known scientists, such as Lord Kelvin.

Called to Serve

Stokes was a devout man throughout his life. He wrote books and papers and spoke at scientific societies, emphasizing his faith.

Concept Check

1. In what way did George Stokes's research connect with forces?
2. How did George Stokes express his belief in God?

Science and Technology

Extend

Neuroprosthetics

Many of us know or have seen people with prosthetics. In Canada, Terry Fox ran across the country with a prosthetic leg to raise money for cancer research. In science, Stephen Hawking communicates his theories of the Universe with complex computer voice synthesizers operated by his eye movements.

Traditional prosthetics replace missing limbs or provide assistance to those experiencing paralysis with devices that provide minimal feedback to the user. Patients with prosthetic arms, for example, report accidentally breaking dishes or crushing cans because they are not able to gauge how much pressure they are using when they pick items up. Too little pressure causes the items to slip and fall out of their grasp.

Scientists have been successful in connecting the prosthetic directly to the patient's brain. In the case of cochlear, or inner ear, implants, patients do not have a perfect replacement, but they can have a telephone conversation and perhaps hear a university or college lecture—even if they may not be able to appreciate music as much as they would like.

Limb replacement may be trickier because scientists have not been able to precisely identify which neurons in the brain are responsible for specific movements. However, early experiments have proven promising. Patients can identify tingling sensations and vibrations, which allow them enough feedback to have control. This is still far away from the complex network of sensations we may feel from stroking a cat's fur, for example, but it's a start.

Imagine a person confined to a wheelchair and paralyzed from the neck down. Even the simplest of actions, such as taking a sip of water, are outside this person's abilities. With a neuroprosthetic, a net of neurosensors implanted in the person's brain may allow him or her to concentrate and move a robotic arm to pick up a cup and move it to his or her lips. This technology is in development today and will open up a world of independence to those with paralysis.

Concept Check

1. What benefits do neuroprosthetics offer above traditional prosthetics?
2. How can neuroprosthetics change people's lives?

Chapter 13 Review

Assess/Reflect

Study Guide

Lesson 1

1. Work is done when an object moves in the direction of the force acting on it.
2. Power is the measure of the amount of work done per second.
3. Machines help make work easier by changing the strength or direction of a force, or the distance over which a force is applied. Machines also increase power by increasing the speed at which work can be done. Power is the rate at which work is done.
4. To calculate mechanical advantage, you divide the output force by the input force. Efficiency can be calculated by dividing the work output (WO) by the amount of work input (WI).

Lesson 2

1. An inclined plane is a simple machine with a slanted surface. A screw is an inclined plane wrapped around a post. A wedge is an inclined plane with two inclined planes attached back to back.
2. The length of the slanted plane represents the effort force. The height of the raised end represents the resistance force. To calculate the mechanical advantage of an inclined plane, you divide the length of the slanted plane by the height of the raised end.

Lesson 3

1. In a first-class lever, the fulcrum is positioned between the effort force and the resistance force. In a second-class lever, the resistance force is found between the fulcrum and the effort force. In a third-class lever, the effort force is found between the fulcrum and the resistance force.
2. The effort arm is the distance between the fulcrum and the point where effort force is applied. The resistance arm is the distance between the fulcrum and the point where resistance force is exerted. To find the mechanical advantage of a lever, you divide the length of the effort arm by the length of the resistance arm.
3. A wheel and axle is a modified lever made up of an axle that goes through the center of a wheel. The axle is the fulcrum, and the outside of the wheel is the arm of the lever. Effort force is usually applied to the wheel. The axle exerts resistance force.
4. A fixed pulley changes the direction of the input force. A movable pulley reduces the amount of input force needed to lift an object.
5. A compound machine is a combination of two or more simple machines. Scissors are an example of a compound machine made of two first-class levers that share the same fulcrum.

Vocabulary Check

Match the term in the first column to its definition in the second column.

1. effort force
2. efficiency
3. resistance force
4. mechanical advantage
5. fulcrum

A. the point on which a lever rotates
B. force associated with a machine that resists change
C. number of times a machine multiplies the effort put into it
D. measure of the actual work of a machine compared to the work put into it
E. force that must be applied to a machine to cause change

Multiple Choice

Choose the best answer.

6. A scientist is comparing work input with work output for a machine. What is the scientist studying?
   A. efficiency
   B. power
   C. mechanical advantage
   D. resistance effort

7. What is one disadvantage of an inclined plane?
   A. It decreases the power available to do work.
   B. It does not have a fulcrum.
   C. It increases the distance over which work is done.
   D. Its efficiency is difficult to improve.

8. Which type of simple machines make up a shovel?
   A. inclined plane and screw
   B. lever and wedge
   C. screw and lever
   D. pulley and inclined plane

9. A woman used 60 W of power to do 480 J of work. How long did it take her to do the work?
   A. 8 sec
   B. 40 sec
   C. 260 sec
   D. 420 sec

Check Point

Answer the following questions.

10. A student is using a shovel to dig a ditch. How could the student increase the power available to him without increasing the amount of work to be done?

11. How much effort force would you have to use to lift a 800-N load using a pulley system with a mechanical advantage of 4?

12. Rank the machines in the table according to efficiency. Use calculated quantities to support your answer.

Machine	Work Input	Work Output
Machine A	30 J	19 J
Machine B	22 J	20 J
Machine C	40 J	15 J

13. A wedge is a modified inclined plane. How would you calculate its mechanical advantage?

14. A student makes a model of a simple machine. In the model, the resistance force is located between the fulcrum and the effort force. What is the student modeling?

Energy Transferred and Transformed

Scripture Spotlight

On the first day of Creation, God created the Sun. This was the first form of energy for Earth. God gave us the ability to find other ways to use energy. By learning about energy, you can better understand how God cares for us. In this chapter, you will read the following passages.

Exodus 3:1–3 (p. 526)
Revelation 3:15–16 (p. 539)

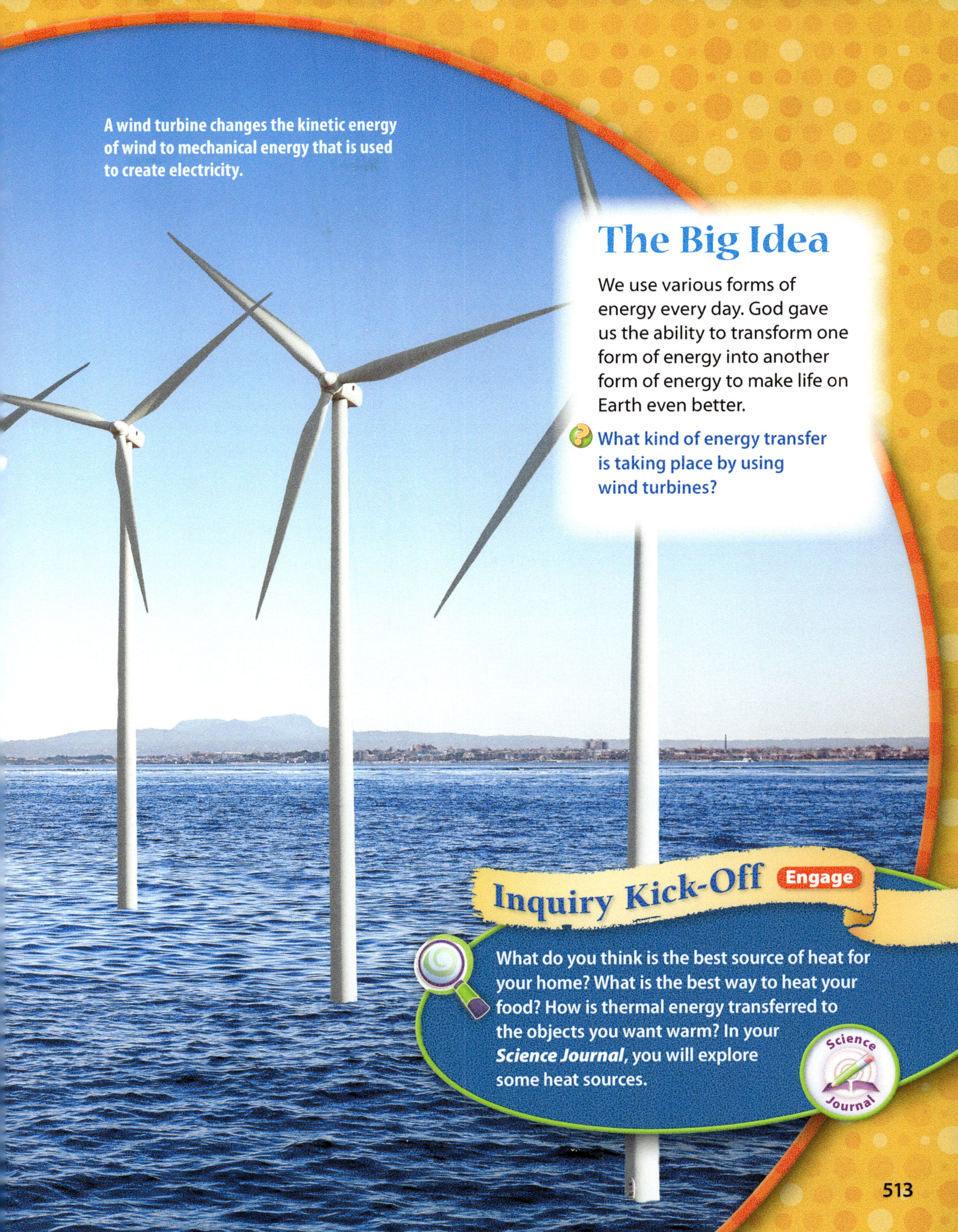

A wind turbine changes the kinetic energy of wind to mechanical energy that is used to create electricity.

The Big Idea

We use various forms of energy every day. God gave us the ability to transform one form of energy into another form of energy to make life on Earth even better.

What kind of energy transfer is taking place by using wind turbines?

Inquiry Kick-Off

Engage

What do you think is the best source of heat for your home? What is the best way to heat your food? How is thermal energy transferred to the objects you want warm? In your ***Science Journal***, you will explore some heat sources.

Lesson 1

Objectives

- Distinguish between kinetic and potential energy.
- Calculate kinetic energy and gravitational potential energy.
- Describe forms of energy.

Vocabulary

gravitational potential energy

chemical energy

nuclear energy

electrical energy

Essential Question

What Are the Forms of Energy?

Energy is a property of all substances and is associated with things such as mechanical motion, light, electricity, thermal energy, sound, and nuclear and chemical reactions. Look around the classroom; how many different examples of energy can you see? How are these examples of energy different? An avalanche roaring down the side of a mountain and a gallon of gasoline both have energy. Think of the energy in the moving mass of snow and the energy in a gallon of gasoline. How are they the same? How are they different?

Kinetic and Potential Energy Explain

You have learned that energy can be categorized as either *kinetic energy* or *potential energy*. The word *kinetic* comes from the Greek word *kinetikos*, which means "to move or put in motion." Kinetic energy is the energy related to the motion of an object. Potential energy is the energy associated with the position of an object relative to other objects. Let's look more closely at each of these types of energy.

A lot of potential energy is converted to kinetic energy during this avalanche.

What gave the snow its potential energy?

Consider a baseball that is traveling at 145 km/hr with respect to a batter. The moving ball has kinetic energy because it can do work on another object if the two collide. For example, if the ball hits the batter, the collision will cause the ball to be deflected. This means the ball's velocity changes, so the batter must have exerted a force on the ball (by Newton's second law). By Newton's third law, the ball exerted the same amount of force on the batter but in the opposite direction. The batter and the ball are both compressed slightly in the collision, so the ball applied this force over a small distance. Applying a force over a distance is the definition of *work*, so we conclude that the ball did work on the batter. The batter knows the ball did work on him because he feels the pain of his muscles being compressed and other unpleasant sensations. How would the kinetic energy of the ball change if the ball were traveling at 75 km/hr instead of 145 km/hr?

Faith Connection

Compare potential and kinetic energy with the power of the Holy Spirit being present with us or active within us and through us, depending on our choice.

What is happening in this picture in terms of kinetic and potential energy?

Kinetic Energy

Recall that kinetic energy is the energy of motion. An object that is moving has kinetic energy. There are many forms of kinetic energy: things can vibrate, they can spin, or they can move from place to place. To keep matters simple, we will focus on the kinetic energy that causes objects to move from place to place. This kind of kinetic energy depends upon two variables: the mass (m) of the object and the speed (v) of the object. The equation for kinetic energy is:

$$\text{kinetic energy} = \frac{1}{2} \times m \times v^2$$

This equation reveals that the kinetic energy of an object is directly proportional to the square of its speed. The kinetic energy is dependent upon the square of the speed. For a twofold increase in speed, how much would kinetic energy increase?

Explore-a-Lab

Structured Inquiry

How is potential and kinetic energy converted when a ball bounces?

You will need graph paper, a table-tennis ball, and a meterstick. Working with a partner, drop a table-tennis ball from a height of 25 cm onto a hard surface. Measure the height of each subsequent bounce. Create a table to record the results. Make a graph of the data. Repeat the lab, but drop the ball from 50 cm and then 100 cm. Draw a simple diagram to show where the potential energy is the greatest and where the kinetic energy is the greatest. How are potential energy and kinetic energy converted when a ball bounces? Which drop height produced the greatest number of bounces? When the ball hits the surface, what force causes the ball to lose some of its kinetic energy?

Potential Energy

Look at the skier standing at the top of a hill in the image below. Does she have kinetic energy with respect to the hill? No, but she has the *potential* to gain kinetic energy because she can increase her speed by sliding down the hill. In her initial position at the top of the hill, she has *potential energy*. As long as she stays on top of the hill, her potential energy does not change.

At the top of the hill, the skier has potential energy.

Water held behind a dam is another example of potential energy. The water can be released at any time to flow downhill, which increases its kinetic energy so that it can do work by turning a waterwheel or electric generator.

Another form of potential energy is a barrel of oil. Under the right conditions, this barrel of oil can create kinetic energy. For the skier, the potential energy is due to her position with respect to Earth. What determines the potential energy of the barrel of oil?

How does the skier's position give her potential energy? You know she has potential energy because she can convert her position to kinetic energy by sliding down the hill. What causes her to move downward? What would happen if the skier and the hill were on the Moon? Could she still convert her position to kinetic energy? Would her potential energy still be the same? At the bottom of the hill on the Moon, how would her kinetic energy differ?

This example illustrates why potential energy requires at least two objects. If our skier were all alone in outer space, no gravity would be acting on her, so she would not have the potential to move. She would have zero gravitational potential energy. **Gravitational potential energy** is the energy an object, such as the skier, possesses as a result of the force of gravity acting on it. Gravitational potential energy depends on mass and position relative to Earth. What are other situations in which the mass and the distance between two objects determine their gravitational potential energy?

At the bottom of the hill, the skier's potential energy has become kinetic energy.

What energy does the skier have when she is halfway down the hill?

Calculating Kinetic Energy and Gravitational Potential Energy

The kinetic energy and potential energy of an object are measured relative to something else. Kinetic energy depends on the speed of the object relative to another object. Potential energy depends on the position of one object relative to another object.

The kinetic energy K of an object with mass m that is moving at a speed v is:

$$K = \frac{1}{2}mv^2$$

Based on this equation, which has more influence on kinetic energy—mass or velocity? You can use the equation to calculate the kinetic energy of a 20-kg dog running down the street straight past you at 10 m/s with respect to you. The mass of the dog is $m = 20$ kg, and its speed is $v = 10$ m/s. Plug these numbers into the equation for kinetic energy to solve:

$$\begin{aligned} K_{\text{dog}} &= \frac{1}{2}mv^2 \\ &= \frac{1}{2} \times 20\text{ kg} \times (10\text{ m/s})^2 \\ &= \frac{1}{2} \times 20\text{ kg} \times (10\text{ m/s}) \times (10\text{ m/s}) \\ &= 1000\text{ J} \end{aligned}$$

The dog's kinetic energy with respect to you is 1000 J.

Science Journal

Check out your *Science Journal* for a Structured Inquiry that demonstrates how potential energy can be converted to kinetic energy.

Extend

Recall that the skier had gravitational potential energy because of her mass and her position relative to Earth. When calculating the gravitational potential energy of an object, its weight is expressed in newtons. To determine newtons, you multiply the mass of an object in kilograms by the acceleration due to gravity. You will remember that, according to Newton's law of universal gravitation, the acceleration due to gravity is 9.8 m/s^2. Therefore, near the surface of Earth, the gravitational potential energy U of an object with mass m at a height h above a reference position is

$$U \text{ (gravitational potential energy)} = m \text{ (mass)} \times g \text{ (acceleration of gravity)} \times h \text{ (height)}$$

Imagine you lift a 2.0-kg book a distance of 1.0 m from the floor to a table top. If the floor is 0.0 m, then the table top would be at 1.0 m, so the height to use in the calculation would be:

$$\begin{aligned} h &= \text{final height} - \text{initial height} \\ &= 1.0\text{ m} - 0.0\text{ m} \\ &= 1.0\text{ m} \end{aligned}$$

Now use m = 2.0 kg and h = 1.0 m in the equation for gravitational potential energy:

$$
\begin{aligned}
U &= m \times g \times h \\
&= 2.0\ \text{kg} \times 9.8\ \text{m/s}^2 \times 1.0\ \text{m} \\
&= 2.0 \times 9.8 \times 1.0\ \text{kg} \times \text{m/s}^2 \times \text{m} \\
&= 20\ \text{N} \times \text{m} \\
&= 20\ \text{J}
\end{aligned}
$$

The potential energy of the lifted book is 20 J. What if the book were set on the window sill three floors (10 m) above the ground? How would the potential energy relative to the ground differ from the potential energy relative to the floor?

While gravity acts between every object in the Universe, only for very large masses, such as planets and stars, is gravity noticeable in any meaningful way. Typically, we only consider gravitational potential energy when at least one object is a large body such as a planet or star. The gravitational potential energy of the skier due to her position with respect to Earth is noticeable. Why is the gravitational potential energy between you and a classmate not noticeable?

Explain why these books have gravitational potential energy.

Explore-a-Lab

Guided Inquiry

How can you use energy transfer to move an object?

Work in groups of four to make a toy to move a ball. Use your knowledge of kinetic energy and energy transfer to create a toy to propel a ping-pong ball at least five inches. You can use tinker toys, rubber bands, paper cups, craft sticks, pipe cleaners, balloons, tape, string, straws, and any other materials your teacher makes available to create your transport device. Draw your design. If you have to modify your design, note those changes on your drawing as well. What machines are included in your device? Share your device with the rest of the class and show how it works. Record how far you moved the ball and compare it to other groups' designs. What design moved the ball the farthest? What part of the design made it the most successful?

Record your work for this inquiry. Your teacher may also assign the related Guided Inquiry.

Investigating Potential Energy

How can you measure gravitational potential energy?

Materials
- balance
- marble
- plastic wrap
- modeling clay (soft)
- metric ruler
- graph paper
- steel ball bearing

Procedure

1. Use the balance to determine the mass of the marble and steel ball bearing in kg (you will need to convert from grams). **Record** the masses in your ***Science Journal***.
2. Place a sheet of plastic wrap on the floor. Place a 15 cm square of soft clay about 2 cm thick on the plastic wrap.
3. Drop the marble from a height of 0.5 m. **Measure** and record the depth of the dent that it makes.
4. **Calculate** and record the gravitational potential energy of the marble before it was dropped using the equation below.

   $$U = m \text{ (mass)} \times g \text{ (acceleration of gravity)} \times h \text{ (height)}$$
5. Repeat Steps 3–4 dropping the marble from heights of 1 m and 2 m.
6. **Construct a graph** of your data. Plot the heights from which you dropped the ball on the x-axis and the average depth of the corresponding dents on the y-axis.
7. Repeat Steps 3–6 with the steel ball bearing.

Create Explanations

1. How can you measure gravitational potential energy?
2. How is the ball's kinetic energy related to the gravitational potential energy that the ball had before it was dropped?
3. Why does the size of the crater depend on the potential energy of the object?
4. How could you drop the marble and the steel ball bearing so that they make a crater that has the same depth?

Other Forms of Energy Explain

There are many forms of energy, but all are forms of kinetic energy or potential energy, or some combination of the two. The table shows the various forms of energy and gives their composition in terms of kinetic energy and potential energy.

Forms of Energy and Their Composition	
Energy form	**Energy composition**
Mechanical energy	Kinetic and potential energy
Light	Kinetic energy
Sound	Kinetic and potential energy
Chemical energy	Potential energy
Nuclear energy	Potential energy
Elastic energy	Potential energy
Thermal energy	Kinetic and potential energy
Electric energy	Kinetic energy when power is being used Potential energy when no power is being used

Chemical energy is the energy stored in the positions of atoms and molecules in compounds. It is transformed to kinetic energy by pulling atoms or molecules completely apart so that they don't come back together again. To transform chemical energy, a small amount of energy is used to separate some of the atoms and molecules from each other. Once they break their bonds, the atoms and molecules fly apart at high speeds. This is similar to what would happen if you were pulling on a rope attached to a wall, and someone suddenly cut the rope. As the atoms and molecules fly away from each other, they release much more energy in the form of kinetic energy than it took to separate them from each other. Some examples of chemical energy being released are the explosion of dynamite and the burning of gasoline in the engine of a car.

Into what forms of energy is the chemical energy of the fuel in this fire converted?

Nuclear energy is similar to chemical energy, but the potential energy does not come from the relative position of atoms and molecules. Instead, the potential energy in nuclear energy comes from the relative position of the particles that are inside the nucleus of atoms. The forces holding these particles together are much, much stronger than the forces of molecular bonds, so much more kinetic energy is released when they fly apart.

You already know that *thermal energy* is the average mechanical energy of atoms or molecules. In a hot room, the air molecules move faster, on average, than those in a cold room. This movement can be motion in a straight line, motion due to spinning, motion due to molecules vibrating, or any combination of these motions. Solids and liquids also contain thermal energy. What kinds of motion contribute to thermal energy in solids and liquids? Why do you think thermal energy is said to contain both kinetic *and* potential energy if it is just the total mechanical energy of small particles?

Water below Earth's surface can be heated by hot rock. As a result of the transfer of thermal energy, the water turns into a gaseous state and erupts from the surface as a geyser.

Electrical energy is kinetic energy if power is being used, but it is potential energy if power is not being used. In a new battery, negative charges are concentrated at one end, and positive charges are concentrated at the other end. Negative charges repel each other, so they want nothing more than to move away from each other (and likewise for positive charges). This means that the charges have the potential to move, so they have electric potential energy. If you connect the battery to a circuit with a light bulb, the charges move through the circuit. The kinetic energy of the moving electrons can do work, for example, by bumping into the atoms of the filament of a light bulb, which raises the temperature of the filament enough to make it emit light. Why do we say that electrical energy is kinetic energy if power is being used and is potential energy if power is not being used?

Concept Check Assess/Reflect

Summary: What are the forms of energy? Energy comes in two basic forms—kinetic energy and potential energy. All other forms of energy are either kinetic energy, potential energy, or a combination of the two. Kinetic energy is the energy due to motion of one object relative to another. An object of mass m moving at speed v has kinetic energy of:

$$K = \frac{1}{2}mv^2$$

Potential energy is due to the position of one object relative to another. For objects near the surface of Earth, their gravitational potential energy is:

$$U = mgh$$

where h is the height of the object with respect to a reference position. Kinetic energy and potential energy can be transformed from one into the other. All forms of energy are either kinetic energy or potential energy. For example, nuclear and chemical energy are forms of potential energy, thermal energy is combined kinetic and potential energy, and electrical energy is kinetic energy if power is being used and is potential energy if power is not being used.

1. How is potential energy different from kinetic energy?
2. Why are chemical and nuclear energy considered forms of potential energy?
3. Is it possible to convert mechanical energy into nuclear potential energy? Why or why not?

Lesson 2

Essential Question

How Is Energy Transformed?

Objectives

- Explain how energy transforms from one form to another.
- Demonstrate the law of conservation of energy.
- Discuss when and how energy is transferred.

Vocabulary

energy transformation
system
surroundings
law of conservation of energy
energy transfer

Suppose you jump over a small fence on the way to school. Every second of every day, energy is being changed from one form into another all around us. How does the Sun's energy change? What energy change happens in plants? If you eat a large bowl of pasta and then run a mile, what energy changes take place in your body? Would any form of life be able to survive without the ability to change energy? Why or why not? What other energy changes are taking place around you as you read this?

Transforming Energy Explain

Energy transformation occurs when energy changes from one form to another. How would you go about studying the transformation of energy? For example, suppose you want to investigate the energy transformations that occur when sunlight hits a leaf. What is the first thing you would do? Actually, you may already have done it, perhaps without thinking about it. The first thing you do is define what it is you are studying. In this case, it's the leaf. A scientist would call the leaf the system. A **system** consists of all the things that are being studied. Everything outside the system is called the **surroundings**. To keep track of what happens to energy, scientists often investigate the interactions between a system and its surroundings.

The energy of sunlight powers many processes on Earth.

Into what other forms of energy can this solar energy be changed?

Consider a leaf as the system. You know that it absorbs sunlight. In doing so, energy is transformed from light energy to chemical energy and thermal energy. In the case of the leaf, the process of photosynthesis uses light energy to create complex molecules from simple molecules. Light energy that is not used in photosynthesis is transformed into thermal energy.

If you consider yourself as a system, what forms of energy are you absorbing from your surroundings? What forms of energy are leaving you and entering the surroundings?

Examples of Energy Transformation

In the previous lesson, you learned that potential energy transforms into kinetic energy when a skier slides down a hill. Even though potential energy is transformed or changes into kinetic energy, the total energy of the skier–Earth system does not change—only the form of the energy changes. With the skier at the top of the hill, the system has a certain amount of potential energy. When she arrives at the bottom of the hill, the system has the same amount of energy, but it is in the form of kinetic energy.

Faith Connection

How does God transform the lives of people? How can you help?

Roller Coaster

Another example of energy transforming back and forth from potential energy to kinetic energy is a roller coaster going over a track. On the top of a hill, the roller coaster has potential energy. As the roller coaster coasts down into a valley between hills, this potential energy is transformed into kinetic energy. As it coasts up the next hill, the roller coaster slows down, and the kinetic energy is transformed back into potential energy. What other processes transform potential energy into kinetic energy? Where does the original energy come from? Where has it gone when the roller coaster stops?

At what point on the tracks does this roller coaster have the most kinetic energy?

Explore-a-Lab

Structured Inquiry

How can you measure potential and kinetic energy?

Gather three different-sized marbles, a balance, a ruler, a milk carton, and foam tubing. Create an inclined plane with the tubing, resting the upper part on books or a chair so it is at a constant height. Cut the milk carton in half and place it at the bottom of the tube to catch the marbles. Find and record the mass of each marble. Place one marble at the top of the inclined plane and release it. Record how far the marble pushes the carton. Repeat this process two more times and find the average distance. Repeat the process for the other marbles. Which marble had the most potential and kinetic energy? How did mass affect the amount of energy each marble had? Change the height of your inclined plane and repeat the experiment. How did your results change? What adjustments can you make so all of the marbles exhibit the same amount of kinetic energy?

Scripture Spotlight

Read **Exodus 3:1–3.** What energy transformation usually occurs when wood burns? How can you tell that transformation was not occurring in this passage?

Blender

Now consider electrical energy that is used to run a blender. Before you turn on the blender, the electrical energy is in the form of potential energy. This potential energy comes from the presence of significantly more negative charges in the wire than positive charges. Because negative charges repel each other, they will move away from each other if possible. This possibility happens when the blender is turned on. The negative charges can now move through the switch and into the electrical motor. As they pass through the motor, the repulsive force between the negative charges is used to push on the motor, causing it to spin. In this way, the potential energy that the negative charges had before the blender was turned on is transformed into the kinetic energy of the spinning motor.

Into what form of energy is the kinetic energy of the moving gasoline particles converted by the pistons?

Automobiles

In cars, potential energy also is transformed into kinetic energy. The potential energy comes from the relative position of the atoms and molecules in the gasoline. In the piston of the motor, a spark gives some of the molecules in the gasoline vapor the energy to escape each other's attractive force. These molecules fly apart at high speed, so they have high kinetic energy. They then hit other molecules in the fuel, causing an explosive chain reaction. Some of these vapor molecules with high kinetic energy hit the piston and cause it to move, which in turn starts a sequence of events that makes the car move. The end result is that the potential energy in the gasoline is transformed into the kinetic energy of the moving car. Besides moving the piston, what else does the chemical energy do? What other forms does it take?

Nuclear Power Stations

In nuclear power stations, nuclear energy is converted into electrical energy. Recall that nuclear energy is due to the relative position of neutrons and protons inside the nucleus of atoms in the nuclear fuel. Hitting these nuclei with neutrons gives them the energy they need to split apart, and in the process they release more neutrons. This reaction would become an explosive chain reaction, such as what occurs in a nuclear bomb, if a material were not used to absorb some of the neutrons, creating a controlled nuclear reaction. The fragments of the split nuclei fly apart at high speeds and warm the water around them. The thermal energy in the water is then used to drive turbines that produce electricity. What is similar about all these examples of energy transformation? What is different?

Conservation of Energy Explain

In all processes of energy transformation, no energy is created or lost. Actually, it is not possible to create or destroy energy; it can only be transformed into different forms. This is called the **law of conservation of energy**. Mathematically, we can write the law of conservation of energy as:

$$\text{final energy} = \text{initial energy}$$

The final energy is the energy of the system after the transformation, and the initial energy is the energy of the system before the transformation.

Although the law of conservation of energy is very important and very powerful, it is actually just an accounting system. It is a bit like a parent who knows that her child has 12 blocks. When the parent cleans up the child's room, she may find only ten blocks. Because the parent knows the number of blocks is conserved, she knows that two blocks are missing. So she looks under the bed and behind the dresser until the missing blocks are found. Conservation of energy is the same idea. If we know how much energy a system starts with, we know how much it will end up with after an energy transformation is done. So we can figure out what happened if some energy is missing.

Bar Diagrams

When applying the law of conservation of energy, it is sometimes useful to use a tool called a *bar diagram* to keep track of the various forms of energy. In a bar diagram, the heights of the bars represent the energy. On the right of the bar diagram, a bar represents the total energy. On the left of the bar diagram, bars represent the different forms of energy that contribute to the total energy. Each form of energy has its own bar. The bar diagram below is for a system that has ¼ kinetic energy and ¾ potential energy.

System with ¼ Kinetic Energy and ¾ Potential Energy

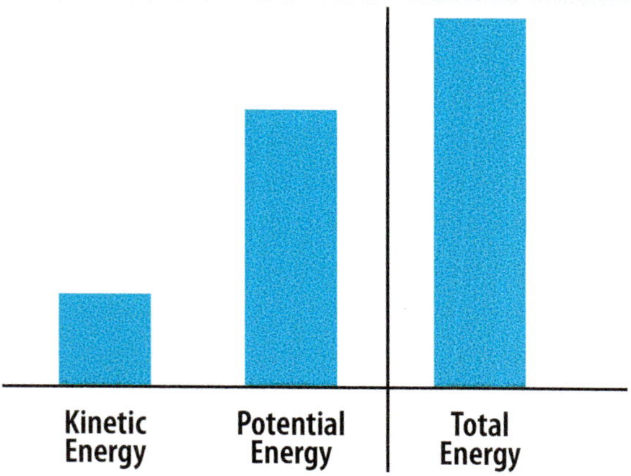

The total height of the bars on the left should be the same as the height of the bar on the right.

Now let's use a bar diagram to apply conservation of energy to the roller coaster–Earth system that was discussed earlier. When the roller coaster is at the top of the track, it moves slowly, so it has a small amount of kinetic energy. Being high on the track, it has a relatively large amount of potential energy. Adding these two types of energy together gives the total energy.

Roller Coaster at Top of Track

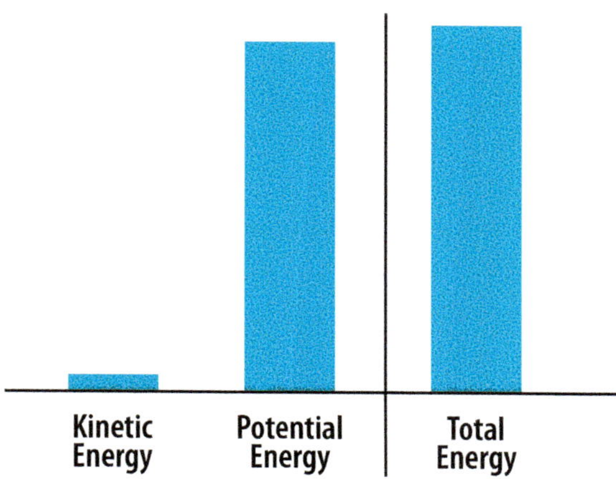

The total energy of the roller coaster at the top of the track is made up of mostly potential energy and a little bit of kinetic energy.

Math in Science

Create a bar graph similar to the one on the previous page that shows the kinetic energy, potential energy, and total energy for the roller coaster when it is halfway down the track.

Calculating Energy Changes

Bar diagrams are helpful for getting an overall view of how energy is transformed yet conserved. But the real power of conservation of energy is that it lets you figure out things that nature may be hiding from you. For example, suppose you are standing on the top of a small hill, and your friend is skiing down a different hill toward your hill at 10 m/s. You are 10 m vertically higher than your friend. You wonder whether your friend is going to be able to slide all the way up the hill to reach you. Luckily, you remember conservation of energy and so you pull out your mobile phone to do a quick calculation.

Your friend's initial energy is kinetic energy, so:

$$\text{initial energy} = \text{kinetic energy}$$

You learned how to calculate kinetic energy in the previous lesson. Apply the same formula here:

$$\begin{aligned}\text{kinetic energy} &= \tfrac{1}{2} \times \text{mass} \times \text{speed}^2 \\ &= \tfrac{1}{2} mv^2\end{aligned}$$

In this case, let's say your friend's mass is $m = 40$ kg. Her speed is $v = 10$ m/s. Plug these numbers into the formula for kinetic energy:

$$\begin{aligned}\text{kinetic energy} &= \tfrac{1}{2} \times \text{mass} \times \text{speed}^2 \\ &= \tfrac{1}{2} \times 40\ \text{kg} \times (10\ \text{m/s})^2 \\ &= \tfrac{1}{2} \times 40\ \text{kg} \times 10\ \text{m/s} \times 10\ \text{m/s} \\ &= 2{,}000\ \text{J}\end{aligned}$$

This means your friend's initial energy is 2,000 J. When your friend slides as far up the hill as she can, her final energy will be all potential energy, so:

$$\text{final energy} = \text{potential energy}$$

You also learned how to calculate potential energy in the previous lesson. Apply the same formula here:

$$\begin{aligned}\text{potential energy} &= \text{mass} \times \text{acceleration of gravity} \times \text{height} \\ &= mgh\end{aligned}$$

In this equation, m is again your friend's mass, g is the acceleration due to gravity (9.8 m/s^2), and h is the vertical height up to which she slides. We don't know h, but we can find the product of mass × acceleration of gravity (mg). Let's find that and then use conservation of energy to find the height h. The product mg is:

$$\begin{aligned} mg &= 40 \text{ kg} \times 9.8 \text{ m/s}^2 \\ &= 392 \text{ N} \end{aligned}$$

So the potential energy of your friend after she slides up the hill is

$$\begin{aligned} \text{potential energy} &= mgh \\ &= 392 \text{ N} \times h \end{aligned}$$

This means that your friend's final energy is 392 N times the height up to which she slides. According to the law of conservation of energy, her final energy must be the same as her initial energy. You just figured out that her initial energy is 2000 J, and her final energy is $392 \text{ N} \times h$, so you can write

$$\begin{aligned} \text{final energy} &= \text{initial energy} \\ 392 \text{ N} \times h &= 2000 \text{ J} \end{aligned}$$

To find the height h, divide each side of this equation by 392 N.

$$\begin{aligned} \text{final energy} &= \text{initial energy} \\ \frac{392 \text{ N} \times h}{392 \text{ N}} &= \frac{2000 \text{ J}}{392 \text{ N}} \\ \overbrace{\frac{392 \text{ N}}{392 \text{ N}}}^{=1} \times h &= \frac{2000 \text{ J}}{392 \text{ N}} \\ h &= \frac{2000}{392} \frac{\text{J}}{\text{N}} \\ h &= 5.1 \text{ m} \end{aligned}$$

So your friend will only slide up to a height of 5.1 m. She'll have to walk the rest of the way to get up to your height. Notice that, in the last step of this calculation, the units J/N changed to m. Can you convince yourself that this is correct?

Work and Energy Efficiency

Now that you know that energy is always conserved in energy transformations, you might wonder why we constantly try to conserve energy. What is the point if energy is never created or destroyed? To understand, consider again our skier as she slides down the hill, but this time do not neglect air resistance and the friction between her skis and the snow. These forces act on the skier in the direction opposite to her motion and so do negative work on her. When she gets to the bottom of the hill, her kinetic energy is therefore less than the potential energy she had at the top of the hill. The missing energy went into increasing the thermal energy of the air and the snow. If we add this thermal energy to the final kinetic energy, we find that energy is still conserved:

final kinetic energy + thermal energy = initial potential energy

This brings us back to the question: Why should we constantly try to conserve energy? To answer this question, ask yourself which form of energy is more efficient for doing work: kinetic energy or thermal energy?

An important characteristic of thermal energy is that the motion of the particles in the system is random. If all the particles move in the same way, as in a gust of wind, then this motion does not contribute to thermal energy. To use thermal energy to do work, we can put a hot gas in a piston. As the hot gas particles move freely in the cylinder, some collide with the bottom of the piston. They push the piston out of the cylinder, and thus do some work. But most of the hot gas particles actually just collide with the sides of the cylinder, which do not move. These particles do no work—they just heat up the piston as the kinetic energy is changed to thermal energy. This thermal energy escapes the system and can do no work to the system. Where does this thermal energy go?.

Now suppose that all the hot gas particles could move toward the piston. When they collide with the piston, all the particles would contribute to making the piston move. What effect would this have on the piston? What kind of energy do the gas particles have?

This example shows why kinetic energy is more efficient at doing work than thermal energy, which is much more diffuse so it can't do much work.

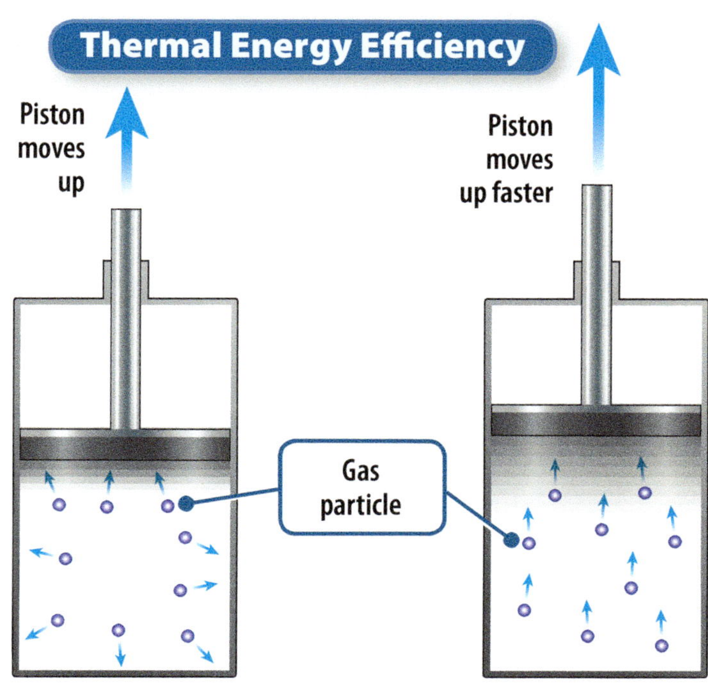

Piston engines are inefficient because energy is lost for various reasons, including to cool the system and to move the pistons. Note that the image on the right is hypothetical to show how the engine is actually inefficient. No engine system can make all of the gas particles move in the same direction at the same time.

Explore-a-Lab

Structured Inquiry

Where does the energy go?

You will need some string and two 1-L water bottles filled halfway with water.

1. Tie a single piece of string about 1 m long between two supports (chairs or desks would work). This support string will support the two water bottles.
2. With other pieces of string, hang the water bottles from their necks so they are about 30 cm below the support string. The water bottles should be about 50 cm away from each other on the support string and about 50 cm away from each support.
3. With one water bottle not moving, pull the other water bottle away from the support string (like a swing on the playground) about halfway up. This is the release height.
4. Let the water bottle swing down so that it starts swinging back and forth perpendicular to the support string.

What happens to the other water bottle as the first one swings? How is the swinging of one water bottle related to the swinging of the other water bottle? Do they ever swing at the same time? If so, do they swing to original release height at the same time? How does this demonstrate conservation of energy?

Transferring Energy Explain

You learned that energy can be classified according to its form, such as chemical energy or nuclear energy, and that energy can change forms. Energy can also be classified according to whether it moves from one place to another without changing form. This is known as **energy transfer**. Energy is often transferred between a system and its surroundings. If this happens, the system is not isolated from its surroundings.

A system that is isolated exchanges no energy with its surroundings. Energy is conserved in an isolated system because no energy can enter or leave the system. In this case, the initial energy E_i of the system is equal to the final energy E_f:

$$E_f = E_i$$

If a system is not isolated, then energy can enter or leave the system. In this case, you must include the energy transferred into or out of the system when applying conservation of energy:

$$E_f = E_i \pm E_{transfer}$$

Use the plus sign if energy is transferred into the system (think "added to the system") and the minus sign if energy is transferred out of the system (think "subtracted from the system").

Record your work for this inquiry. Your teacher may also assign the related Guided Inquiry.

Energy Conservation

How can you demonstrate the law of conservation of energy?

Materials

- flexible ruler
- toy car
- meterstick
- stopwatch
- masking tape

Procedure

1. Working with a team, sketch ideas on how to use the law of conservation of energy while changing one form of energy into another by using a ruler to move a toy car. At this point, you should not be experimenting with the car and ruler.
2. Using your team's sketches, **predict** how you can maximize the amount of energy transferred in the experiment you designed. Consider the different **variables** you need to control or monitor.
3. With your team, agree how you will **measure** the amount of energy put in to start the movement of the car and how to determine the amount of energy that actually moved the car.
4. Conduct your trial and **record** your results in your ***Science Journal***. Repeat with at least two more trials.

Analyze Results

Create a graph showing the amount of energy put in and the outcome. Explain how much energy you think you conserved. Where did the lost energy go?

Create Explanations

1. How can you demonstrate the law of conservation of energy?
2. What variables did you control to make your experiment as accurate as possible?
3. How could you improve your experiment?

How can you transfer energy into or out of a system? One way is for the system to do work on something in the surroundings. This is a mechanical transfer of energy. Can you think of other ways to transfer energy?

Light energy, for example, transfers energy from one place to another at the speed of light. What about sound energy? Can sound transfer energy from one place to another? How can you tell? Electrical energy also transfers itself from one place to another. You know this because high-power cables transfer the electrical energy from dams and nuclear reactors to our homes. We know the energy has arrived in our homes because we can transform it into other forms of energy. What is your favorite way to transform electrical energy in your home?

Nuclear and chemical energy are not transferable forms of energy. To transfer chemical and nuclear energy, they must first be transformed into a transferable form of energy, such as electrical energy. Why not transform it into light energy? To obtain electrical energy, nuclear potential energy in the nuclei of atoms is first transformed into the kinetic energy of the split nuclear fragments that fly apart at high speeds. These fast-moving particles collide with water molecules, which makes the water molecules move faster. In other words, they heat up the water. Finally, the water gains enough thermal energy to become high-pressure steam, which is used to force a turbine to spin, creating electrical energy.

Summary: How is energy transformed? Energy can be neither created nor destroyed; it can only be transformed from one form to another. Every time energy is transformed, the quantity of energy present before the transformation is the same as the quantity of energy present after the transformation. This is the law of conservation of energy. For an isolated system, no energy is exchanged with the surroundings, so the system must always have the same total energy. If you know the energy of an isolated system at one point in time, then you know its energy for all times. Forms of energy that are transferrable (light and sound) can move from one place to another without changing forms. Forms of energy that cannot move without changing form (chemical and nuclear) are not transferrable forms of energy.

1. Why is conservation of energy so useful for scientists?
2. If your body is defined as the system, in what ways is energy transferred between it and its surroundings?
3. How can you demonstrate that sound transfers energy from one place to another?
4. If you were studying the transformation of energy of a skier as she slides down a hill, what would you define as the system?

Essential Question

What Is Thermal Energy?

By touching an object with your bare hands, you can tell whether it's hot or cold. What gives matter the property of being hot or cold? And why is this property transmitted to your hands when you touch something? Is there any matter that does not transmit its heat to the surrounding matter? If you placed a candle in the middle of an empty freezer, will the freezer get warmer or will the cold put out the flame? Can the temperature get cold enough to freeze the flame?

Thermal Energy and Random Motion Explain

A thimblefull of air contains billions of molecules, all zipping around with different kinetic energies. How can you possibly describe the energy of all these molecules? One way to tackle this problem is by calculating the average energy of all the particles in a system. The **kinetic theory of matter** does just that: it describes matter as composed of large numbers of very small particles that are constantly moving about randomly.

Objectives

- Describe thermal energy in terms of the kinetic theory of matter.
- Explain how temperature, thermal energy, and heat are related.
- Explain the effects of thermal energy on the properties of matter.

Vocabulary

kinetic theory of matter
absolute zero
thermal equilibrium
thermal conductivity
specific heat capacity

Like all matter, these marshmallows are made of moving particles.

What affect does the flame have on the particles that make up the marshmallows?

In a gas or liquid, the particles can move from one location to another, spin, and/or vibrate. In a solid, the particles cannot move from place to place or spin, but they can vibrate. Each particle that moves from place to place or that spins has kinetic energy. If the particles vibrate, they have both kinetic and potential energy. As you know, the *thermal energy* of a substance or a system is the total energy associated with the random movement of all the particles that make up the substance or the system.

Another important property of thermal energy is that it depends on the size of the system that you choose. If your system is a thimbleful of water at room temperature, its thermal energy is about 1200 J. This means that, in theory, enough energy is in the water to move a 1200-kg object a distance of 1 m (1200 kg is the mass of a small car!). If you pour out half of the water, then the thermal energy of the water in the system is divided in half—600 J—because you have half the number of particles in the thimble.

Now consider the thermal energy of a roomful of air. Each air molecule moves randomly about the room, bumping occasionally against other molecules. In addition to this motion, the molecules can spin and/or vibrate. To find the total thermal energy of the air in the room, you would have to add these three contributions to the total kinetic energy from all the air molecules in the room! Fortunately, this impossible task is not necessary because simpler ways exist to calculate the thermal energy. To understand how this is done, you must understand the difference between thermal energy, temperature, and heat.

Temperature and Thermal Energy

Thermal energy is measured in joules. The thermal energy of a substance depends on its temperature. You know that *temperature* is a measure of how hot or cold a substance or a system is. Temperature is measured in degrees. How hot or cold something is depends on the average kinetic energy of its particles. Therefore, temperature is a measure of the average kinetic energy of the particles of a substance or a system.

Several different temperature scales exist. The Fahrenheit and the Celsius scales are the most commonly used. Zero degrees Celsius (0°C) is the temperature at which water freezes at sea level, and 100°C is the temperature at which water boils at sea level. How do you measure the temperature of systems that are colder than 0°C? What is the coldest temperature possible? As the temperature of a system is lowered, the kinetic energy of the particles in the system lowers. Eventually, the

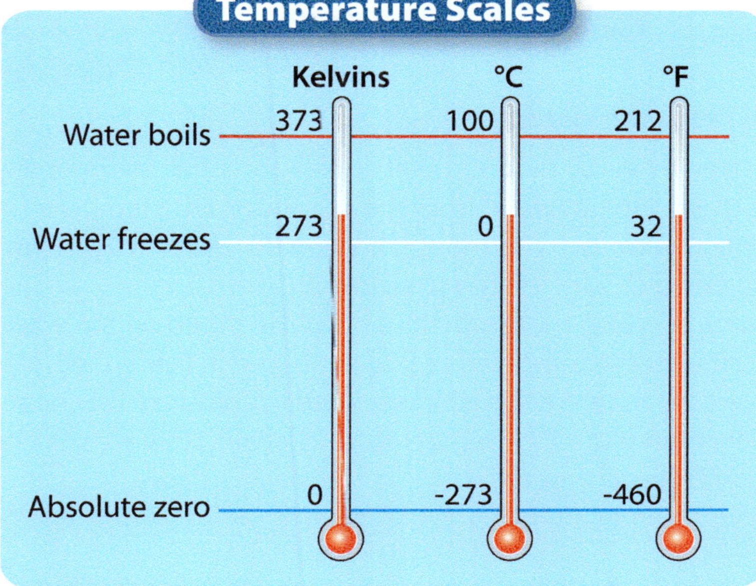

The Celsius scale is based on the boiling and freezing points of water.

How does that differ from the kelvin scale?

particles will have almost zero motion. On the Celsius scale, this happens at –273°C. Because it is not possible to have a temperature less than –273°C, another temperature scale called the kelvin scale defines this temperature to be **absolute zero**. This shift of the zero to the bottom of the temperature scale is the only difference between the Celsius scale and the kelvin scale. This means that a temperature difference is the same in each scale. For example, the temperature difference between an initial temperature of 50°C and a final temperature 80°C is

final temperature – initial temperature = 80°C – 50°C = 30°C

Let's find this temperature difference on the kelvin scale. To convert a temperature in Celsius to a temperature in kelvin, add 273°C and then change the °C to K (note that there is no degree sign in front of the *K*), so

initial temperature in kelvins = initial temperature in Celsius + 273°C
= 50°C + 273°C
= 323 K

final temperature in kelvin = final temperature in Celsius + 273°C
= 80°C + 273°C
= 353 K

Subtracting the initial temperature in kelvin from the final temperature in Kelvin gives

final temperature in kelvin – final temperature in Celsius
= 353 K – 323 K = 30 K

Because we calculated the *difference* between two temperatures, the results are the same in Celsius and kelvins.

Explore-a-Lab

Structured Inquiry

How hot is the water?

You need a Celsius thermometer and a Fahrenheit thermometer. Fill a glass with hot water from the tap. Record the temperature of the water with each thermometer every minute for 10 or 15 min. Make one plot with the temperature in Celsius on the *y*-axis and the time at which the temperature was measured on the *x*-axis. Do the same for the temperature you measured in Fahrenheit. Now convert the temperature you measured in Celsius to kelvins and make a third plot of the temperature in kelvin. How are the graphs similar? How do they differ?

If temperature differences are the same in the Celsius and kelvin scales, why bother with the kelvin scale? The advantage of the kelvin scale is that a temperature of a system using the kelvin scale is directly proportional to the thermal energy of the system. For example, if you heat a system from 100 K to 200 K, then its thermal energy doubles. To find how the thermal energy of a system at 200°C compares with that at 100 °C, you have to first convert these temperatures to the kelvin scale.

Heat Explain

In what direction does heat travel between these pancakes and the plate?

Heat is something we experience every day. But what is heat? And how is it connected with temperature? These questions baffled humankind for years and gave rise to some curious theories.

What happens when you place four hot pancakes on a plate? After a while, the plate warms up, and the pancakes cool down. What does this tell you about the thermal energy of the plate and of the pancakes? And where did the energy come from to change the temperature of the plate? The plate is in contact with the pancake, the table, and the surrounding air. Perhaps the table or the air gave up some thermal energy to warm the plate. But we know from experience that you can put a plate at room temperature on a table, and the plate's temperature doesn't change. This means its thermal energy doesn't change in this situation. Therefore, the energy to warm the plate holding the four pancakes must have come

from the pancakes. This would also help explain why the pancakes cooled. They transferred some of their thermal energy to the plate. Recall that thermal energy that is transferred between substances at different temperatures is called *heat*.

Now place some ice cubes on the plate and ask the same questions. What happens to the plate? Of course, the plate cools, and the ice cubes warm and melt. Did the ice cube transfer "cold" to the plate? No, the plate cooled because it transferred some of its thermal energy to the ice cubes. Heat is always transferred from hotter substances to colder substances. In this sense, temperature may be thought of as the ability of a substance to transfer thermal energy to its surroundings. Why can't heat transfer from cold substances to hot substances?

Scripture Spotlight

Read **Revelation 3:15–16.** Does it use the scientific meaning of *cold* and *hot*? How can you ensure that your church is not lukewarm?

Heat and Temperature

Are heat and temperature the same thing? Consider placing two pancakes on the same plate as in the earlier pancake example. Will the final temperature of the plate be the same as when it contained four pancakes? The plate does not get as hot because two pancakes contain half the thermal energy of four pancakes, so they have only half as much heat available to transfer to the plate. From this example, you learn that the amount of heat a system can transfer depends on the size of the system. For example, imagine a bathtub full of water at 35°C. When the water cools, let's say that the temperature of the air in the bathroom increases by 1 °C. If the bathtub had been half full of water, what would have been the change in air temperature?

Now consider what happens if, after being heated by the pancakes, you cut the plate into two equal-size pieces. Will the temperature of each piece also be divided by two? Why or why not? What does this tell you about how heat and temperature differ?

Thermal Equilibrium

The temperature of a system tells us how well the system can heat other systems. A system at a high temperature is better for heating than a system at low temperature. When two systems are put in contact, the high-temperature system heats the cold-temperature system. The heating stops when the two systems are at the same temperature because each system then has the same ability to heat. At this point, the two systems are in **thermal equilibrium**. This means that the same amount of heat is transferred in both directions between the systems, so overall no heating occurs.

Record your work for this inquiry. Your teacher may also assign the related Guided Inquiry.

Heat and Temperature

What makes molecules move?

Procedure

Materials
- glowstick
- two plastic-foam cups
- one thermometer
- ½ cup ice
- hot and cold water

1. Remove a glowstick from its wrapper. Bend the glowstick until you hear a snap. This will activate the glowstick.
2. Place hot water in one plastic-foam cup and ice water in another plastic-foam cup.
3. Start with the cup containing the hot water. **Measure** and **record** the temperature of the water. Place the glowstick in the hot water. Record your observations.
4. After 5 min, again measure and record the temperature of the hot water.
5. Measure and record the temperature of the ice water. Place the glowstick in the water. Record your observations.
6. After 5 min, measure and record the temperature of the ice water. **Calculate** the difference between the starting and ending temperature.

Cup	Initial Temperature (°C)	Ending Temperature (°C)	Change in Temperature (°C)	Glowstick Observation
Hot water				
Ice water				

Analyze Results

Create a graph showing how the temperatures changed. What caused these results? How did this affect your observations?

Create Explanations

1. What makes molecules move?
2. Is there a difference in the glow of the glowstick in the hot water and the glowstick in the cold water? Why?
3. What is the relationship between heat and molecular movement? How did you come to this conclusion?

Thermal Conductivity Explain

If you have ever walked barefoot around your home, you may have noticed that some surfaces feel colder on your feet than others. For example, rugs feel much warmer to your feet than tiles. However, both the rug and the tiles are at the same temperature. How do we know that they are both at the same temperature? And if they are at the same temperature, why do the tiles feel colder than the rug?

This example highlights an important property of substances, which is the ability to conduct heat. This property of a substance is called its **thermal conductivity**. Substances with high thermal conductivity quickly transfer heat from hotter parts of the substance to cooler parts. What kitchen material should have a high thermal conductivity?

Good thermal conductors are often denser materials, such as stone or steel. In these materials, the atoms and molecules are closer together, so heat is usually transferred more quickly between them. Materials that are less dense, such as gases, are usually poor thermal conductors. Gas particles are very far apart and so must travel far to transfer heat to a neighboring particle. Rugs, for example, trap a lot of air in their fibers, which is one reason that rugs are poor thermal conductors.

This floor probably feels cold on the person's feet.

Which would have a higher thermal conductivity—this floor or one made of wood? Why?

Explore-a-Lab
Structured Inquiry

What makes a good thermal conductor?

Gather samples of ceramic tile, carpet, glass, aluminum, cotton, oven mitt, stone, concrete, plastic, and rubber. Predict which of these items will conduct the most thermal energy by ranking these items from 1–10, with 1 being the most conductive. Hold your hand on each item for 30 sec and record how well it conducted heat using your ranking system. Which items conducted the most heat? Were your predictions correct? Explain why the best conductors conducted heat.

Specific Heat Capacity Explain

Another important property of a substance is the amount of heat required to change its temperature. The heat required to change the temperature of one kilogram of a substance by one degree Celsius is called the **specific heat capacity** of the substance. Because heat is a form of energy, the specific heat capacity has units of joules per kilogram per °C (J/kg °C).

Water, for example, has a relatively high specific heat capacity. Copper has a relatively low specific heat capacity. If you warm 10 kg of water for about 10 min on a gas stove, the temperature of the water will increase by about 15°C. If you do the same with a 10-kg copper block, its temperature will increase by about 160°C! It takes much more heat to change the temperature of water than it does to change the temperature of copper, so water has a higher specific heat capacity.

Specific Heat of Common Materials

Substance	Specific Heat Capacity (J/kg °C)
Brass	385
Copper	385
Glass	837
Air	1,000
Ice (268 K)	2090
Water	4186
Wood	1700

By using the specific heat capacity, we can determine the amount of heat required to change the temperature of a substance. All we need to know is the mass of the substance. The equation is:

$$\underbrace{E}_{\text{heat transferred to substance}} = \underbrace{m}_{\text{mass}} \times \underbrace{c}_{\text{specific heat capacity}} \times \underbrace{(T_f - T_i)}_{\text{final temperature minus initial temperature}}$$

In this equation, E is the heat transferred to the substance (measured in joules), m is the mass of the substance, and c is its specific heat capacity. The temperature of the substance before heating it, the initial temperature, is T_i. After heating the substance, its final temperature is T_f.

Let's use this equation to find how much heat is required to change the temperature of 1.0 kg of wood from 20°C to 50°C. From the table, we see that c = 1700 J/kg°C for wood. The initial temperature is $T_i = 20°C$, and the final temperature is $T_f = 50°C$. The mass of the piece of wood is m = 1.0 kg. Inserting these values into the equation above gives the following result:

$$
\begin{aligned}
E &= m \times c \times (T_f - T_i) \\
&= 1.0\,\text{kg} \times 1700\,\text{J/kg°C} \times (50°\text{C} - 20°\text{C}) \\
&= 51{,}000\,\text{J}
\end{aligned}
$$

It takes 51,000 J of heat to change the temperature of a 1-kg piece of wood from 20°C to 50°C. How much would it take to change the temperature of the same piece of wood from 50°C to 80°C?

Explore-a-Lab

Structured Inquiry

Where does the heat go?

Follow these steps below to see how different materials change temperature at different rates.

1. Place a pot of water with about 1 L of water at room temperature on a hot plate. Record the temperature of the water.
2. Turn the hot plate on high and heat the water for 3 min and then record the temperature of the water again.
3. Cool the pot by running tap water over it for a while.
4. Place a block of aluminum or steel in the pot and fill the pot with water to the same level as it was before. Let the pot, water, and block come to room temperature. The block of aluminum or steel should be about 5 cm × 5 cm × 5 cm. If you don't have these materials, a stone of about the same size will do the trick. Note that the water in the pot should cover the metal or stone.
5. Turn the hot plate on high and heat the water for 3 min and then record the temperature of the water again.

How does the temperature of the water plus block compare with the temperature of the water alone? How can you explain any difference?

Math in Science

Suppose we add 51,000 J of heat to 1 kg of water at 20°C. What would the final temperature of the water be? The specific heat capacity of water is 4180 J/kg°C. Solve the equation for specific heat capacity for the final temperature by following these steps:

1. Divide each side of the equation by the product $m \times c$. This is done below, except for the last step, which you should fill in.

$$\frac{E}{m \times c} = \frac{m \times c \times (T_f - T_i)}{m \times c}$$

$$\frac{E}{m \times c} = \frac{m \times c}{m \times c} \times (T_f - T_i)$$

$$\frac{E}{m \times c} = ?$$

2. Using the result you got from Step 1, add the initial temperature (T_i) to each side of the equation. This should give you an equation with only the final temperature (T_f) on the right-hand side.
3. Using the result you got from step 2, plug in the values for heat ($E = 51{,}000$ J), mass ($m = 1$ kg), specific heat capacity ($c = 4180$ J/kg°C), and initial temperature ($T_i = 20$ °C) to find the final temperature of the water.
4. Evaluate your answer. Does it seem reasonable? Why or why not?

Thermal Energy and States of Matter Explain

How do thermal energy, temperature, and heat affect the properties of matter? What happens if you leave a pot of water heating on the stove for a long time? Does the same thing happen if you leave a pot of water in the freezer for a long time? In the first case, the water is eventually transformed into steam, and, in the second case, it is transformed into ice. These are examples of changes of state. The "state" of matter can be solid, liquid, or gas. Why does matter change state, and what role does thermal energy play in this change?

Changes of State

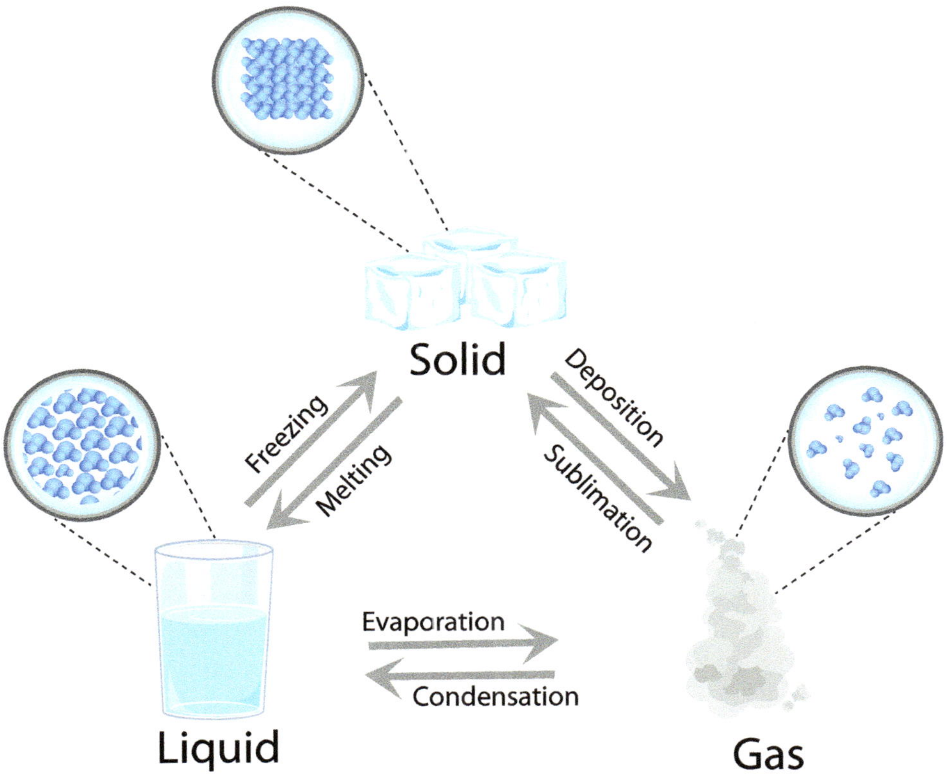

Solid

Freezing

Melting

Deposition

Sublimation

Liquid

Evaporation

Condensation

Gas

Heating a substance causes its temperature to rise. The particles in the substance therefore begin to move more rapidly. For example, heating water increases the water's temperature, which means the average kinetic energy of the water molecules increases. Add enough heat, and the average kinetic energy of the water molecules will be enough to break the bonds that hold them together in the liquid. At this point, the molecules vaporize by leaving the liquid and becoming steam molecules. This is an example of a phase change from liquid to gas.

Now consider heating a solid such as ice. As more heat is added, the ice molecules vibrate faster and faster until the kinetic energy of the molecules is sufficient to break the bonds that hold them together in the form of ice. The molecules then become water molecules (i.e., liquid). This is an example of a change of state from solid to liquid, which is called *melting*.

Changes of state also occur when a substance loses heat. The phase change from gas to liquid is called *condensation*. The phase change from liquid to solid is called *freezing*. Can a substance change state by going from solid to gas? Or from gas to solid?

For any given substance, its changes of state occur at specific temperatures and pressures. The temperature at which a substance melts is called the *melting point*. For vaporization, freezing, and condensation, we refer to the vaporization, freezing, and condensation points respectively. For example, ice changes to water at 0°C, so its melting point is 0°C.

Explore-a-Lab

Structured Inquiry

How much energy does it take to melt ice?

Place a liter of warm water in a pot. Measure its temperature. Measure the mass of an ice cube and then place it in the warm water. As soon as the ice cube is completely melted, measure the temperature of the water again. Use the equation for specific heat capacity to calculate the energy lost by the water to melt the ice cube.

Concept Check

Assess/Reflect

Summary: What is thermal energy? Thermal energy is the total energy associated with the random motion of all the particles in a system. Heat is thermal energy that is transferred from one substance or system to another. Temperature is a measure of the average kinetic energy per particle in a system. Heat flows from a substance with higher temperature to a substance with lower temperature. When the two substances are at the same temperature, no heat is transferred, and the substances are at thermal equilibrium. Two important properties of substances are thermal conductivity and specific heat capacity. The thermal conductivity of a substance tells us how fast the substance transfers heat from hotter parts to colder parts of the substance. The specific heat capacity tells us how much heat is needed to change the temperature of 1 kg of the substance by 1°C. Depending on how much thermal energy matter contains, it can be in different phases, such as gas, liquid, or solid.

1. Which has more thermal energy, a pot of boiling water or an iceberg? Why?
2. When two blocks of aluminum at 20°C are placed in contact, does heat move from one block to the other? Why or why not?
3. You add 1000 J of heat to 10 g of copper, 10 g of glass, and 10 kg of water. For which substance does the temperature change the most? For which does the temperature change the least? Why?

Lesson 4

Essential Question

How Is Thermal Energy Transferred?

Thermal energy can be transferred in several ways. In fact, we witness this every day. The Sun heats Earth. Put your hand above a candle, and warm air rises to heat your hand. Why does the warm air rise and not just go out in all directions? A block of ice sitting on hot concrete is warmed and melts. Can you think of other examples of thermal energy transferring from one system to another?

Heating and Random Motion Explain

You know that thermal energy is the energy associated with the random motions of all the particles in a system. How is random motion related to heating? To understand this, imagine you have a container divided into two compartments by a wall, as in the illustration below. Each compartment contains air, but the air in the right compartment is hotter than the air in the left compartment. What does this tell you about the average speed of the air particles in the right compartment compared with that of the air particles in the left compartment? What happens if you remove the wall separating the two compartments?

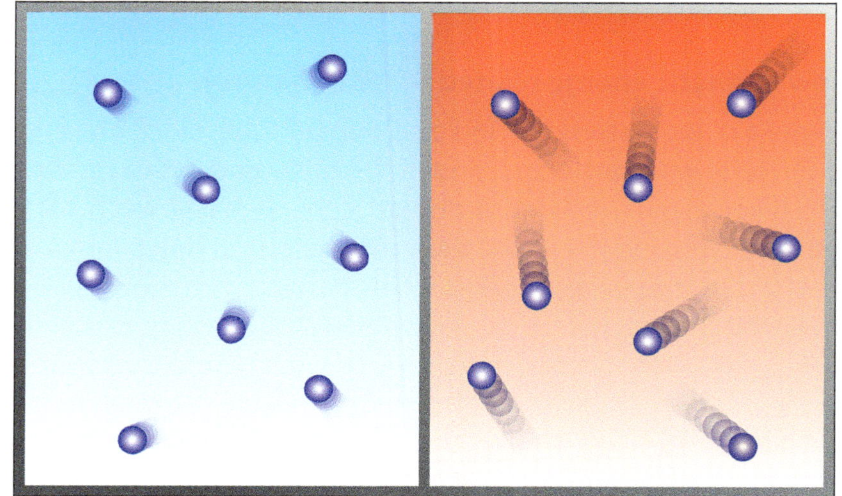

The container is separated into two compartments. The air in the right compartment is at a higher temperature than the air in the left compartment.

Objectives

- Describe how thermal energy is transferred.
- Compare and contrast the transfer of thermal energy by conduction, convection, and radiation.
- Distinguish between thermal conductors and insulators.
- Describe ways in which thermal energy is used.

Vocabulary

conduction

radiation

convection

Because the particles are moving about in random directions, they will inevitably start to cross into the opposite compartment and collide with the particles there. Most collisions between gas particles will result in the speedier particle slowing down and the slower particle speeding up. After enough of these collisions, the average speed of the particles from the two compartments will be the same. The average speed of the particles from the left-hand compartment will have increased and that of the particles from the right-hand compartment will have decreased. What does this tell you about the temperature of the gas in the container?

Notice that the thermal energy of the particles from the left-hand compartment has increased, whereas the thermal energy of the particles from the right-hand compartment has decreased. This means that thermal energy was transferred from the particles that initially had more thermal energy to the particles that initially had less thermal energy. This example shows how the transfer of thermal energy occurs in a gas because of the random motion of the gas particles.

Now imagine placing two solids at different temperatures in thermal contact. What will happen? Because the particles of a solid vibrate, they can collide with their neighbors. Using the same reasoning as for a gas, these collisions will spread the available thermal energy around to all the particles in the system so that, on average, each particle has its fair share of thermal energy. By describing systems in terms of microscopic particles that move about in random manners, the kinetic theory of matter gives us an understanding of how thermal energy is transferred on a macroscopic scale.

Explore-a-Lab

Structured Inquiry

 **How is kinetic energy related to temperature?**

Create a crude piston by using a cardboard tube. Block one end of the tube with cardboard, and fit the other end with a piston, which you can make from cardboard, cork, or plastic foam. Fill the tube with a few marbles and push the piston down into the tube (but not so far as to prevent the marbles from moving). Now shake the tube in directions perpendicular to the long axis of the tube (why perpendicular?) and observe what happens to your piston. Can you explain what happens? How are the marbles like gas particles? How is their motion analogous to temperature?

Conduction

What happens if you hold an iron poker in a fire for several minutes? You will likely notice that after several minutes, the end of the poker you are holding gets hotter. How does the thermal energy generated by the fire transfer through the length of the poker? The poker gets hotter because of conduction. **Conduction** is the transfer of thermal energy as heat through a substance without the substance moving as a whole. The kinetic theory of matter helps us understand how conduction occurs.

Thermal energy moves throughout this metal bar by conduction.

The huge number of collisions between the hot, gas particles of the fire and the cold particles in the tip of the iron poker increases the average thermal energy of the particles in the tip of the poker. Within the poker, the particles with increased thermal energy collide with neighboring particles farther away from the fire that have (on average) less thermal energy. These collisions transfer some of the thermal energy from the particles nearer the fire to particles farther from the fire. The particles nearer the fire may then be resupplied with thermal energy from the fire, and so the process continues. The result is that thermal energy is transferred through the length of the poker to the end that you are holding.

Heating by conduction can occur within a single object or between objects that are in thermal contact. For example, placing your feet on a hot-water bottle heats your feet by conduction. Heating by conduction requires a continuous path of substance between a warmer zone and a colder zone. This is why thermos bottles consist of inner and outer bottles separated by an evacuated space. Heating by conduction is not possible through an evacuated space, so the hot liquid in the inner bottle takes longer to transfer its thermal energy to the cooler outer bottle. Eventually, however, the hot liquid in the thermos bottle does cool down. How does this happen?

Radiation

Conduction is not the only way in which thermal energy can be transferred. For example, the Sun heats Earth by a process called radiation. **Radiation** is the transfer of energy by electromagnetic waves, such as infrared waves. Unlike conduction, radiation can transfer thermal energy through empty space. As a result of radiation, the Sun cools and Earth is heated, so thermal energy is transferred.

Radiation also explains why a hot liquid in a thermos bottle eventually cools. The hot liquid first transfers thermal energy by conduction to the inner bottle, which in turn transfers thermal energy by radiation to the outer bottle. The outer bottle then transfers the thermal energy to the surroundings. Why is the outer bottle often coated with a mirror material?

What forms of thermal energy transfer are taking place in this picture?

Convection

A third way to transfer thermal energy is by convection. **Convection** is heating by moving a mass of warm substance from a warm area to a cooler area. Heating by convection occurs in fluids. Consider how thermal energy from a portable room heater spreads throughout a room. The heater heats the nearby air, which becomes less dense and rises. Cooler air near the ceiling descends to replace the warmer air that has risen, creating a circular flow of air that transports the thermal energy of the heater to distant parts of the room. Convection currents in a pot of boiling water with macaroni are what make the macaroni circulate between the top and bottom of the pot. Convection currents in the atmosphere arise from the same principle and contribute greatly to weather and other climatic phenomena. Can you think of other planet-size systems where convection currents might exist?

Thermal Energy Transfer		
Heat Transfer	**Definition**	**Example**
Conduction	Transfer of thermal energy through a substance without the substance moving as a whole Must have matter present	A spoon heating when placed in hot water
Radiation	Transfer of thermal energy by electromagnetic waves Radiation does not require matter	Sunlight warming the Earth
Convection	Transfer of thermal energy by motion of a hot substance from a hot region to a cold region Must have matter present	Circulation of hot and cold material; movement of boiling water in a pan

Faith Connection

The Word of God is best spread from person to person. Is this most like conduction, convection, or radiation?

Thermal Conductors and Insulators

Recall that the ability of a material to transfer thermal energy by conduction is called *thermal conductivity*. Thermal conductivity is measured in watts per meter per degree kelvin (W/m·K). Notice that materials that conduct electricity are good thermal conductors as well. This is because the electrons that are free to move and conduct electricity can also move and conduct thermal energy.

Thermal Conductivity	
Material	**Conductivity (W/m·K)**
Air	0.026
Aluminum	237
Copper	401
Plastic foam	0.029

A material that is a poor thermal conductor is called a *thermal insulator*. The table shows that air is a poor thermal conductor, so it is a good thermal insulator. Compared with solids, air is much less dense. This means that air molecules have to move much farther before they can interact with another molecule and transfer thermal energy. The result is that thermal energy is transferred very slowly in air. In fact, many materials, both natural and human-made exploit this property of air to be good thermal insulators. For example, mammals have fur that traps air near the body, which slows thermal energy transfer from the body to the surroundings, keeping the animal warmer.

Uses of Thermal Energy Explain

How is thermal energy harnessed in our society? Humans have long used thermal energy to cook food and heat metals. Long before science explained the workings of thermal energy, humans discovered how to harness thermal energy to do work. Devices that do this are called *heat engines*. Heat engines were built before the time of Christ, but it was not until science developed an understanding of thermal energy in the nineteenth century that their use became widespread.

Modern heat engines include the internal combustion engine found in cars, the jet engine, and steam turbines. All these engines convert thermal energy into motion. To do this, a gas is heated in an enclosed space. As the gas expands, it does work. In an internal combustion engine, the work is done by the hot gas pushing on the piston and forcing it to move. After the piston has moved the desired distance, the gas, which has now cooled, is released into the environment, and the cycle begins again.

Steam turbines are well suited for generating electricity and are often used in electrical power plants. Steam turbines are also used for sea transportation. In a steam turbine, water is heated to produce high-pressure steam. In one type of steam turbine, the high-pressure steam is released so that it hits the blades of the turbine, causing the turbine to spin. After passing the blades of the turbine, the steam is cooler and can be condensed back into water. The energy required to create the high-pressure steam often comes from nuclear energy, natural gas, or coal.

These cooling towers cool steam so that it condenses to form liquid water.

Record your work for this inquiry. Your teacher may also assign the related Guided Inquiry.

Cold Stuff

Which insulator works best to prevent energy transfer?

Materials

- cotton
- steel wool
- plastic container or bowl (750 m)
- stop watch
- 3 plastic juice bottles (small)
- thermometer

Procedure

1. Fill one of the plastic juice bottles with cotton, and fill a second bottle with steel wool. The third bottle will be filled with air.
2. Fill the plastic bowl with 500 mL of ice water. Insert the thermometer into the bottle filled with cotton. Do not let the thermometer touch the bottom of the bottle.
3. Once the thermometer is inserted in the bottle, wait for 2 min and **record** the initial temperature of the thermometer.
4. Now place the bottle with the cotton and the thermometer into the bowl of ice water and start the stopwatch. Hold the bottle down so it is submerged as much as possible in the ice water, but don't let water spill into the bottle.
5. While holding the bottle in the water, **measure** and record the insulator's temperature every 30 sec for 5 min. Then remove the bottle from the water.
6. Repeat Steps 2–5, this time using the bottle filled with steel wool.
7. Repeat Steps 2–5, this time using the bottle filled with air.
8. **Construct a graph** of your data.

Create Explanations

1. Which insulator works best to prevent energy transfer?
2. Where did the heat inside the containers go as they were cooling?
3. What other materials do you think might make good insulators?
4. What materials would make poor insulators?

Explore-a-Lab

Structured Inquiry

 **Why do some materials conduct heat better than others?**

Use a plastic knife to place a small amount of cold soft margarine on the end of a metal spoon handle and on a plastic spoon handle. Make sure to place the same amount of margarine on each spoon. Place the bowls of both spoons in a beaker of hot water. Start the stopwatch. Carefully watch the margarine on the spoons. Record the time for the margarine to melt and slide down a spoon. Continue observing until the margarine on the other spoon melts and slides down and record the time. Construct a bar graph that compares the data for each spoon. Why did the margarine melt faster on one of the spoons? How could you design this activity so that the time it takes the butter to melt on each spoon is about the same amount of time? Try your idea and see if it works.

Concept Check

Assess/Reflect

Summary: How is thermal energy transferred? The kinetic theory of matter explains how the random motion of particles in a system transfers thermal energy from hotter parts of the system to colder parts of the system. Three ways for thermal energy to transfer are by conduction, radiation, and convection. Conduction is the transfer of thermal energy through a substance without the substance moving as a whole. Radiation is the transfer of thermal energy by electromagnetic waves. Convection is the transfer of thermal energy by motion of a substance from a hot region to a cold region. Substances that easily transfer thermal energy by conduction are called conductors. Substances that do not easily transfer thermal energy by conduction are called insulators.

1. Why does the random motion of particles lead to the transfer thermal energy from hot regions to cold regions and never vice versa?
2. Why are materials of lower density typically poorer thermal conductors compared to materials of higher density?
3. Describe the method(s) of heat transfer that occur when you heat a pot of water on a gas stove.
4. What type of engine would be more efficient: an electrical engine or a gasoline engine? Why?
5. Would convection work if there were no gravity? Why or why not?

Careers in Science

Extend

Ocean Engineer

Have you heard someone say that "space is the final frontier"? That might be true for some explorers, but here on Earth, the oceans are the final frontier. Much of the oceans are unexplored or too dangerous to study. But thanks to the relatively new field of ocean engineering, we know more about the oceans than ever before. And we are able to study them safely and less expensively.

Ocean engineers solve problems that come up when working in an ocean environment. They design ways to use the ocean's resources, including the energy from ocean waves and tides. They do all of this while protecting the oceans from human activities that can damage or pollute the ocean environment.

Ocean engineers design and build many types of equipment. This equipment includes vehicles that cruise the oceans, such as tugboats, aircraft carriers, ships, and submarines; offshore oil rigs that help drill oil from deep below the ocean's surface; vehicles that are used for deep sea exploration; underwater robots that are used to explore areas that humans cannot get to; and sonar that locates underwater features, such as mountains and valleys, volcanoes, or even sunken ships, such as the *Titanic*.

Other ocean engineers study the ocean water itself. They try to understand how waves, currents, and even the salt in the water affect the equipment used in the ocean.

Ocean engineers take courses in many different subjects. They study math and science, plus engineering courses in mechanical, electrical, chemical, and acoustical, or sound engineering. They often work with ocean scientists from other disciplines such as biology, geology, and chemistry.

Concept Check

1. Why are vehicles that can study the deep sea important to ocean studies?
2. How might the work of an ocean engineer help protect the animals living in the ocean?

Tsunami Warning Systems

Science and Technology

Extend

Tsunamis, or seismic sea waves, are large ocean water waves that form when an up-and-down movement on the ocean floor occurs as the result of an earthquake. About 35 earthquakes occur every day somewhere on Earth. Many of these occur on the ocean floor. Some tsunamis form far out at sea, and others form closer to shore. In any case, because ocean waves increase in height as they approach shore, tsunamis have the ability to grow very large in height. The size of tsunamis makes them extremely dangerous when they break on land.

Governments around the world have established tsunami warning system (TWS). Their purpose is to help reduce the loss of life and property in the event that a tsunami reaches land. The longer people have before a tsunami hits, the greater their chances are of reaching safety.

The TWS consists of stations positioned around the world, satellites in space, and specially designed buoys called DART. DART stands for Deep Ocean Assessment and Reporting of Tsunamis. The system works in the following way. First, the DART surface buoy is set afloat. It is connected to the ocean floor by a cable and an anchor. A second piece of equipment called a tsunameter is placed on the ocean floor a short distance from the DART. Both are able to communicate with each other and with a satellite in space. When an earthquake on the ocean floor occurs, the tsunameter detects the up-and-down movement of the ocean floor. At the surface, the wave generated by the up-and-down movement causes the DART to move up-and-down as well. The DART records the change in sea level caused by the wave. Both sets of information are sent to a satellite that then sends the data to the TWS. The arrival times of the waves as they pass the DART buoys allow the TWS to calculate how fast the wave is moving and how long it will take a tsunami to reach land.

The National Tsunami Warning Center in Palmer, Alaska, provides warnings for the U.S. mainland, Canada, Puerto Rico, and the U.S. Virgin Islands. The Pacific Tsunami Warning Center in Ewa Beach, Hawaii, provides warnings for Hawaii and other American territories in the Pacific Ocean. The International Tsunami Information Center provides warnings for 25 member countries in the Pacific basin.

Concept Check

1. Why are there so many tsunami warning centers around the Pacific Ocean basin?
2. Why are both the buoy and the tsunameter necessary for tsunami detection?

Chapter 14 Review Assess/Reflect

Study Guide

Lesson 1

1. Kinetic energy is the energy an object has because of its motion. Potential energy is the energy an object has because of its position.
2. The gravitational potential energy U of an object is calculated by using the formula $U = m \times g \times h$, where m is the mass of the object, $g = 9.8$ m/s^2 is the acceleration of gravity at the surface of Earth, and h is the height of the object with respect to a reference position that you are free to choose. The kinetic energy K of an object is calculated by using the formula $K = \frac{1}{2} \times m \times v^2$, where m is the mass of the object and v is its speed.
3. Forms of energy include mechanical energy, light energy, electrical energy, sound energy, elastic energy, chemical energy, nuclear energy, and thermal energy. Each form of energy is either kinetic energy or potential energy or a combination of both. See the chart on page 521 for information about each energy type.

Lesson 2

1. Energy transforms from one form to another when a system changes. Examples include a roller coaster that, by rolling down a track, transforms its potential energy at the top of the track to kinetic energy. Other examples include the conversion of chemical energy in car engines and the conversion of light energy in leaves.
2. The law of conservation of energy states that energy cannot be created or destroyed. It is possible to change only its form. This law can be demonstrated by using bar diagrams to track forms of energy or by using the formulas for calculating kinetic and potential energy.
3. Transforming energy refers to energy changing from one form to another, such as a roller coaster changing potential energy into kinetic energy as it descends a hill. Transferring energy refers to energy that can move from one place to another without changing form, such as light energy or electrical energy.

Lesson 3

1. The kinetic theory of matter describes matter as composed of large numbers of small particles that are constantly moving in a random manner. Thermal energy is simply the sum of the mechanical energy of each particle in a system.
2. Heating is the transfer of thermal energy from a region with higher temperature to a region with colder temperature. Temperature is a measure of the average kinetic energy of all the particles in a system. Thermal energy is the total kinetic energy due to the random motion of all the particles in a system.
3. Adding thermal energy to a solid can cause it to change to a liquid. Adding thermal energy to a liquid can cause it to change to a gas. Removing thermal energy from a gas can cause it to change to a liquid. Removing thermal energy from a liquid can cause it to change to a solid.

Lesson 4

1. Heating is the transfer of thermal energy through particles colliding with each other. Thermal energy is transferred from regions with higher temperature to regions with lower temperature.
2. Conduction is the transfer of thermal energy as heat through a substance without the substance moving as a whole. Convection is heating by transferring matter from one place

to another. Radiation is the transfer of thermal energy by electromagnetic waves. Transferring heat by conduction and convection requires matter. Radiation can transfer energy though a vacuum; no matter is needed.

3. Good thermal conductors rapidly transfer thermal energy. Thermal insulators transfer thermal energy very slowly.
4. Thermal energy is used for cooking and for heating our homes. It is also used in heat engines, which include car engines, jet engines, and the steam turbines often used to generate electricity.

Vocabulary Check

Classify the energy forms as kinetic energy, potential energy, or a combination of both.

1. Thermal energy: __________
2. Chemical energy: __________
3. Nuclear energy: __________

Fill in the blank:

4. Sunlight hits your skin and warms you. This is an example of __________.
5. Energy __________ occurs when energy moves from one place to another without changing form.

Multiple Choice Questions

6. The thermal energy in a liter of water is about 1.2 million J. If you pour out 2/3 of the water, the thermal energy of the water left is
   A. 0.8 million J
   B. 0.4 million J
   C. 1.2 million J
   D. 0.3 million J
7. A 0.02-kg arrow is shot at a speed of 50 m/s. What is its kinetic energy?
   A. 250 J
   B. 500 J
   C. 5 J
   D. 25 J
8. With respect to the ground, the potential energy of a 10-kg monkey at the top of a 30-m high tree is
   A. 1470 J.
   B. 2280 J.
   C. 2940 J.
   D. 53.40 J.
9. Knowing the specific heat capacity of different substances, which substance requires the most heat to change the temperature?
   A. ice (2090 J/kg°C)
   B. glass (837 J/kg°C)
   C. water (4186 J/kg°C)
   D. wood (1700 J/kg°C)

Check Point

10. If you were a football player, which collision would you rather avoid and why—one with a 70-kg opponent running towards you at 10 m/s or one with a 100-kg opponent running toward you at 8 m/s?
11. If you are floating in space far from any other object, do you have gravitational potential energy? Why or why not?
12. Into what forms of energy is the chemical energy of the food you eat converted to in the body?
13. Which would you expect to be a better thermal conductor—a metal spoon or a wooden spoon? Explain your reasoning.
14. You have two identical cups, one containing a dense liquid and one containing a liquid that is less dense. You heat both for the same amount of time on your stove. Which liquid would you expect to be hotter? Explain your reasoning.

Using the Glossary

The Glossary is like a small dictionary. It lists important science terms from your book. The words and terms are listed in alphabetical order. Each word is followed by its meaning. Some words have more than one meaning. This Glossary gives the scientific meaning. That is how the word is used in this book. After the meaning is a page number that tells on what page the word is first used.

absolute magnitude How bright a star would be if all the stars were the same distance from Earth. (p. 363)

absolute zero A temperature of zero degrees on the kelvin scale that is equivalent to –273°C. (p. 537)

action-reaction force pair Two forces that are equal but act in opposite directions and on different objects. (p. 464)

aerobic bacteria Bacteria that require oxygen to grow. (singular: aerobic bacterium) (p. 27)

AIDS (acquired immune deficiency syndrome) An immune disease, caused by HIV, in which the immune system can no longer protect the body against invasion by pathogens or opportunistic infections. (p. 245)

air sac An air-filled cavity present in the body of birds that makes bones lighter. (p. 121)

allele Each form of a gene for a particular trait. (p. 149)

anaerobic bacteria Bacteria that grow only in the absence of oxygen. (singular: anaerobic bacterium) (p. 27)

annelid The name given to a segmented worm. (p. 68)

antenna A jointed, movable, sensory structure on the head of arthropods. (plural: antennae) (p. 81)

antibiotic A chemical that kills bacteria or prevents them from reproducing. (p. 28)

antibody A type of protein produced by a B cell that helps the immune system by sticking to a pathogen and marking it for destruction. (p. 220)

antigen A specific marker that each pathogen contains that only one type of killer T cell can recognize. (p. 220)

anus The lowest part of the large intestine through which solid waste passes from the body. (p. 270)

aphelion The point when Earth is farthest from the Sun. (p. 293)

apparent magnitude The brightness of a star as it appears to an observer on Earth. (p. 362)

arthropod An invertebrate that has a segmented body, an exoskeleton, and jointed appendages. (p. 80)

associative learning A type of learning where stimuli and events are found to be correlated. (p. 128)

asterism A pattern of easily recognizable stars. It is smaller than a constellation. (p. 361)

asteroid belt A region of the Solar System between the orbits of Mars and Jupiter that is a collection of millions of small rocky objects orbiting the Sun. (p. 309)

astronomical unit (AU) A unit of distance used to measure distances within our Solar System and based on the average distance between Earth and the Sun. One AU is equal to 150,000,000 km. (p. 297)

asymmetry When the body is shaped irregularly and cannot be split into equal parts. (p. 59)

autoimmune disease A condition in which the immune system targets a person's cells, tissues, or organs by mistake. (p. 224)

autotroph An organism that makes its own food using energy from sunlight or from inorganic compounds. (p. 25)

average speed The total distance traveled divided by the total time it took to travel that distance. (p. 441)

balanced forces Equal forces that act in opposite directions. (p. 459)

barbule The part of a feather that has tiny hooks to hold the barbs together. (p. 120)

barred spiral galaxy A type of galaxy that has arms that spiral outward from a large central bar. (p. 371)

big bang theory A theory that the Universe formed about 14 billion years ago as the result of a cosmic explosion. It marks the beginning of the Universe. (p. 381)

bilateral symmetry When the right and left body parts of an organism are arranged the same way. (p. 58)

bile A fluid secreted by the liver that helps break large fat particles into smaller pieces. (p. 269)

binary fission A form of reproduction where a single-celled organism grows in size and then splits into two complete organisms. (p. 25)

binomial nomenclature A two-part naming system in which the genus name is first, followed by the species name. (p. 17)

birth control A way to reduce the chance of pregnancy. (p. 207)

bivalve A mollusk that has a shell of two halves, or valves. (p. 75)

black dwarf A white dwarf star that has cooled and that no longer gives off light. (p. 368)

black hole The massive remains of a supernova with a gravitational pull so strong that not even light can escape its gravity. (p. 369)

blueshift The effect where the light from an object in space moving toward the observer at a very high speed appears bluer that it would otherwise. (p. 383)

budding A form of asexual reproduction where the offspring grows out of the body of the parent. (p. 181)

bullying An intentional act that causes harm to others and may involve verbal harassment, verbal or nonverbal threats, physical assault, stalking, or other methods of coercion, such as manipulation, blackmail, or extortion. (p. 193)

buoyant force The upward force of a fluid or a gas on an object. (p. 471)

carrier An individual having a specified gene that is not expressed in its phenotype. (p. 163)

Centers for Disease Control and Prevention (CDC) An organization in the United States that monitors many kinds of diseases nationwide and collects data on possible disease threats. (p. 250)

centripetal force The force that makes an object move in a curved path. (p. 471)

chemical digestion The breakdown of food particles into molecules by digestive juices. (p. 263)

chemical energy The energy stored in the positions of atoms and molecules in compounds. (p. 521)

chromatophore In mollusks, a tiny bag of pigment encircled by muscles that changes color depending on the state of the muscle. (p. 92)

chyme Food broken down and mixed with acids and enzymes in the stomach. (p. 267)

cilia Short, hairlike structures of some protists used for movement and searching for food. (singular: cilium) (p. 34)

circumpolar constellation A constellation that appears to circle the North Pole due to Earth's rotation and that is visible all year long. (p. 361)

cnidarian The group of invertebrates that have radial symmetry and tentacles with stinging cells. Cnidarians include: corals, sea anemones, sea jellies, hydras, and the Portuguese man-of-war. (p. 65)

codominance A pattern of inheritance where multiple alleles can have more than one dominant trait expressed. (p. 159)

codon Each sequence of three bases (such as ACG, CGT, CAC, etc.) in a DNA strand. (p. 142)

cognition The ability to gain knowledge and comprehension. (p. 190)

common name A word used to describe many similar species. (p. 17)

communicable disease An infectious disease that can spread from person to person. (p. 226)

Communicable Disease Control Division (CDCD) The Canadian agency that monitors many kinds of diseases and collects data on possible disease threats. (p. 250)

complete dominance A trait that requires only one dominant allele. (p. 152)

complete metamorphosis A transformation in insects that involves four stages: egg, larva, pupa, and adult. (p. 83)

complex inheritance Patterns of inheritance that differ from simple patterns of inheritance. Traits result from the influence of multiple factors. (p. 156)

compound eye Many individual, specialized receptors fused together in invertebrates. Each receptor works like a lens. (p. 89)

condom A barrier method of birth control, using a thin sheath (usually made of latex) worn on the penis. (p. 207)

conduction The transfer of thermal energy as heat through a substance without the substance moving as a whole. (p. 549)

constriction A method of killing prey where the snake wraps its body around an animal of prey and squeezes until the prey suffocates. (p. 115)

contact force A force that involves contact between the object exerting the force and the object being changed by the force. (p. 461)

convection The motion of warm material that rises, cools off, and sinks again, producing a continuous circulation of material and transfer of heat. (p. 551)

courtship A behavior that involves or results in communication. (p. 131)

crater A depression on the surface of a planet or moon caused by impact with a meteroite, an asteroid, a comet, or another object in space. (p. 343)

dark matter The missing matter that cannot be seen but is theorized to be present in much of the Universe. (p. 385)

dialysis The repeated removal of wastes and excess salt and water from the blood of a kidney patient. (p. 280)

diffraction The bending of waves around a barrier or through an opening. (p. 421)

digestion The process in which the organs from the digestive system break down foods into individual molecules so that the foods can be absorbed into the bloodstream and transported to the cells. (p. 263)

digestive system The group of organs that work together to process food for use by the body. It includes the mouth, esophagus, stomach, liver, gallbladder, pancreas, and the intestines. (p. 263)

direct contact The transmission of a pathogen from person to person. (p. 233)

domain A taxonomic category above the kingdom level. (p. 20)

dominant allele An allele whose trait is expressed in the phenotype when the organism has the allele. (p. 150)

Doppler effect The apparent change in the frequency of a wave as the distance between the wave source and the observer changes. (p. 412)

echinoderm A spiny-skinned marine animal with radial symmetry, including sea stars, sand dollars, brittle stars, sea cucumbers, and sea urchins. (p. 77)

eclipse When one object partially or totally blocks sunlight from another object. (p. 345)

efficiency The measure of the actual work of a machine compared with the work put into it. (p. 488)

effort arm The part of a lever that is the distance between the fulcrum and the point where effort force is applied. (p. 498)

effort force The force that must be applied to a machine to cause the machine to change the strength or direction of a force or the distance over which a force is applied. Also called the input force. (p. 491)

elastic collision A type of collision in which both kinetic energy and momentum are conserved, such as when a tennis ball is dropped onto a hard surface. (p. 454)

elastic force The force a solid exerts to return to its original shape after it has been deformed. (p. 470)

electrical energy Kinetic energy if power is being used or potential energy if power is not being used. (p. 523)

electrical force The force that acts between electrically charged objects. (p. 472)

ellipse A flattened circle. (p. 293)

elliptical galaxy A type of galaxy that is shaped like a large ellipse. (p. 371)

endoskeleton An internal skeleton. Usually made of calcium or silica. (p. 77)

energy transfer Energy that is classified according to whether it moves from one place to another without changing form. (p. 532)

energy transformation Occurs when energy changes from one form to another. (p. 524)

enzyme A substance that speeds up chemical reactions in the body. (p. 263)

equinox The moment between the solstices when Earth moves into a position where neither end of Earth's axis points toward or away from the Sun and the length of day and night are equal. (p. 337)

estrogen The female sex hormone. (p. 185)

excretion The process by which wastes are removed from the body. (p. 276)

excretory system The group of organs responsible for removing wastes from the body. (p. 276)

exoskeleton An external skeleton. Usually made from chitin or calcium. (p. 81)

experimental science A type of science that deals with events in which the beginning conditions are known and can be duplicated. (p. 378)

fall equinox The day when the Sun rises exactly in the east because neither pole tilts toward or away from the Sun (September 22 or September 23 in the northern hemisphere). (p. 339)

filoplume A hairlike feather that helps birds sense things in their environment or is decorative. (p. 120)

flagella Whiplike structures of some animal-like protists used for movement. (singular: flagellum) (p. 34)

foci The two points of an elliptical orbit. (singular: focus) (p. 294)

frame of reference The position of an object in relation to other objects in the area. (p. 442)

fulcrum The point on which a lever pivots or rotates. (p. 498)

fusion A process that occurs when nuclei of lightweight elements, such as hydrogen, combine to form heavier elements, such as helium. (p. 365)

galaxy A vast group of stars, dust, and gas held together by gravity. (p. 370)

gallbladder A small sac-like organ that stores bile from the liver that is not immediately needed for digestion. (p. 269)

genetic engineering The ability to select a gene from one kind of organism and insert it into the genetic make-up of another. (p. 166)

genetically modified organism (GMO) An organism that expresses a trait received from the DNA of another kind of living thing. (p. 167)

genetics The study of how traits are inherited. (p. 142)

genotype The combination of alleles that an organism has for a particular trait. (p. 150)

geocentric theory The idea that all objects in space move around Earth. (p. 312)

gravitational potential energy The energy an object possesses as a result of the force of gravity acting on it. (p. 517)

habituation A type of learning in which a continued stimulus eventually results in no response being given to that stimulus. (p. 128)

heliocentric theory The idea that the Sun is the center of the Universe and that the planets orbit the Sun in circular orbits. (p. 312)

hemizygous When a parent organism carries only one allele for a certain trait. (p. 163)

herpetology The study of amphibians and reptiles. (p. 111)

heterotroph An organism that cannot make its own food but must eat other living things. Includes herbivores, carnivores, and scavengers. (p. 25)

heterozygous The state in which the two alleles are different. (p. 150)

highland A tall, rugged area on the Moon's surface that appears as a light-colored area to us when we look at the Moon. (p. 343)

historical science A type of science that studies past events that cannot be repeated and for which there is no direct experimental data. It involves making inferences based on the results of these events. (p. 378)

HIV (human immunodeficiency virus) The virus that causes AIDS. (p. 245)

homozygous The state in which both alleles are the same. (p. 150)

hyphae Thin filaments that make up the body of a fungi. (singular: hypha) (p. 40)

I

immune response The third line of defense against a pathogen, where a specific T cell reacts to an antigen of a pathogen. (p. 220)

incest Sexual abuse when the abuser is a family member. (p. 195)

incomplete dominance A form of inheritance where alleles for a specific trait are neither dominant nor recessive. Both are expressed, resulting in a blend of characteristics. (p. 156)

incomplete metamorphosis A transformation in insects that involves three stages: egg, nymph, and adult. (p. 83)

index of refraction The way in which a material refracts light, calculated by dividing the speed of light in a vacuum by the speed of light in the medium. (p. 420)

indirect contact The transmission of a pathogen from person to object to person. (p. 233)

inelastic collision A type of collision in which momentum is conserved, but some of the kinetic energy converts to other forms of energy, such as what happens when a ball of clay is dropped onto a hard surface. (p. 455)

infectious disease A disease that is caused by pathogens that get into the body and cause problems. (p. 226)

inflammatory response The second line of defense against a pathogen, where a damaged cell releases a chemical causing the area under attack to swell and increase blood flow. (p. 219)

infrasound Sound that has a frequency of less than 20 Hz. (p. 408)

input force (IF) The amount of force applied to a machine and used in calculating mechanical advantage. (p. 487)

instantaneous speed The speed at a given moment. (p. 441)

intensity The amount of energy, measured in decibels, that is transported per unit time past a given point of the medium through which the sound travels. (p. 409)

interference The result when two or more light waves meet or overlap. (p. 425)

irregular galaxy A type of galaxy that has an irregular shape that does not fit easily into other classification schemes. (p. 371)

K

karyotype A picture of chromosomes arranged by their size and shape. (p. 145)

kidney Main organ of the excretory system used to remove wastes and excess fluids from the body and to maintain the body's water balance. (p. 277)

kinetic friction The force that tends to stop the motion of one object when two solid surfaces in contact are moving past one another. (p. 468)

kinetic theory of matter A theory that describes matter as composed of large numbers of very small particles that are constantly moving about randomly. (p. 535)

kingdom A group of similar phyla. (p. 19)

Kuiper Belt A disk-shaped region beyond Neptune, from about 30 to 55 AU from the Sun, composed of millions of icy and rocky objects. (p. 305)

law of conservation of energy Law stating that it is not possible to create or destroy energy and that energy can only be transformed into different forms. Final energy = initial energy. (p. 527)

law of reflection Law stating that a beam of light will reflect off a surface at the same angle at which it struck the surface. (p. 419)

liver A large organ under the rib cage that filters harmful substances from the blood before passing it to the rest of the body and that aids in the digestive process. (p. 269)

Local Group A cluster of nearby galaxies that includes the Milky Way and the Magellanic Clouds. (p. 375)

longitudinal wave A wave that vibrates parallel to the motion of the wave's movement. (p. 399)

lunar eclipse An eclipse that occurs when Earth is between the Sun and the Moon, casting its shadow on the Moon. (p. 346)

macrophage A phagocytic tissue cell that is derived from a monocyte and that works slowly to surround and break down pathogens. (p. 219)

magnetic force An attraction or repulsion exerted on moving electrical charges and on certain metals such as iron and nickel. (p. 472)

main sequence The region on the H-R diagram that extends from the upper left to the lower right where most stars are found. (p. 366)

mantle The fleshy tissue that covers and protects the mollusk's internal body parts. (p. 71)

maria Smooth, dark areas visible on the Moon that were formed by ancient lava flows that hardened on the Moon's surface. (singular: mare) (p. 343)

mechanical advantage (MA) The number of times a machine multiplies the effort put into it and how effective the machine will be at performing the work desired. (p. 487)

mechanical digestion The physical tearing and grinding of food into smaller pieces. (p. 263)

mechanical force A force in which the object exerting the force is in physical contact with the object being pushed or pulled. Also called contact force. (p. 472)

mechanical wave A wave that cannot transmit energy in a vacuum and that requires matter to transport its energy from one location to another. (p. 403)

medium The matter through which waves move. (p. 397)

meiosis A type of cell division that reduces the number of chromosomes in each resulting cell to half the original number found in the parent cell. Meiosis produces the gametes, egg and sperm. (p. 183)

meteorite A meteor that lands on Earth's surface. (p. 310)

microorganism A life form too small to see without magnification. (p. 24)

Milky Way Galaxy The galaxy that contains the Sun and our Solar System. (p. 370)

mollusk An animal that has a soft, unsegmented body; bilateral symmetry; and a body divided into a head, foot, and mantle. Mollusks include gastropods, bivalves, cephalopods and similar animals. (p. 70)

molt To shed an exoskeleton and replace it with a new one. (p. 81)

momentum The property that a mass has due to its motion. It is calculated by finding of the product of the mass and the velocity of the object. (p. 452)

multiple alleles When more than two allele forms exist for a trait. (p. 157)

mutation Any change in the genetic code. (p. 143)

mycelium A tangled mass of hyphae. (p. 40)

mycology The study of fungi. (p. 42)

neap tide The smallest daily tidal range caused by the gravitational effect of the Sun lessening the gravitational effect of the Moon when the Sun, the Moon, and Earth form a right angle relative to one another at the first-quarter and third-quarter moons. (p. 350)

nephron A filtering unit in the kidney. (p. 277)

net force The combination of all the forces acting on an object. (p. 458)

neutron star A very dense, very small star that is a remnant of a supernova. (p. 368)

noncontact force A force, such as magnetic force, that does not involve contact between the force and the object that is being changed by the force. (p. 461)

noninfectious disease A disease such as cancer or emphysema, that is not caused by disease-causing organisms, or pathogens. (p. 226)

nonoptical telescope An instrument that uses invisible radiation to observe distant objects. (p. 316)

nuclear energy The energy that comes from the relative position of the particles that are inside the nucleus of atoms. (p. 522)

nuclear force A powerful force that holds protons and neutrons in place within the core of an atom. (p. 473)

ocelli Simple, light-sensitive cells of an invertebrate used for seeing. (singular: ocellus) (p. 89)

olfaction The chemical senses of taste and smell in invertebrates. (p. 91)

opaque object An object that does not allow light to pass through it. (p. 429)

operculum The protective bony cover that lies over the soft, feathery surfaces of gills. (p. 104)

opportunistic infection An infection caused by microorganisms commonly present in or on the human body or found in the environment that wait for the immune system to be damaged before striking. (p. 245)

optical telescope An instrument that uses visible light to magnify and observe distant objects. (p. 313)

orbital plane The plane defined by a planet's orbital path around the Sun. (p. 335)

output force (OF) The force applied by the machine to an object and used in calculating mechanical advantage. (p. 487)

pancreas A gland located behind the stomach that aids in the neutralization of acids in the small intestine and that releases enzymes into the small intestine to help with the digestion of protein. (p. 269)

pandemic A disease outbreak that affects a very large group of people spread over a wide geographic area. (p. 250)

parallax The apparent change in position of an object when viewed from different locations, or points of view. (p. 364)

parasite An organism that feeds on other organisms. (p. 45)

pasteurization The process of heating a substance to destroy or slow the growth of microorganisms in it. (p. 236)

penumbra The lighter part of the Moon's shadow that falls on part of Earth during a solar eclipse. (p. 346)

perihelion The point where Earth is closest to the Sun. (p. 293)

phenotype The genetic characteristics of an organism that can be seen and measured. (p. 150)

pheromone A chemical released by one individual that affects the behavior of another individual of the same species. (p. 91)

photon A particle, or bundle of electromagnetic waves, that travels in wavelike patterns at the speed of light (300,000 km/s). (p. 415)

phytoplankton Small photosynthetic organisms that float and drift near the ocean's surface. (p. 33)

poikilotherm An endothermic animal, such as a hummingbird, whose internal temperature varies considerably. (p. 117)

polygenic Quality of some traits where several genes are involved, and those genes have multiple alleles. (p. 163)

power The rate at which work is done, measured in watts. (p. 484)

primary sex characteristic Any physical or psychological trait that is associated with sexual reproduction. (p. 186)

protist A eukaryotic organism that cannot be classified as a fungus, a plant, or an animal. (p. 32)

protoplanet A relatively small, round object in space that has some structured interior layers and that, according to one theory, eventually combined with many other protoplanets to form the planets in our Solar System. (p. 386)

protozoan Single-celled eukaryotes that move and are heterotrophs. (p. 32)

pseudopodia Extensions of an amoeba used to move around and to capture and engulf food. (singular: pseudopodium) (p. 34)

pulsar A fast-spinning neutron star that sends out a pulsating beam of radiation. (p. 368)

Punnett square A table used to help predict the likelihood of genotypes that can be passed from parents to offspring. (p. 152)

qualitative A description that uses no numbers. (p. 439)

quantitative A description that uses numbers. (p. 439)

quasar A source of light that is billions of light-years away in space. (p. 369)

radial symmetry When the body parts of an organism are arranged in a circle around a center point. (p. 59)

radiant energy Energy transmitted by electromagnetic waves. (p. 415)

radiation The transfer of energy by electromagnetic waves. (p. 550)

radula A toothed, or file-like, structure used for feeding by gastropods and other mollusks. (p. 71)

rape Forcing sexual acts on individuals without their permission. (p. 196)

recessive allele An allele whose trait is expressed in the phenotype only if the organism's alleles are both for the recessive trait. Always hidden by the dominant allele. (p. 150)

rectum A lower part of the large intestine in which solid waste accumulates before being pushed out through the anus. (p. 270)

red giant A star with a mass similar to that of the Sun that has used up all the hydrogen in its core, causing the core to contract and the core temperature to increase, triggering the outer layers to expand and cool. (p. 368)

redshift The color shift of a galaxy toward the red end of the light spectrum as it moves away from the observer. (p. 383)

reference point Another object that appears to stay in place when tracking a moving object. (p. 438)

reflecting telescope A type of optical telescope that uses curved and flat mirrors to collect and focus light onto an eyepiece lens. (p. 314)

refracting telescope A type of optical telescope that uses glass lenses to collect and focus light onto an eyepiece lens. (p. 314)

regeneration The process of growing new body parts. (p. 63)

regolith The material that covers the surface of a planet or moon. (p. 343)

resistance arm The part of a lever that is the distance between the fulcrum and the point where resistance force is exerted. (p. 498)

resistance force The force applied by a machine that resists a change in the strength or direction of a force or in the distance over which a force is applied. Also called the output force or load force. (p. 491)

retrograde Refers to Venus's daily rotation on its axis (clockwise rotation) in a direction that is the exact opposite from the rotation of Earth and all the other planets in our Solar System (counterclockwise rotation). (p. 298)

retrovirus A type of virus where the genetic material of the virus is transmitted from its RNA to the host cell's DNA, which is a reverse flow of genetic information. (p. 246)

risk taking Acting in ways that may have unpleasant or dangerous results. (p. 192)

saliva A fluid that contains digestive enzymes that begin the process of breaking down carbohydrates into smaller sugars. (p. 265)

salivary gland Gland in the mouth that releases saliva. (p. 265)

saprophyte An organism, such as a fungus, that gets its nourishment from dead and decaying organisms. (p. 42)

scale One of many hard plates that cover some fish and reptiles. (p. 102)

scientific name A two-word name derived from binomial nomenclature that is used to describe a specific organism. It is made up of the genus name and the species name. (p. 17)

scute A bony plate that makes up the protective shells of turtles, tortoises, and terrapins. (p. 114)

secondary sex characteristic Any physical or psychological trait that is associated with sexual differentiation. (p. 186)

selective breeding A method of getting offspring with new abilities that are desired by having parents with certain traits reproduce. (p. 166)

semiplume A type of feather that is under the outer feathers and provides insulation and shape for the bird. (p. 120)

sensitization A type of learning in which a repeated stimulus eventually results in a better response being given to that stimulus. (p. 128)

sex-linked trait A trait that is determined by the alleles on a sex chromosome. (p. 161)

sexual abstinence A decision to avoid all kinds of intimate sexual behavior until marriage. (p. 200)

sexual feeling Attraction to another person. (p. 199)

sexual harassment Unwanted sexual advances. (p. 195)

sexually transmitted disease (STD) An infection that can spread from person to person through sexual contact. (p. 207)

sidereal day The time Earth takes to make one complete rotation compared with a distant star. (p. 329)

simple inheritance Inheritance that involves one set of alleles that produce only two kinds of phenotypes. (p. 152)

slope On a line graph, the ratio between the change in the y-value (rise) and the x-value (run). (p. 452)

solar day The time Earth takes to make one complete rotation in relation to the Sun. (p. 329)

solar eclipse An eclipse that occurs when the Moon comes between Earth and the Sun and the Moon blocks our view of the Sun because its shadow falls on part of Earth. (p. 346)

solstice The moment when the North Pole points directly toward or away from the Sun. (p. 337)

space probe A spacecraft that travels beyond Earth's orbit to explore the Solar System and transmit data back to Earth. (p. 320)

spawning A method of reproduction in which aquatic organisms release sex cells in the water. (p. 102)

species A group of organisms that resemble one another and can mate to produce fertile offspring. (p. 17)

specific heat capacity The heat required to change the temperature of one kilogram of a substance by one degree Celsius. (p. 542)

spiral galaxy A type of galaxy that has arms that spiral outward from a central bulge, similar to the arms of a pinwheel. (p. 371)

sponge An invertebrate with a hollow body, many small pores, and a large opening at the top. (p. 62)

spongin The flexible, fibrous substance that makes up the body of a sponge. (p. 63)

spring equinox The day when the Sun rises exactly in the east because neither pole tilts toward or away from the Sun (March 20 or March 21 in the northern hemisphere). (p. 339)

spring tide The largest daily tidal range caused by the combined gravitational effect of the Sun and the Moon when the Sun, the Moon, and Earth are aligned in space at the Full Moon and the New Moon. (p. 350)

static friction The friction between two surfaces when they are not moving relative to each other. (p. 468)

summer solstice The day when the Sun appears to move through its highest path in the sky (June 21 or June 22 in the northern hemisphere). (p. 338)

supergiant A star more massive than the Sun that has the ability to fuse heavier elements together in its core. (p. 368)

supernova When fusion stops in a supergiant star and its outer layers explode. (p. 368)

surroundings Everything outside the system. (p. 524)

symmetry Describes the balanced distribution of an animal's parts around an axis. (p. 58)

system Consists of all the things that are being studied. (p. 524)

taxonomy The science of classifying living things. (p. 19)

testosterone The male sex hormone. (p. 185)

thermal conductivity The ability of a substance to conduct heat, indicating how fast a substance transfers heat from hotter parts to colder parts of the substance. (p. 541)

thermal equilibrium A condition that occurs when the same amount of heat is transferred in both directions between two systems that are put in contact, so overall no heating occurs. (p. 539)

tidal range The difference between high tides and low tides. (p. 350)

translucent material A material that allows some wavelengths of light to pass through it. (p. 430)

transparent material A material that allows all wavelengths of light to pass through it. (p. 430)

transverse wave A wave in which the motion of the particles is perpendicular to the motion of the wave. (p. 397)

tube foot Attached to the end of the water-vascular system, it allows an echinoderm to move. (plural: tube feet) (p. 77)

ultrasound Sound that has a frequency greater than 20,000 Hz. (p. 408)

umbra The darkest part of the Moon's shadow that falls on part of Earth during an eclipse. (p. 346)

unbalanced forces Forces that have a net force that is not equal to zero and that cause a change in an object's motion. (p. 459)

unprotected sex Sexual intercourse that does not involve a contraceptive. (p. 207)

urea A nitrogen waste in urine. (p. 277)

ureter A narrow tube that carries urine out of the kidney to the urinary bladder. (p. 277)

urethra A tube through which urine is removed from the body. (p. 277)

urinary bladder A sac-like organ where urine is stored until it can be removed from the body. (p. 277)

vector An animal that carries disease-causing pathogens and spreads them to humans. (p. 253)

venom A poison used to immobilize or kill prey. (p. 115)

villi Fingerlike projections along the lining of the small intestine that aid in the absorption of nutrients. (singular: villus) (p. 267)

virus A piece of genetic material coated with protein. (p. 228)

water-vascular system A system of water-filled canals inside an echinoderm. (p. 77)

white dwarf A star with a mass similar to that of the Sun that has used up all of its helium and that no longer produces energy. (p. 368)

winter solstice In the northern hemisphere, the day when the Sun appears to reach its lowest or southernmost point in the sky (December 21 or December 22). (p. 339)

zygote The initial cell of a new organism produced by the merging of an egg and sperm during sexual reproduction. (p. 183)

Index

F

G

H

M

N

O

P

Q

R

S

Photo Credits

CVR (Main Image) Comstock/Thinkstock, (Background) Chad Baker/Thinkstock, (Banner) (1) Fotorich01/Shutterstock, (2) Janprchal/Shutterstock, (3) Courtesy of Larry Blackmer, (4) Steshkin Yevgeniy/Shutterstock, (5) Lucarelli Temistocle/Shutterstock, (6) NASA Images/NASA, (7) Matej Pavlansky/Shutterstock Barnett/Shutterstock; **SS 0** Comstock/Thinkstock; **SS 5** Ase/Shutterstock; **SS 7** Stocktrek Images/Thinkstock; **SS 8** NASA/JAXA; **SS 9** razlomov/Shutterstock; **SS 10** Jakub Krechowicz/Shutterstock; **12** dibrova/Shutterstock; **14–15** ggw1962/Shutterstock; **17** (t) Gertjan Hooijer/Shutterstock; **17** (b) iStockphoto/Thinkstock; **20** iStockphoto/Thinkstock; **21** (r) kldy/Shutterstock; **22** Joyce Mar/Shutterstock; **24** Blend Images/Thinkstock; **25** Sebastian Kaulitzki/Shutterstock; **27** Alila Medical Media/Shutterstock; **28–29** NOAA/PMEL/EOI; **30** Sarah2/Shutterstock; **32** Jubal Harshaw/Shutterstock; **33** EthanDaniels/Shutterstock; **34** Lebendkulturen.de/Shutterstock; **36–37** iStockphoto/Thinkstock; **38** Blamb/Shutterstock; **39** iStockphoto/Thinkstock; **40** Kichigin/Shutterstock; **42** iko/Shutterstock; **43** Dziewul/Shutterstock; **44** Hugh Lansdown/Shutterstock; **45** Margrit Hirsch/Shutterstock; **46** Berna Namoglu/Shutterstock; **47** catolla/Shutterstock; **48** Georgios Kollidas/Shutterstock; **49** Regien Paassen/Shutterstock; **52–53** David Rose/Shutterstock; **54** Lynne Carpenter/Shutterstock; **57** Joe Belanger/Shutterstock; **58** Tischenko Irina/Shutterstock; **59** Ethan Daniels/Shutterstock; **60–61** Vitaly Korovin/Shutterstock; **61** Dennis Sabo/Shutterstock; **66** JonMilnes/Shutterstock; **67** Juan Gaertner/Shutterstock; **68** Photocrea/Shutterstock; **70** Ethan Daniels/Shutterstock; **72** Steven Maltby/Shutterstock; **74** Dorling Kindersley RF/Thinkstock; **75** Circumnavigation/Shutterstock; **76** Vitaly Korovin/Shutterstock; **80** (l) Kheng Guan Toh/Shutterstock; **84** (l) fivespots/Shutterstock, (r) Milos Luzanin/Shutterstock; **89** Liew Weng Keong/Shutterstock; **91** Catalin Petolea/Shutterstock; **92** Kjersti Joergensen/Shutterstock; **93** USBFCO/Shutterstock; **94** Library of Congress; **95** Bluehand/Shutterstock; **98–99** Srdjan Draskovic/Shutterstock; **101** Tischenko Irina/Shutterstock; **103** bluehand/Shutterstock; **104** (l) Dorling Kindersley RF/Thinkstock, (r) iStockphoto/Thinkstock; **106** Comstock Images/Thinkstock; **108** Ryan M. Bolton/Shutterstock; **109** iStockphoto/Thinkstock; **111** iStockphoto/Thinkstock; **112** Heiko Kiera/Shutterstock; **114** Matt Jeppson/Shutterstock; **115** (l) Stockbyte/Thinkstock, (r) iStockphoto/Thinkstock; **116** iStockphoto/Thinkstock; **117** (t) Larry Blackmer Collection; **118** Alexonline/Shutterstock; **121** Dorling Kindersley RF/Thinkstock; **123** Janelle Lugge/Shutterstock; **124** eZeePics Studio/Shutterstock; **126** Kerstiny/Shutterstock; **129** Acon Cheng/Shutterstock; **132** iStockphoto/Thinkstock; **134** Butterfly Hunter/Shutterstock; **135** iStockphoto/Thinkstock; **138** Little_Desire/Shutterstock; **141–142** Tribalium/Shutterstock; **143** Luka Skywalker/Shutterstock; **145** (b) somersault1824/Shutterstock; **147** Alila Medical Media/Shutterstock; **149** michaeljung/Shutterstock; **150** (l) Sanjay Deva/Shutterstock, (r) gosphotodesign/Shutterstock; **152** (t) Creatista/Shutterstock, (b) Nchlsft/Shutterstock; **155** (t) iStockphoto/Thinkstock, (b) Ron Chapple Studios/Thinkstock; **156** Meaofoto/Shutterstock; **158** (tl) iStockphoto/Thinkstock, (tr) Linn Currie/Shutterstock, (bl) Dasha Petrenko/Shutterstock, (br) cath5/Shutterstock; **159** Kenneth Sponsler/Shutterstock; **160** ducu59us/Shutterstock; **161** Alila Medical Media/Shutterstock; **163** iStockphoto/Thinkstock; **165** Sofiaworld/Shutterstock; **172** rook76/Shutterstock; **173** Mandy Godbehear/Shutterstock; **176–177** Mandy Godbehear/Shutterstock; **178–179** Fuse/Thinkstock; **180** Denis Nata/Shutterstock; **181** Tiago Sá Brito/Shutterstock; **183** Daisy Daisy/Shutterstock; **184** (l to r) Jason Stitt/Shutterstock, (2) Andresr/Shutterstock, (3) Jezper/Shutterstock, (4) Alex Mit/Shutterstock, (5) Sebastian Kaulitzki/Shutterstock, (6) Sebastian Kaulitzki/Shutterstock, (7) Sebastian Kaulitzki/Shutterstock, (8) Jenya Baz/Shutterstock; **187** Pedalist/Shutterstock; **189** lithian/Shutterstock; **190** Sergiyn/Shutterstock; **197** Creatas/Thinkstock; **199** Dragon Images/Shutterstock; **203** Aleksandr Markin/Shutterstock; **205** Monkey Business Images/Shutterstock; **207** Amir Ridhwan/Shutterstock; **210** wavebreakmedia/Shutterstock; **211** Dmitry Melnikov/Shutterstock; **213** Mandy Godbehear/Shutterstock; **214** Rob Byron/Shutterstock; **216** Masterfile RF; **217** Leonello Calvetti/Shutterstock; **221** Dorling Kindersley RF/Thinkstock; **222** LSkywalker/Shutterstock; **225** Piotr Marcinski/Shutterstock; **226** Zurijeta/Shutterstock; **227** (t) iStockphoto/Thinkstock, (b) Paket/Shutterstock; **228** RioPatuca/Shutterstock; **229** (t) Michelangelus/Shutterstock, (b) iStockphoto/Thinkstock; **231** Sarah2/Shutterstock; **232** RAJ CREATIONZS/Shutterstock; **234** Ingram Publishing/Thinkstock; **238** Syda Productions/

Shutterstock; **239** wavebreakmedia/Shutterstock; **240** Hemera/Thinkstock; **242** iStock/Thinkstock; **243** iStockphoto/Thinkstock; **248** iStockphoto/Thinkstock; **251** iStockphoto/Thinkstock; **253** iStockphoto/Thinkstock; **254** Getty Images News/Thinkstock; **256** Andrea Danti/Shutterstock; **257** iStockphoto/Thinkstock; **260** spotmatik/Shutterstock; **262** Hemera/Thinkstock; **264** leonello calvetti/Shutterstock; **265** Alila Medical Media/Shutterstock; **266** leonello calvetti/Shutterstock; **267** (l) Hemera/Thinkstock, (r) leonello calvetti/Shutterstock; **269** leonello calvetti/Shutterstock; **270** leonello calvetti/Shutterstock; **271** Liquidlibrary/Thinkstock; **277** iStockphoto/Thinkstock; **282** Tyler Olson/Shutterstock; **284** Aleksander Popovski/Nedley Health Systems; **285** Noam Armonn/Shutterstock; **288–289** mpanch/Shutterstock; **290–291** Igor Zh./Shutterstock; **292** Oko Laa/Shutterstock; **298** (t) Luis Stortini Sabor aka CVADRAT/Shutterstock, (b) NASA/JPL; **299** Lee Prince/Shutterstock; **300** Stocktrek Images/Thinkstock; **301** NASA/JPL/USGS; **302** (t) NASA Planetary Photojournal, (b) Milagli/Shutterstock; **303** NASA/JPL; **304** Diego Barucco/Shutterstock; **305** MichaelTaylor/Shutterstock; **306** a. v. ley/Shutterstock; **307** NASA/JPL; **310** Shalygin/Shutterstock; **311** Kenneth V. Pilon/Shutterstock; **312–313** fstockfoto/Shutterstock; **316** Neo Edmund/Shutterstock; **317** Harrison H. Schmitt/NASA; **318** NASA **319** NASA; **320** NASA/JPL; **321** NASA/JPL–Caltech; **322** iurii/Shutterstock; **323** (bkgd) ilyianne/Shutterstock, (inset) Stocktrek Images/Thinkstock; **332–333** oorka/Shutterstock; **334** janprchal/Shutterstock; **339** INSAGO/Shutterstock; **341** BlueRingMedia/Shutterstock; **344** Orla/Shutterstock; **349** Rafael Pacheco/Shutterstock; **350** Neil A. Armstrong/NASA; **352** NASA/JAXA; **358** Nicku/Shutterstock; **359** iStockphoto/Thinkstock; **356–357** Stocktrek Images/Thinkstock; **358** ESA/AURA/Caltech/NASA; **359** Brian Maudsley/Shutterstock; **366** Zhabska Tetyana/Shutterstock; **367** Digital Vision/Thinkstock; **369** Hemera/Thinkstock; **370** iStockphoto/Thinkstock; **371** (top to bottom) Stocktrek Images/Thinkstock, (2) Wolfgang Kloehr/Shutterstock, (3) Stocktrek Images/Thinkstock, (4) Stocktrek Images/Thinkstock; **374** (bkgd)Stocktrek Images/Thinkstock, (t) NASA/JPL–Caltech, (c) Courtesey of TRW, (b) ESA/Hubble; **376** (t to b) NASA/ESA/T.M.Brown, (2) NASA/JPL/Caltech, (3) NASA/ESA/Hubble Heritage Team, (4) NASA/JPL; **377** iStockphoto/Thinkstock; **378** NASA; **379** NASA; **381** Designua/Shutterstock; **384–385** iStockphoto/Thinkstock; **388** Sanford Roth/Science Source; **389** (bkgd) iStockphoto/Thinkstock, (inset) NOAO/Science Source; **392–393** bikeriderlondon/Shutterstock; **394–395** EpicStockMedia/Shutterstock; **396** Willyam Bradberry/Shutterstock; **399** Eliks/Shutterstock; **402** iStockphoto/Thinkstock; **403** Jag_cz/Shutterstock; **404** Joe Gough/Shutterstock; **405** iStockphoto/Thinkstock; **408** iStockphoto/Thinkstock; **409** Away/Shutterstock; **410** Eduard Kyslynskyy/Shutterstock; **411** Kzenon/Shutterstock; **412** Designua/Shutterstock; **414** Mopic/Shutterstock; **420** funkyfrogstock; **421** GIPhotoStock/Science Source; **425** Sybille Yates/Shutterstock; **430** Christopher Hall/Shutterstock; **432** iStockphoto/Thinkstock; **433** Digital Vision/Thinkstock; **436–437** Ipatov/Shutterstock; **438** Peter Gudella/Shutterstock; **439** peresanz/Shutterstock; **442** Peter Bernik/Shutterstock; **444** Ivsanmas/Shutterstock; **447** iStockphoto/Thinkstock; **448** Hemera/Thinkstock; **454** Sashkin/Shutterstock; **455** stockshoppe/Shutterstock; **457** iStockphoto/Thinkstock; **461** Marlon Lopez MMG1 Design/Shutterstock; **466** Aletia/Shutterstock; **469** (t)maxik/Shutterstock, (b) AlexRoz/Shutterstock; **470** Joggie Botma/Shutterstock; **471** Jeff Whyte/Shutterstock; **476** Courtesy Ryan Pierce/Southern Adventist University; **477** Margoe Edwards/Shutterstock; **480–481** Bikeworldtravel/Shutterstock; **482** GIRODJL/Shutterstock; **484** Michael Stokes/Shutterstock; **485** iStockphoto/Thinkstock; **487** DJTaylor/Shutterstock; **488** IM_photo/Shutterstock; **489** U.S. Federal Trade Commission; **490** Frederic Legrand/Shutterstock; **491** Federico Rostagno/Shutterstock; **493** Somchai Som/Shutterstock; **494** kreatorex/Shutterstock; **496** MilanB/Shutterstock; **499** (t) pryzmat/Shutterstock, (b) wavebreakmedia/Shutterstock; **500** foodiepics/Shutterstock; **503** fotum/Shutterstock; **506** Pavel L Photo and Video/Shutterstock; **508** Science Source; **509** The Washington Post/Getty Images; **512–513** artjazz/Shutterstock; **514** My Good Images/Shutterstock; **515** Todd Taulman/Shutterstock; **519** Alexeysun/Shutterstock; **521** Kalmatsuy/Shutterstock; **522** janaph/Shutterstock; **524** Igor Zh./Shutterstock; **525** Racheal Grazias/Shutterstock; **526** Iaroslav Neliubov/Shutterstock; **535** Phil McDonald/Shutterstock; **538** scphoto60/Shutterstock; **541** Jan Vancura/Shutterstock; **545** snapgalleria/Shutterstock; **549** mady70/Shutterstock; **550** Jason Patrick Ross/Shutterstock; **551** BlueRingMedia/Shutterstock; **553** Robert Neumann/Shutterstock; **556** NOAA; **557** T.P. Walker/NOAA.